DISCARDED

Benchmark Papers in Geology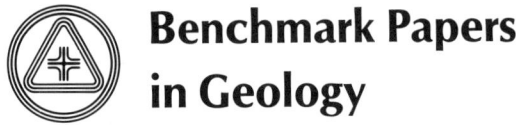

Series Editor: Rhodes W. Fairbridge
Columbia University

A selection from the published volumes in this series

Volume
- 2 RIVER MORPHOLOGY / *Stanley A. Schumm*
- 16 GEOCHEMISTRY OF WATER / *Yasushi Kitano*
- 25 ENVIRONMENTAL GEOLOGY / *Frederick Betz, Jr.*
- 37 STATISTICAL ANALYSIS IN GEOLOGY / *John M. Cubitt and Stephen Henley*
- 39 BEACH PROCESSES AND COASTAL HYDRODYNAMICS / *John S. Fisher and Robert Dolan*
- 59 KARST GEOMORPHOLOGY / *M. M. Sweeting*
- 73 CHEMICAL HYDROGEOLOGY / *William Back and R. Allan Freeze*

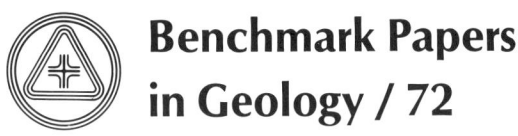

Benchmark Papers in Geology / 72

A BENCHMARK® Books Series

PHYSICAL HYDROGEOLOGY

Edited by

R. ALLAN FREEZE
University of British Columbia

and

WILLIAM BACK
U.S. Geological Survey

Hutchinson Ross Publishing Company

Stroudsburg, Pennsylvania

Copyright ©1983 by **Hutchinson Ross Publishing Company**
Benchmark Papers in Geology, Volume 72
Library of Congress Catalog Card Number: 82-2976
ISBN: 0-87933-431-2

All rights reserved. No part of this book covered by the copyrights hereon may be reproduced or transmitted in any form or by any means—graphic, electronic, or mechanical, including photocopying, recording, taping, or information storage and retrieval systems—without written permission of the publisher.

85 84 83 1 2 3 4 5
Manufactured in the United States of America.

LIBRARY OF CONGRESS CATALOGING IN PUBLICATION DATA
Main entry under title:
Physical hydrogeology.
   (Benchmark papers in geology; 72)
   Includes indexes.
   1. Hydrogeology. I. Freeze, R. Allan. II. Back,
William, 1925-   III. Series.
GB1003.2.P49   551.48   82-2976
ISBN 0-87933-431-2   AACR2

Distributed worldwide by Van Nostrand Reinhold Company Inc.,
135 W. 50th Street, New York, NY 10020.

CONTENTS

Series Editor's Foreword	ix
Preface	xi
Contents by Author	xiii
Introduction	1

PART I: PHYSICS OF GROUNDWATER FLOW

Editors' Comments on Papers 1 Through 8 — 8

1   DARCY, H.: Determination of the Laws of Flow of Water Through Sand — 14
Translated from *Les fontaines publiques de la ville de Dijon,* Victor Dalmont, Paris, 1856, pp. 590-594

2   SLICHTER, C. S.: Theoretical Investigation of the Motion of Ground Waters — 21
U.S. Geol. Survey Annual Rept. **19** (part 2): 295, 297, 329-332 (1899)

3   MEINZER, O. E.: The Occurrence of Ground Water in the United States; with a Discussion of Principles — 26
U.S. Geol. Survey Water-Supply Paper 489, 1923, pp. 1-8

4   HUBBERT, M. K.: The Theory of Ground-Water Motion — 36
Jour. Geology **48**:785-819 (1940)

5   MAASLAND, M.: Soil Anisotropy and Land Drainage — 71
Drainage of Agricultural Lands, J. N. Luthin, ed., American Society of Agronomy, Madison, Wisc., 1957, pp. 216-228

6   MEINZER, O. E.: Compressibility and Elasticity of Artesian Aquifers — 85
Econ. Geology **23**:263-291 (1928)

7   JACOB, C. E.: On the Flow of Water in an Elastic Artesian Aquifer — 114
Am. Geophys. Union Trans. **21**:574-586 (1940)

8   COOPER, H. H., JR.: The Equation of Groundwater Flow in Fixed and Deforming Coordinates — 127
Jour. Geophys. Research **71**:4785-4790 (1966)

PART II: WELL AND AQUIFER HYDRAULICS

Editors' Comments on Papers 9 Through 18 — 134

9   THEIS, C. V.: The Relation Between the Lowering of the Piezometric Surface and the Rate and Duration of Discharge of a Well Using Ground-Water Storage — 141
Am. Geophys. Union Trans. **16**:519-524 (1935)

Contents

10 BOULTON, N. S.: The Drawdown of the Water-Table under Non-Steady Conditions near a Pumped Well in an Unconfined Formation — 147
Inst. Civil Engineers Proc. **3** (part 3):564–579 (1954)

11 HANTUSH, M. S., and C. E. JACOB: Non-Steady Radial Flow in an Infinite Leaky Aquifer — 163
Am. Geophys. Union Trans. **36**:95–100 (1955)

12 NEUMAN, S. P., and P. A. WITHERSPOON: Applicability of Current Theories of Flow in Leaky Aquifers — 169
Water Resources Research **5**:817–829 (1969)

13 FERRIS, J. G., D. B. KNOWLES, R. H. BROWN, and R. W. STALLMAN: Theory of Aquifer Tests — 182
U.S. Geol. Survey Water-Supply Paper 1536-E, 1962, pp. 69, 144–153

14 STALLMAN, R. W.: Use of Numerical Methods for Analyzing Data on Ground Water Levels — 193
Internat. Assoc. Sci. Hydrology Pub. **41**:227–231 (1956)

15 TYSON, H. N., JR., and E. M. WEBER: Ground-Water Management for the Nation's Future—Computer Simulation of Ground-Water Basins — 198
Am. Soc. Civil Engineers Proc., Jour. Hydraulics Div. **90**:59–77 (1964)

16 PINDER, G. F., and J. D. BREDEHOEFT: Application of the Digital Computer for Aquifer Evaluation — 217
Water Resources Research **4**:1069–1093 (1968)

17 WALTON, W. C., and T. A. PRICKETT: Hydrogeologic Electric Analog Computers — 243
Am. Soc. Civil Engineers Proc., Jour. Hydraulics Div. **89**:67–91 (1963)

18 TODD, D. K.: Ground-Water Flow in Relation to a Flooding Stream — 268
Am. Soc. Civil Engineers Proc., Jour. Hydraulics Div. **81**(separate):628-1-628-20 (1955)

PART III: REGIONAL GROUNDWATER FLOW

Editors' Comments on Papers 19 Through 32 — 290

19 CHAMBERLIN, T. C.: Requisite and Qualifying Conditions of Artesian Wells — 297
U.S. Geol. Survey Annual Rept. **5**:131–141 (1885)

20 DARTON, N. H.: Preliminary Report on Artesian Waters of a Portion of the Dakotas — 308
U.S. Geol. Survey Annual Rept. **17**(part 2):609–617 (1897)

21 FIEDLER, A. G., and S. S. NYE: Geology and Ground-Water Resources of the Roswell Artesian Basin, New Mexico (abstract) — 317
U.S. Geol. Survey Water-Supply Paper 639, 1933, p. 1

22 BENNETT, R. R., and R. R. MEYER: Geology and Ground-Water Resources of the Baltimore Area (abstract) — 318
Maryland Dept. Geology, Mines and Water Resources Bull. **4**:1–4 (1952)

23 LOHMAN, S. W.: Geology and Artesian Water Supply of the Grand Junction Area, Colorado (abstract) — 322
U.S. Geol. Survey Prof. Paper 451, 1965, pp. 1–4

24	NEWCOMB, R. C.:	Effect of Tectonic Structure on the Occurrence of Ground Water in the Basalt of the Columbia River Group of the Dalles Area, Oregon and Washington (abstract) *U.S. Geol. Survey Prof. Paper 383-C,* 1969, p. 1	326
25	COOLEY, M. E., J. W. HARSHBARGER, J. P. AKERS, and W. F. HARDT:	Regional Hydrogeology of the Navajo and Hopi Indian Reservations, Arizona, New Mexico, and Utah (abstract) *U.S. Geol. Survey Prof. Paper 521-A,* 1969, pp. 1-2	327
26	TÓTH, J.:	A Theoretical Analysis of Groundwater Flow in Small Drainage Basins *Jour. Geophys. Research* **68**:4795-4812 (1963)	328
27	FREEZE, R. A., and P. A. WITHERSPOON:	Theoretical Analysis of Regional Groundwater Flow. 2. Effect of Water-Table Configuration and Subsurface Permeability Variation *Water Resources Research* **3**:623-634 (1967)	346
28	WALTON, W. C.:	Leaky Artesian Aquifer Conditions in Illinois *Illinois Water Survey Rept. Inv. 39,* 1960, pp. 5-6, 17-25, 27	358
29	DOMENICO, P. A., and M. D. MIFFLIN:	Water from Low-Permeability Sediments and Land Subsidence *Water Resources Research* **1**:563-576 (1965)	370
30	POLAND, J. F., and G. H. DAVIS:	Land Subsidence due to Withdrawal of Fluids *Geol. Soc. America Rev. Engineering Geol.* **2**:187, 238-239, 252-262 (1969)	384
31	LEGRAND, H. E., and V. T. STRINGFIELD:	Development of Permeability and Storage in the Tertiary Limestones of the Southeastern States, USA *Internat. Assoc. Sci. Hydrology Bull.* **11**:61-73 (1966)	398
32	DAVIS, S. N.:	Porosity and Permeability of Natural Materials *Flow Through Porous Media,* R. J. M. De Wiest, ed., Academic Press, New York, 1969, pp. 53-58, 65, 70, 80-81, 86-89	411

Author Citation Index	423
Subject Index	427
About the Editors	431

SERIES EDITOR'S FOREWORD

The philosophy behind the Benchmark Papers in Geology is one of collection, sifting, and rediffusion. Scientific literature today is so vast, so dispersed, and, in the case of old papers, so inaccessible for readers not in the immediate neighborhood of major libraries that much valuable information has been ignored by default. It has become just so difficult, or so time consuming, to search out the key papers in any basic area of research that one can hardly blame a busy person for skimping on some of his or her "homework."

This series of volumes has been devised, therefore, as a practical solution to this critical problem. The geologist, perhaps even more than any other scientist, often suffers from twin difficulties—isolation from central library resources and immensely diffused sources of material. New colleges and industrial libraries simply cannot afford to purchase complete runs of all the world's earth science literature. Specialists simply cannot locate reprints or copies of all their principal reference materials. So it is that we are now making a concerted effort to gather into single volumes the critical materials needed to reconstruct the background of any and every major topic of our discipline.

We are interpreting "geology" in its broadest sense: the fundamental science of the planet Earth, its materials, its history, and its dynamics. Because of training in "earthy" materials, we also take in astrogeology, the corresponding aspect of the planetary sciences. Besides the classical core disciplines such as mineralogy, petrology, structure, geomorphology, paleontology, and stratigraphy, we embrace the newer fields of geophysics and geochemistry, applied also to oceanography, geochronology, and paleoecology. We recognize the work of the mining geologists, the petroleum geologists, the hydrologists, and the engineering and environmental geologists. Each specialist needs a working library. We are endeavoring to make the task of compiling such a library a little easier.

Each volume in the series contains an introduction prepared by a specialist (the volume editor)—a "state of the art" opening or a summary of the object and content of the volume. The articles, usually some twenty to fifty reproduced either in their entirety or in significant extracts, are selected in an attempt to cover the field, from the key papers of the last century to fairly recent work. Where the original works are in foreign languages, we have endeavored to locate or commission translations. Geologists, because

Series Editor's Foreword

of their global subject, are often acutely aware of the oneness of our world. The selections cannot therefore be restricted to any one country, and whenever possible an attempt is made to scan the world literature.

To each article, or group of kindred articles, some sort of "highlight commentary" is usually supplied by the volume editor. This commentary should serve to bring that article into historical perspective and to emphasize its particular role in the growth of the field. References, or citations, wherever possible, will be reproduced in their entirety—for by this means the observant reader can assess the background material available to that particular author, or, if desired, he or she too can double check the earlier sources.

A "benchmark," in surveyor's terminology, is an established point on the ground that is recorded on our maps. It is usually anything that is a vantage point, from a modest hill to a mountain peak. From the historical viewpoint, these benchmarks are the bricks of our scientific edifice.

RHODES W. FAIRBRIDGE

PREFACE

This is the first of two volumes that attempt to bring together the classic literature in the field of hydrogeology. This volume emphasizes physical hydrogeology; a companion volume treats the chemical aspects.

We hope that this collection will be of use to our hydrogeological colleagues, both as a historical document and as a teaching aid in graduate programs in hydrogeology. This volume may also be a valuable introduction to the subject for researchers from other fields within the earth sciences and for researchers from the broader set of interdisciplinary environmental sciences.

Hydrogeology is itself an interdisciplinary subject, and the literature is dispersed across a wide variety of outlets. Many of our selections are from government publication series, but a larger number are from technical journals. Some of these journals are primarily scientific in their emphasis; others have a strong engineering component. The original raison d'être for most of the work reported in this volume was the desire to improve technology for water-resource development. In retrospect, it is now clear that this same body of fundamental hydrogeological knowledge has more diverse applications in such endeavors as the design of agricultural irrigation and drainage systems, the design of subsurface nuclear waste repositories, the analysis of slope-stability problems, the appraisal of geothermal resources, and the interpretation of the genesis of petroleum reservoirs and economic mineral deposits.

We solicited the help of several of our colleagues in the preparation of this volume. Without holding them in any way responsible for any errors of commission or omission, which remain ours alone, we wish to thank the following hydrogeologists for their thoughtful advice: John D. Bredehoeft, Stanley N. Davis, Robert N. Farvolden, Francis R. Hall, Harry E. LeGrand, T. N. (Nari) Narasimhan, Shlomo P. Neuman, and Jozsef Tóth. We are also grateful to Hilton H. Cooper, John G. Ferris, and Gerald Meyer who reviewed the editors' contributions to the manuscript. The senior author is indebted to Martin Engi and Maurice Pryce for their help on the translation of Darcy.

R. ALLAN FREEZE
WILLIAM BACK

CONTENTS BY AUTHOR

Akers, J. P., 327
Bennett, R. R., 318
Bredehoeft, J. D., 217
Brown, R. H., 182
Boulton, N. S., 147
Chamberlin, T. C., 297
Cooley, M. E., 327
Cooper, H. H., Jr., 127
Darcy, H., 14
Darton, N. H., 308
Davis, G. H., 384
Davis, S. N., 411
Domenico, P. A., 370
Ferris, J. G., 182
Fiedler, A. G., 317
Freeze, R. A., 346
Hantush, M. S., 163
Hardt, W. F., 327
Harshbarger, J. W., 327
Hubbert, M. K., 36
Jacob, C. E., 114, 163
Knowles, D. B., 182

Legrand, H. E., 398
Lohman, S. W., 322
Maasland, M., 71
Meinzer, O. E., 26, 85
Meyer, R. R., 318
Mifflin, M. D., 370
Neuman, S. P., 169
Newcomb, R. C., 326
Nye, S. S., 317
Pinder, G. F., 217
Poland, J. F., 384
Prickett, T. A., 243
Slichter, C. S., 21
Stallman, R. W. 182, 193
Stringfield, V. T., 398
Theis, C. V., 141
Todd, D. K., 268
Tóth, J., 328
Tyson, H. N., Jr., 198
Walton, W. C., 243
Weber, E. M., 198
Witherspoon, P. A., 169, 346

ARTESIAN WELL AT PRAIRIE DU CHIEN, WISCONSIN.

Source: T. C. Chamberlin, 1885, Requisite and Qualifying Conditions of Artesian Wells, *U.S. Geol. Survey Annual Rept.* **5:**Frontispiece.

PHYSICAL HYDROGEOLOGY

INTRODUCTION

SELECTION CRITERIA

In his book on the history of science, Kuhn (1962) suggested that scientific thought is advanced through alternating periods of revolution and evolution. Revolutionary ideas are those that arise outside the mainstream of evolutionary development and deflect the mainstream into new and rewarding channels of investigation. Within this context, we first selected those papers that we believe have had a revolutionary impact on the science of hydrogeology; second, we tried to choose papers from the evolutionary periods of development that have had recognized and lasting impact. Taken together, these are the papers without which any historical perspective of the field would be incomplete.

In this rapidly expanding field, many important developments occurred as the result of the independent but collective work of a large number of researchers. In such cases, representative examples from these larger suites are presented in order to guide the reader through the historical development of the science without encountering major conceptual gaps. In the editors' comments preceding each section, we alert the reader to the many important papers that could not be included in these volumes and to those review articles that provide a more complete synthesis than can be attempted here.

This collection represents a selection from the tens of thousands of hydrogeological publications that have appeared since the birth of the science in the late nineteenth century. Any selection process is fraught with danger; it is always subjective, and it is open to personal bias. In order to focus on the essential areas and keep the volume size manageable, it was necessary to exclude a wide variety of important topics, such as: water management, unsaturated flow, phreatophytes, geophysical exploration, theory of flow through fractured rocks, geothermal resources, and the inverse problem. Some of these (such as water management and unsaturated flow) deserve Benchmark volumes of their own. Others (such as fracture flow and the inverse problem) represent topics of current research, making the selection of benchmark papers premature. For this reason, we have also excluded

Introduction

papers that appeared after 1970. We also omitted papers with a civil engineering or theoretical mechanics emphasis and in doing so, omitted translations of the late nineteenth century European contributions of Dupuit, Boussinesq, Forchheimer, and the Thiems, father and son.

With the exclusion of the early European contributions, the book becomes unabashedly North American. The near absence of foreign literature is similarly reflected in the reference lists of the papers themselves and in the reference lists of most current North American research articles. This lack of scientific interaction across cultural and linguistic boundaries is a historical fact of hydrogeology that is not mirrored in the more basic sciences and for which we have no ready explanation. It is certainly not due to the absence of important work in other countries. As De Wiest's translation of Polubarinova-Kochina (1962) revealed, there is a body of work in the Soviet Union that grew from the same nineteenth century European roots and paralleled much of the twentieth century North American research. The Netherlands, in particular, has provided important research on agricultural drainage, and similarly important contributions have come in recent years from Israel, Great Britain, France, Norway, India and Japan.

Throughout most of this century, the United States Geological Survey (U.S.G.S.) has played a leading role in the development of hydrogeology in North America. Although the theoretical contributions of its staff are well represented in this volume, the massive effort of the U.S. Geological Survey and the many state agencies in field hydrogeological studies is probably somewhat underrepresented. This is due to the fact that individual conceptual breakthroughs seem to be more easily identified as benchmarks than are originally conceived field programs integrating a variety of concepts and methods. In addition it is difficult to meaningfully excerpt lengthy field reports. Despite these difficulties, we have included excerpts and abstracts from several of the classic field studies. We recognize that field studies have had a tremendous impact on the emergence of hydrogeology as a scientific discipline in North America. In summary, the result of our selection process is a volume that emphasizes physically based groundwater research within a hydrogeological context, carried out in twentieth-century North America. The papers have been grouped under three headings: Physics of Groundwater Flow, Well and Aquifer Hydraulics, and Regional Groundwater Flow. The organization is by topic rather than by date of publication, but the papers within each section are ordered more or less sequentially from early concepts to more recent developments. All geochemical topics, including mass transport and seawater intrusion, are treated in the companion volume on *Chemical Hydrogeology,* edited by William Back and R. Allan Freeze (1983).

HISTORICAL PERSPECTIVE

In the context of Kuhn's vision of the history of science, we believe that there have been three revolutions in physical hydrogeology: the historic set of experiments carried out by Darcy (1856) in Dijon, France; the transient well-hydraulics analysis of C. V. Theis in 1935 (Paper 9); and the introduction of large digital computers in the early 1960s. Darcy provided the empirical law on which all subsequent work has been based, and Theis provided methodology for both the measurement of in situ hydrologic properties of geologic formations and the prediction of the response of groundwater systems to pumping. Digital computers provide the means for assessment of groundwater resources on a regional scale within the context of the full hydrologic cycle. Paper 1 is an excerpt from Darcy (1856); Theis (Paper 9) begins Part II. Computer models appear at the ends of both Parts II and III.

The study of groundwater problems has grown steadily in scale. Darcy's law developed from a laboratory column experiment; the earliest applications of the theory were in the design of filter beds and individual wells; the Theis breakthrough opened the door to studies at aquifer scale rather than well scale; and the computer age has seen the scope of study grow from individual aquifers to aquifer-aquitard systems, groundwater basins, and full watersheds. Part I deals largely with the theory of groundwater flow at one point or in a small elemental volume; Part II deals with flow to wells; and Part III deals with regional flow systems.

Hydrogeology, as its name implies, is historically rooted in both geology and hydrology. For those who wish to learn more about these roots, booklength accounts are available in both sciences. Adams (1954) traced the historical development of the geological sciences; Biswas (1970) did the same for the hydrological sciences. Rouse and Ince (1957) included much of hydrological interest in their book on the history of hydraulics. All three provide a wealth of detail from the contributions of the early Egyptians and the Greek and Roman philosophers through the emergence of the earth sciences in Europe during the eighteenth and nineteenth centuries. Several shorter articles on the history of hydrogeology are by Baker and Horton (1936), discussing the ideas of the early philosophers; Ferris and Sayre (1955), tracing the development of quantitative groundwater hydrology from

Introduction

Darcy through to 1950; and Bredehoeft et al. (1982), outlining the growth of ideas with respect to the regional circulation of groundwater.

Chow (1964) provides a concise discussion of the history of hydrology. He divided the time since the year 1800 into four periods. The Period of Modernization (1800–1900) is described as the grand era of experimental hydrology; the Period of Empiricism (1900–1930) is noted as a time when modern quantitative hydrology was still immature. The Period of Rationalization (1930–1950) saw the emergence of many great hydrologists who used rational analysis rather than empiricism to solve hydrologic problems; and the Period of Theorization (1950–present) has seen the innovative application of mathematical analysis in hydrology. Using this classification, Darcy's experiments fall into the Period of Modernization; Theis's contributions belong to the Period of Rationalization; and the growth of computer-simulation technology is a feature of the Period of Theorization.

More coherent treatment of the science can be found in other hydrogeology textbooks. Those that have found greatest favor among hydrogeologists are (listed from earliest to most recent): Tolman (1937), Todd (1959), Davis and De Wiest (1966), Walton (1970), Domenico (1972), and Freeze and Cherry (1979).

REFERENCES

Adams, F. D., 1954, *The Birth and Development of the Geological Sciences,* Dover, New York, 506p.

Back, W., and R. A. Freeze, eds., 1983, *Chemical Hydrogeology,* Benchmark Papers in Geology, vol. 73, Hutchinson Ross, Stroudsburg, Pa., 432p.

Baker, M. N., and R. E. Horton, 1936, Historical Development of Ideas Regarding the Origin of Springs and Groundwater *Am. Geophys. Union Trans.* **17:**395–400.

Biswas, A., 1970, *History of Hydrology,* North-Holland, Amsterdam, 336p.

Bredehoeft, J. D., W. Back, and B. B. Hanshaw, 1982, Regional Ground Water Flow Concepts in the United States: Historical Perspective, in *Proceedings of Symposium on Recent Trends in Hydrogeology,* Geological Society of America Special Paper 189, pp. 297–316.

Chow, V. T., 1964, Hydrology and Its Development, in *Handbook of Applied Hydrology,* V. T. Chow, ed., McGraw-Hill, New York, 21p.

Darcy, H., 1856, *Les fontaines publiques de la ville de Dijon,* Victor Dalmont, Paris.

Davis, S. N., and R. J. M. De Wiest, 1966, *Hydrogeology,* Wiley, New York, 463p.

Domenico, P. A., 1972, *Concepts and Models in Groundwater Hydrology,* McGraw-Hill, New York, 405p.

Ferris, J. G., and A. N. Sayre, 1955, The Quantitative Approach to Ground-Water Investigations, *Econ. Geology,* 50th anniv. issue, pp. 714–747.

Freeze, R. A., and J. A. Cherry, 1979, *Groundwater,* Prentice-Hall, Englewood Cliffs, N.J., 604p.

Kuhn, T. S., 1962, *The Structure of Scientific Revolutions,* University of Chicago Press, Chicago, 172p.
Polubarinova-Kochina, P. Ya., 1962, *Theory of Groundwater Movement,* R. J. M. De Wiest, trans., Princeton University Press, Princeton, N.J., 613p.
Rouse, H., and S. Ince, 1957, *History of Hydraulics,* Iowa State University, Institute of Hydraulic Research, Ames, Iowa.
Todd, D. K., 1959, *Ground Water Hydrology,* Wiley, New York, 336p.
Tolman, C. F., 1937, Ground Water, McGraw-Hill, New York, 593p.
Walton, W. C., 1970, *Groundwater Resource Evaluation,* McGraw-Hill, New York, 664p.

Part I

PHYSICS OF GROUNDWATER FLOW

Editors' Comments on Papers 1 Through 8

1 **DARCY**
Excerpt from *Determination of the Laws of Flow of Water Through Sand*

2 **SLICHTER**
Excerpt from *Theoretical Investigation of the Motion of Ground Waters*

3 **MEINZER**
Excerpt from *The Occurrence of Ground Water in the United States; with a Discussion of Principles*

4 **HUBBERT**
Excerpt from *The Theory of Ground-Water Motion*

5 **MAASLAND**
Excerpt from *Soil Anisotropy and Land Drainage*

6 **MEINZER**
Compressibility and Elasticity of Artesian Aquifers

7 **JACOB**
On the Flow of Water in an Elastic Artesian Aquifer

8 **COOPER**
The Equation of Groundwater Flow in Fixed and Deforming Coordinates

It is often said that mathematics is the language of physics, and this relationship holds with respect to the physics of groundwater flow. The physics of flow through porous media is embodied in a flow equation that distills into one short mathematical notation all the essential truths of groundwater flow. The solutions to this equation, obtained in the context of boundary value problems that represent idealized hydrogeological environments, are the basis of almost all the quantitative methodology in common use today in the field of groundwater hydrology.

The equation for the flow of water through saturated porous media in two or three dimensions has been known in its steady-state form since the work of Forchheimer (1886, 1898) in the late nineteenth century. His understanding was based on an analogy with the heat-flow equation. Theis (Paper 9) invoked the same analogy in 1935 in presenting a solution to the transient form of the flow equation, although he did not present the fundamental differential equation itself. It was not until 1940, with the work of Jacob (Paper 7), that the transient equation was first developed from basic physical principles; it was not until 1966 that Cooper (Paper 8) laid to rest some important challenges that had arisen with respect to Jacob's development.

The purpose of this section is to chart the scientific march from Darcy's experiments to the Jacob-Cooper formulation of the transient groundwater flow equation. In principle, the development of the equation requires: (1) the definition of a hydraulic potential; (2) the development of a constitutive law that relates the flow rate to the potential gradient; (3) an understanding of the properties of the medium that control the flow (hydraulic conductivity, porosity, and compressibility); (4) an understanding of the concept of effective stress; and (5) an application of the continuity relationship for conservation of mass. From the current viewpoint, these five items read like a logical sequence of steps, but historically, they were developed independently by a number of different scientists and not at all in the sequential order listed above. The hydraulic potential was not fully understood until the work of Hubbert (Paper 4) in 1940, long after the experimental derivation of the constitutive law by Darcy (Paper 1) in 1856. Hubbert was also the first to critically examine the hydraulic conductivity parameter in detail; and Maasland (Paper 5) provided a clear summary of the effects of anisotropy on flow. Meinzer (Paper 6) first recognized the importance of aquifer compressibility in the groundwater storage term, and he is also responsible (Paper 3) for one of the best early discussions of porosity, drawing upon Slichter's earlier study of the interstitial geometry of packings of spheres. The combination of Darcy's law and the steady-state continuity relationship leads to the steady-state flow equation as shown by Slichter (Paper 2). The combination of Darcy's law, the effective stress principle, and the transient continuity relationship leads to the transient flow equation presented by Jacob (Paper 7) and Cooper (Paper 8).

Looking at some of the papers in greater detail, Paper 1 is a translation of a portion of the appendix to Darcy's report describing his experimental apparatus and the results. Because we could not find an earlier translation of the entire pertinent section (although Fried, 1965, translated a portion of it), in the interest of continuity we have prepared a new translation of the complete section. There are several points of interest. It is clear from the context of the main body of the report that Darcy's experiments aimed at improving the design of filter sands for water purification, and were apparently performed at a hospital. The experimental column was of larger diameter and the sand somewhat coarser than one might expect. In addition to the law bearing his name, Darcy also developed the equations for a falling head permeameter. The two components of head, due to pressure and elevation, were clearly understood and appear as separate terms in his writings. Of historical interest is the statement that in some runs he set up pressures less than atmospheric at the base of the column. It is not clear from the manuscript how this condition was maintained or how the water exited from a basal zone that ought to have been unsaturated or tension saturated. This latter point is niggling of course; the fact is that the experiments carried out by Darcy with the help of his assistant, Ritter, in Dijon, France in 1855 and 1856 represent the beginning of groundwater hydrology as a quantitative science.

During the latter half of the nineteenth century, several European hydraulic engineers such as Boussinesq, Dupuit, Forchheimer, and the two Thiems made important contributions to the development of the science. Dupuit (1863) developed a linear constitutive law, similar to Darcy's, based on hydraulic theory rather than experimental evidence. He also produced the first formal mathematical analysis of a groundwater hydraulics problem, that of radial flow toward a pumped well in an unconfined aquifer. The assumptions he invoked in this analysis, namely, that the hydraulic gradient is equal to the slope of the water table and that it is invariant with depth, have come to be known as the Dupuit assumptions, and methods based on these assumptions are still in wide use today.

If Gauss was the "prince of mathematicians," then surely Forchheimer was the prince of groundwater hydraulics. Forchheimer (1886, 1898) was the first to note the analogy between groundwater flow and heat flow, and the first to use the Laplace equation for the description of steady-state groundwater flow. He also introduced the concept of a flow net and first applied conformal mapping and the method of images. He clarified the Dupuit assumptions and recognized that steady-state flow in unconfined aquifers under the Dupuit assumptions would obey Laplace's equation with respect to the square of the hydraulic head rather than the hydraulic head itself.

Dupuit's formula for the discharge from a well in an unconfined aquifer required advance knowledge of the radius of the zone of influence at steady state. Adolph Thiem (1870) carried out extensive field investigations to clarify the controls on the radius of influence.

His son Gunther Thiem (1906) was the first to recognize that Dupuit's equations could be applied at any two points on the profile of the cone of depression around a well. This realization led to the first inverse application of a solution to the steady-state flow equation and hence, to the first use of pumping tests as a practical tool for *in situ* measurement of the hydraulic properties of geologic formations.

During the latter half of the nineteenth century, important developments also occurred in North America. There was little interchange of information between the United States and Europe and as a result many of the European breakthroughs were duplicated in America. C. S. Slichter of the U.S. Geological Survey, working twelve years after Forchheimer and apparently unaware of his work, utilized the same heat-flow literature to arrive at Laplace's equation, flow-net construction, and conformal mapping. Paper 2 is an excerpt from Slichter's 1898 report in which he develops Laplace's equation and discusses the flow-net approach.

During the early decades of the twentieth century the unquestioned North American leader in hydrogeology was Oscar Meinzer of the U.S. Geological Survey. His two classic *Water-Supply Papers* (Meinzer, 1923; Paper 3) are still reprinted and widely used today. His major contribution to the understanding of the physics of groundwater flow came in his 1928 paper on the compressibility of artesian aquifers (Paper 6) wherein he invoked the effective stress principle. Largely on the basis of his work on the Dakota sandstone, Meinzer recognized that the water in a confined aquifer supports part of the overlying load and that aquifers compact when fluid pressure is decreased. Although the basic concept of effective stress must be ascribed to Terzaghi (1925), Meinzer's realization that the concept applied as well to aquifers as to laboratory soil samples is a major breakthrough in its own right.

With Darcy's law established and the two components of groundwater storage, porosity and compressibility, clarified by Meinzer, the stage was set for development of the transient flow equation from first principles rather than from a heat-flow analogy. Such a development was presented in 1940 by Jacob (Paper 7).

In the same year, Hubbert published his classic treatise on the theory of groundwater motion (excerpted here as Paper 4). In it Hubbert clarified the fundamental nature of both hydraulic potential and the coefficient of permeability. On the basis of the latter analysis we now differentiate between permeability, a function of porous media alone, and hydraulic conductivity, the proportionality constant in Darcy's law that is a function of both fluid and media.

As attempts to apply the flow equations intensified, it became clear that actual hydrogeologic environments are both heterogeneous and anisotropic with respect to hydraulic conductivity. One of the clearest expositions of the effects of anisotropy on flow is found in Maasland (Paper 5). In a section of his paper following our excerpt, Maasland also outlined the relationships between stratified heterogeneous systems and homogeneous anisotropic systems. He traced the concept of anisotropy back to Versluys (1915), and among the other early contributions, he singled out those of Dachler (1933), Vreedenburgh (1936), and Muskat (1937).

During the early 1960s doubts were voiced as to the validity of Jacob's development of the groundwater flow equation. The doubts revolved around the fact that the effective stress law he invoked assumed that only vertical stresses exist. A full analysis should have dealt with the interaction between a three-dimensional stress field and a three-dimensional fluid flow field. In addition, stresses lead to strains, and in low-permeability compacting media the rate of movement of the grains of the porous media may approach the rate of movement of the water flowing through them. To treat this condition it was suggested that a Lagrangian coordinate system might be better suited to the problem than an Eulerian one and that the Darcy velocity might better be considered as the differential velocity of the water relative to the grains. Finally, the question arose as to whether hydraulic conductivity ought to be considered a function of effective stress in transient analyses of compressible systems.

On investigation, hydrogeologists discovered that Biot (1941, 1955), a physicist working in a petroleum research institute, had already coupled a three-dimensional stress field with the fluid-flow field. His work was probably not fully understood by hydrogeologists until Verruijt (1969) provided an interpretation of Biot's notation and placed the work into the context of earlier groundwater developments. In the meantime, De Wiest (1966) improved the Jacob equation with respect to the variation of hydraulic conductivity with effective stress but not with respect to the storage side of the equation. It was Cooper (Paper 8) who clarified the relationship between the development of the flow equation in fixed coordinates and deforming coordinates. Cooper concluded that Jacob's equation was correct for almost all practical applications.

REFERENCES

Biot, M. A., 1941, General Theory of Three-Dimensional Consolidation, *Jour. Appl. Physics* **12:**155–164.
Biot, M. A., 1955, Theory of Elasticity and Consolidation for a Porous Anisotropic Solid, *Jour. Appl. Physics* **26:**182–185.

Dachler, R., 1933, Über Sickerwasserstromungen in geschichtetem Material, *Wasserwirtschaft* **2**:13-16.

De Wiest, R. J. M., 1966, On the Storage Coefficient and the Equation of Groundwater Flow, *Jour. Geophys. Research* **71**:1117-1122.

Dupuit, J., 1863, *Études théoriques et pratiques sur le mouvement des eaux dans les canaux découverts et à travers les terrains perméables*, Dunod, Paris, 304p.

Forchheimer, P., 1886, Über die Ergiebigkeit von Brunnenanlagen und Sickerschlitzen, *Zeitschr. Architekten und Ingenieurvereins (Hannover)* **32**:539-564.

Forchheimer, P., 1898, Grundwasserspiegel bei Brunnenanlagen, *Zeitschr. Architekten und Ingenieurvereins (Hannover)* **50**:629-635, 645-648.

Fried, J., 1965, The Reproduction of Darcy's Law and Its English Translation, *Water Resources Bull.* **1**:4-11.

Meinzer, O. E., 1923, Outline of Ground Water Hydrology with Definitions, *U.S. Geol. Survey Water-Supply Paper 494*, 71p.

Muskat, M., 1937, *The Flow of Homogeneous Fluids Through Porous Media*, McGraw-Hill, New York, 763p.

Terzaghi, K., 1925, *Erdbaumechanic auf Bodenphysikalischer Grundlage*, Denticke, Vienna, 399p.

Thiem, A., 1870, Die Ergiebigkeit artesicher Bohrlöcher, Schachtbrunnen, und Filtergallerien, *Jour. Gasbeleucht. Wasserversorgung.* **14**:450-567.

Thiem, G., 1906, *Hydrologische Methoden*, Gephardt, Leipzig, 56p.

Verruijt, A., 1969, Elastic Storage of Aquifers, in *Flow Through Porous Media*, R. J. M. De Wiest, ed., Academic Press, New York, pp. 331-376.

Versluys, J., 1915, De onbepaalde vergelijking der permanente beweging van het grondwater, *Geologie en Mijnbouw* **1**:349-360.

Vreedenburgh, C. G. F., 1936, On the Steady Flow of Water Percolating Through Soils with Homogeneous-anisotropic permeability, *Internat. Conf. Soil Mechanics and Found. Engineering*, **1**:222-225.

1

DETERMINATION OF THE LAWS OF THE FLOW OF WATER THROUGH SAND

H. Darcy

This excerpt was translated expressly for this Benchmark volume by R. Allan Freeze, University of British Columbia, from pp. 590–594 of Les fontaines publiques de la ville de Dijon, Victor Dalmont, Paris, *647p.*

THE PUBLIC FOUNTAINS OF THE CITY OF DIJON

Development and application of the principles to follow and the formulae to use in questions of the

DISTRIBUTION OF WATER

Supplemented by an appendix relating to the water supplies of several cities, the filtration of water, and the manufacture of cast-iron, lead, sheet-metal, and bitumen pipes.

by

HENRY DARCY
Inspector-General of Bridges and Highways

PARIS
VICTOR DALMONT, PUBLISHER

Successor to Carilian-Goeury and V^{or} Dalmont
Printer to the Imperial Corps of Bridges, Highways, and Mines
Quai des Augustins, 49
1856

Flow of Water Through Sand

I approach now an account of the experiments that I have carried out at Dijon together with Engineer Charles Ritter, to determine the laws of flow of water through sand. These experiments have been repeated by Chief Engineer Baumgarten.

The apparatus used (Plate 24, Figure 3, see page **20**) consisted of a vertical column 2.50 m high, formed from a piece of pipe 0.35 m in internal diameter, and closed at each end by a bolted endplate.

In the interior, 0.20 m above the bottom, there is a horizontal grating, intended to support the sand, and which divides the column into two compartments. This partition is formed by the superposition from bottom to top of a grating of prismatic iron bars of 0.007 m thickness, a grating of cylindrical bars of 0.005 m diameter, and finally a screen with 0.002 m spacing. The spacing of bars in each of the gratings is equal to their thickness, and the two gratings are arranged so that their bars are oriented perpendicular to one another.

The upper chamber of the column receives water from a tube connected to the hospital water system, and from which a tap allows regulation of the discharge at will; the lower chamber opens through a tap into a measuring basin 1.0 m on a side.

The pressure at the two ends of the column is measured by U-tube mercury manometers; and finally, each of the compartments is furnished with an air vent, which is essential for the operation of the apparatus.

The experiments have been carried out with a silicious sand from the Saône, composed as follows:

> 58% sand passing the 0.77 mm screen
> 13% sand passing the 1.10 mm screen
> 12% sand passing the 2.00 mm screen
> 17% fine gravel, shell fragments, etc.

It has void space of approximately 38%.

The sand was poured and packed into the column, which had previously been filled with water so that the voids of the soil mass would contain no air, and the height of the sand was measured only at the end of each series of experiments after the passage of water had suitably packed it.

Each experiment consisted of establishing a specified pressure in the upper chamber of the column by the adjustment of the inflow tap; then, when it was established by means of two observations that the flow had become essentially uniform, the outflow from the filter during a certain time was noted, and the mean outflow per minute was calculated from it.

Under weak heads, the almost total quiescence of the mercury in the manometer permitted measurement to the nearest millimeter, representing 26.2 mm of water; when operated under strong pressures, the inflow tap was almost entirely open, and then the manometer, in spite of the diaphragm with which it was furnished, exhibited continual oscillations. Nevertheless, the strong oscillations were only random, and it was possible to measure the mean height of the mercury to the nearest 5 millimeters, that is, to know the water pressure to the nearest 1.30 m [sic].

All these manometer fluctuations were due to the water-hammer effects from the continual turning on and off of the numerous standpipes at the hospital where the experimental apparatus was located.

All the pressures have been referred to the level of the lower face of the filter, and no account has been taken of the friction in the upper part of the column, which was evidently negligible.

The table of experiments [Table 1] as well as their graphical representation, demonstrate that the discharge through each filter increases proportionally with the head.

For the filters that have been utilized, the discharge per second per square meter is related very roughly to the head by the following relations:

First series: $Q = 0.493P$
Second series: $Q = 0.145P$
Third series: $Q = 0.126P$
Fourth series: $Q = 0.123P$

Denoting I as the head per meter of thickness of the filter, these formulae become the following:

First series: $Q = 0.286I$
Second series: $Q = 0.165I$
Third series: $Q = 0.216I$
Fourth series: $Q = 0.332I$

The differences between the values of the coefficient Q/I arise because the sand employed has not been consistently homogenized. For the second series, it has not been washed; for the third series, it was washed; for the fourth, it was very well washed and of a slightly coarser grain size.

It appears then, that for identical sands, one can assume that the discharge is directly proportional to the head and inversely proportional to the thickness of the layer traversed.

In the preceding experiments, the pressure below the filter has always been equal to that of the atmosphere; it was interesting to investigate whether the law of proportionality that has come to be recognized between the discharges and the heads that produce them still holds when the pressure under the filter was greater than or less than atmospheric pressure: such is the goal of the new experiments carried out on the 17th and 18th of February 1856 by M. Ritter.

These experiments are reported in the following table [Table 2]: column 4 gives the pressure above the filter; column 5 the pressure below the filter, sometimes greater than and sometimes less than the weight P of the atmosphere; column 6 presents the difference in pressure; and finally, column 7 indicates the ratios between the flow rate and the differences in pressures existing above and below the filter. The thickness of the sand layer was equal to 1.10 m.

The constant values of the ratios in column 7 testify to the truth of the law already stated: it can be noted in passing that the pressures above and below the filter range between very wide limits: below the filter the pressure varied from $P + 9.88$ to $P - 3.60$, and above the filter from $P + 12.88$ to $P + 2.98$.

Thus, denoting the thickness of the sand layer by e, its surface area by s, atmospheric pressure by P, and the height of water above the layer by h, we will have $P + h$ for the pressure to which the upper end will be subjected; and further, if $P + h_o$ is the pressure supported by the lower surface, k is a coefficient depending on the permeability of the layer, and q is the flow rate, we have

$$q = k\frac{s}{e}[h + e \pm h_o]$$

which reduces to

$$q = k\frac{s}{e}(h + e)$$

when $h_o = 0$, or when the pressure below the filter is equal to atmospheric pressure.

It is easy to determine the law that governs the decline in the height of water h above the filter; as a matter of fact, if dh is the distance that this level falls during time dt, then the rate of decline will be $-dh/dt$; and the above equation then gives for this rate the expression

$$\frac{q}{s} = v = \frac{k}{e}(h + e).$$

Thus we have

$$-\frac{dh}{dt} = \frac{k}{e}(h + e)$$

or

$$\frac{dh}{h + e} = -\frac{k}{e}dt$$

and

$$\ln(h + e) = C - \frac{k}{e}t.$$

If the value h_o corresponds to the time t_o and h to any time t, it follows that

$$\ln(h + e) = \ln(h_o + e) - \frac{k}{e}(t - t_o). \tag{1}$$

If we now replace $h + e$ and $h_o + e$ by qe/sk and $q_o e/sk$, it follows that

$$\ln q = \ln q_o - \frac{k}{e}(t - t_o) \tag{2}$$

and the two equations (1) and (2) give, respectively, the law of decline in height above the filter, and the law of variation in flow rate starting from time t_o.

If k and e were unknown, it can be seen that it would require two preliminary experiments in order to make the unknown ratio k/e vanish from the second term.

Table 1 Table of experiments carried out at Dijon on October 29 and 30 and November 2, 1855.

Number of the Experiment	Duration (min)	Mean flow rate (l/min)	Mean pressure	Ratio between volumes and pressure	Observations
1st Series, with a sand thickness of 0.58 m.					
1	25	3.60	1.11	3.25	The sand has not been washed.
2	20	7.65	2.36	3.24	
3	15	12.00	4.00	3.00	The manometer column experienced only slight movements.
4	18	14.28	4.90	2.91	
5	17	15.20	5.02	3.03	
6	17	21.80	7.63	2.86	
7	11	23.41	8.13	2.88	Very appreciable oscillations
8	15	24.50	8.58	2.85	
9	13	27.80	9.86	2.82	Strong manometer oscillations
10	10	29.40	10.89	2.70	
2nd Series, with a sand thickness of 1.14 m.					
1	30	2.66	2.60	1.01	The sand is not washed.
2	21	4.28	4.70	0.91	
3	26	6.26	7.71	0.81	
4	18	8.60	10.34	0.83	Very strong oscillations
5	10	8.90	10.75	0.83	
6	24	10.40	12.34	0.84	
3rd Series, with a sand thickness of 1.71 m.					
1	31	2.13	2.57	0.83	Washed sand
2	20	3.90	5.09	0.77	
3	17	7.25	9.46	0.76	Very strong oscillations
4	20	8.55	12.35	0.69	
4th Series, with a sand thickness of 1.70 m.					
1	20	5.25	6.98	0.75	Washed sand with a slightly coarser grain size than the preceding.
2	20	7.00	9.95	0.70	Weak oscillations as a result of the partial obstruction of the manometer opening
3	20	10.30	13.93	0.74	

Table 2

Number of the Experiment	Duration	Mean Flow Rate	Mean Pressure		Difference in Pressures	Ratio between volumes and pressures	Observations
			Above the Filter	Below the Filter			
1	2	3	4	5	6	7	8
	min	ℓ/min	m	m	m		
1	15	18.8	$P + 9.48$	$P - 3.60$	13.08	1.44	Strong oscillations in the upper manometer.
2	15	18.3	$P + 12.88$	$P\ \ 0$	12.88	1.42	″
3	10	18.0	$P + 9.80$	$P - 2.78$	12.58	1.43	″
4	10	17.4	$P + 12.87$	$P + 0.46$	12.41	1.40	Weak
5	20	18.1	$P + 12.80$	$P + 0.49$	12.35	1.47	Rather Weak
6	16	14.9	$P + 8.86$	$P - 0.83$	9.69	1.54	Almost none
7	15	12.1	$P + 12.84$	$P + 4.40$	8.44	1.43	Very strong
8	15	9.8	$P + 6.71$	$P\ \ 0$	6.71	1.46	Very weak
9	20	7.9	$P + 12.81$	$P + 7.03$	5.78	1.37	Very strong
10	20	8.65	$P + 5.58$	$P\ \ 0$	5.58	1.55	Almost none
11	20	4.5	$P + 2.98$	$P\ \ 0$	2.98	1.51	″
12	20	4.15	$P + 12.86$	$P + 9.88$	2.98	1.39	Quite strong The cause of these oscillations was explained by this time.

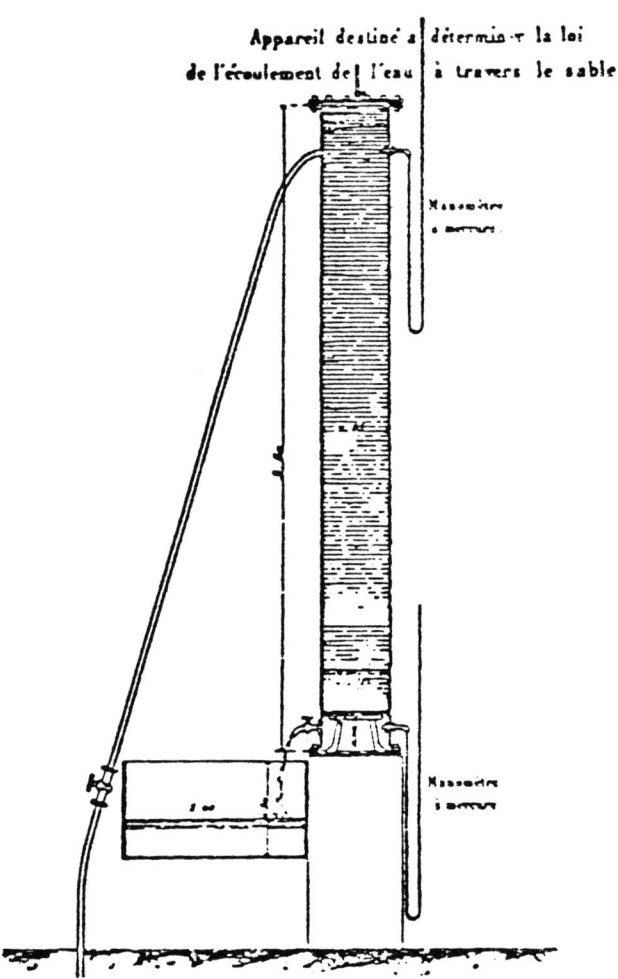

Figure 3

THEORETICAL INVESTIGATION OF THE MOTION OF GROUND WATERS

BY

CHARLES S. SLICHTER

CONTENTS.

	Page.
INTRODUCTION AND SUMMARY	301
CHAPTER I. Laws of the rectilinear flow of ground water through a soil	305
Application and illustration of the formulas obtained above	323
CHAPTER II. General laws of the flow of ground waters	329
CHAPTER III. Motion of ground water in horizontal planes	333
Method of conjugate functions or conformal transformation	344
CHAPTER IV. Motion of ground water in vertical planes	351
CHAPTER V. Flow of artesian wells and their mutual interference	358
Wells in a region in which the ground waters have a constant velocity in a general direction	368
Mutual interference of two wells	371
Mutual interference of three wells	375
Mutual interference of a large number of wells arranged in a row	378
APPENDIX. Bibliography of papers on the motion of ground waters and related topics	381

LIST OF TABLES.

	Page
TABLE I. Relation of porosity m to face angle θ	312
II. Constants for various porosities of an ideal soil	326
III. Coefficients of viscosity for water for various temperatures	328
IV. Capacity of wells of various sizes, extending 200 feet into sandstone of a given kind, flowing under 10-foot head	362
V. Capacity of the wells of Table IV, allowing in each case for the friction of 200 feet of pipe same size as well	363
VI. Drop in pressure at various distances from a 6 inch well which is lowered 10 feet by pumping	364
VII. Capacity of 6-inch well extending 100 feet in material of various kinds, for various heads of pressure	367
VIII. Mutual interference of two 6-inch wells placed at various distances apart	374
IX. Mutual interference of three 6-inch wells placed various distances apart	376
X. Mutual interference of a row of wells whose members are various distances apart	380

[*Editors' Note:* In the original, material precedes this excerpt.]

GENERAL LAWS OF THE FLOW OF GROUND WATERS.

1. Experimenters have usually claimed that the velocity of the flow of a liquid in a given direction through a column of soil is proportional to the difference in pressure at the ends of the column and inversely proportional to the length of the column. This law is often referred to as "Darcy's law,"[1] and may be expressed by a formula, as follows:

$$v = k \frac{p}{h}, \qquad (1)$$

in which v is the velocity, p is the difference in pressure at the ends of the column of soil, and h is the length of the column. The pressure at the ends of the column must be taken just inside of the soil itself, and not in the liquid outside of the surface of the soil. There appears to be a sudden reduction in pressure as the liquid enters the soil, just as there is a loss of head due to the orifice of influx in case of water entering an ordinary water main.[2] The constant k depends upon the size of the soil grains, the porosity of the soil, and the viscosity of the liquid. The form of this constant was investigated from a theoretical standpoint in Chapter I. (See Chapter I, equation (47).) Darcy and others have determined its value experimentally for different soils, and have suggested various forms of experimental filters suited for this purpose.[3]

Darcy's law is entirely analogous to the well-known law of Poiseuille for the flow of a liquid through a capillary tube,[4] and, like this latter law, it fails to hold for high pressure gradients; for example, the law may be true for a pressure variation not exceeding 1 gram per length of 10 centimeters, but may fail to hold for a pressure variation of 10 grams per length of 1 centimeter. The limits within which the law holds, and the modifications which a second approximation require, can be determined only by exhaustive experiments on a wide range of materials.

---

[1] H. Darcy, Les fontaines publiques de la ville de Dijon, Paris, 1856. Darcy-Bazin, Recherches hydrauliques, Paris, 1865.

[2] Allen Hazen, Some physical properties of sands and gravels: Report Mass. State Board of Health, 1892, p. 541. Consult also bibliography in the appendix.

[3] Hazen, loc. cit., and Otto Lueger, Wasserversorgung der Städte, Stuttgart, p. 125.

[4] The most important memoirs on the flow of liquids and gases through capillary tubes are the following: Poiseuille, Recherches sur le mouvement des liquids dans les tubes de très-petits diamètres: Acad. Sciences, Savants estrangers, 1842, Vol. IX, p. 433. Graham, On the motions of gases: Phil. Trans. Royal Soc. London, 1846, p. 573; 1849, p. 349. O. E. Meyer, Ueber die innere Reibung der Gase: Pogg. Annalen, Vol. CXXVII, 1866, p. 253; Vol. CXLVIII, 1873, p. 1. M. Couette, Étude sur le Frottement des liquids.

Equation (1) may also be written

$$v = k \frac{\partial p}{\partial h}.$$

2. If a fluid in a porous medium is not required to move in a single direction, but has liberty to move in any direction, then if at a given point, P, the motion is in direction s we may write

$$v = k \frac{\partial p}{\partial s}, \qquad (2)$$

in case the motion is steady, using the term "steady motion" in its usual sense in hydrodynamics, as descriptive of motion which does not vary with the lapse of time. The pores of a soil are arranged in such a haphazard manner that we are at liberty to suppose that at any particular point the fluid is free to move in any desired direction. The flow of a fluid in a soil, then, is not different from the flow of a hypothetical fluid which we may suppose to replace both the fluid in the soil and the soil itself, which hypothetical fluid has the law of flow for steady motion given by the equation

$$v = k \frac{\partial p}{\partial s}.$$

In case no external attractive forces are acting, the equations of motion in three dimensions of the hypothetical fluid are

$$\begin{aligned} u &= \frac{dx}{dt} = k \frac{\partial p}{\partial x}, \\ v &= \frac{dy}{dt} = k \frac{\partial p}{\partial y}, \\ w &= \frac{dz}{dt} = k \frac{\partial p}{\partial z}. \end{aligned} \qquad (3)$$

The equation of continuity is

$$\frac{\partial u}{\partial x} + \frac{\partial v}{\partial y} + \frac{\partial w}{\partial z} = 0, \qquad (4)$$

the same as in the case of ordinary liquids. The equation of continuity holds for both perfect and viscous liquids, and is simply the mathematical statement that a given mass of the liquid does not change its volume during the given motion—that is, that the liquid is incompressible. If we substitute from (3) into (4) we obtain

$$\frac{\partial^2 p}{\partial x^2} + \frac{\partial^2 p}{\partial y^2} + \frac{\partial^2 p}{\partial z^2} = 0. \qquad (5)$$

3. This equation (equation 5) is the familiar equation occurring in nearly all branches of applied mathematics, known as Laplace's equation. The function p, which satisfies this differential equation, is called

a *potential function*, and from equations (2) and (3) it is observed that the velocity in any direction is the differential coefficient of the potential function with respect to that direction. In the hydrodynamics of perfect fluids the function that possesses this property is called the "velocity potential," but in the hydrodynamics of the hypothetical fluid now under consideration the velocity potential (omitting the constant k) is identical with the pressure function. The coincidence of the pressure function with the potential function is a matter of great interest and importance, and presents a striking analogy with the mathematics of electrical action.

If, as some claim, the velocity of ground water is not directly proportional to pressure, but varies in an arithmetical progression of the form

$$v = kp + c,$$

then equations (3) above must be modified by the addition of a constant, but equation (5) will be unaltered. Thus Laplace's equation is satisfied in either case.

It seems remarkable that the fact that the solution of any problem in the motion of ground waters depends upon the solution of the differential equation (5) has not been pointed out before. The existence of this function is made the basis of nearly all the work in the following pages.

4. If we assume that an external impressed force, as gravity, acts upon the liquid in the z direction with constant intensity g, and if the density of the fluid be ρ, then the expression for w becomes

$$w = k \left(\frac{\partial p}{\partial z} + \frac{\rho g\, dz}{dz} \right). \tag{6}$$

Therefore the equations of motion with gravity acting are

$$u = \frac{dx}{dt} = k \frac{\partial p}{\partial x},$$

$$v = \frac{dy}{dt} = k \frac{\partial p}{\partial y}, \tag{7}$$

$$w = \frac{dz}{dt} = k \frac{\partial p}{\partial z} + k\rho g.$$

These hold, of course, only for the case of steady motion. Equations (4) and (5) are unchanged by the new hypothesis.

5. We have, therefore, shown that a problem in the steady motion of ground waters is mathematically analogous to a problem in the steady flow of heat or electricity, or to a problem in the steady motion of a perfect fluid. The unknown function p must be determined from the partial differential equation (5), subject to the boundary conditions present in each particular problem undertaken. The function p is

necessarily finite, continuous, and single valued at all points of the hypothetical liquid, and the general methods usual in potential theory apply. In a determinate problem the value of p must be given over a closed boundary, or along a given axis in space, from which we shall be able to derive the value of p at every point of the region expressed as a function of x, y, and z. After p has been determined by the appropriate solution of (5) for a given problem, the equation $p = c$ will give, for different values of c, a series of surfaces upon each one of which the pressure has the constant value assigned. These are the *equipotential or equipressural surfaces*. Orthogonal to this series of surfaces there exists a series of lines in space, along which the particles of liquid actually move. These lines are the *lines of flow*.

Equations (7) introduce an interesting series of problems, in which, however, the boundary conditions are somewhat more difficult to handle than in the case of equations (3).

The equations above written involve the assumption that the medium in which the flow is taking place is everywhere of the same structure—as, for example, a soil or rock which does not essentially change, from place to place, the size or character of its pores. Such is usually the case within the same geological formation, but in stratified material the constant k is apt to be different for vertical motion from what it is for horizontal motion.

[*Editors' Note:* Material has been omitted at this point.]

THE OCCURRENCE OF GROUND WATER IN THE UNITED STATES

WITH A DISCUSSION OF PRINCIPLES.

By Oscar Edward Meinzer.

INTRODUCTION.

The writer has planned and partly prepared a series of six papers on ground water in the United States. These papers are to deal with (1) occurrence, (2) origin, discharge, and quantity, (3) movement and head, (4) quality, (5) recovery and use, and (6) ground-water provinces. The present paper is the first of the series.

The writer is indebted to many colaborers for assistance in preparing this paper, especially to the following members of the United States Geological Survey: M. R. Campbell, who read the entire paper; E. W. Shaw, C. E. Van Orstrand, A. F. Melcher, and C. K. Wentworth, who examined Chapter I; W. C. Alden, who examined Chapters II and IV with special reference to their statements regarding glacial geology; E. S. Larsen, who examined a part of Chapter II; G. W. Stose, who examined Chapter III; T. W. Stanton, who examined Chapter IV; L. W. Stephenson, who examined the parts of Chapter IV that relate to the Coastal Plain; C. W. Cooke, who furnished many unpublished data on the geology of the Coastal Plain; Miss M. G. Wilmarth, who gave valuable help in compiling the geologic sections; D. G. Thompson and Miss Norah E. Dowell, who furnished original data on the mechanical composition and porosity of various sedimentary materials; B. H. Lane, who made valuable criticisms of the text; and Martin Solem, who had charge of the preparation of the illustrations.

Chapter I. PRINCIPLES OF OCCURRENCE.

ROCKS AS RECEPTACLES OF WATER.

The rocks that form the crust of the earth are in few places, if anywhere, solid throughout. They contain numerous open spaces, called voids or interstices, and these spaces are the receptacles that hold the water that is found below the surface of the land and is recovered in part through springs and wells. There are many kinds of rocks, and they differ greatly in the number, size, shape, and arrangement of their interstices and hence in their properties as containers of water. The occurrence of water in the rocks of any region is therefore determined by the character, distribution, and structure of the rocks it contains—that is, by the geology of the region. Most rocks have numerous interstices of very small size, but some are characterized by a few large openings, such as joints or caverns. In most rocks the interstices are connected, so that the water can move through the rocks by percolating from one interstice to another; but in some rocks the interstices are largely isolated, and there is little opportunity for the water to percolate. The interstices are generally irregular in shape, but different types of irregularities are characteristic of different kinds of rocks. The differences in rocks with respect to their interstices result from the differences in the minerals of which they are composed and from the great diversity of geologic processes by which they were produced or later modified.

POROSITY OF ROCKS.

DEFINITION OF TERM.[1]

The porosity of a rock is its property of containing interstices. Some authors have used the term to refer only to minute interstices, which they call pores, but in comparison with the size of the earth itself even the largest openings are no more than pores, and the term "porosity" is much more useful if it is made to apply to all openings, instead of only to openings having an arbitrary limit of size. Porosity is expressed quantitatively as the percentage of the total volume of the rock that is occupied by interstices or that is not occupied by solid rock material. A rock is said to be saturated when all its interstices are filled with water. In a saturated rock the porosity is practically the percentage of the total volume of the rock that is occupied by water.

---

[1] See Gregory, H. E., and others, Military geology and topography, p. 114, New Haven, Yale Univ. Press, 1918 (chapter on water supply prepared chiefly by O. E. Meinzer).

CONDITIONS CONTROLLING POROSITY.

The porosity of a sedimentary deposit depends chiefly on (1) the shape and arrangement of its constituent particles, (2) the degree of assortment of its particles, (3) the cementation and compacting to which it has been subjected since its deposition, (4) the removal of mineral matter through solution by percolating waters, and (5) the fracturing of the rock, resulting in joints and other openings. Well-sorted deposits of uncemented gravel, sand, or silt have a high porosity, regardless of whether they consist of large or small grains. If, however, the material is poorly sorted small particles occupy the spaces between the larger ones, still smaller ones occupy the spaces between these small particles, and so on, with the result that the porosity is greatly reduced (fig. 1, A and B). Boulder clay, which is an

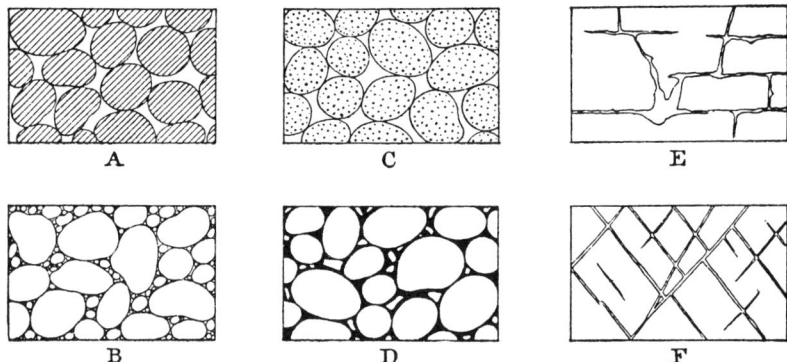

FIGURE 1.—Diagram showing several types of rock interstices and the relation of rock texture to porosity, A, Well-sorted sedimentary deposit having high porosity; B, poorly sorted sedimentary deposit having low porosity; C, well-sorted sedimentary deposit consisting of pebbles that are themselves porous, so that the deposit as a whole has a very high porosity; D, well-sorted sedimentary deposit whose porosity has been diminished by the deposition of mineral matter in the interstices; E, rock rendered porous by solution; F, rock rendered porous by fracturing.

unassorted mixture of glacial drift containing particles of great variety in size, may have a very low porosity, whereas outwash gravel and sand, derived from the same source but assorted by running water, may be highly porous. Well-sorted uncemented gravel may be composed of pebbles that are themselves porous, so that the deposit as a whole has a very high porosity (fig. 1, C). Well-sorted porous gravel, sand, or silt may gradually have its interstices filled with mineral matter deposited out of solution from percolating waters, and under extreme conditions it may become a practically impervious conglomerate or quartzite of very low porosity (fig. 1, D). On the other hand, relatively soluble rock, such as limestone, though originally dense, may become cavernous as a result of the removal of part of its substance through the solvent action of percolating water (fig. 1, E). Furthermore hard, brittle rock, such as limestone, hard sandstone, or most igneous and metamorphic rocks, may acquire

4   OCCURRENCE OF GROUND WATER IN THE UNITED STATES.

large interstices through fracturing that results from shrinkage or deformation of the rocks or through other agencies (fig. 1, F). Solution channels and fractures may be large and of great practical importance, but they are rarely abundant enough to give an otherwise dense rock a high porosity.

POROSITY OF GRANULAR DEPOSITS.

RELATION OF POROSITY TO ARRANGEMENT OF GRAINS.

The most common type of water-bearing materials consists of deposits composed of fragments of rock that were more or less rounded by wear before they were deposited. In such deposits the water exists in the irregular spaces that remain between these fragments or grains. To investigate the water-bearing characteristics of such material Slichter[2] first made a theoretical study of the most simple case—an "ideal soil" consisting of spherical grains of equal size.

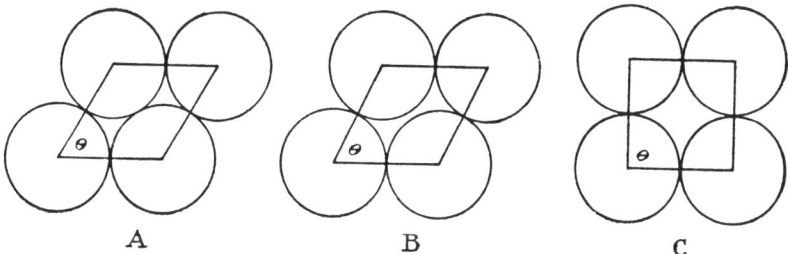

FIGURE 2.—Sections of four contiguous spheres of equal size. A, Most compact arrangement; B, less compact arrangement; C, least compact arrangement.

He described the conditions with these simple assumptions, as follows:

In order to study the nature of the pores, we may separate out from the mass of the soil eight contiguous grains in such manner that the lines joining their centers form an equilateral parallelopiped or rhombohedron, as represented in Plate I, A,[3] in which the white rods mark the position and direction of two of the pores. By studying the properties of the pores of this rhombohedron we may arrive at the properties to be assigned to the pores of the entire mass of soil, since this rhombohedron constitutes the element of volume, or the unit element, which, if repeated, will give the entire mass of soil.

If the grains of soil are arranged in the most compact manner possible, each grain will touch surrounding grains at twelve points, and the element of volume will be a rhombohedron having face angles equal to 60° and 120° (fig. 2, A). If the grains are not arranged in the most compact manner the rhombohedron will have its face angles greater than 60° (fig. 2, B), and each sphere will touch other spheres in but six points but will nearly touch in six other points. The most open arrangement of the soil grains which is possible with the grains in contact is had when the rhombohedron is a cube (fig. 2, C).

---

[2] Slichter, C. S., Theoretical investigation of the motion of ground water: U. S. Geol. Survey Nineteenth Ann. Rept., pt. 2, pp. 305-328, 1899.

[3] The numbers of the illustrations cited have been changed to conform to those of the present paper.—O. E. M.

A. UNIT ELEMENT OF A GROUP OF SPHERES ARRANGED IN THE MOST COMPACT MANNER POSSIBLE.

Face angles 60° and 120°.

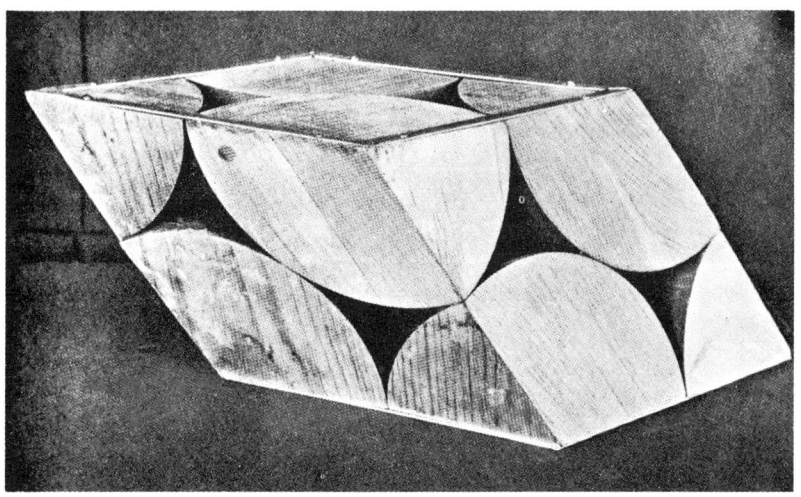

B. UNIT RHOMBOHEDRON FORMED BY PASSING PLANES THROUGH THE CENTERS OF EIGHT CONTIGUOUS SPHERES IN THE MOST COMPACT ARRANGEMENT OF A GROUP OF SPHERES.

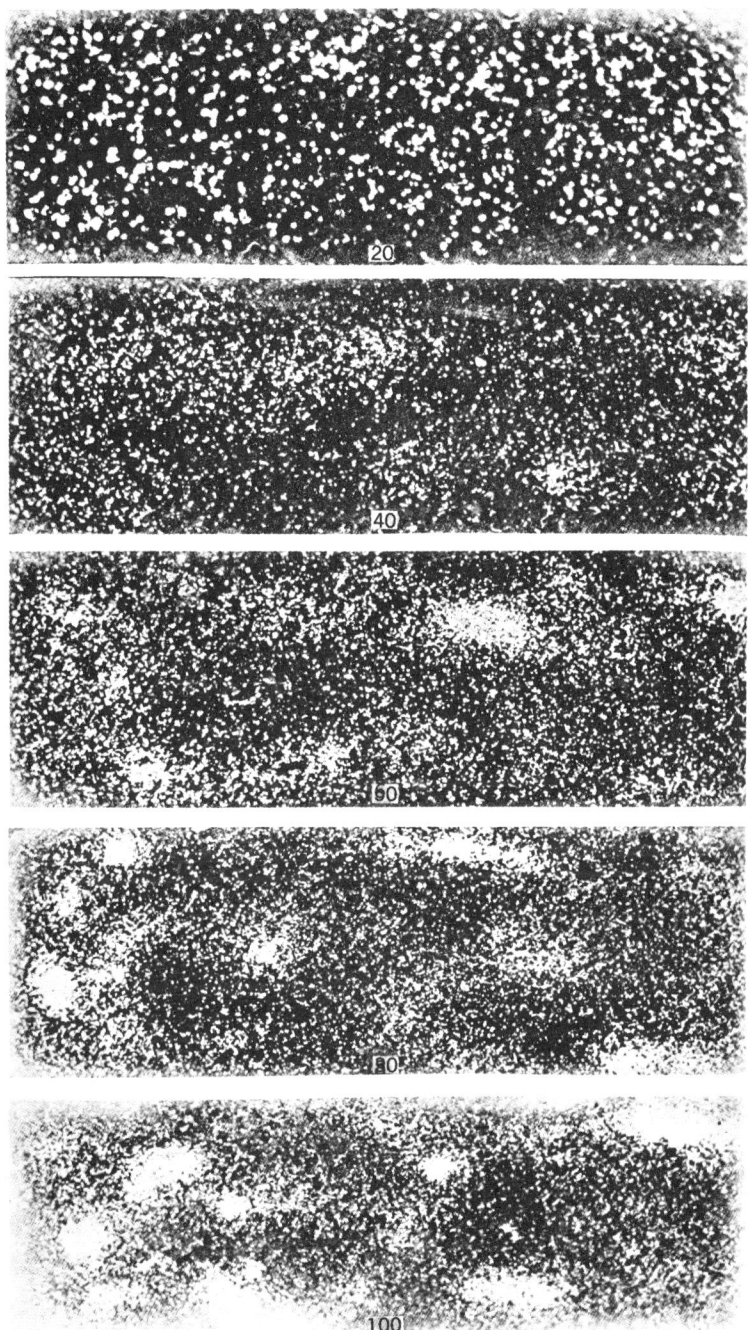

SANDS USED IN KING'S EXPERIMENT OF WATER-YIELDING CAPACITIES.
Natural size.

Plate I, *B*, shows the rhombohedron formed by joining the centers of the spheres of Plate I, *A*.

If we imagine a soil made up of particles arranged so that the lines joining their centers form cubes, the percentage of open space to the whole space, or the so-called porosity, can be found by dividing (1) the difference between the volume of a sphere and the volume of the circumscribed cube by (2) the volume of the circumscribed cube, which gives a porosity of 47.64 per cent. If the particles are arranged as compactly as possible, as in Plate I, *A*, the percentage of pore space can be found by dividing (1) the difference between the volume of a sphere and the volume of a rhombohedron whose acute face angles are 60° and whose edges equal the diameter of the sphere by (2) the volume of this rhombohedron, which gives a porosity of 25.95 per cent. This fact is shown nicely by considering that the pieces of eight different spheres which make the rhombohedron of Plate I, *B*, can be placed together so as to make a perfect sphere. It is plain that the eight pieces would make a complete sphere even if the face angle θ had not the value 60° but had any other value up to 90°. If we measure the porosity of a soil composed of grains of nearly uniform size, we shall find a large variation in the results, depending largely upon the manner in which the soil was packed; but usually the porosity will lie within these limits.

The pores through such an ideal soil are capillary tubes of approximately triangular cross section. The pore enlarges slightly in area as it follows the surfaces of the spherical soil grains and then diminishes again to its former value.

RELATION OF POROSITY TO SIZE OF GRAINS.

It will be noted that the size of the grains does not enter into Slichter's calculations. If other conditions are the same, a material will have the same porosity whether it consists of large or small grains. Thus, although there is wide range in the porosity of each of the four principal types of granular deposits—gravel, sand, silt, and clay—there is probably no great difference in the average porosity of the different groups. On the whole, silt and clay are about as porous as sand and gravel.

RELATION OF POROSITY TO SHAPE OF GRAINS.

Natural sedimentary deposits differ from the "ideal soil" investigated by Slichter in being made up of grains that are not perfect spheres and that are not all of the same size. The shapes of the grains differ considerably, according to the character of the minerals of which they are composed and the shapes of the original fragments; also according to the kind and amount of breaking up and wear they received before they were deposited. Irregularity in shape results in a larger possible range in porosity. To some extent the irregularities tend to counteract one another, but it is believed that the porosity of many deposits is increased by the irregular, angular shapes of its constituent particles.

RELATION OF POROSITY TO DEGREE OF ASSORTMENT.

Variety in size of grain, or the degree of assortment, is of fundamental importance with respect to the porosity of a deposit. A deposit composed of large grains of uniform size has a high porosity, and a deposit composed of small grains of uniform size has an equally high porosity; but a deposit composed of a mixture of grains of these two sizes has a much lower porosity. If small grains are added to a deposit of large grains the small grains will occupy the interstices between the large ones, thereby reducing the amount of void space (fig. 1, A and B). If in the opening between the spheres shown in figure 1, A, a small sphere is placed, this small sphere will occupy space that would otherwise have been empty or occupied by water—that is, it will reduce the porosity to the extent of its own volume. If large grains are added to a deposit of small grains, they will fill with solid rock the spaces which they occupy and which would otherwise be occupied by an aggregation of the small grains with their intervening interstices. Those interstices will thus be replaced by solid rock, and the porosity will be correspondingly reduced. This is illustrated in figure 3, in which a represents a deposit of nearly uniform grains and b represents the same deposit with the addition of a large grain or rock (R) that contains no interstices. Obviously, the large rock has displaced a number of interstices and reduced the porosity of the deposit.

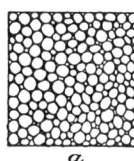 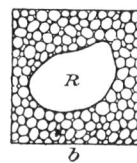

FIGURE 3.—Diagrams showing reduction in porosity caused by addition of a large grain (R) to an aggregation of small grains.

The amount of variation in size of grain, or the degree of assortment of a deposit, can be quantitatively expressed by means of a mechanical analysis in which are given the proportions of the sample that consist of grains of specified sizes. The mechanical analyses of eight samples examined by Hazen[4] are given in the following table and are graphically represented in figure 4.

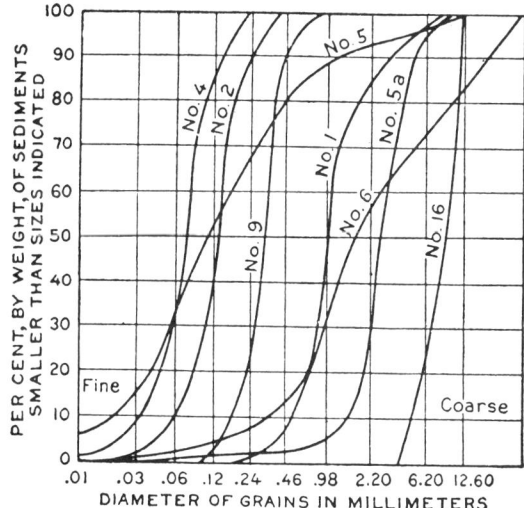

FIGURE 4.—Diagram showing mechanical composition of materials used in experiments by Hazen. The lines representing diameters of the particles are spaced according to the logarithms of these diameters.

---

[4] Hazen, Allen, Experiments upon the purification of sewage and water at the Lawrence experiment station: Massachusetts State Board of Health Twenty-third Ann. Rept., for 1891, pp. 429–431, 1892.

PRINCIPLES OF OCCURRENCE.

Mechanical composition of materials used in experiments by Hazen.

Diameter of grains (millimeters).	Per cent of total sample by weight.							
	No. 3.	No. 4.	No. 2.	No. 9.	No. 6.	No. 1.	No. 5a.	No. 16.
Less than 12.6	99				83	100	100	98
Less than 6.2	96				73	97	95	27
Less than 2.2	92				57	85	31	0
Less than 0.98	89			100	32	53	4	
Less than 0.46	80		100	91	13	7	2	
Less than 0.24	67	100	90	26	7	1.5	1.5	
Less than 0.12	51	85	43	3	4	0	1.0	
Less than 0.06	33	35	10	0	2		.5	
Less than 0.03	16	10	2		.5		0	
Less than 0.01 (organic)	6	1	0		0			
Effective size of grain in millimeters[a]	.02	.03	.06	.17	.35	.48	1.40	5.00
Uniformity coefficient	9.0	2.3	2.3	2.0	7.8	2.4	2.4	1.8
Porosity (per cent by volume)	36	44	42	42	32	40		45

[a] The effective size of grain, as defined by Hazen, is the diameter of a grain of such size that 10 per cent of the sample (by weight) consists of smaller grains and 90 per cent of larger grains.

In order to have a simple quantitative expression of the degree of uniformity in size, an arbitrary quantity is used—the uniformity coefficient. This is the ratio of the diameter of a grain that has 60 per cent (by weight) of the sample finer than itself to the diameter of a grain that has 10 per cent finer than itself. This coefficient can be obtained from a mechanical analysis such as those given above, or it can conveniently be obtained from a curve representing a mechanical analysis, such as those shown in figure 4. In explanation of the uniformity coefficient and its relation to the graphic representation of mechanical analyses, as in figure 4, Hazen makes the following statements:

> For study and comparison the results have been plotted and are shown in the accompanying diagram [fig. 4], the height of a curve at any point showing the per cent of material finer than the size indicated at the bottom of the diagram. The lines representing the diameters are spaced according to the logarithms of the diameters of the particles, as in this way materials of corresponding uniformity in the range of sizes of their particles give equally steep curves, regardless of the absolute sizes of the particles, thus greatly facilitating a comparison of different materials. This scale also shows adequately every grade of material from 0.01 to 10 millimeters in a small space, and without unduly extending any portion of the scale. * * * If all the grains of a sand were absolutely of the same size, the uniformity coefficient would be 1; with most comparatively even-grained sands the coefficient ranges from 2 to 3; with No. 6 and No. 5, the figures are about 8 and 9, respectively; and some extremely uneven sands have coefficients as high as 20 or 30; but the data in regard to the action of such materials is as yet limited.

In regard to the relation of the uniformity coefficient to the porosity, Hazen[5] makes the following statement:

> The amount of open space depends upon the shape and uniformity in size of the particles of sand and is independent of their absolute size. The materials which have the sharpest rise on the diagram [fig. 4], indicating the greatest uniformity in

---
[5] Op. cit., p. 432.

size, have the greatest open space, while the sands having a more gradual rise pack more closely; the finer particles occupy the spaces between the larger stones, greatly reducing the open space.

Obviously, the uniformity coefficient is an index to the porosity. The larger this coefficient the smaller the porosity. The relation for the samples whose mechanical analyses are given in the preceding table is shown in figure 5.

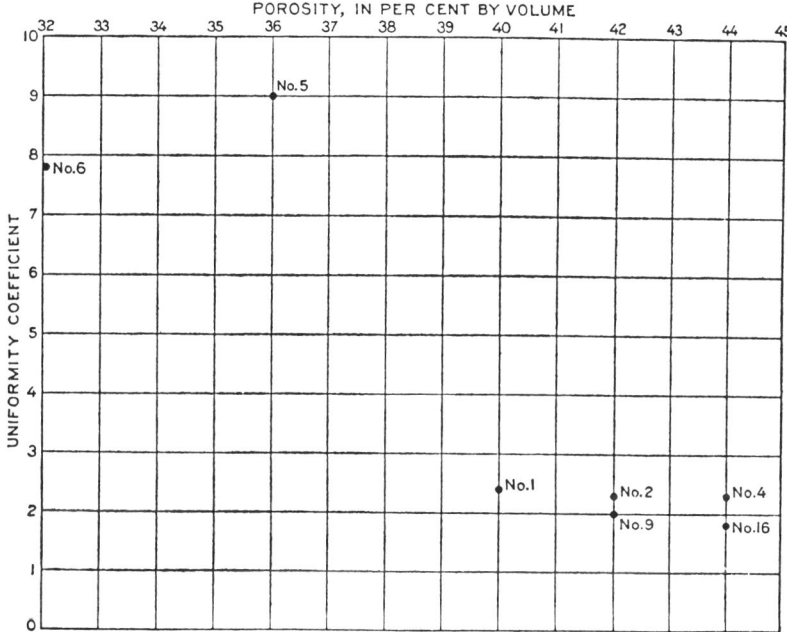

FIGURE 5.—Diagram showing relation between the porosity and the uniformity coefficients of materials used in experiments by Hazen.

[Editors' Note: Material has been omitted at this point.]

THE THEORY OF GROUND-WATER MOTION

M. KING HUBBERT
Columbia University

ABSTRACT

The existing analytical treatments of ground-water flow have mostly been founded upon the conception, borrowed from the flow of ideal frictionless fluids of classical hydrodynamics, that ground-water motion is derivable from a velocity potential. Another widely held view is that pressure is such a function with the fluid flow in the direction of the negative pressure gradient. In the present paper it is shown that a more exceptionless analytical theory results if a potential whose value at a given point is defined to be equal to the work required to transform a unit mass of fluid from an arbitrary standard state to the state at the point in question is employed. Denoting this function by Φ, it is shown that the differential equation of fluid flow in an isotropic medium is given by $\mathbf{q} = -\sigma \operatorname{grad} \Phi$, where $\mathbf{q}$ is the flow vector whose magnitude is equal to the volume of fluid crossing a unit of area normal to the flow direction in unit time, and σ a specific conductivity parameter depending upon both the properties of the fluid and the medium. This is an expression of Darcy's law and is physically, as well as mathematically, analogous to Ohm's law in electricity and leads to the same deductions in analogous situations.

It is shown that $\sigma = k\rho/\eta$, where k is the permeability parameter depending upon the geometrical properties of the medium only and ρ and η are the density and viscosity, respectively, of the fluid.

The remainder of the paper is devoted to deducing the consequences of Darcy's law as just expressed, with particular regard for the practical problems of ground-water hydrology.

INTRODUCTION

Since the pioneer experiments of Darcy[1] on the flow of water through filter sands, a succession of important analytical treatments have appeared in which the flow of ground water has been discussed

[1] Henry Darcy, *Les Fontaines publiques de la ville de Dijon* (Paris: Victor Dalmont, 1856).

as a problem in field theory of mathematical physics. Outstanding among these analytical treatments is the now almost classical study of Slichter[2] and the recent treatments of Dachler[3] and of Muskat.[4]

While the importance of these and similar works cannot be overestimated, they have all fallen short in one or more important respects of the goal of establishing a theory of ground-water motion which is both free from internal contradictions and in conformity with all the fundamental principles of physics which ground-water motion must satisfy. In these treatments ample precaution has been taken not to violate the principle of the conservation of matter, but much less care has been exercised with respect to the equally inviolable first and second laws of thermodynamics.

In the present paper an attempt will be made to establish the theory of ground-water motion upon such a basis that the deduced consequences will be in more complete conformity with the principles of physics than has been the case heretofore. In doing this the entire subject will have to be re-examined from first principles because the contradictions to be avoided are inherent in the fundamental conceptions employed. The results we shall achieve will, in the main, resemble those of the existing treatments, but significant differences will arise where the neglect of energy relationships is nonallowable.

The present treatment will, as far as possible, be logically complete; yet the reader may occasionally feel the need for amplification of fundamental principles that can here be discussed only briefly. For this he is referred to standard treatises of potential theory, thermodynamics, hydrodynamics, dynamic meteorology, heat conduction, and electrodynamics. While many treatises of these various subjects exist, among those that the author has employed extensively are: Planck, *Theory of Heat* and *Thermodynamics;* Gibbs, "On the Equilibrium of Heterogeneous Substances" (in *Collected Works*); Kellogg, *Foundation of Potential Theory;* Coffin, *Vector Analysis;*

[2] C. S. Slichter, "Theoretical Investigation of the Motion of Ground Water," *19th Ann. Rept. U.S. Geol. Surv.*, Part II (1897–98), pp. 295–384.

[3] Robert Dachler, *Grundwasserströmung* (Vienna: J. Springer, 1936).

[4] Morris Muskat, *Flow of Homogeneous Fluids through Porous Media* (New York: McGraw-Hill, 1937).

Ewald, Pöschl, and Prandtl, *The Physics of Solids and Fluids;* Prandtl and Tietjens, *Hydro- and Aerodynamics* (2 vols.); V. Bjerknes, "Dynamic Meteorology and Hydrography" (*Carnegie Inst. Wash. Pub. 88* [1911]); Abraham and Becker, *Classical Electricity and Magnetism;* Page and Adams, *Principles of Electricity.*

For the geological background of ground-water hydrology no better source exists than the publications of the Division of Ground Water of the United States Geological Survey, particularly Meinzer, "The Occurrence of Ground Water in the United States with a Discussion of Principles" (*U.S. Geol. Surv. Water-Supply Paper 489* [1923]). A recent treatment of the same subject is Tolman, *Ground Water.*

THE DARCY EXPERIMENT

The general problem of ground-water motion is principally this: Suppose we are given the complete geometry (at least statistically) of an underground region extending from the earth's surface to some arbitrary depth, and we know the rate of the flow of water across the boundary of the region at all points, what is the nature of the flow at every point in its interior? A variation of this problem that is commonly encountered in practice consists in being given the geometry and the state of flow throughout a region in space and being required to anticipate the changes in the flow system which will result from certain specified alterations to be imposed upon it.

Stated thus broadly, both these problems appear formidable, and indeed are so until we develop certain necessary relationships which will enable us to deal with them. In order to acquire the necessary analytical tools for use in more general problems, we arbitrarily choose a simple flow system which we investigate exhaustively. Having done this, we then shall return with the knowledge thus gained to problems of a more complex nature.

The motion of ground water differs from the more familiar problems of fluid flow in that in this case the flow, instead of occurring in open basins or channels, takes place through an intricately branched network of open spaces interpenetrating a skeletal solid framework. There are, therefore, two distinct and essential elements to every such flow problem: the properties of the fluid, and the properties of the solid framework or *medium*. The flow properties of the fluid are sufficiently determined for most purposes by its viscosity, η, and its density, ρ.

The two more important properties of the medium are its *porosity* and its *permeability*. The porosity, ϵ, may be defined as the ratio of the free space, or voids, to the total volume. In rocks this ranges from zero to more than 50 per cent. The permeability of the medium pertains to the facility with which fluids flow through it. For the present only a qualitative definition is possible. Of two media through which the same fluid is made to flow under identical conditions, we may say that the one through which the flow is more rapid has the greater permeability. A medium may be said to be *isotropic* with respect to permeability if it is equally permeable to flow in all directions.

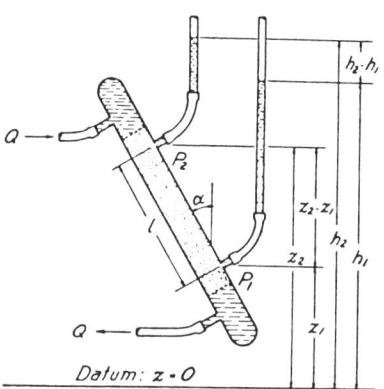

FIG. 1.—Apparatus for studying the flow of liquid through a permeable material.

Now, to begin our analysis, we choose a region of space occupied by a medium which is homogeneous and isotropic with respect to porosity and permeability. We assume, moreover, that the solid framework is insoluble and chemically inert with respect to the fluid passing through it, and ideally rigid. For the region to be investigated we choose a volume whose shape is that of a right circular cylinder of cross-sectional area A and length l, where both A and l are large, compared with the small irregularities or grain-size of the medium. We require that the flow through the volume be rectilinear and in a direction parallel to the axis of the cylinder.

These conditions may be approximately realized experimentally if we construct the apparatus shown in Figure 1. This consists of a large metal cylinder mounted so as to pivot about a horizontal axis perpendicular to its own axis. The mid-section of the cylinder between two screens is filled with sand or other suitable granular material of uniform grade. The ends of the cylinder are closed, but near

THE THEORY OF GROUND-WATER MOTION

each end inlet and outlet hose connections are attached. At points P_1 and P_2 in the same axial line and distance l apart two manometer openings are tapped into the mid-section filled with the permeable material. By means of flexible rubber tubing these are connected with permanently mounted open-topped manometers.

We choose a standard datum of elevation above which the elevations of the points P_1 and P_2 are z_1 and z_2, respectively. We let the angle of tilt of the large cylinder be α, where α is zero when the cylinder is vertical with P_2 above P_1.

Water is made to flow through the system at a total discharge rate of Q units of volume per unit of time. It rises to elevations h_1 and h_2 above the standard datum in the two manometer tubes terminated at P_1 and P_2, respectively.

We now wish to investigate the relations among the several variables of the system. We adopt the convention that the total discharge, Q, is positive when the flow is directed from P_1 to P_2, and negative when from P_2 to P_1. By varying Q we establish the following relationships: When

$$\left. \begin{array}{llll} Q = 0, & h_2 = h_1 & \text{or} & h_2 - h_1 = 0, \\ Q > 0, & h_2 < h_1 & \text{or} & h_2 - h_1 < 0, \\ Q < 0, & h_2 > h_1 & \text{or} & h_2 - h_1 > 0. \end{array} \right\} \quad (1)$$

We also observe that this relationship is one of proportionality and that

$$Q \propto -(h_2 - h_1). \qquad (2)$$

We next investigate the effect of changing the angle of tilt α. Letting Q remain constant, we tilt the system so that α assumes all possible values from zero to 180°, and we discover that the results expressed by equations (1) and (2) are in nowise influenced by the value of α, being exactly the same whether the system is upright, inclined, horizontal, or upside down.

Two other important variables of the system whose effects require investigation are the area of cross section A and the length l. This could be done by rebuilding the apparatus and varying first A and then l; but to do so is hardly necessary, since the answers are implicit in the experiment already described.

Consider first the variation of A. Since the direction of the flow is axial, no change will be produced if we insert a thin axial partition dividing the cylinder into two halves of area $A/2$ each and discharge $Q/2$ in each half. In this case the manometer readings, which remain unchanged, are registering the flow in only one-half of the system. Still further partitioning would give us a series of n equal parallel tubes, each conducting water at a rate Q/n, with still no change in the manometer readings, although only the flow Q/n in a tube whose cross section is A/n is being recorded. From this it is clear that a reduction in cross section of the flow system has no influence on the manometer readings provided the discharge across that area is varied at the same rate. This leads us to the conclusion that it is the ratio Q/A which must be kept constant if the manometer readings are not to vary. The total discharge divided by the cross-sectional area is obviously equal to the discharge per unit of area or to the *specific discharge q*. Consequently,

$$\frac{Q}{A} = q \propto -(h_2 - h_1) \tag{3}$$

or, for a given value of the manometer readings,

$$Q = qA, \tag{4}$$

where the specific discharge, q, is the constant of proportionality.

In a similar manner we investigate the effect of varying the length l between the bases of the two manometer tubes. The height difference $h_2 - h_1$ is the reaction to the discharge q through the length l. Since the flow is uniform over every cross section of the system, then for length dl there must correspond a fall in h, of dh, such that

$$\frac{dh}{dl} = \frac{h_2 - h_1}{l}, \tag{5}$$

and these quantities must be proportional to q. Or, if $(h_2 - h_1)$ is kept constant,

$$Q \propto \frac{1}{l}. \tag{6}$$

Now, if we combine the results of equations (2)–(6) into a single expression, we obtain

$$Q = -KA \cdot \frac{h_2 - h_1}{l} ; \qquad (7)$$

or, by employing the specific discharge q and the differential expression dh/dl,

$$q = \frac{Q}{A} = -K \cdot \frac{dh}{dl}, \qquad (8)$$

where K is a constant of proportionality.

With the exception of the variation of the angle of tilt, α, which, as we have seen, produces no effect, the experiment just described is essentially that performed originally by Henry Darcy, whose results are better expressed in his own words:

> Ainsi, en appelant e l'eppaiseur de la couche de sable, P la pression atmosphérique, h la hauteur de l'eau sur cette couche, on aura $P + h$ pour la pression a laquelle sera soumise la base superieure; saient, de plus, $P \pm h_0$ la pression supportié par la surface inferieure, k un coefficient dependant de la perméabilité de la couche, q le volume débité, on a
>
> $$q = k\frac{s}{e}[h + e \pm h_0]$$
>
> qui se réduit à
>
> $$q = k\frac{s}{e}[h + e], \quad \text{quand} \quad h_0 = 0,$$
>
> au lorsque la pression sous le filtre est equal à la pression atmosphérique.[5]

Darcy explains elsewhere that s is the area of cross section. Although the language employed is necessarily somewhat archaic by present standards, it seems clear that his standard datum from which the manometer heights were measured was the lower base of his filter sand. Then $h + e$ is the height in the upper manometer, and h_0 that in the lower, this being either positive or negative, depending upon whether or not a vacuum was employed. It is clear, therefore, that the results announced by Darcy are entirely in accord with those expressed by equations (7) and (8)—a relationship which appropriately has come to be known as "Darcy's law."

[5] *Op. cit.*, pp. 570 ff.

Darcy also noted, significantly, that the relationship was no longer valid for fluid velocities greater than 10–11 cm/sec.

THE PHYSICAL SIGNIFICANCE OF DARCY'S LAW

The relationships embodied in equation (8) form the necessary basis upon which any analytical theory of ground-water motion must rest; but the method of obtaining this equation has been empirical, and, expressed in this primitive form, the equation is of slight usefulness because it expresses only what we have learned already and gives us no insight into the deeper mechanism of fluid flow.

What determines the direction of the fluid flow in the first place? What would be the effect if we changed from a finer to a coarser sand? How would the flow rate be altered if we changed the viscosity or the density of the fluid? These and other similar ones are questions that equation (8) does not answer; yet they are questions to which it is highly important that answers be known.

Let us consider first the question of what determines the direction of the fluid motion, that is, whether the flow in Figure 1 is to be directed from P_1 to P_2 or from P_2 to P_1. We obviously cannot say that elevation is the determining factor, because, as we have seen already, if the flow is initially from a higher to a lower elevation, an inversion of the system changes the flow from a lower to a higher elevation without affecting in the slightest degree either its rate or the readings of the manometers.

If it is not elevation, perhaps it is the fluid pressure that is the determining factor in the flow, with the flow always directed away from regions where the pressure is higher and toward those where it is lower. In fact, the great majority of all writers upon this subject have stated that this is so, and many have employed equations of the form

$$q = -K' \cdot \frac{dp}{dl} \qquad (9)$$

as a statement of Darcy's law, presumably under the impression that equations (8) and (9) are physically equivalent statements.

Whether or not this is so may readily be determined in the following manner: At any point P in a flow system whose elevation above the standard datum is z, we terminate a manometer tube in which the liquid rises to a height h above the datum. The pressure at the point P is determined by the height of the liquid column above P and is given by

$$p = \rho g(h - z) + p_0, \qquad (10)$$

where p is the pressure, ρ the density of the liquid, g the acceleration due to gravity, and p_0 the pressure of the atmosphere. Now, if we apply equation (10) to the points P_1 and P_2 of Figure 1, we obtain

$$\left. \begin{array}{l} p_1 = \rho g(h_1 - z_1) + p_0, \\ p_2 = \rho g(h_2 - z_2) + p_0, \end{array} \right\} \qquad (11)$$

and the difference between the pressures at the two points is

$$p_2 - p_1 = \rho g(h_2 - h_1) - \rho g(z_2 - z_1). \qquad (12)$$

Now we already know that as $(h_2 - h_1) \to 0$, $q \to 0$; but in that case p_1 is less than p_2 by the quantity $\rho g(z_2 - z_1)$, which gives us a case where the flow is zero when one of the two pressures is greater than the other. Next, suppose we make h_2 slightly greater than h_1, which will correspond to flow from P_2 to P_1, but still not great enough for $(h_2 - h_1)$ to equal $(z_2 - z_1)$. In this case the pressure p_2 will still be less than p_1, and we shall have flow from a region of lower to one of higher pressure. Then if we invert the system by rotation through 180° about a horizontal axis, the pressure at P_2 will become greater than that at P_1 without the rate of flow or the manometer reading being changed. Consequently, we must conclude that a liquid can with equal facility be made to flow from a region of higher to one of lower pressure, or from a region of lower to one of higher, quite arbitrarily.

To investigate equation (9) we need only to express equation (12) differentially:

$$\frac{dp}{dl} = \rho g \cdot \frac{dh}{dl} - \rho g \cdot \frac{dz}{dl}. \qquad (13)$$

But

$$\frac{dz}{dl} = \cos \alpha ,\qquad(14)$$

so that when we insert this into equation (13) we obtain

$$\frac{dp}{dl} = \rho g \cdot \frac{dh}{dl} - \rho g \cos \alpha .\qquad(15)$$

When this is substituted into equation (9), it gives

$$q = -K'\rho g \left(\frac{dh}{dl} - \cos \alpha\right) ,\qquad(16)$$

which is manifestly not equivalent to equation (8) under any condition except when the second term, that of cos α, is negligible, compared with the first. This is only true either when cos α is approximately zero, corresponding to nearly horizontal flow, or when dh/dl is very large, compared with unity, the maximum value of cos α.

In the experiment of Figure 1, and in ground-water motion in general, the direction of flow may have any inclination from vertical to horizontal, so that the assumption of cos $\alpha = 0$ is not tenable. Moreover, dh/dl is ordinarily very much smaller than unity, the maximum value of cos α. Hence, in problems of ground-water flow, equation (9) is not only not equivalent to Darcy's law, as expressed by equation (8), but is not even a valid approximation to it. Consequently, the employment of equation (9) or any equivalent statement wherein the rate of flow is assumed to be proportional to the pressure gradient, when dealing with ground-water problems, is to be ruled out on the grounds of being physically erroneous.

When dealing with the flow of gases, on the other hand, dh/dl, or its equivalent, may be large, compared with cos α, in which case the employment of equation (9) becomes an allowable approximation.

What, then, does determine the direction of the flow? What we still seek to find is some physical quantity, capable of measurement at every point in a flow system, whose properties are such that the flow always occurs from regions in which the quantity has higher values to those in which it has lower, regardless of the direction in space. What we have demonstrated so far is that neither elevation nor pressure is such a quantity.

Formally, the manometer height h satisfies this condition entirely, but to adopt it empirically without further investigation would be like reading the length of the mercury column of a thermometer without knowing that temperature was the physical quantity being indicated. The quantity we seek, therefore, is evidently the hidden physical quantity whose magnitude is indicated by the height h of the liquid in the manometer tube, measured from a standard datum.

The clue to this quantity is to be found if we direct our attention momentarily to an apparently irrelevant experiment. Suppose we take a simple pendulum, start it swinging with a wide amplitude, and then leave it entirely free from outside disturbance. We record its motion by means of time exposures taken periodically, the time of each exposure being made equal to the period of the pendulum. We shuffle the photographs and ask a second person to arrange them in chronological order. This is illustrated by Figure 2, where A, B, C, D, and E are prints selected at random, and the problem is to determine the chronological order in which the exposures were made.

By inspection we would say that the order was D, B, E, A, and C, because this is the order of decreasing amplitude of swing, and our experience tells us that an isolated pendulum always swings with decreasing amplitude—never the reverse.

We have, therefore, an experimental process which only goes in a single direction, and it is this unidirectional characteristic that we wish to consider. This we may best do by taking into account the energy transformations of the process. To set the pendulum swinging in the first place, we had to give it an initial supply of potential energy. As it swung back and forth, there were periodical transformations of potential energy to kinetic energy and back to potential energy again, the sum of the two comprising the mechanical energy of the system and remaining constant for constant amplitude of swing. But we have noted already that the amplitude continuously diminished; consequently, there must have been a continuous dissipation of the mechanical energy initially supplied to the system. Also, ultimate equilibrium corresponds to a state of rest with the pendulum at its lowest possible position, that is to say, with the kinetic energy of the system equal to zero, and the potential energy the minimum possible compatible with the constraints of the system.

We know, moreover, that the progressive decrease in the amplitude of swing is due to frictional resistances; and, by the principle of the conservation of energy, we know that the mechanical energy lost by the system reappears as heat at the temperature of the surroundings. Now, if we could transform this thermal energy back to mechanical energy again without other permanent changes, then we could remotivate the pendulum, allowing it to use over and over again the same energy, which would be a form of perpetual motion.

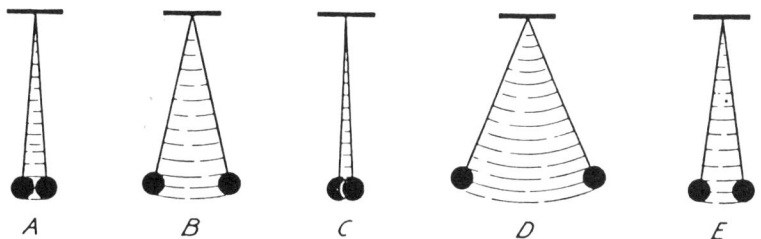

Fig. 2.—Unidirectional and irreversible transformation of isolated mechanical system.

Our purely negative experience in this connection leads to the conclusion that such a process is impossible, and a generalization of this fact so as to apply to all manner of processes gives us the Second Law of Thermodynamics, which states, in essence, that any material transformation which involves friction, or its equivalent, is unidirectional and irreversible in character, meaning that, once the process has taken place, by no method whatsoever can it be undone again. In other words, if we have any initial configuration of matter in an isolated system, and this configuration undergoes spontaneous change accompanied by friction, it is quite impossible ever to restore the matter contained within that system to its initial configuration and at the same time leave the material configuration external to that system undisturbed.

It is this method of reasoning that we now wish to employ with respect to our problem of fluid flow. The flow of a fluid is a mechanical process; and, when the fluid has viscosity and the flow occurs through the small passages of porous rocks, friction is not only present but is one of the dominant influences in the process. Consequently, if such flow occurs at all, it must be accompanied by an irreversible transformation of mechanical to thermal energy through the mechanism of fluid friction. Furthermore, like the pendulum, final equilibrium must correspond to a configuration in which the kinetic energy is zero; and the potential energy, the minimum possible compatible with the constraints of the system. Hence, to have flow, there must have been supplied an initial store of mechanical energy, and the changes that occur thereafter must always be in the direction of a decrease by dissipation into heat of this mechanical energy.

So far this gives the generalized direction of the process but not the geometrical direction of the flow in space. The latter is obtained, however, when we consider that the mechanical energy of the system is associated with the elements of its mass or volume and that these occupy particular positions in space. The direction of the flow in space must therefore be away from regions in which the mechanical energy per unit of mass is higher and toward regions in which it is lower.

It is not to be inferred that the energy which a system possesses in consequence of the fact that an element of mass occupies a given position resides upon or within the mass element. For example, the potential energy $U = gz$ of a unit mass due to the earth's gravitational field is an energy of the unit mass and the earth considered as a single system; yet we may unambiguously refer this energy to the unit mass at a specified position. This same is true of the fluid system with which we now deal.

The mechanical energy of the fluid per unit of mass is, therefore, evidently the physical quantity we set out to find; and hereafter we shall refer to it as the *potential* of the fluid in question, there being as many distinct potentials as there are distinct fluids to be dealt with.

Our problem now resolves itself into the determination of the fluid potential, or mechanical energy per unit of mass, of the given fluid at any arbitrary point in space. To obtain this we note first that energy is a relative quantity and is measurable by the amount of work required to effect any given transformation from some arbitrary initial state to a specified final state. The potential of a fluid at a specified point is, accordingly, the work required to transform a unit of mass of the fluid from an arbitrarily chosen standard state to the state at the point under consideration. For the standard state it is convenient to employ an elevation of zero, a pressure of 1 atmosphere, and a velocity of zero (relative to the earth's surface). Let the fluid in its final state at the point P be characterized by an elevation z, a pressure p, and a velocity v. We may also let V_0 be the volume per unit mass, or specific volume, of the fluid in its standard state, and V be that for the final state. Let the corresponding densities be ρ_0 and ρ. We may also note that the density is the reciprocal of the specific volume: $V = 1/\rho$.

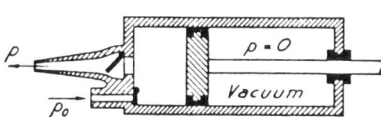

FIG. 3.—Pump for transforming liquid from standard to final state.

We wish to find the work required to transform a unit mass of the fluid from the initial to the final state, and to do this we imagine a pump constructed along the lines indicated by Figure 3. This consists of a cylinder with frictionless piston, on the front of which is the fluid chamber and on the back a perfect vacuum. Inlet and outlet valves are provided. We then imagine the transformation to be effected by the following successive steps:

1. Under standard conditions we slowly withdraw the piston and charge the cylinder with unit mass of the fluid. The work done by the piston on the fluid is then

$$w_1 = -p_0 V_0. \tag{17}$$

2. Next we lift the pump with its fluid contents to the point P of elevation z. The work expended for this is

$$w_2 = +gz + m_p gz, \tag{18}$$

where gz is the work required to lift the unit mass of the fluid, and $m_p gz$ that required to lift the pump alone.

3. The contents of the cylinder are injected into the system at point P. The work required for this is

$$w_3 = + \int_V^{V_0} p \cdot dV + pV. \qquad (19)$$

The first term to the right of equation (19) is the work of compression of the fluid in order to raise its pressure from p_0 to p before it can be injected. The pV term is the work of injection against the pressure p.

4. The fluid is accelerated from a velocity of zero to that of v, requiring an amount of work

$$w_4 = + \frac{v^2}{2}. \qquad (20)$$

5. The cylinder is returned to its initial position at zero elevation, thus completing the cycle. This requires an amount of work

$$w_5 = - m_p g z. \qquad (21)$$

The sum of these separate amounts of work is the potential Φ of the fluid at the point P. Performing the addition and canceling out terms that repeat with opposite signs gives us

$$\Phi = gz - p_0 V_0 + \int_V^{V_0} p\, dV + pV + \frac{v^2}{2}. \qquad (22)$$

In this, the first and last terms on the right-hand side are the gravitational potential energy and the kinetic energy, respectively. The significance of the other three terms is best visualized by means of the "indicator diagram" of Figure 4, in which the pressure in the cylinder is plotted against piston displacement for both compressible and incompressible fluids. If the fluid is incompressible, a condition satisfied approximately by liquids under ordinary pressure ranges,

$$\int p \cdot dV = 0 \quad \text{and} \quad V = V_0. \qquad (23)$$

In this case the pressure-volume work reduces to $(p - p_0)V$, and equation (22) simplifies to

$$\Phi = gz + (p - p_0)V + \frac{v^2}{2}. \tag{24}$$

By a mathematical transformation we can convert equation (22) into another form whose physical significance may not be im-

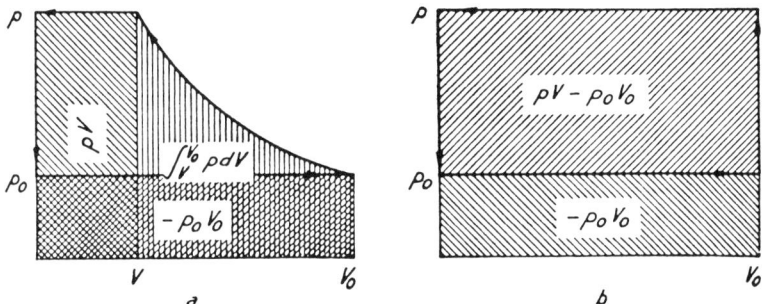

FIG. 4.—Pump indicator diagrams: a, for gases; b, for liquids

mediately apparent but which will prove to be of great usefulness later. To effect this we make use of the fact that

$$d(pV) = p \cdot dV + V \cdot dp,$$

or

$$\int p \cdot dV = \int d(pV) - \int V \cdot dp. \tag{25}$$

Then, as a definite integral, this becomes

$$\left. \begin{array}{l} \int_V^{V_0} p \cdot dV = \int_{pV}^{p_0V_0} d(pV) - \int_p^{p_0} V \cdot dp = p_0V_0 - pV \\ \qquad\qquad + \int_{p_0}^{p} V \cdot dp. \end{array} \right\} \tag{26}$$

Substituting this value for $\int_V^{V_0} pdV$ into equation (22) transforms that equation into

$$\Phi = gz + \int_{p_0}^{p} V \cdot dp + \frac{v^2}{2}, \tag{27}$$

which, when we substitute $1/\rho$ for V, becomes

$$\Phi = gz + \int_{p_0}^{p}\frac{dp}{\rho} + \frac{v^2}{2}. \tag{28}$$

A graphical interpretation of equation (26) is readily afforded by noting that the area enclosed by the indicator diagram of Figure 4, *a*, is equal to the work performed by the pump per cycle and that this is given by $\int_{p_0}^{p} V \cdot dp$ or by $\int_{p_0}^{p}\frac{dp}{\rho}$.

Equation (28) is an expression for the fluid potential at a point P in the most general form that we shall require. In order for the value of Φ so defined to be unique, however, it is necessary to stipulate that the density of the fluid must be a function of the pressure only, for otherwise the value of $\int_{p_0}^{p} V dp$ will be indeterminate. For liquids this condition is automatically complied with to the extent that their densities may be regarded as constant. For gases it is satisfied by isothermal conditions, which for many problems is a satisfactory approximation, and also by adiabatic conditions, which are of great importance in meteorology.

If the fluid flows without friction, as is approximately the case for liquids of small viscosity and gases, when moving in large open spaces, then each element of mass retains its mechanical energy undiminished, and along any given path of fluid flow each unit of mass will have the same energy and potential. For that case

$$\Phi = gz + \int_{p_0}^{p}\frac{dp}{\rho} + \frac{v^2}{2} = \text{constant}, \tag{29}$$

which is a generalization of a celebrated theorem announced in 1738 by Daniel Bernoulli relating the elevation, pressure, and velocity along a given flowline of a fluid in frictionless flow.

If the flow is not frictionless, as is pre-eminently the case with ground water and similar fluid flow, then the mechanical energy initially possessed by an element of the fluid must continuously be dissipated as this element traverses its path of flow. Consequently, the value of Φ in this instance must continuously decrease in the direction of the flow.

Another consequence of friction in ground-water motion is to damp out any large velocities, the flow velocity of ground water being rarely as great as 1 cm/sec and commonly of the order of a few centimeters per day. For our purposes we can accordingly neglect the kinetic energy term $v^2/2$ as being negligible in comparison with the other terms of the equation for the potential. This simplifies the expression for the potential down to

$$\Phi = gz + \int_{p_0}^{p} \frac{dp}{\rho}, \qquad (30)$$

which, for liquids, reduces still farther to

$$\Phi = gz + \frac{p - p_0}{\rho}. \qquad (31)$$

Now, in order to be able to measure Φ at any point P of elevation z and pressure p in a liquid flow system, suppose we return to our earlier device of terminating at the point P a manometer tube into which the liquid rises to the height h above the standard datum. The pressure p in terms of the manometer reading and the elevation z is then given by equation (10). When this is substituted into equation (31), we obtain the remarkably simple result

$$\Phi = gz + \frac{[\rho g(h - z) + p_0] - p_0}{\rho} = gh. \qquad (32)$$

Hence the fluid potential is indeed the hidden physical quantity we originally set out to discover, since its magnitude is indicated by the height h of the manometer and is numerically equal to h multiplied by the acceleration due to gravity.

With this background, let us return now to the results of the Darcy experiment as expressed empirically by equation (8). From equation (32)

$$\frac{dh}{dl} = \frac{1}{g} \cdot \frac{d\Phi}{dl}, \qquad (33)$$

which, when employed in conjunction with equation (8), gives the two equivalent alternative expressions of Darcy's law:

$$q = -K \cdot \frac{dh}{dl} = -K \cdot \frac{1}{g} \cdot \frac{d\Phi}{dl}. \tag{34}$$

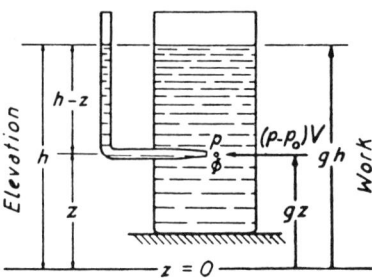

FIG. 5.—Fluid potential at any point inside a body of static liquid

ANALYSIS OF THE PARAMETER K

Our next task is to analyze the parameter K. In the experiment of Figure 1 we investigated certain variables whose effects are stated explicitly in equation (8), but in performing this experiment we arbitrarily kept all other factors constant except the particular ones under investigation. The experiment, however, was incomplete in that a number of possible variables were not taken into consideration. What, for example, would have been the effect of a change of the density or the viscosity of the fluid, of the coarseness of the sand, or of other possible variables of the system?

Manifestly, all these other variables were included in the stipulation "all else remaining constant"; and their combined effects are lumped together into the "constant" of proportionality K, which is accordingly, therefore, not a true constant but a variable parameter whose value depends upon those of the lumped variables. It is important that we find these hidden variables so that we can include them explicitly in our flow equation, enabling us to eliminate K entirely. To do this we must delve somewhat more deeply than heretofore into the mechanics of fluid flow.

In what follows, two different points of view—the microscopic and the macroscopic—distinguished from one another by the size scale adopted, will prove useful. When we employ the microscopic point of view, we shall be concerned with fluid elements that are large compared with the kinetic irregularities of molecular motion but smaller than the open passages of the medium. When we employ the macroscopic point of view, the fluid elements we speak of shall be large enough that the irregularities of flow due to the medium need not be considered, but only the statistical resultant.

In order to determine what the various factors are that influence the rate of flow, and the part they play, we need to know what the forces are that act upon the fluid elements and upon what these depend. To determine these we shall stipulate that the flow through the sand in the apparatus of Figure 1 be kept *steady*, that is, not changing with time, which will be accomplished if we maintain q constant at any arbitrarily chosen value. Then we adopt the microscopic point of view in order to examine the flow in some detail.

When seen on this scale, the flow will consist of the passage of the fluid through an intricate, branching, three-dimensional network of interstices, analogous to the two-dimensional flow of a river through a complex of small islands. If we choose a particular point, fixed with respect to the framework, then a fluid particle passing through this point will follow a definite path, or *streamline*, and every other particle passing through the same point will transverse the same path and assume the same velocities at corresponding points along the path. Consequently, to every fixed point within the flow system there corresponds a particular fluid velocity, both in magnitude and direction, and through each point there passes a particular streamline. We may accordingly think of the flow as being represented by a family of streamlines, one passing through each point, or by a *field of velocity vectors*, one terminated at each fixed point occupied by the fluid.

In addition to this, around any fixed point we may take a small volume element, also fixed with respect to the framework. While the fluid continuously flows through such an element, at every instant the volume element is occupied by a particular element of the fluid. The forces acting upon this element are of two kinds: those that act upon its surface, or *surface* forces, and those that act upon the mass within the volume element, or *body* forces. If dV is the volume of the element, then $\rho \cdot dV$ will be its mass, and by Newton's laws of motion the sum of the applied forces is equal to the product of the mass by its acceleration. Then

$$\rho \cdot dV \cdot \mathbf{a} = \mathbf{f}_a + \mathbf{f}_b, \qquad (35)$$

where the vector $\mathbf{a}$ is the acceleration and the vectors $\mathbf{f}_a$ and $\mathbf{f}_b$ the surface and body forces, respectively.

Now, by the D'Alembert principle, we can introduce a fictitious force

$$\mathbf{f}_i = -\rho \cdot dV \mathbf{a}, \qquad (36)$$

which is the reaction due to inertia of the body to the applied forces $\mathbf{f}_a$ and $\mathbf{f}_b$. Then we have

$$-\mathbf{f}_i = \mathbf{f}_a + \mathbf{f}_b, \quad \text{or} \quad \mathbf{f}_i + \mathbf{f}_a + \mathbf{f}_b = 0, \qquad (37)$$

which tells us that the moving element of fluid is in a state of dynamic equilibrium under these three forces in a manner quite analogous to the more familiar static equilibrium.

A more useful classification of the forces acting upon the fluid element is upon the basis of function. On this basis we may speak of a *driving* force $\mathbf{f}_d$, a resistive force arising from frictional resistance due to the fluid viscosity, $\mathbf{f}_r$, and a reactive force due to the inertial reaction to acceleration $\mathbf{f}_i$. We may think of $\mathbf{f}_d$ as being the independently variable force which supplies the energy to the fluid element. Both $\mathbf{f}_r$ and $\mathbf{f}_i$ are dependent variables, depending only upon the state of motion of the fluid and being zero when the fluid motion is zero. The inertial force, $\mathbf{f}_i$, neither contributes energy to nor subtracts it from the system, whereas $\mathbf{f}_r$ is the energy-dissipating force.

From this it is clear that $\mathbf{f}_d$ and $\mathbf{f}_r$ are the principal forces from the point of view of function, while $\mathbf{f}_i$ plays the role of a superimposed modifying influence.

In order to discover the factors which determine $\mathbf{f}_i$ and $\mathbf{f}_r$, the microscopic viewpoint is necessary. The driving force, $\mathbf{f}_d$, can also be deduced from this viewpoint, but to do so is unduly complicated.

We shall accordingly investigate the forces f_i and f_r acting upon a microscopic element and shall then extend our results by integration to a macroscopic volume element before deriving f_d. On a microscopic scale, however, when f_i and f_r are known, f_d, being the sum of these two, is uniquely determined.

In order to investigate the forces due to inertia and to viscosity, let us erect co-ordinate axes with the x-axis parallel to the direction

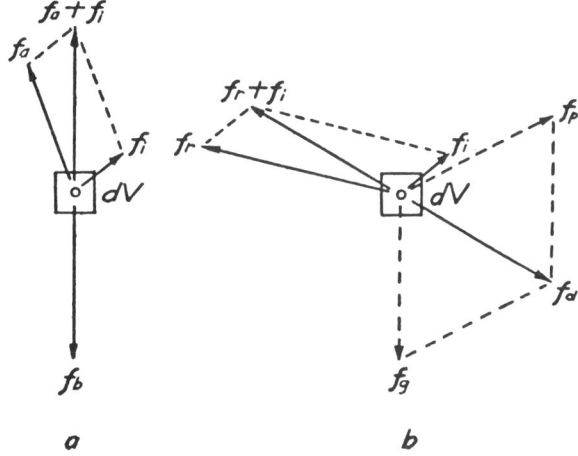

Fig. 6.—The forces which act upon a microscopic fluid element classified a as surface force, body force, and inertial force and b as driving force, resistive force, and inertial force.

of fluid flow. We let the volume element dV have lengths of side dx, dy, and dz. We let u, v, and w be the components of the velocity parallel to the x-, the y-, and the z-axes, respectively.

Taking first the force due to inertia, we resolve the force and the acceleration upon which it depends into components parallel to the separate axes. Taking the x-component,

$$a_x = \frac{du}{dt} = \frac{\partial u}{\partial t} + \frac{\partial u}{\partial x} \cdot \frac{\partial x}{\partial t} + \frac{\partial u}{\partial y} \cdot \frac{\partial y}{\partial t} + \frac{\partial u}{\partial z} \cdot \frac{\partial z}{\partial t}, \qquad (38)$$

where $\partial u/\partial t$ signifies the change of u with time at a particular point, $\partial u/\partial x$ the change of u with x at a particular time, and $\partial x/\partial t$ the change of the x-co-ordinate with time, of a particular particle, and so for the other terms. But

$$\left.\begin{aligned} \frac{\partial x}{\partial t} &= u, \\ \frac{\partial y}{\partial t} &= v = 0, \\ \frac{\partial z}{\partial t} &= w = 0, \end{aligned}\right\} \quad (39)$$

and, for steady motion,

$$\frac{\partial u}{\partial t} = 0. \quad (40)$$

Then, in this special case equation (38) simplifies to

$$a_x = u \cdot \frac{\partial u}{\partial x}, \quad (41)$$

and the force component is

$$f_{ix} = -\rho \cdot dV \cdot u \cdot \frac{\partial u}{\partial x}. \quad (42)$$

Similar expressions obtain for the components parallel to each of the other axes, but all that concerns us is the *form* of equation (42), so that we do not need to consider the matter in greater detail.

Now let us consider the force due to viscosity. Upon the y and z faces of the volume-element shear stresses will act. Let τ_{yx} be the shear stress parallel to the x-axis upon the y-face whose outward directed normal is toward the negative end of the y-axis. Upon the opposite face let the corresponding stress be $\tau_{yx} + (\partial \tau_{yx}/\partial y)dy$. Then the net force parallel to the x-axis produced by these two stresses will be

$$f_{r y x} = \left(\tau_{yx} + \frac{\partial \tau_{yx}}{\partial y} \cdot dy \right) dxdz - \tau_{yx} \cdot dxdz = +\frac{\partial \tau_{yx}}{\partial y} \cdot dV. \quad (43)$$

Complete elaboration would provide additional terms, but again they would all be of the form of the right-hand term of equation (43).

Our next problem is to relate the shear stress acting upon the surface of a fluid element with the state of its flow. By Newton's law of viscosity the rate of shear in a viscous fluid is proportional to the intensity of the shear stress acting upon it. If γ_{yx} is the change of

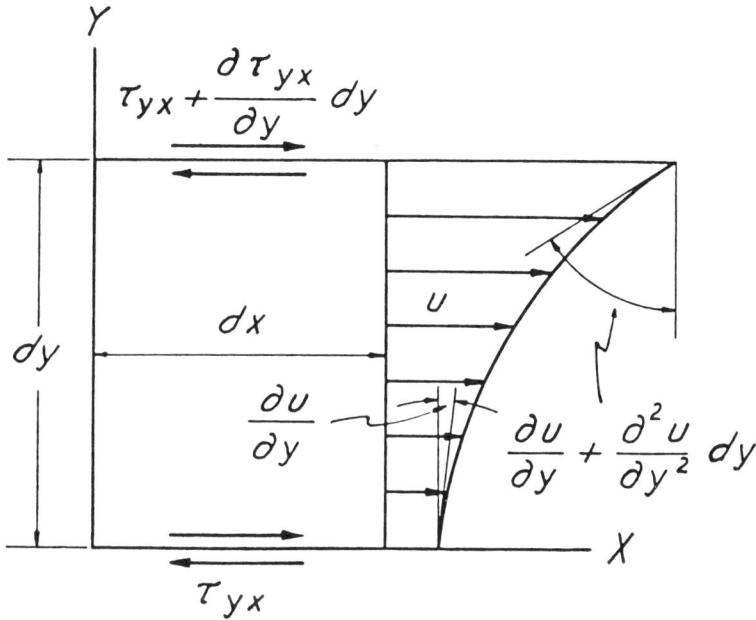

Fig. 7.—Manner of dependence upon the fluid velocity gradients of shearing stresses and unbalanced frictional forces acting upon microscopic fluid element.

an original right angle in the fluid with sides parallel to the x- and y-axes, then $d\gamma_{yx}/dt$ is the time rate of the shear, and

$$\tau_{yx} = \eta \cdot \frac{d\gamma_{yx}}{dt}, \tag{44}$$

where the constant of proportionality η is the viscosity of the fluid. But

$$d\gamma_{yx} = \frac{\left(u + \frac{\partial u}{\partial y} \cdot dy\right) dt - u\,dt}{dy} = \frac{\partial u}{\partial y} \cdot dt, \tag{45}$$

from which we obtain

$$\frac{d\gamma_{yx}}{dt} = \frac{\partial u}{\partial y},\qquad(46)$$

which, when substituted into equation (44), gives

$$\tau_{yz} = \eta \cdot \frac{\partial u}{\partial y}.\qquad(47)$$

Now if we differentiate this with respect to y,

$$\frac{\partial \tau_{yz}}{\partial y} = \eta \cdot \frac{\partial^2 u}{\partial y^2},\qquad(48)$$

which can then be substituted into equation (43), with the result that

$$f_{ryz} = \eta \cdot \frac{\partial^2 u}{\partial y^2} \cdot dV.\qquad(49)$$

What we are interested in finding is the manner in which the inertial and resistive forces vary with the specific discharge q and with a change of scale of the medium. For generality, suppose that we have two *geometrically similar* media of different-length scales, and through these we cause two different fluids to undergo *kinematically similar* flow. Geometrical similarity requires that all corresponding lengths of the two systems must have a constant ratio

$$\frac{l_2}{l_1} = \frac{d_2}{d_1},\qquad(50)$$

where d_1 and d_2 are corresponding grain diameters of the two systems, these being taken as characteristic lengths.

Kinematic similarity requires the two flow systems to be exact replicas of one another except for the length and time scales. All streamlines must be geometrically similar, and all corresponding velocities must have identical directions and proportional magnitudes. Since the specific discharge q has the dimensions

$$[q] = \left[\frac{\frac{\text{Volume}}{\text{Area}}}{\text{Time}}\right] = [LT^{-1}] = [\text{Velocity}],\qquad(51)$$

this represents a generalized velocity. Consequently, in kinematically similar flow all corresponding velocities must bear to each other the ratio

$$\frac{v_2}{v_1} = \frac{q_2}{q_1}. \tag{52}$$

Finally, since forces act at all points upon a fluid in motion, in order for two flows to be kinematically similar, it is necessary for all corresponding forces of the two systems to be similar, that is, to have identical directions and proportional magnitudes:

$$\frac{f_{i2}}{f_{i1}} = \frac{f_{r2}}{f_{r1}} = \frac{f_{d2}}{f_{d1}}. \tag{53}$$

This also necessitates that the corresponding force-equilibrium triangles of the two systems must be similar, so that

$$f_{i1} : f_{r1} : f_{d1} :: f_{i2} : f_{r2} : f_{d2} \tag{54}$$

or

$$\frac{f_{i1}}{f_{r1}} = \frac{f_{i2}}{f_{r2}}. \tag{55}$$

The manner in which the terms of equation (55) depend upon the velocity and the length scales is given by equations (42) and (49). For either of the two flow systems

$$\frac{f_i}{f_r} \propto \frac{f_{ix}}{f_{ryz}} = \frac{\rho u \cdot \dfrac{\partial u}{\partial x} \cdot dV}{\eta \dfrac{\partial^2 u}{\partial y^2} \cdot dV}. \tag{56}$$

In this

$$u \propto q, \quad \text{and} \quad \frac{\partial u}{\partial x} \propto \frac{q}{d}. \tag{57}$$

Similarly,

$$\frac{\partial^2 u}{\partial y^2} = \frac{\partial}{\partial y}\left(\frac{\partial u}{\partial y}\right) \propto \frac{q}{d^2}. \tag{58}$$

Substituting these results into equation (56) gives

$$\frac{f_i}{f_r} \propto \frac{\rho \cdot \frac{q^2}{d}}{\eta \cdot \frac{q}{d^2}} = \frac{qd}{\frac{\eta}{\rho}} = R. \tag{59}$$

Since equations (56)–(59) are identical for the two kinematically similar systems, then it follows that the quantity R defined by equation (59) must be the same for each. R is known as the Reynolds' number of the system, so called in honor of Osborne Reynolds,[6] whose pioneer studies first disclosed its significance.

It is important that the generality of equation (59) be appreciated, for, since it applies equally well to either of two kinematically similar systems, then

$$\frac{q_1 d_1}{\frac{\eta_1}{\rho_1}} = \frac{q_2 d_2}{\frac{\eta_2}{\rho_2}} = R, \tag{59a}$$

even though the length scales, the discharges, and the viscosity and density of the fluids may differ widely in the two cases.

Conversely, any two states of flow through geometrically similar frameworks will be kinematically similar when both have the same value of Reynolds' number; but for the same fluid in the same framework, since R is proportional to q, this condition cannot strictly be satisfied for different velocities of flow. Practically, it is satisfied for very small values of R, since, while both f_i and f_r tend to zero as q tends to zero, f_i decreases faster than f_r, and their ratio, which is proportional to R, tends also to zero as q and R tend to zero—the so-called "creeping motion" of flow. When f_i is negligible compared with f_r, then our force equilibrium becomes simply

$$\mathbf{f}_r = -\mathbf{f}_d \tag{60}$$

for every volume element within the flow system.

[6] "An Experimental Investigation of the Circumstances Which Determine Whether the Motion of Water Shall Be Direct or Sinuous and of the Law of Resistance in Parallel Channels," *Phil. Trans. Royal Soc. London*, Vol. 174 (1883), pp. 935-82; also *Papers on Mechanical and Physical Subjects* (Cambridge: University Press, 1901), Vol. II, pp. 51-105.

Now, if we agree to keep the flow velocities small enough so that the forces due to inertia are negligible and so that, therefore, equation (60) is valid, we can integrate f_r over all the microscopic volume elements of a macroscopic volume. The resultant will be $\mathbf{F}_r$, the resistive force acting upon the macroscopic element of volume. To perform this integration we observe that at every point f_r is directed parallel to the streamline, and, at least in the majority of cases, opposite to the direction of motion. Also, at each point f_r is proportional in magnitude to the velocity vector v.

Suppose the macroscopic flow is uniform and rectilinear. Then, if we choose an axis parallel to this macroscopic flow direction, we can resolve the microscopic velocity at every point into an axial component $v \cos \theta$ and a normal component $v \sin \theta$, where θ is the angle the streamline makes with this axis. Since the net fluid motion normal to the axis is zero, then there must be an equal number of normal components in all directions, so that they cancel one another completely.

The axial components, on the contrary, all have the same direction; and their average value, $\bar{v}$, over a macroscopic volume element is to be obtained by integration:

$$\bar{v} = \iiint_{\epsilon \cdot \Delta V} \frac{v \cos \theta \cdot dV}{\epsilon \cdot \Delta V} = \frac{q}{\epsilon}. \tag{61}$$

The product $\epsilon \cdot \Delta V$ is the fraction of a macroscopic volume that is occupied by the fluid, and division by this quantity of the integral of equation (61) is necessary to obtain the average velocity. Otherwise we would have the *sum* of the axial components.

Now, bearing in mind that f_r is proportional to v, it, too, can be resolved into radial and axial components and integrated. For the radial components

$$\iiint_{\epsilon \cdot \Delta V} f_r \sin \theta \cdot dV \propto \iiint_{\epsilon \cdot \Delta V} v \sin \theta \cdot dV = 0. \tag{62}$$

This leaves only the axial components whose algebraic sum or integral over the volume $\epsilon \cdot \Delta V$ is the macroscopic resistive force $\mathbf{F}_r$:

$$\iiint_{\epsilon \cdot \Delta V} f_r \cos\theta \cdot dV = \mathbf{F}_r \propto -\eta \cdot \frac{q}{d^2} \cdot \epsilon \cdot \Delta V. \tag{63}$$

Now, if we let $1/N$ be the constant of proportionality, we may write for the resistive force:

$$\mathbf{F}_r = -\frac{1}{N} \cdot \eta \cdot \frac{q}{d^2} \cdot \epsilon \cdot \Delta V. \tag{64}$$

If we divide this by $\rho\epsilon \cdot \Delta V$, we shall obtain

$$\frac{\mathbf{F}_r}{\rho\epsilon \cdot \Delta V} = -\frac{1}{N} \cdot \frac{\eta}{\rho} \cdot \frac{q}{d^2}, \tag{65}$$

which is the resistive force per unit of fluid mass.

On the macroscopic scale the two forces which act upon the fluid contained in a volume element ΔV are the driving force and the resistive force, and these must be equal in magnitude and opposite in direction, so that

$$\mathbf{F}_r = -\mathbf{F}_d. \tag{66}$$

Also, the directions of both of these forces must be parallel to the macroscopic direction of flow, $\mathbf{F}_d$ in the direction of motion and $\mathbf{F}_r$ in the opposite direction.

We have already related $\mathbf{F}_r$ to the macroscopic motion of the fluid; let us now investigate the driving force, $\mathbf{F}_d$. Let us take a macroscopic element of volume consisting of a prism whose length is Δl and whose area of cross section is ΔA, oriented with its axis parallel to the direction of macroscopic flow. The driving force acting upon the fluid within this volume element will be the resultant of the surface forces upon its exterior and the body forces upon the mass within its interior. In this case the surface forces consist of normal stresses only, since the shear stresses are expended against the rigid framework of the medium throughout the interior of the body and are not transmitted to distances appreciably greater than the mean grain diameter d. Their effect is accordingly, on a macroscopic scale, similar to a body force and gives rise to the resistive force $\mathbf{F}_r$, which has already been evaluated. The normal stress acting upon the surface of the macroscopic volume element is the hydrostatic fluid pressure p. For the body force we have only the attraction due to gravity.

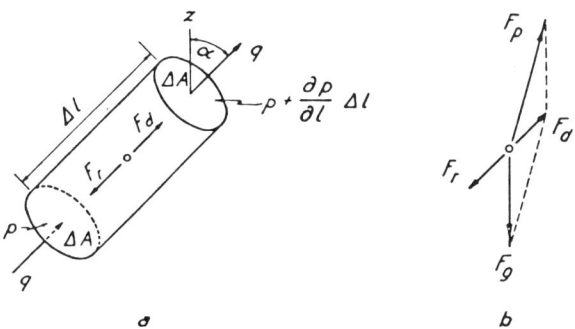

Fig. 8.—*a*, Forces acting upon macroscopic fluid volume element in direction parallel to flow; *b*, relation between forces parallel to flow and the total forces due to pressure gradient and gravity.

The total driving force must be the sum of the net force upon the volume element due to pressure and of the force due to gravitational attraction. Since the fluid is free to flow in any direction, this resultant must be in the direction of the fluid flow, and we need only consider the components of the other two forces in that direction.

For the component of the force due to pressure, let p be the pressure upon the upstream end of the prism, and $p + (\partial p/\partial l)\Delta l$ that upon the downstream end. We define the direction of the flow by the angle a which it makes from the upward directed vertical. Then in the direction of the flow a the component of the driving force due to pressure is

$$F_{pa} = p\epsilon \cdot \Delta A - \left(p + \frac{\partial p}{\partial l} \cdot \Delta l\right)\epsilon \cdot \Delta A = -\frac{\partial p}{\partial l} \cdot \epsilon \cdot \Delta V, \quad (67)$$

where $\epsilon \cdot \Delta A$ is the fraction of ΔA that is occupied by the fluid, and $\epsilon \cdot \Delta V$ the volume of fluid within the prism.

If we divide equation (67) by $\rho\epsilon\cdot\Delta V$, we obtain

$$\frac{F_{pa}}{\rho\cdot\epsilon\cdot\Delta V} = -\frac{1}{\rho}\cdot\frac{\partial p}{\partial l}, \tag{68}$$

which is the component of the force per unit of mass of the fluid in the direction a produced by the change of macroscopic pressure in that direction. It is numerically equal to the rate of increase with distance of the pressure and oppositely directed. That is, the force due to pressure alone is always directed from regions where the pressure is higher toward those where it is lower.

The component in the direction a of the force exerted by gravity acting upon the fluid contained within the volume ΔV is

$$F_{ga} = -g\cos a\cdot\rho\epsilon\cdot\Delta V = -g\cdot\frac{\partial z}{\partial l}\cdot\rho\epsilon\cdot\Delta V. \tag{69}$$

The negative sign here is necessary to allow for the fact that for upward flow $\cos a$ is positive, while for downward flow it is negative. In the expression $\partial z/\partial l$, z is the vertical co-ordinate, or the elevation above the standard datum.

Now, adding equations (67) and (69) gives us the total driving force acting upon the fluid in the volume element ΔV,

$$\mathbf{F}_d = \left(-g\cdot\frac{\partial z}{\partial l} - \frac{1}{\rho}\cdot\frac{\partial p}{\partial l}\right)\rho\epsilon\cdot\Delta V; \tag{70}$$

and, if we divide this by $\rho\epsilon\cdot\Delta V$, we obtain

$$\frac{\mathbf{F}_d}{\rho\epsilon\cdot\Delta V} = -g\cdot\frac{\partial z}{\partial l} - \frac{1}{\rho}\cdot\frac{\partial p}{\partial l}, \tag{71}$$

which is the total driving force acting upon each unit of mass of the fluid.

If we turn now to equations (30) and (31) and differentiate them with respect to l, and then compare the results with equation (71), we shall find that

$$-g\cdot\frac{\partial z}{\partial l} - \frac{1}{\rho}\cdot\frac{\partial p}{\partial l} = -\frac{\partial \Phi}{\partial l}, \tag{72}$$

so that

$$\frac{F_d}{\rho\epsilon \cdot \Delta V} = -\frac{\partial \Phi}{\partial l}. \qquad (73)$$

Hence, the total driving force per unit of mass of the fluid is numerically equal to the rate of increase with distance of the fluid potential and is oppositely directed.

Since the sum of the macroscopic forces per unit of mass must be zero, then, by adding the expressions for these forces as given by equations (65) and (73), we obtain

$$-\frac{1}{N} \cdot \frac{\eta}{\rho} \cdot \frac{q}{d^2} - \frac{\partial \Phi}{\partial l} = 0,$$

which, when solved for q, gives

$$q = -Nd^2 \cdot \frac{\rho}{\eta} \cdot \frac{\partial \Phi}{\partial l} = -Nd^2 \cdot \frac{\rho}{\eta} g \cdot \frac{\partial h}{\partial l}. \qquad (74)$$

When this is compared with Darcy's law as expressed by equation (34), it is clear that equations (34) and (74) are equivalent and that

$$K = Nd^2\rho \cdot \frac{1}{\eta} \cdot g, \qquad (75)$$

so that the five factors to the right of equation (75) are evidently the quantities we originally set out to discover which were lumped together as the parameter K. Of these, all have been defined except the factor N, which we employed as a factor of proportionality when relating the resistive force to the fluid velocity. N is simply a dimensionless numerical coefficient whose value depends upon the geometrical shape of the internal structure of the medium through which the flow occurs. For two geometrically similar media the values of N would be the same; for dissimilar media, such as rounded versus angular grain shapes, the values of N would be different. The dimensions of N are therefore

$$[N] = [\text{angle}] = [LL^{-1}] = [L^0]. \qquad (76)$$

Since its effect is determined solely by experiment, no more precise definition of N is required.

That the five quantities to the right in equation (75) are the correct ones, that they occur to their proper powers, and that no other essential quantities have been omitted may be demonstrated by a simple dimensional check. We rearrange equation (74) as follows:

$$\frac{q}{\frac{\partial h}{\partial l}} = -Nd^2\rho \cdot \frac{1}{\eta} \cdot g,$$

where the terms to be investigated are segregated to the right. If the equation is correct, then the dimensions of the right-hand term must be the same as those of the term to the left.

The dimensions of the separate factors are:

$$\left.\begin{aligned} [q] &= [LT^{-1}], \\ \left[\frac{1}{\frac{\partial h}{\partial l}}\right] &= [L^0], \\ [N] &= [L^0], \\ [d^2] &= [L^2], \\ [\rho] &= [ML^{-3}], \\ \left[\frac{1}{\eta}\right] &= [M^{-1}L^{+1}T^{+1}], \\ [g] &= [LT^{-2}]. \end{aligned}\right\} \quad (77)$$

Introducing these values into equation (74) gives the dimensional expression

$$\left[\frac{q}{\frac{\partial h}{\partial l}}\right] = \left[Nd^2\rho \cdot \frac{1}{\eta} \cdot g\right], \qquad (78)$$

or

$$\left.\begin{aligned} [M^0L^{+1}T^{-1}] &= [L^0L^2M^{+1}L^{-3}M^{-1}L^{+1}T^{+1}L^{+1}T^{-2}] \\ &= [M^0L^{+1}T^{-1}]. \end{aligned}\right\} \quad (79)$$

The fact that these dimensions balance is interpreted to mean that all the factors contained in K are given in the right-hand side of equation (78). Of these, only N might be further broken down into additional components, such as porosity and other geometrical parameters. In practice there is no need for such dissection, since N and d^2 are more conveniently lumped together as a single property of the medium. Consequently, we may regard the equations (74):

$$q = -Nd^2 \cdot \frac{\rho}{\eta} \cdot \frac{\partial \Phi}{\partial l} = -Nd^2 \cdot \frac{\rho}{\eta} \cdot g \cdot \frac{\partial h}{\partial l},$$

as equivalent and complete expressions of Darcy's law in the most general form we shall require. This tells us that, in addition to the factors investigated already, q varies directly as a factor N depending upon the internal shape of the medium, as the square of the grain size or other suitable length scale indicating coarseness, as the density of the fluid, and inversely as the viscosity of the fluid.

We come now to the problem of a more precise definition of *permeability* than the qualitative one given earlier. Of the five quantities N, d^2, ρ, $1/\eta$, and g of equation (74), the first two, N and d^2, are properties solely of the medium; the second two, ρ and η, are properties solely of the fluid; the last, g, is a property of the earth's gravitational field and may be taken, for present purposes, as constant. It would be possible to select from these five quantities any combination which includes the properties of the medium, N and d^2, and to make of this a single lumped parameter which could then be called the "coefficient of permeability." As will be discussed later, this is essentially what has been done already, so that numerous dimensionally unlike quantities, all called "permeability," are currently in use.

To avoid this sort of confusion, we ask what it is that we wish a coefficient of permeability to signify. We have already noted that the concept pertains to the facility with which a given rock or material transmits fluids. Presumably, then, permeability is to be taken as a property of the medium alone. If so, then a coefficient of permeability for a given medium should not change value when different fluids are employed. To satisfy this requirement we lump the factors Nd^2, depending upon the medium only, into the single factor k and then write Darcy's law as

$$q = -k \cdot \frac{\rho}{\eta} \cdot \frac{\partial \Phi}{\partial l}, \qquad (80)$$

where k is the coefficient of permeability of the medium. For any given medium its value is obtained by an experiment analogous to that of Figure 1, where all quantities except k are measured by experiment, and then equation (80) is solved for k.

The dimensions of k are

$$[k] = [Nd^2] = [L^2]. \qquad (81)$$

For purposes of mathematical analysis we are frequently concerned with the flow rate without regard to the parts played separately by the properties of the fluid and those of the medium. In this case it is more convenient to lump the factors k, ρ, and $1/\eta$ together into a single parameter σ, which then simplifies Darcy's law to

$$q = -\sigma \cdot \frac{\partial \Phi}{\partial l}, \qquad (82)$$

a form which is physically, as well as mathematically, analogous to Ohm's law in electricity when applied to extended media. In the flow of electricity the factor analogous to σ is called the *specific electrical conductivity;* here we shall call σ the *specific fluid conductivity.* From its definition it is clear that the specific fluid conductivity is a property both of the medium and of the fluid. For a homogeneous liquid under isothermal conditions the density and viscosity are constant. In this case the specific conductivity will vary proportionately to the permeability of the medium.

[Editors' Note: Material has been omitted at this point.]

Soil Anisotropy and Land Drainage

Marinus Maasland

IF AT ANY POINT of a soil the hydraulic conductivity (of dimensions LT^{-1}) is the same in every direction, the soil is said to be *isotropic*. If the soil medium is isotropic and, moreover, has the same hydraulic conductivity at all points, the soil is said to be *homogeneous-isotropic*. If the hydraulic conductivity at a point in a soil medium is *not* the same in every direction, and if this variation of hydraulic conductivity with direction is of the same nature over the whole medium, the soil is said to be *homogeneous-anisotropic*. Thus in homogeneous-anisotropic soil the hydraulic conductivity is a function of direction, but not of the coordinates. Laplace's equation is assumed to be valid for groundwater flow in the steady state through homogeneous-isotropic soils; that is,

$$\nabla^2 \phi = 0$$

where $\phi = (k'\gamma g/\mu)(p/\gamma g + z) = (k'\gamma g/\mu)h = kh.$

Here k' is the soil permeability as used by Muskat (1937) and others, μ is the viscosity of the soil water, γ is the density of the soil water, g is the acceleration of gravity, p is the pressure at a point in the soil, z is the vertical coordinate of the point where the pressure is p, k is equal to $(k'\gamma g/\mu)$, and $h = (p/\gamma g + z)$.

The lumped constant k, the hydraulic conductivity [Muskat (1937, p. 294) uses $\bar{k}$ where we use k], represents physically the infiltration rate, say in feet per day, for a water-saturated vertical column of soil whose upper surface is kept covered with a thin layer of water. Physically, h is the height above (or below) some arbitrary level at which groundwater would stand in a test pipe inserted in the soil. In such a test pipe $p/\gamma g$ is the distance of the water level above the pipe bottom and z (the value of z may be positive or negative) is the distance from the point to the reference level. It may be remarked that the water level in such a test pipe will correspond to the level of the water table only if the groundwater is in static equilibrium. The most pertinent characteristic of the hydraulic head h is that the expression $k\, \delta h/\delta s$ gives the quantity of water per unit area per unit time passing through an area perpendicular to the direction s.

Thus far, in the study of the flow of groundwater, attention has been directed mainly to the problem of the flow in isotropic homogeneous soils. Actual measurements, however, either in the field or in the laboratory, indicate that in nature we are usually concerned with anisotropic soils.

When the problem involves components of flow along more than one direction with different hydraulic conductivities, the anisotropy can be taken into account by applying a transformation of the coordinates as outlined in Section V.1. This is illustrated further by the treatment of the problems of flow towards ditches and tile drains, and of flow towards small cavities, in V.4, and V.5, respectively. It will be shown that the analytical problem in the transformed coordinate system is equivalent to one of flow in an isotropic medium. Therefore, from the analytical point of view, one returns in the treatment of such anisotropic systems, after the modification of the geometry, to the solution of problems for isotropic systems; a complete discussion of the latter will implicitly include at the same time the solution for similar problems in which anisotropy is to be taken into account. Hence, in most cases, it will suffice to consider the problem from the beginning as one in an isotropic medium, and only at the very end introduce the appropriate transformation of coordinates if the effects of anisotropy are to be studied.

In Section V.2 it is shown that there is a connection between microstratification and anisotropy. A relationship is then derived between the hydraulic conductivities of regularly alternating homogeneous-isotropic layers and apparent anisotropy. The results of some model experiments are also given.

Factors causing anisotropy and some measured anisotropy quotients are mentioned in V.3.

Muskat (1937) has shown that anisotropy may cause a material diminution in the rate at which oil enters a well, penetrating only partly into an anisotropic bed. He presents calculations and graphs which emphasize the importance of the effect. Muskat (1937, p. 283) is sometimes quoted as stating that the flow towards a partially penetrating well will become almost exactly radial at a distance from the well equal to twice the sandthickness. It should be pointed out however, that—as implied by Muskat— this applies to isotropic beds only. The critical distance referred to is, in case of anisotropy, to be multiplied by the square root of the ratio of horizontal and vertical hydraulic conductivity (k_h and k_v respectively). The result is that the effect of partial penetration is accentuated; the lateral extent of this disturbing influence is increased, as usually $k_h/k_v > 1$. The minimum distance from the well at which the flow is radial, is an important factor in determining the suitable location for the installation of piezometers for observing and measuring drawdown, if these measurements are used for calculating the transmissibility (the thickness of the sandbed times the horizontal hydraulic conductivity). The effect of anisotropy on wells is analyzed fully by Muskat and is not discussed here.

Shih te Yang (1949) derives relationships between the hydraulic conductivity components in different directions at a certain point. He elaborates on the applicability of Mohr's circle to the problem; "Mohr's circle" is usually used in a graphical method in soil mechanics for determining forces and stresses in an earth body. This work is not discussed further.

1. THEORY OF FLUID FLOW THROUGH ANISOTROPIC MEDIA

Pertinent theory on fluid flow in anisotropic media has been given by Versluys (1915), Samsioe (1931), Dachler (1933 or 1936, pp. 133–138), Vreedenburgh (1935, 1936, 1937), Muskat (1937), Ferrandon (1948, 1954), Scheidegger (1954), and Maasland and Kirkham (1955). The theoretical results will depend on the validity of Darcy's law. Since Darcy's law is valid, not only for water but also (at low pressure gradients) for air (Muskat, 1937; Kirkham, 1946), the results for water may also be used for air. In the following formulae the hydraulic conductivity k is used; we could also have used the permeability k' for expressing and deriving the results as anisotropy is a property of the flow medium. There are five principal results; they, for convenience, will be expressed as theorems.

a. General Theorems on Anisotrpy

Theorem I. A porous medium, consisting of any number of arbitrarily directed sets of parallel, elementary flow tubes, can always be replaced by an equivalent, fictitious, porous medium of equal size with three, mutually perpendicular, uniquely directed systems of pore tubes. In this fictitious medium, the net flow per unit area is the same in every direction as in the actual medium, provided that the hydraulic head (gauge pressure—in the case of air flow) is the same everywhere in the fictitious medium as in the actual medium.

The three unique, mutually perpendicular directions are called the *principal directions* x, y, z, of anisotropy; and there correspond to these directions, for the most general case, three unique values of hydraulic conductivity k_x, k_y, and k_z. For the special case that $k_x = k_y$, and with k_z not being equal to k_x or k_y, the principal directions x and y are replaced by a principal plane perpendicular to the direction z, which latter direction then remains alone as a unique principal direction.

Theorem I, as applied to water movement in porous media, is apparently due to Versluys (1915), who derived the above theorem for any combination of arbitrarily directed sets of parallel, non-intersecting capillaries.

Versluys assumed that—for viscous flow—the results of that derivation can also be applied to such intersecting elementary flow tubes as occur in soils. The analogy between the equations of flow through porous media and the equations of flow through capillaries is discussed by Scheidegger (1953, 1955).

Scheidegger (1953) proposed a theory of flow through porous media based upon the statistics of disordered phenomena. It is assumed in his "hypothesis of complete disorder" that the geometric conditions for the motion of a fluid particle prevailing at a spot in a particular porous medium are entirely uncorrelated with those of any other spot of that particular piece of material. This hypothesis, as compared to capillaric flow, corresponds to the opposite limit case. It is shown that the statistical theory gives a connection between the macroscopic hydraulic conductivity and porosity, and the average of the microscopic reciprocal "resistance." From this theory it follows that if certain factors are set equal to zero, the motion is the same as described by Darcy's law. If those factors are not set equal to zero, then a new quantity must be introduced, the "dispersivity," a constant of the porous medium.

The above theory is developed further by Scheidegger (1955) to make it applicable to any type of microscopic flow equations. A theorem is then proved, stating that the flow through porous media is described by the superposition of two effects; firstly, one corresponding to the average flow through a set of small channels, and, secondly, a dispersivity effect. The individual particles of the fluid do not only move along the streamlines resulting from Darcy's law but they are also dispersed sideways.

It is known from experimental studies that the deviations from Darcy's law are generally small: Dachler (1936), Muskat (1937). In the theory and applications presented in this section it will be assumed that Darcy's law is valid and that the dispersivity effect may be neglected.

Ferrandon (1948, 1954) proposed the tensor theory of hydraulic conductivity which is also discussed by Irmay (1951) and by Scheidegger (1954). In the case of non-isotropy, the simple, single factor k (scalar) becomes a symmetric tensor $\bar{k}$ (the "hydraulic conductivity-tensor") having nine components, six of which are different (Irmay, 1951). For a homogeneous soil, the number of variables can be reduced to three (i.e., the hydraulic conductivities in the principal directions) by suitably orienting the coordinate axes. The fact that the hydraulic conductivity of an anisotropic porous medium can be represented as a symmetric tensor leads immediately to the above theorem and to the conclusion that, in general, the potential gradient and the flow direction do not coincide. They do coincide along the principal directions of anisotropy.

The corresponding theorem occurs in electricity in the case of an anisotropic dielectric. See, for example, Smythe (1939, pp. 20–21) for a concise treatment. Proof of Theorem I will not be given here.

Theorem II. The effect of an anisotropy in the hydraulic conductivity is equivalent to the effect of shrinkage or expansion of the coordinates of a point in the flow system. That is, one can, by suitably shrinking or expanding the coordinates of each point in an anisotropic medium, obtain an equivalent, homogeneous, isotropic system.

A proof of Theorem II, for two-dimensional flow, is given by Dachler (1933, 1936) and by Samsioe (1931); and, for three-dimensional flow, by Vreedenburgh (1936). Maasland and Kirkham (1955) give a simpler proof and at the same time a general one by modifying, as follows, a treatment given by Muskat (1937).

Let v_x, v_y, v_z and k_x, k_y and k_z be, respectively, the velocities and hydraulic conductivities for the principal directions (Theorem I), in a homogeneous anisotropic medium. Let h be the hydraulic head at a point (x, y, z). Then, Darcy's law may be written,

$$v_x = -k_x \frac{\delta h}{\delta x}, \qquad v_y = -k_y \frac{\delta h}{\delta y}, \qquad v_z = -k_z \frac{\delta h}{\delta z}. \tag{V.1}$$

The equation of continuity is

$$\delta v_x/\delta x + \delta v_y/\delta y + \delta v_z/\delta z = 0. \tag{V.2}$$

From Equations (V.1) and (V.2) we obtain, instead of Laplace's equation, the result

$$k_x \delta^2 h/\delta x^2 + k_y \delta^2 h/\delta y^2 + k_z \delta^2 h/\delta z^2 = 0. \tag{V.3}$$

Now let k_0 be an arbitrary constant having the dimensions of k_x, k_y and k_z; and let x', y', and z' be defined by

$$x' = (k_0/k_x)^{1/2} x, \tag{V.4}$$

$$y' = (k_0/k_y)^{1/2} y, \tag{V.5}$$

$$z' = (k_0/k_z)^{1/2} z. \tag{V.6}$$

From Equations (V.3), (V.4), (V.5) and (V.6) it follows at once that

$$\delta^2 h/\delta x'^2 + \delta^2 h/\delta y'^2 + \delta^2 h/\delta z'^2 = 0, \tag{V.7}$$

which is Laplace's equation for a homogeneous isotropic medium, whence Theorem II is true. The shrinkages and expansions are given by Equations (V.4), (V.5) and (V.6), equations which differ from those given by Muskat (1937) by the arbitrary factor k_0 (in connection with this, see Section V.1.b).

Theorem III. The hydraulic conductivity, k, for the equivalent homogeneous *isotropic medium into which the anisotropic medium may be expanded or shrunk (Theorem II) is related to the hydraulic conductivities of the actual anisotropic system by the relation*

$$k = (k_x k_y k_z / k_0)^{1/2}, \tag{V.8}$$

where k_0 is the arbitrary constant, and k_x, k_y, and k_z are the hydraulic conductivities for the principal directions (Theorem I) of the actual anisotropic medium.

A proof of Theorem III may be found in Vreedenburgh (1936). Maasland and Kirkham (1955) give a shorter and simpler proof which follows.

Consider, in an anisotropic medium, an elementary tetrahedron $ABCD$ (Fig. 1), its surfaces ACD, ADB, and ABC to be mutually perpendicular to colinear axes x, y, z and x', y', z', respectively.

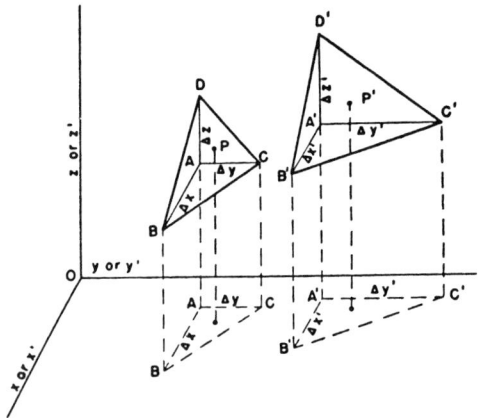

Fig. 1. Transformation of a tetrahedron (solid lines) $ABCD$ into $A'B'C'D'$ by means of Equations (4), (5) and (6). The figure is drawn for $(k_x/k_0)^{1/2} = 4/3$, $(k_y/k_0)^{1/2} = 1/2$ and $(k_z/k_0)^{1/2} = 3/4$. The broken line triangles are projections of the bases of the tetrahedrons in the x, y plane.

Let the surface BCD (at least in the limit as the area BCD approaches zero) be part of an equipotential surface, and let BCD include a point P. Next, consider a tetrahedron $A'B'C'D'$, which results by application of Equations (V.4), (V.5) and (V.6) in $ABCD$. Point P will now be transformed to P' lying on a triangular surface $B'C'D'$; the other sides of the transformed tetrahedron will be $A'C'D'$, $A'D'B'$ and $A'B'C'$. Observe that the latter three surfaces will be respectively parallel, but in general, not equal to ACD, ADB and ABC; and that the fourth surface $B'C'D'$ will, in general, be neither equal to nor parallel to BCD; but notice that the hydraulic head h at P and P' will be the same.

Now observe that the amount of fluid which goes through BCD per unit time must be equal to that which goes through $B'C'D'$ per unit time. (If this were not so, the transformation would not be equivalent.) Observe, again, by the law of conservation of mass, that the quantity of fluid which goes through BCD per unit time is equal to the sum of the quantities which goes through ACD, ADB, and ABC per unit time; and that the quantity of fluid, which goes through $B'C'D'$ per unit time, is equal to the sum of the quantities, which goes through $A'C'D'$, $A'D'B'$ and $A'B'C'$ per unit time.

Therefore, letting F_{ACD}, etc., be the quantity of flow per unit time through surfaces ACD, etc., we may write

$$F_{ACD} + F_{ADB} + F_{ABC} = F_{A'C'D'} + F_{A'D'B'} + F_{A'B'C'} \qquad (V.9)$$

But, by Darcy's law and taking $AB = \Delta x$, etc., (Fig. 1), we have, very nearly, and exactly in the limit, the results

$$F_{ACD} = -\tfrac{1}{2} k_x \, \Delta y \, \Delta z \, \delta h/\delta x \qquad (V.10)$$

$$F_{ADB} = -\tfrac{1}{2} k_y \, \Delta x \, \Delta z \, \delta h/\delta y \qquad (V.11)$$

$$F_{ABC} = -\tfrac{1}{2} k_z \, \Delta x \, \Delta y \, \delta h/\delta z \qquad (V.12)$$

Likewise in the transformed isotropic system, which could, it is to be remembered, exist in actuality, and for which the hydraulic conductivity is k, we can write

$$F_{A'C'D'} = -\tfrac{1}{2} k \, \Delta y' \, \Delta z' \, \delta h/\delta x' \qquad (V.13)$$

$$F_{A'D'B'} = -\tfrac{1}{2} k \, \Delta x' \, \Delta z' \, \delta h/\delta y' \qquad (V.14)$$

$$F_{A'B'C'} = -\tfrac{1}{2} k \, \Delta x' \, \Delta y' \, \delta h/\delta z' \qquad (V.15)$$

Now, by Equations (V.4), (V.5), (V.6), we have

$$\Delta x' = (k_0/k_x)^{1/2} \, \Delta x, \qquad (V.16)$$

$$\Delta y' = (k_0/k_y)^{1/2} \, \Delta y, \qquad (V.17)$$

$$\Delta z' = (k_0/k_z)^{1/2} \, \Delta z, \qquad (V.18)$$

and these last three equations imply also (since $\Delta x \to \delta x$, $\Delta y \to \delta y$ etc.), that

$$\delta h/\delta x' = (k_x/k_0)^{1/2}(\delta h/\delta x), \qquad (V.19)$$

$$\delta h/\delta y' = (k_y/k_0)^{1/2}(\delta h/\delta y), \qquad (V.20)$$

$$\delta h/\delta z' = (k_z/k_0)^{1/2}(\delta h/\delta z). \qquad (V.21)$$

Putting Equations (V.16) to (V.21) in (V.13) to (V.15), we obtain

$$F_{A'C'D'} = -\tfrac{1}{2} k [k_0 k_x/(k_y k_z)]^{1/2} \, \Delta y \, \Delta z \, \delta h/\delta x, \qquad (V.22)$$

II. Theory of Land Drainage

$$F_{A'D'B'} = -\tfrac{1}{2}k[k_0 k_y/(k_x k_z)]^{1/2}\,\Delta x\,\Delta z\,\delta h/\delta y, \quad (V.23)$$

$$F_{A'B'C'} = -\tfrac{1}{2}k[k_0 k_z/(k_x k_y)]^{1/2}\,\Delta x\,\Delta y\,\delta h/\delta z. \quad (V.24)$$

Putting Equations (V.10) to (V.12) and (V.22) to (V.24) in (V.9), we see that the resulting equation can only be true if (V.8) is true—which completes our proof.

Theorem IV. *If the square root of the directional hydraulic conductivity (that is, the hydraulic conductivity in the flow direction) is plotted in all the corresponding directions at a point of an anisotropic medium, then one obtains an ellipsoid; this ellipsoid is called the ellipsoid of direction.*

A proof of this theorem is given by Vreedenburgh (1936). Muskat (1937, pp. 225–227) and Scheidegger (1954) derived pertinent formulae. Here, the theorem will be proved by extending a treatment given by Muskat (1937).

The direction of the streamlines in an anisotropic system will, in general, not coincide with the direction of the normal to the equipotentials. The angle θ between the streamline and the normal to the equipotential at a point O in an anisotropic system, is given by (see e.g., Wylie 1951, p. 432, Equation 11):

$$\cos\theta = \frac{\mathbf{v}\cdot\nabla h}{|\mathbf{v}||\nabla h|} = \frac{k_x(\delta h/\delta x)^2 + k_y(\delta h/\delta y)^2 + k_z(\delta h/\delta z)^2}{|\mathbf{v}||\nabla h|}, \quad (V.25)$$

where $\mathbf{v}$ is the vector velocity of the fluid at point (x, y, z) and ∇h the operational vector with components $\delta h/\delta x$, $\delta h/\delta y$, $\delta h/\delta z$. The resultant velocity along the streamline is then

$$|\mathbf{v}| = k_r\,|\nabla h|\,\cos\theta, \quad (V.26)$$

in which k_r is the hydraulic conductivity in direction r, making the angles θ_x, θ_y, θ_z with coordinate axes. Substitution of Equation (V.25) in (V.26) gives

$$k_r = \frac{|\mathbf{v}|}{|\nabla h|\cos\theta} = \frac{|\mathbf{v}|^2}{\mathbf{v}\cdot\nabla h}$$

$$= \frac{k_x^2(\delta h/\delta x)^2 + k_y^2(\delta h/\delta y)^2 + k_z^2(\delta h/\delta z)^2}{k_x(\delta h/\delta x)^2 + k_y(\delta h/\delta y)^2 + k_z(\delta h/\delta z)^2}. \quad (V.27)$$

We have also the relations

$$v_x = v\cos\theta_x = -k_x(\delta h/\delta x),$$
$$v_y = v\cos\theta_y = -k_y(\delta h/\delta y), \quad (V.28)$$
$$v_z = v\cos\theta_z = -k_z(\delta h/\delta z),$$

in which v_x, v_y and v_z are the velocities of the fluid in the direction of coordinate axes x, y, z. Substitution of Equation (V.28) in (V.27) gives

$$k_r = \frac{\cos^2 \theta_x + \cos^2 \theta_y + \cos^2 \theta_z}{\cos^2 \theta_x/k_x + \cos^2 \theta_y/k_y + \cos^2 \theta_z/k_z},$$

or

$$1/k_r = \frac{\cos^2 \theta_x}{k_x} + \frac{\cos^2 \theta_y}{k_y} + \frac{\cos^2 \theta_z}{k_z}, \quad (V.29)$$

which for two-dimensional flow problems reduces to

$$1/k_r = \frac{\cos^2 \alpha}{k_x} + \frac{\sin^2 \alpha}{k_y}; \quad (V.30)$$

α being the angle between the directional hydraulic conductivity and the x axis.

If now in any direction $(\theta_x, \theta_y, \theta_z)$ in an anisotropic medium, a radius $OP = r = (k_r)^{1/2}$ is drawn from point O, then the coordinates of P are $x = r \cos \theta_x$, $y = r \cos \theta_y$ and $z = r \cos \theta_z$. It follows from Equation (V.29) that the geometrical locus of P is the ellipsoid

$$x^2/k_x + y^2/k_y + z^2/k_z = 1 \quad (V.31)$$

which is the equation of an ellipsoid with semi-axes $k_x^{1/2}$, $k_y^{1/2}$, $k_z^{1/2}$, respectively, in the principal directions of anisotropy. Vreedenburgh (1936) calls Equation (V.31) the ellipsoid of direction.

Scheidegger (1954) erroneously concludes from Equation (V.29) that the plotting of the *inverse* square root of the directional permeability will yield an ellipsoid. It is shown here that an ellipsoid is obtained by plotting the square root of directional permeability.

Theorem V. The equipotentials in an anisotropic medium are conjugate to the flow lines with regard to the ellipsoid (V.31).

This theorem is due to Vreedenburgh (1936). A simple proof can be given as follows.

A diameter D and a diametral plane M are said to be conjugate with regard to an ellipsoid if D contains the centers of the sections parallel to the diametral plane M, and M bisects the chords parallel to D. (Diameters and diametral planes go through the center of the ellipsoid; chords are straight lines which intersect the ellipsoid at two points.)

Let the flow line D have the directional components $\cos \theta_x$, $\cos \theta_y$ and $\cos \theta_z$; let M be the plane $Ax + By + Cz = 0$. Then D and M are conjugate with regard to the ellipsoid (V.31) if (see, e.g., Osgood and Groustein, 1927)

$$A : B : C = \cos \theta_x/k_x : \cos \theta_y/k_y : \cos \theta_z/k_z.$$

It follows that—if M is conjugate to D—the equation of M must be

$$x \cos \theta_x/k_x + y \cos \theta_y/k_y + z \cos \theta_z/k_z = 0. \qquad \text{(V.32)}$$

Let the normal n to the equipotential surface at a point O have the directional components $\cos \phi_x$, $\cos \phi_y$ and $\cos \phi_z$. Then

$$\begin{aligned} v_x &= -k_x(\delta h/\delta n) \cos \phi_x \\ v_y &= -k_y(\delta h/\delta n) \cos \phi_y \\ v_z &= -k_z(\delta h/\delta n) \cos \phi_z \end{aligned} \qquad \text{(V.33)}$$

From Equations (V.28) and (V.33) it follows that

$$\cos \phi_x : \cos \phi_y : \cos \phi_z = \frac{\cos \theta_x}{k_x} : \frac{\cos \theta_y}{k_y} : \frac{\cos \theta_z}{k_z} \qquad \text{(V.34)}$$

It is seen from this equation that Equation (V.32) is the equation of the tangent-plane to the equipotential surface at point O, which completes our proof.

It follows from this theorem that, in general, the potential gradient and the flow lines do not have the same direction. They do have the same direction if the potential gradient coincides with one of the principal directions of anisotropy.

b. Two-Dimensional Flow Problems in Anisotropic Soil

The equipotentials in two-dimensional flow problems are cylindrical surfaces. Only one diametral section of the ellipsoid of direction (Equation V.31) need then be considered, i.e., the section in the plane normal to the generators of the equipotentials. A diametral section of an ellipsoid is usually an ellipse. This ellipse will be called the ellipse of direction.

The formulae needed for transforming a two-dimensional anisotropic flow problem into a fictitious isotropic one, can readily be obtained from the formulae derived in V.1.a. for the three-dimensional flow problem. Let the coordinate axes x, y coincide with the principal directions of anisotropy, and let the hydraulic conductivities in these directions be k_h and k_v respectively. The transformation of the anisotropic medium into a fictitious isotropic one is then achieved by applying Equations (V.4) and (V.5).

The hydraulic conductivity k for the equivalent homogeneous isotropic system is

$$k = (k_h k_v)^{1/2}. \qquad \text{(V.35)}$$

It is observed that k_0 does not occur in Equation (V.35) which may be explained as follows. k_0 is an arbitrary factor which can be introduced in any isotropic or anisotropic flow problem according to the following rule: expand or shrink a flow medium by a factor n; the rate of flow will then be the same in the expanded medium as in the original medium if the hydraulic conductivity of the expanded medium is equal to the hydraulic conductivity of the original medium divided by the expansion factor n. The hydraulic head h is assumed to be the same everywhere in the expanded medium as in the original soil. The rule follows immediately from Darcy's law:

$$Q = -k_s S(\delta h/\delta s),$$

in which

k_s = hydraulic conductivity in flow direction s.

S = surface area normal to the flow direction s.

S in this formula is replaced by Sn^2 and s by sn. It follows that Q is not affected by the expansion if k_s is divided by n.

For a two-dimensional system we write (Darcy):

$$Q = -k_s L(\delta h/\delta s)$$

Both L and s have the dimension of length and must be multiplied by n for expansion of the medium. Q is the rate of flow through a strip of unit width (the width of the strip—which is measured in the direction normal to x, y plane—is not affected by the expansion). It is observed from the above formula that, in this case, the hydraulic conductivity k_s is not affected by the expansion or shrinkage.

The factor k_0 is useful for transforming anisotropic systems that consist of two or more layers (each of which have different hydraulic conductivities), into fictitious isotropic systems. It is used for the matching of boundaries between layers (see Section V.2.f), and sometimes for the matching and the determination of the boundary conditions in transformed systems.

Figure 2 shows the direction of flow and the direction of the normal n to the equipotential h at a point O of an anisotropic medium. The ellipse of direction

$$x^2/k_h + y^2/k_v = 1 \qquad (V.36)$$

is transformed into the circle: $(x'^2 + y'^2 = k_0)$ in Figs. 2C and 2D. It is observed that in Fig. 2D the flow lines are normal to the equipotentials. By Darcy's law, and taking $dh = \delta h$, etc., (Fig. 2-A) we have, very nearly, and exactly in the limit, the results

$$v_x = v_\alpha \cos \alpha = -k_h\, \delta h/\delta x = -k_h(\delta h/\delta n) \sin \phi,$$

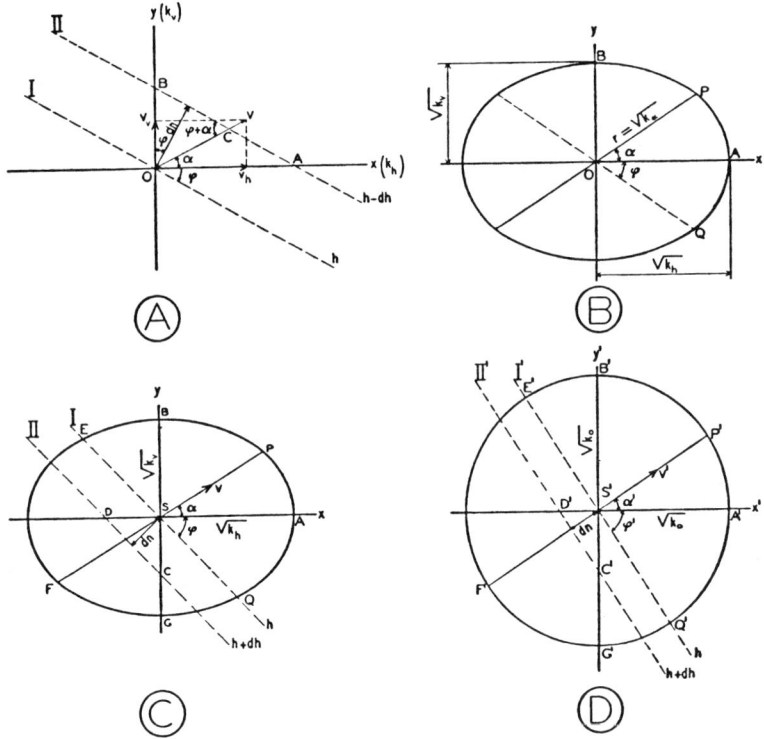

FIG. 2. Two-dimensional flow in an anisotropic medium. 2A: Dependence of the direction of flow on the relative magnitudes of k_h and k_v. 2B: The ellipse of direction: $x^2/k_h + y^2/k_v = 1$. 2C and 2D: Transformation of an anisotropic system into a fictitious isotropic system.

and

$$v_y = v_\alpha \sin \alpha = -k_v \, \delta h/\delta y = -k_v (\delta h/\delta n) \cos \phi; \quad (V.37)$$

α being the angle between the flow direction and the x axis, and ϕ the angle between the normal to the equipotential line through point O and the y axis. It follows from Equation (V.37) that

$$v_y/v_x = \tan \alpha = (k_v/k_h) \cot \phi,$$

or

$$\tan \alpha \tan \phi = k_v/k_h. \quad (V.38)$$

This equation gives the relationship between directions of the flow lines and equipotentials in a two-dimensional anisotropic system and the ratio of the hydraulic conductivities in the principal directions of anisotropy.

It follows from Equation (V.38) that the flow lines and equipotential lines are conjugate with regard to the ellipse (Equation V.36).

Equation (V.30) for the directional hydraulic conductivity and Equation (V.36) for the ellipse of direction can be derived as follows. We define

$$v_\alpha = -k_\alpha \, \delta h/\delta s, \qquad (V.39)$$

and it is observed from Fig. 2-A (since $d\dot{n} \to \delta n$, $dh \to \delta h$, etc.) that

$$\delta h/\delta s = \delta h/OC = (\delta h/\delta n) \sin(\alpha + \phi). \qquad (V.40)$$

Substitution of Equation (V.40) in (V.39) gives

$$k_\alpha = -v_\alpha/(\delta h/\delta n)(\sin\alpha \cos\phi + \cos\alpha \sin\phi). \qquad (V.41)$$

Substitution of Equation (V.37) in (V.41) gives

$$1/k_\alpha = \frac{\cos^2\alpha}{k_h} + \frac{\sin^2\alpha}{k_v} \qquad (V.30)$$

which can be written as

$$\frac{(k_\alpha^{1/2}\cos\alpha)^2}{k_h} + \frac{(k_\alpha^{1/2}\sin\alpha)^2}{k_v} = 1,$$

or

$$x^2/k_h + y^2/k_v = 1; \qquad (V.36)$$

which is the ellipse of direction.

Any radius of the ellipse of direction is equal to the square root of the hydraulic conductivity in the direction of the radius. The hydraulic conductivity (k_α) in any direction can be computed from Equation (V.30). The relationship between the direction of flow lines and equipotentials is given by Equation (V.38).

[*Editors' Note:* Material has been omitted at this point.]

REFERENCES

[*Editors' Note:* Only the references cited in the preceding excerpt are reproduced here.]

DACHLER, R. 1933. Über Sickerwasserströmungen in geschichtetem Material. Wasserwirtschaft 2:13-16.

DACHLER, R. 1936. Grundwasserströmung. Julius Springer, Vienna.

FERRANDON, J. 1948. Les lois de l'écoulement de filtration. Genie Civil. 125:24-28.

FERRANDON, J. 1954. Mécanique des terrains perméables. Houille Blanche 9:466-480.

IRMAY, S. 1951. Darcy law for non-isotropic soils. Assoc. Intern. Hydrol. Sci. U.G.G.I., Assemblée Gen. Bruxelles 2:178.

KIRKHAM, DON. 1946. Proposed method for field measurement of permeability of soil below the water table. Soil Sci. Soc. Amer. Proc. 10:58–68.
MAASLAND, M. AND KIRKHAM, D. 1955. Theory and measurement of anisotropic air permeability in soil. Soil Sci. Soc. Amer. Proc. 19:395–400.
MUSKAT, M. 1937. The Flow of Homogeneous Fluids Through Porous Media. McGraw-Hill, New York; or reprinted 1946, J. W. Edwards, Ann Arbor, Mich.
OSGOOD, W. R. AND GROUSTEIN, W. C. 1927. Plane and Solid Geometry. Macmillan, New York.
SAMSIOE, A. F. 1931. Einfluss von Rohrbrunnen auf die Bewegung des Grundwassers. Zeitsch. angew. Math. und Mech. 11:124–135.
SCHEIDEGGER, A. E. 1953. Statistical hydrodynamics in porous media. Jour. Appl. Phys. 25:994–1001.
SCHEIDEGGER, A. E. 1954. Directional permeability of porous media to homogeneous fluids. Geofisica Pura e Applicata. Milano 28:75–90.
SCHEIDEGGER, A. E. 1955. General statistical hydrodynamics in porous media. Geofisica Pura e Applicata. Milano 30:17–26.
VERSLUYS, J. 1915. De onbepaalde vergelijking der permanente beweging van het grondwater. Verh. Geol.-Mijnbouw. Genoot. Ned. en Kolonien. Geol. Serie 1:349–360.
VREEDENBURGH, C. G. F. 1935. Over de stationnaire waterbeweging door grond met homogeen anisotrope doorlaatbaarheid. Ingen. in Ned. Indië 11:140–143.
VREEDENBURGH, C. G. F. 1936. On the steady flow of water percolating through soils with homogeneous-anisotropic permeability. Proc. Intern. Conf. Soil Mech. and Foundation Eng. 1:222–225.
VREEDENBURGH, C. G. F. 1937. De parallelstroming door grond bestaande uit evenwijdige regelmatig afwisselende lagen van verschillende dikte en doorlaatbaarheid. Ingen. in Ned. Indië 8:111–113.
WYLIE, C. R. 1951. Advanced Engineering Mathematics. Ed. 1. McGraw-Hill, New York.
YANG, SHIH TE. 1949. Seepage toward a well by the relaxation method. Thesis. Harvard Univ. Library, Cambridge, Mass.

ERRATUM

Page 223, line 3 should read: "Putting Equations (V. 10) to (V. 12) and (V. 22) to (V. 24) in (V. 9), we . . . ."

Reprinted from *Econ. Geology* 23:263-291 (1928)

COMPRESSIBILITY AND ELASTICITY OF ARTESIAN AQUIFERS.[1]

OSCAR EDWARD MEINZER.[2]

CONTENTS.

Definition of the problem	263
Previous investigations	265
Evidence of compressibility	266
Evidence from laboratory tests	266
Evidence from subsidence of surface	268
Evidence from excess of discharge over recharge	269
Evidence from lag in the decline in head and local character of changes in hydraulic gradient	271
Evidence from tides in wells produced by ocean tides	272
Evidence from fluctuations in head produced by railroad trains	276
Evidences of elasticity	277
Quantitative estimates of compression and expansion	280
Importance of correlative factors	284
Elasticity of the water	285
Solids precipitated out of solution	285
Gases liberated from solution	286
Sand, silt, and clay removed with the water	287
Leakage of wells	287
Leakage of confining beds	288
Significance of compressibility and elasticity in problems of hydrology	289

DEFINITION OF THE PROBLEM.

THE pore space in an artesian aquifer is filled with water that is under hydrostatic pressure. Thus the artesian water exerts a force that acts against the weight of the overlying rocks. In ex-

---

[1] Published with the permission of the Director, United States Geological Survey, and presented in part before the Geological Society of America, Dec. 31, 1927.

[2] When this manuscript was received by the editor, W. L. Russell's manuscript on "The Origin of Artesian Pressure" was in press. When Mr. Russell's paper appeared in No. 2, Mr. Meinzer's manuscript was in press. They are, therefore, independent papers published in the order of their receipt by the editor.—EDITOR.

treme cases this hydrostatic force may be virtually as great as the weight of the rocks but generally it is much less. When wells are drilled to the aquifer and water is withdrawn the hydrostatic pressure is reduced (Fig. 1).

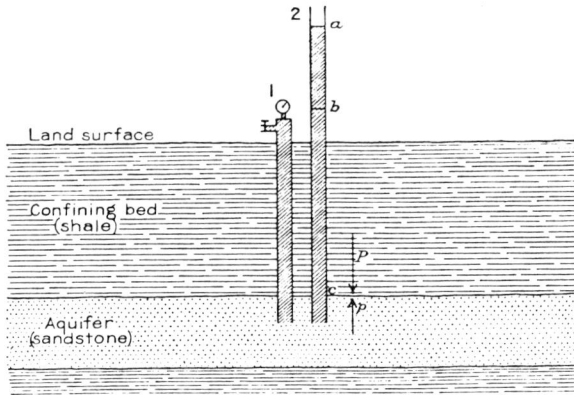

Fig. 1. Ideal section showing pressure relations in an artesian aquifer. The hydrostatic pressure of the artesian water is measured by means of a pressure gage in well No. 1 and by means of a column of water in well No. 2. The water exerts an upward pressure against the overlying confining bed equal to p, which is measured by the weight of a column of water ac in height. The confining bed exerts a downward pressure P due to its weight. As the specific gravity of the shale that forms the confining bed is much greater than that of water, the downward pressure P is greater than the upward pressure p. Only the difference, $P-p$, is borne by the sandstone that forms the aquifer. When the valve is opened in well No. 1, water escapes and the hydrostatic pressure is reduced. The upward pressure of the water is then decreased to the quantity p', which is measured by the weight of a column of water bc in height. The pressure upon the sandstone is accordingly increased by the quantity $p-p'$, which is measured by a column of water ab in height.

It has generally been assumed that the formations which constitute the aquifers are incompressible and inelastic and that, therefore, changes in the hydrostatic pressure do not produce any changes in the pore space of the rock. There are, however, several lines of evidence that the artesian water, especially in strata of sand or soft sandstone, supports a part of the load of the overlying rock and that the aquifers are compressed when the artesian pressure is decreased and expanded when it is increased. The evidence indicates that there are only very moderate amounts of contraction and expansion as a result of the changes of pressure produced by the operation of wells. Nevertheless, the amounts are more than can be attributed to the elasticity or other volume change in the water itself and they are large enough to affect radically the conclusions in regard to the recharge, movement, and discharge of the water in the artesian aquifers.

PREVIOUS INVESTIGATIONS.

Various investigations have been made of the compressibility of rock materials such as sand, silt, clay, and marl,[3] and of shrinkage produced by the drainage of such materials.[4] These shrinkage investigations have been conducted largely in Germany in connection with problems of subsidence of the surface and they have been concerned chiefly with molecular forces in clays and marls.

The specific subject of the compressibility and elasticity of artesian aquifers and their relation to artesian pressure has received but little attention, and, so far as the writer has information, no effort has hitherto been made to study the subject intensively by the application of critical data. In 1890, Hay[5] suggested, in a paper which he read before the Kansas Academy of Science, that the flow of certain artesian wells in that State may be due to "rock pressure," that is, to the pressure of the rocks overlying plastic artesian aquifers upon the water itself. In 1906, Gregory,[6] adopted the theory of rock pressure as a par-

---

[3] Sorby, H. C., "On the Application of Quantitative Methods to the Study of the Structure and History of Rocks, *Geol. Soc. London Quart. Jour.*, vol. 64, p. 214, 1908. Terzaghi, Charles, "Principles of Soil Mechanics: Elastic Behavior of Sand and Clay," *Eng. News-Record*, vol. 95, pp. 742, et seq., 1925. Hedberg, H. D., "The Effect of Gravitational Compaction on the Structure of Sedimentary Rocks," *Amer. Assoc. Petrol. Geol. Bull.*, vol. 10, pp. 1035–1072, 1926.

[4] Young, L. E., and Stoek, H. H., "Subsidence Resulting from Mining," Univ. Illinois Eng. Exper. Sta. Bull. 91, pp. 47–49, 1916. See also bibliography, pp. 180–205.

[5] Hay, Robert, "Artesian Wells in Kansas and Causes of their Flow," *Amer. Geologist*, vol. 5, pp. 296–301, 1890.

[6] Gregory, J. W., "The Dead Heart of Australia," pp. 288–289. John Murray, London, 1906. The various theories relating to artesian water that were advanced

tial explanation of the flowing wells in central Australia, and emphasized the bearing of the theory upon the question of the depletion of the artesian water supply. The principle of compressibility and elasticity of artesian aquifers was inferred by Veatch [7] in 1906, in his discussion of the effects on the head of the artesian water produced by ocean tides, the rise of rivers and lakes, the passing of trains, and the erection and burning of buildings. The principle of compressibility was more definitely stated by Fuller,[8] in 1908, as follows: "Pressure may be exerted [on the artesian water] by virtue of the weight of the overlying materials alone. This is probably not a common factor, but extensive sinkings of the ground have followed the pumping of water from mines and the withdrawal of the support afforded by hydrostatic pressure, indicating that the rock pressure on the water must have been considerable, at least locally." Apparently the practical importance of the subject has not been appreciated except by Gregory.

EVIDENCE OF COMPRESSIBILITY.

Evidence from Laboratory Tests.—It is generally recognized that deposits of fine-grained material are greatly compressed as they become covered by younger formations through the natural processes of sedimentation. It has been shown by E. W. Shaw that much of the newly deposited material of the Mississippi Delta has a porosity of 80 to 90 per cent.[9] By the time this material has become buried beneath a thousand feet of sediments its porosity will probably be more nearly 35 per cent. Sorby concluded from his investigations that fine-grained material which originally contained as much as 90 per cent. of water may have been made almost solid by the squeezing out of the water, so that

in this book led to considerable discussion between the author and E. F. Pittman, who held very different views.

[7] Veatch, A. C., "Fluctuations of the Water Level in Wells, with Special Reference to Long Island, N. Y.," U. S. Geol. Survey Water-Supply Paper 155, pp. 62, 63, 65–69, 74–75, 1906.

[8] Fuller, M. L., "Summary of the Controlling Factors of Artesian Flows," U. S. Geol. Survey Bull. 319, p. 33, 1908.

[9] Meinzer, O. E., "The Occurrence of Ground Water in the United States, with a Discussion of Principles," U. S. Geol. Survey Water-Supply Paper 489, p. 8, 1923.

in extreme cases shales and slates may occupy only one tenth of the volume which they possessed immediately after they had been deposited.

Tests made by different investigators have shown that in general sand is much less compressible than clay. This difference is due in part to the greater size of the grains but largely to their better rounding. The compressibility is least in materials consisting of uniformly well-rounded grains and greatest in materials of irregular, angular, and elongated particles. It has been shown by Terzaghi [10] that sand with admixture of mica has compressibility comparable to that of clay, and he has, moreover, produced evidence from various sources to show that many clays do contain a considerable percentage of mica or other scalelike particles. In general, the grains of very fine materials are not so well rounded as those of coarser materials, for the reason that the chief natural agencies that produce rounding are not effective on very small particles. This principle has been recognized by Daubreé and others, and was verified by Goldman [11] in his studies of the Catahoula sandstone. Evidence along the same line is furnished by several samples of very fine sand recently obtained from a stratum of the Dakota sandstone in which the grains are largely angular.[12]

The compressibility of sand has been demonstrated by laboratory tests. An experiment was described by King [13] in which an upright cylinder was filled with saturated sand. In this experiment some of the water was squeezed out of the interstices by the pressure produced by the settling of the sand and was forced through a pipe at the bottom of the cylinder and through an at-

[10] Terzaghi, Charles, "Principles of Final Soil Classification," Massachusetts Inst. Technology Publications, vol. 63, pp. 41–43, 1927. Also published in *Public Roads*, vol. 8, No. 3, May, 1927.

[11] Goldman, M. I., "Petrographic Evidence on the Origin of the Catahoula Sandstone of Texas," *Amer. Jour. Sci.*, vol. 29, pp. 271, 272, 1915. See also references given in this paper to Daubreé, Ziegler, Worth, Mackie, Udden, and Früh.

[12] Meinzer, O. E., "Problems of the Soft-water Supply in the Dakota Sandstone, with Special Reference to Conditions in Canton, S. Dak., U. S. Geol. Survey Water-Supply Paper. (In preparation.)

[13] King, F. H., "Principles and Conditions of the Movements of Ground Water," U. S. Geol. Survey, 19th Ann. Rept., Pt. II., pp. 78–80, 1899.

tached hose to a level 6 inches above the water level in the cylinder from which it came. The results of a test recently made by Terzaghi [14] are shown in Figure 2. In this test, pressure was ap-

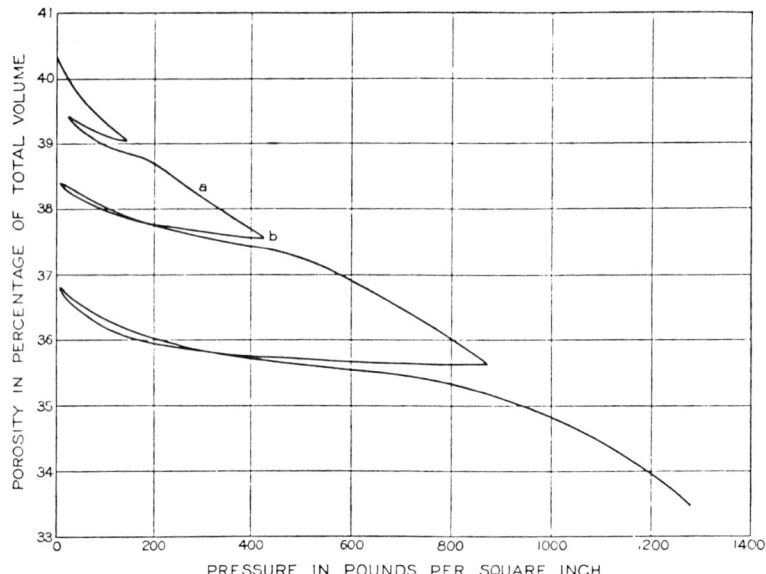

FIG. 2. Diagram showing compressibility and elasticity of compacted sand, according to experiments by Terzaghi (Redrawn from diagram in *Engineering News-Record*, vol. 95, p. 987, 1925). The portion of the curve between *a* and *b* represents approximately the change in pressure upon the Dakota sandstone which resulted from the decline in artesian pressure.

plied to a sample of compacted sand up to about 1,250 pounds per square inch, and the porosity of the sand was decreased by this pressure from about 40 to 34 per cent.

Evidence from Subsidence of Surface.—Subsidence amounting to a maximum of more than 3 feet has recently been described by Pratt and Johnson,[15] in the Goose Creek oil field, on the Gulf Coast in Texas, and has been attributed by them to the extensive extraction of oil, water, gas, and sand from beneath the affected

[14] Terzaghi, Charles, *Eng. News-Record*, vol. 95, pp. 987–990, 1925.

[15] Pratt, W. E., and Johnson, D. W., "Local Subsidence of the Goose Creek Oil Field," *Jour. Geol.*, vol. 34, pp. 577–590, 1926.

area. In a critical paper on the same subject, Snider[16] accepts the explanation that the subsidence was caused by the removal of oil, gas, water, and sand, but shows that the removal of sand was only a very small factor in producing the subsidence and also that drainage of water from the clay strata interbedded with the producing sands can not have been effective in causing shrinkage and subsidence.

Evidence from Excess of Discharge over Recharge.—This subject was first forcibly brought to the attention of the writer in connection with a quantitative study that was made, in colaboration with Herbert A. Hard and Howard E. Simpson, of the artesian water in the upper horizon of the Dakota sandstone in North Dakota.[17] It was estimated in 1923 that in the east-west row of townships under special investigation (T. 129 N.) the rate of discharge from all the artesian wells averaged close to 3,000 gallons a minute in the 38 years since the first well was drilled in 1886. The rate of discharge was estimated to be nearly 10,000 gallons a minute during the peak period at some time between 1905 and 1910, about 5,000 gallons in 1915, and about 2,000 gallons in 1920. A detailed survey made under the direction of Professor Simpson showed that it was almost exactly 1,000 gallons in 1923.

No accurate estimate could be made of the rate of recharge, that is, of the rate of eastward percolation from the intake area into the area of artesian flow in the 6-mile segment of sandstone underlying this row of townships. It was assumed that no appreciable contribution was received from the townships to the north and south, in which there was comparable artesian-well development. According to a rough computation based on inadequate data it appeared that the recharge was much less than the average discharge of nearly 3,000 gallons a minute.

[16] Snider, L. C., " A Suggested Explanation for the Surface Subsidence in the Goose Creek Oil and Gas Field, Texas," *Amer. Assoc. Petrol. Geol. Bull.*, 11, pp. 729–745, 1927.

[17] Meinzer, O. E., and Hard, H. A., " The Artesian-Water Supply of the Dakota Sandstone in North Dakota, with Special Reference to the Edgely Quadrangle," U. S. Geol. Survey Water-Supply Paper 520–E, pp. 90–93, 1925.

The problem of recharge was then approached by another line of reasoning, as follows: During the period from 1915 to 1923, with progressive decrease in the rate of discharge, there was progressive decline in pressure head, indicating that depletion was going on. In 1923, with a discharge of only 1,000 gallons a minute, the head was apparently still declining, which would seem to indicate that the recharge was less than 1,000 gallons a minute. It can not be argued that the rate of recharge decreased on account of decrease in the hydraulic gradient, because the evidence seems to show that at the entrance to the area of artesian flow the hydraulic gradient actually increased from 1915 to 1923. For example, the pressure head dropped farther during this period at Ellendale, N. Dak., than at the margin of the area of artesian flow 10 to 13 miles farther west.

If the rate of discharge in the area of artesian flow has been more rapid than the rate at which water percolated into the sandstone underlying this area, some of the water discharged must have been derived from storage in the sandstone underlying the area. This withdrawal from storage requires a reduction in the interstitial space occupied by water. As will be shown later, the vacated space can not be accounted for by the slight expansion of the water itself nor by any changes in volume that might have resulted from chemical precipitation with the release of pressure. Either the water was replaced in some of the interstices by gas, or else the sandstone was to some extent buoyed up by the artesian pressure within the sandstone, and when this was relieved the sandstone underwent a certain amount of compression in which its total interstitial space was reduced by a volume approximately equal to the volume of the stored water that was discharged. The theory of gas accumulation is believed to be untenable. In his unpublished report Hard states that the gas discharged by some of the wells seems to occur as un unsaturated solution in the artesian waters of the Dakota sandstone and apparently is released from solution by the reduction of pressure incident to the rise of the water to the surface. There is no indication that gas occurs in the sandstone in the gaseous state at the present time. The conclusion, therefore, seems to be inevitable that most of the artesian water that has been discharged from the Dakota sandstone through the flowing wells in this row of townships was taken out of storage in the sandstone underlying the area of artesian flow and that there was a corresponding shrinkage in the total volume of interstitial space. As this is a typical row of townships the same conclusion will probably hold for the rest of North and South Dakota.

Evidence from lag in the decline in head and local character of changes in hydraulic gradient.—If a perfectly tight and inelastic tank were filled with water under great pressure and were then tapped, only a minute quantity of water, equal to the expansion of the water itself, would have to be discharged in order to relieve all the pressure. On the other hand, if the tank were made of elastic material, such as rubber, the discharge would continue, at a constantly diminishing rate, until the strain of the walls of the tank would be almost completely relieved. For this reason pressure tanks used in waterworks are always kept partly filled with air, which furnishes the necessary elasticity.

If the Dakota sandstone were perfectly incompressible any considerable withdrawal of water ought to have resulted in a rapid drop in head and a prompt readjustment of the hydraulic gradient all the way from the area of artesian flow to the outcrop from which the water is derived, hundreds of miles away. This readjustment ought to have occurred with only slight movement of the artesian water. After the readjustment had been made conditions ought to have remained nearly constant, or if the water level at the outcrop had declined there ought to have been a gradual flattening of the hydraulic gradient.

The changes that actually took place seem to have been of very different character. The head did not drop suddenly, but has been declining gradually for about 40 years. The survey made in 1923 showed that there has been much less decline in the water levels in the wells near the western margin of the area of artesian flow than a few miles farther east, suggesting that in all these years there has not been much readjustment of the hydraulic gradient beyond the area in which artesian water was withdrawn (Fig. 3). Moreover, near the west margin of the area of artesian flow, for which data are available, the hydraulic gradient has not flattened out but instead has become steeper.

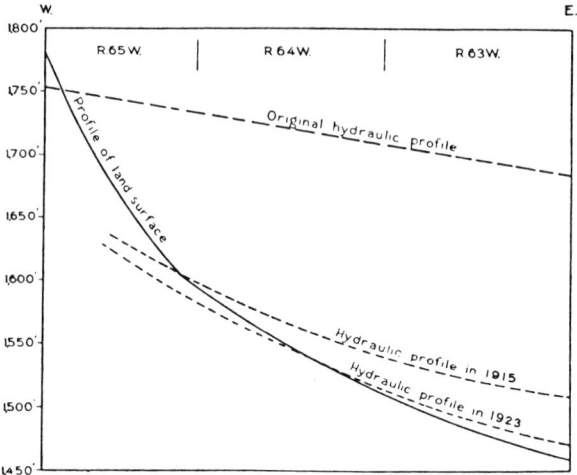

Fig. 3. General east-west section of the area of artesian flow from the Dakota sandstone in the Edgeley quadrangle, N. Dak., showing approximately the original hydraulic gradient, and the gradient in 1915 and 1923. (After Meinzer and Hard, U. S. Geol. Survey Water-Supply Paper 520, Fig. 8.)

In order to reach conclusive results more data are needed as to the head in the large region between the area of flow and the outcrop, but the data at hand certainly point to the conclusion that the Dakota sandstone has not behaved like an incompressible reservoir; that the water discharged by the flowing wells has largely been derived locally from storage and not by transfer of water all the way from the outcrop.

Evidence from Tides in Wells Produced by Ocean Tides.—In numerous places along the sea coast, fluctuations in water level in wells have been observed that are definitely related to the tides in the ocean.[18] In many of these there is no connection between

[18] Veatch, A. C., *op. cit.*, pp. 63–69.

the sea water and the water in the formation that is tapped by the well or else the connection is too indirect to produce the observed fluctuations. As early as 1817 Inglis [19] explained the tidal fluctuation of the head in an artesian well at Bridlington, England, as due to the alternate loading and unloading of ocean water upon the flexible clay bed that overlies the artesian water. This same explanation was given by Veatch for many wells with tidal fluctuations that came under his observation.

An excellent example of such a tidal effect is afforded by a non-flowing well, about 800 feet deep, at Longport, N. J., about 5 miles southwest of Atlantic City, which has been given detailed study by Thompson.[20] The well is tightly cased and passes through a few hundred feet of impermeable clay into a coarse water-bearing sand that is about 80 feet thick. A continuous record for many months, obtained by means of an automatic water-stage recorder over this well, shows that even slight variations and irregularities in the ocean tide are faithfully recorded, with but little lag, in the fluctuations of the water level in the well (Fig. 4).

It shows that the semi-daily range between high and low water in the well varies from about 1 to 3 feet at different times of the month and is a little more than 50 per cent. of the range in the ocean tide.

The facts presented by Thompson show that the increase in artesian pressure in the 800-foot sand when the tide rises is not due to direct communication of the sea water with the artesian water through crevices or interstices in the intervening strata, or as a result of submarine outcrop of this formation. For many years large supplies of water have been pumped from the 800-foot wells in this vicinity, and in 1925 the pumpage averaged about 8 million gallons a day. In a period of 31 years the head dropped from 20 to 25 feet above sea level to as much as 65 feet below

[19] Inglis, Gavin, "On the Cause of Ebbing and Flowing Springs [at Bridlington, Yorkshire]," *Phil. Mag.*, vol. 50, pp. 81–83, 1817.

[20] Thompson, D. G., "Ground-Water Problems on the Barrier Beaches of New Jersey." *Geol. Soc. America Bull.*, vol. 37, p. 466, and Fig. 4, 1926. Also unpublished data.

sea level. At Atlantic City and Longport the 800-foot sand is the only formation to the depth of 2,300 feet to which drilling has been carried that yields a large supply of fresh water. The strata that lie nearer the surface contain only brackish or salty

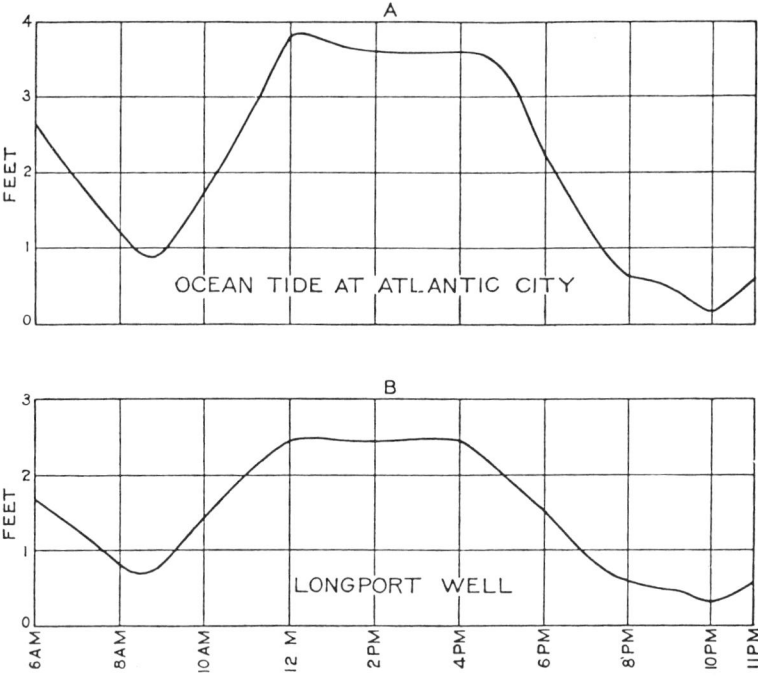

FIG. 4. Diagram showing fluctuation of water level in the 800-foot well near the seashore at Longport, N. J., in relation to the tide in the ocean. These records were made Jan. 22, 1926, when the rising tide was interrupted by a strong northwest wind. (After diagram prepared by Paul Schureman, *Geogr. Review*, p. 481, 1926.)

water, which is effectively shut out from the 800-foot sand by a layer of diatomaceous clay 300 feet thick, as is shown by the fact that the water from the 800-foot wells contains only 6 to 11 parts per million of chloride. Several of these wells have yielded salty water, but it has been proved in each instance that the salty water came through a defective casing from the overlying formations, for when the casings were repaired the wells again furnished fresh water of normal low chloride content. If there were free communication between the sea water and the water in the 800-foot sand through openings in the intervening strata, a noticeable amount of salty water would have been drawn into the 800-foot sand by the protracted and heavy pumping and the consequent great drawdown of the head in the 800-foot sand. Moreover, the salty water in the overlying strata has not suffered much drawdown and now has a much higher head than the water in the 800-foot sand. It is difficult to conceive of any condition that would permit free enough communication of water through the intervening beds to produce the tidal fluctuations in the wells and still maintain this difference in head. The rapid dying out of the tidal fluctuation in the landward direction from the coast, even where the water level in wells is still below sea level, indicates that the fluctuation is not transmitted through the water from some submarine outcrop but is due to local yielding of the overlying clay beds.

There can be no reasonable doubt that a large part of the pressure on the sea bottom produced by the loading of the sea water at high tide is transmitted through the intervening strata upon the water in the 800-foot sand. The tidal fluctuations in the water level in the well record the pressure that is thus transmitted. They indicate that only a part of the pressure produced by the high tide is borne by the sand and overlying strata and that the rest is borne by the water itself.

The water that is squeezed out of the sand between low and high tide to cause the rise in the water level in the Longport and other wells is negligible in quantity. The increased pressure on the water shows that the sand is incompetent to support even a slight additional load without compression but it gives no measure of the compressibility. However, to lower the water level in the well at high tide to its position at low tide would probably require the removal of a large volume of water, and this volume would give some sort of measure of the compression that would result if the sand itself had to support the entire load produced by the rise in sea level. The well at Bridlington was described as overflowing vigorously for 5 hours during each period of high tide, thus indicating considerable compression of the water-bearing material.

Evidence from Fluctuations in Artesian Pressure Produced by Railroad Trains.—Many years ago King[21] discovered that in one of his observation wells situated near a railroad the water level rose whenever a train went by but fell again as soon as the train had passed. He found that a heavy freight train produced a greater rise than a lighter and swifter passenger train and that a locomotive alone did not produce any noticeable effect. This subject is now under investigation by the Geological Survey on wells near railroads in California, Utah, New Mexico, and New Jersey. It is hoped that a quantitative method based on such fluctuations can be developed that will be of use in investigating artesian-water problems and perhaps also in solving other engineering and geophysical problems. In California, H. T. Stearns and T. W. Robinson obtained fluctuations caused by trains in

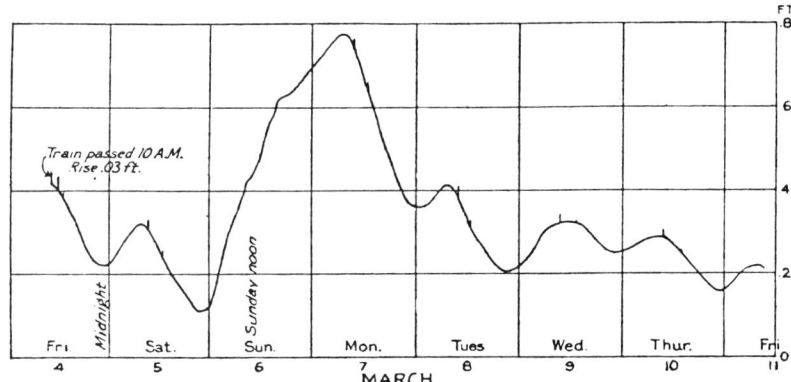

Fig. 5. Automatic record of water level in the Preszler well (200 ft. deep), near Lodi, Calif., March 4 to 11, 1927, showing daily fluctuations due to pumping and abrupt fluctuations due to the passing of trains over a railroad 117 feet from the well. There are two daily trains (except Sundays); one arrives at Lodi at 10 A.M., the other leaves Lodi at 12:25 P.M. Record furnished by H. T. Stearns and T. W. Robinson.

[21] King, F. H., "Observations and Experiments on the Fluctuations in the Level and Rate of Movement of Ground Water on the Wisconsin Agricultural Experiment Station Farm, and at Whitewater, Wisconsin," U. S. Weather Bur. Bull. 5, pp. 67–69, 1892.

three wells that are artesian in character (Fig. 5), but obtained no fluctuations in a well sunk just to the water table.

In Utah, W. N. White did not get even a trace of a rise in a well which he sank just to the water table at a point very near the railroad. In New Mexico, A. G. Fiedler failed to get any rise in a deep artesian well situated near a railroad. Further study will be required before these data can be adequately interpreted. However, they seem to indicate that the rise in the water levels is an artesian phenomenon, and that, beyond a certain limit, neither the artesian aquifers nor the overlying strata are competent to support the weight of the trains but allow the weight to be borne in part by the artesian water itself. If the support given by the water were removed compression of the aquifer would obviously occur.

EVIDENCES OF ELASTICITY.

In testing the pressure in flowing wells in the Dakota artesian basin, Hard [22] found that after a well was closed the pressure, recorded on the gage, would run up quickly at first and would then continue to increase for some time at a constantly diminishing rate. These observations confirmed the reports of earlier investigators. Nettleton,[23] in his investigations in 1890 and 1891, found that the wells differed greatly in regard to the time required for them to recover their full head. In some the recovery was very prompt but in others it was notably slow. In regard to the well at Frederick, S. Dak., Nettleton stated that "the pressure increases for several hours or even days after the flow is shut off, and when opened the flow decreases in the same way until the normal flow is reached, which corresponds somewhat to the time required to gain its maximum pressure." In regard to several other wells with slow recovery he made the following comments:

Plankington, S. Dak.: "When the flow is shut off the pressure quickly runs up to 50 pounds and in three hours it reaches its maximum, 91 pounds

[22] Meinzer, O. E., and Hard, H. A., *op. cit.*, p. 91.
[23] Nettleton, E. S., "Artesian and Underflow Investigation." 52d Cong., 1st sess., S. Ex. Doc. 41, pt. 2, pp. 40–74, 1892.

per square inch." Wolsey, S. Dak.: "If the well has been discharging freely for some time it takes about 18 hours for it to reach its maximum pressure after the water is shut off." Woonsocket, S. Dak.: " The pressure is 125 pounds per square inch when the flow is shut off. . . . The well was allowed to flow freely for 48 hours and was then closed, when it showed a pressure of 85 pounds. . . . A few hours afterwards the closed pressure was 93 pounds." Ellendale, N. Dak.: " When the flow is shut off the pressure rises quickly to 80 pounds and in a few hours it reaches its maximum, which is 115 pounds."

The explanation of the slow recovery seems to be that the water-bearing bed has a certain amount of volume elasticity, that it becomes compressed when the artesian pressure is relieved, and that before the pressure in the well can again reach the pressure that is general in the formation sufficient time has to elapse to allow the water to percolate into the depleted and compressed part of the formation surrounding the well and to expand the interstitial space. So long as the pressure in the shut-down well has not reached the maximum there is obviously a hydraulic gradient from all directions toward the well—a cone of depression in the piezometric surface (Fig. 6). But wherever there is a

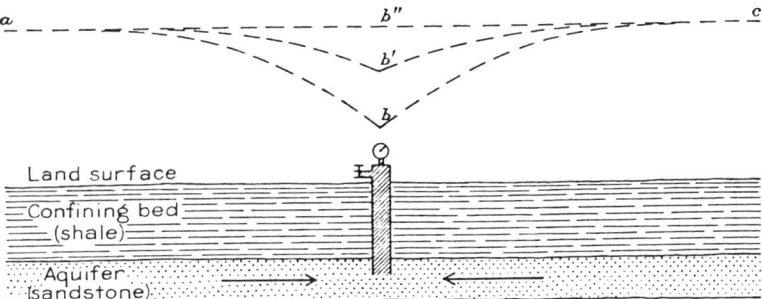

FIG. 6. Ideal section showing gradual recovery of head when an artesian well is closed. The lines ab and cb show the head of the water in the aquifer at all points along the section while the well is flowing. They show that there is a hydraulic gradient from all directions toward the well, as a result of which water percolates toward the well. The lines ab' and cb' show the head some time after the well has been closed. They show that water is still percolating toward the well from all directions although it has no outlet. The lines ab'' and cb'' show the head at a still later time when complete recovery has occurred and there is no longer concentric percolation toward the well.

hydraulic gradient in the water in a permeable rock there is movement of the water in the direction of the gradient. Throughout the period of recovery water is therefore percolating toward the well. But if the well is tightly cased this water has no means of escape. It must be stored in the interstices of the formation in the vicinity of the well. If there are no unfilled interstices at the beginning of the period of recovery, as the conditions seem to require, the inflowing water must, it seems, be stored in space made available by the expansion of the sandstone. If the sandstone were perfectly inelastic the recovery would be nearly instantaneous and would occur without the transfer of any appreciable quantity of water toward the well.

The phenomenon of gradual recovery of head has been observed not only in the wells ending in the Dakota sandstone but to some degree in those of practically every artesian aquifer in regard to which information is available. In Fig. 7 is shown the gradual recovery of the artesian head in the Roswell Artesian

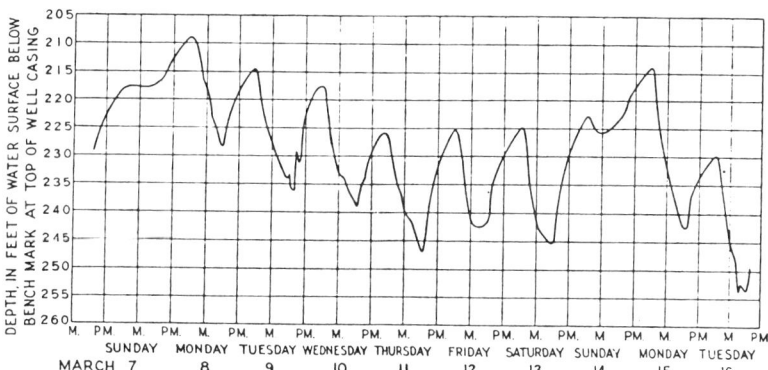

FIG. 7. Automatic record on an observation well, showing fluctuations in head in the Roswell Artesian Basin, N. Mex., produced by variations in the rate of pumping. (After A. G. Fiedler, New Mexico State Engineer, 7th Biennial Rept., Pl. 5, p. 38.) As the pumps that are not operated throughout the night are usually shut down in the evening, the persistent rise in pressure until morning seems to indicate replenishment of storage throughout the night. (The depths below bench mark shown in this diagram range from 20.5 to 26.0 ft., the omission of decimal points being an error.)

Basin at night after certain of the wells have been shut down and a still further recovery on Sundays, when many of the wells are not operated. This seems to prove that there is considerable volume elasticity even in the aquifer of this basin, which consists largely of cavernous limestone.[24] The wells that are not operated throughout the night are generally shut down in the evening. Hence, if there were no elasticity the head should run up about to a maximum long before midnight. The deep well at the Atlantic City waterworks was pumped at approximately uniform rate from the middle of May, 1925, until March 22, 1926, when it was shut down. From March 22 the water level in this well continued to rise for several weeks although there was no decline in pumpage in other parts of the Atlantic City area.[25]

In areas in which there is a seasonal fluctuation in the use of ground water there is likewise a seasonal fluctuation in head. Elasticity is not indicated by the fluctuation itself but by the lag of the fluctuation in head after the fluctuation in draft. Such lag has, for example, been observed by Thompson[26] in the Atlantic City area.

In Terzaghi's laboratory experiments he invariably found that the sand expanded when the pressure was decreased, and he gives approximate values for the expansion coefficient. The expansion was, however, generally much less than the compression that resulted from the preceding application of the pressure (Fig. 2).

QUANTITATIVE ESTIMATES OF COMPRESSION AND EXPANSION.

The foregoing calculations in regard to the Dakota sandstone afford a basis for computing the approximate amount of compression that has taken place. If in the 38 years prior to the time of the investigation the average rate of discharge from the east-west row of townships under consideration was 3,000 gallons a minute

[24] Fiedler, A. G., Report on investigations of the Roswell Artesian Basin, Chaves and Eddy counties, N. Mex., during the year ending June 30, 1926: New Mexico State Engineer, 7th Biennial Rept., pp. 34–43, 1926.
[25] Thompson, D. G., unpublished data.
[26] Thompson, D. G., op. cit., p. 468.

and the average rate of recharge only 500 gallons a minute, the average withdrawal from storage amounted to 2,500 gallons a minute. If the area of depletion consists of the 18 townships described as T. 129 N., Rs. 48-65 W., a withdrawal from storage of 2,500 gallons a minute for 38 years would amount to a layer of water 4.4 inches deep. If the average thickness of the sandstone is estimated to be 60 feet, a compression of 4.4 inches amounts to about .6 of 1 per cent.

The first well at Ellendale, according to the log, reached the Dakota sandstone at a depth of 1,035 feet. The strata above the Dakota are chiefly soft shale, which, with the water they contain, must have a specific gravity of about 2. Hence these strata, owing to their weight, exert a pressure equal to that of a column of water about 2,070 feet high, or 898 pounds to the square inch. According to one report, the original pressure at the surface in this well was 145 pounds to the square inch and according to another it was 175 pounds. If it was 145 pounds the artesian pressure at the top of the Dakota sandstone was 145 pounds plus the weight of a column of water 1,035 feet high, or 594 pounds to the square inch. Then, 594 pounds of the pressure exerted by the strata that overlie the Dakota sandstone was supported by the water in the sandstone, and only 304 pounds by the sandstone itself. If the original artesian pressure at the surface was 175 pounds, the artesian pressure at the top of the sandstone was 624 pounds, and only 274 pounds of the overlying strata was supported by the sandstone. If at any point within the area of artesian flow the head at the surface was as great as the depth to the Dakota sandstone the artesian pressure must have been great enough virtually to float the overlying beds.

In 1923 the water level in the Ellendale well stood almost exactly at the surface. Hence, the head had declined either 145 or 175 pounds to the square inch, according to the report of original pressure that is accepted. If the original pressure was 145 pounds the load placed on the sandstone as a result of the decline in head increased from 304 to 449 pounds, or $47\frac{1}{2}$ per cent.; if the original pressure was 175 pounds the load increased from 274 to 449 pounds, or 64 per cent. For the entire area under consideration the average decline in artesian pressure was probably somewhat less than that at Ellendale, but it is approximately correct to say that according to the computations the compression amounted to .2 of 1 per cent. for each decrease of 100 feet in artesian head.

In the experiments by Terzaghi it was found that, within a moderate range in pressure, the compression proceeded in general about in proportion to the increase in pressure. At least one of Terzaghi's experiments was carried through the range of pressures that are involved in the calculations on the Dakota sandstone. In this experiment, as shown by the diagram in Fig. 2, the pressure was increased from zero to about 425 pounds to the square inch, then decreased nearly to zero, and then increased far beyond 425 pounds. During the increase from 304 to 425 pounds (approximately the increase on the sandstone at Ellendale) the sand was compressed so as to decrease the porosity from about 38.2 to 37.6 per cent.—a decrease in the volume of the sand of about .6 of 1 per cent. The close agreement of these laboratory results with the results previously obtained from the computations on the Dakota sandstone is of course accidental, but it shows that the theory of compressibility presented in this paper is quantitatively reasonable. When the pressure was decreased from 425 pounds the sand expanded, but during the decrease from 425 to 304 pounds the expansion amounted to only about .1 of 1 per cent.

In order to get some idea as to the amount of expansion that is involved in the gradual recovery of head when a flowing well is closed or when pumping on a non-flowing artesian well is stopped, a few rough calculations were made on wells for which considerable exact information was available. One of these calculations was made on the basis of data furnished by Thompson[27] in regard to wells at Avon-by-the-Sea, N. J., which end in a stratum of sand, that lies 420 feet below the surface and is 80 feet thick. On November 12, 1924, Well No. 2, of the city

[27] Thompson, D. G., "Ground Water Supplies in the Vicinity of Asbury Park, N. J." (unpublished).

waterworks, was pumped continuously for 10 hours. The discharge was about 320 gallons a minute at the end of the first hour, 310 at the end of the fifth hour, and 300 at the end of the eighth hour. The discharge apparently continued to decrease during the entire test, but in the last few hours the rate of decrease became very slight. The drawdown amounted to about 100 feet. Well No. 1, situated 200 feet from No. 2, was not pumped, but numerous measurements of the water level in it were made during the test, during a period of one day after the pumping had ceased and again a week later. This well showed a drawdown of somewhat more than 30 feet just before pumping was stopped in well No. 2, a recovery of about 25 feet in 5 hours and about 30 feet in 24 hours (Fig. 8). In a test on April 2, 1925, in which both

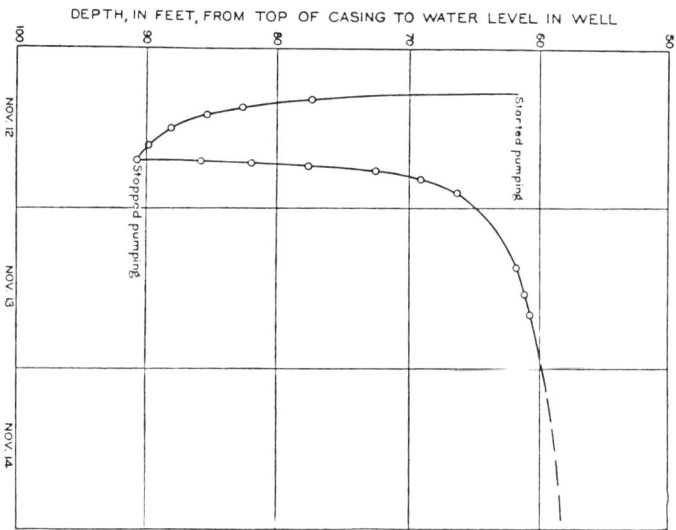

FIG. 8. Diagram showing drawdown and recovery of head in well No. 1 at Avon-by-the-Sea, N. J., as a result of pumping from well No. 2, 200 feet distant, on Nov. 12, 1924, according to data furnished by D. G. Thompson.

wells were pumped simultaneously for 3½ hours, a lowering of the water level of 1.25 feet was observed in the so-called Riggs well, half a mile distant, and a lowering of .42 foot in the Casino well, 800 feet farther away. A part of the lowering may have been due to tidal effect but in part it was almost certainly drawdown due to the pumping in wells Nos. 1 and 2.

The calculations were based on the simple but somewhat inaccurate assumptions that the area of influence covered one square mile and that the rate of percolation into this area after pumping ceased on November 12, 1924, was in proportion to the drawdown in Well No. 1. It was estimated from Figure 8, that the average drawdown in Well No. 1 during the 2-day period after pumping ceased was 4 feet, or one eighth of the drawdown when Well No. 2 was delivering 300 gallons a minute. Therefore, the average rate of percolation into the area of influence during the period should have been one eighth of 300 gallons a minute, or 37½ gallons a minute. This gives a total inflow of 108,000 gallons, or an average of 169 gallons an acre, and an average expansion of .006 inch or .0006 of 1 per cent. in the 80-foot stratum of sand. It was estimated, further, that the average drawdown for the entire area of influence just before pumping ceased was 1 foot. Hence, if the expansion were proportional to the recovery in head it would amount to about .06 of 1 per cent. for an increase of 100 feet in head, or about one third of the compression computed for a decrease of 100 feet in head in the Dakota sandstone.

Computations on other wells give somewhat comparable results. Though the results of these computations are doubtless all greatly in error, they appear to be significant in having the same order of magnitude and one that is entirely reasonable.

IMPORTANCE OF CORRELATIVE FACTORS.

Other changes that might be produced by the variations in pressure and might possibly account for the phenomena that have been discussed are as follows: (1) expansion and contraction of the water itself; (2) precipitation of solids out of solution in the water; (3) escape of gases out of solution; (4) removal of sand, silt, and clay through the wells with the water; (5) leakage through well casings or up through the holes on the outside of the casings, and (6) leakage through the confining beds. In the following paragraphs all of these possible causes are briefly but critically considered, and the conclusion is reached that they are not adequate to account fully for the phenomena that have been attributed to the compressibility and elasticity of the artesian aquifers.

Elasticity of the Water.—The compressibility of water under moderate pressures, such as are involved in artesian aquifers, amounts to about .015 of 1 per cent. for each 100 feet of head.[28] The volume of water that could be stored in a sandstone having a porosity of 35 per cent. through compression of the water by increasing the artesian head 100 feet would therefore be equal to about .005 of 1 per cent. of the volume of the sandstone. As water has virtually perfect volume elasticity, this would also be the increase in volume of water that would result from a lowering of head of 100 feet. For the Dakota sandstone in the row of townships under consideration the increase in volume due to the drop of approximately 300 feet in artesian head must have amounted to about .015 of 1 per cent. of the volume of the sandstone, or about .1 inch. This appears to be of a different order of magnitude from the depth of 4.4 inches of artesian water that were taken from storage according to the previous calculations.

Solids precipitated out of solution.—Some solids dissolved in water under pressure are precipitated out of solution when the pressure is decreased and this precipitation generally causes an increase in aggregate volume of the water and solids. Ten samples of water from the upper horizon of the Dakota sandstone, which is here under consideration, were analyzed by Shepard [29] early in the history of artesian water development. These samples contained an average of 2,261 parts per million of dissolved solids, most of which consisted of soluble sodium salts that would not be precipitated by the change in pressure. They contained an average of only 27 parts per million of calcium and only 20 parts of magnesium. Even if all of these amounts of calcium and magnesium had been precipitated as carbonates, the

[28] Fowle, F. E., Smithsonian physical tables, 7th revised edition, p. 107, 1923.
[29] Shepard, J. H., "The Artesian Waters of South Dakota," South Dakota Agric. Coll. and Exp. Sta., Bull. 41, 1895.

volume of the precipitated material would be only about .002 of 1 per cent. of the volume of the sandstone, or not much more than .01 inch—an amount that is insignificant in comparison with the amount of water that was removed from storage. Moreover, analyses of water from this horizon made in recent years show about the same chemical composition as the earlier analyses, indicating that there has not been much precipitation. Even if the precipitation of solids were quantitatively of some consequence, the process could not account for most of the phenomena that have been presented as evidences of compressibility and elasticity.

Gases Liberated from Solution.—The quantity of any gas that can be held in solution in artesian water decreases with decrease in pressure. If the water in an artesian aquifer is nearly saturated with gas some of the gas may escape out of solution when the artesian pressure is decreased and may occupy some of the interstices of the aquifer in gaseous form. By such a process large amounts of artesian water might be removed from an incompressible aquifer and its place might be taken by the liberated gas. This process may have been operative in some places, but in most artesian aquifers the gas is apparently too meager in quantity to pass out of solution until it approaches the surface in its upward ascent through the well. The available information indicates that release of gas has not been a factor in any of the artesian aquifers that have been considered in this paper, but fuller and more precise data on the quantities of gas delivered by the artesian water would be desirable. At any rate the process of gas liberation would not account for the increase in artesian pressure that occurs when there is loading at the surface, as by ocean tides or railroad trains. As is well known, gas in many wells is liberated from the water as it approaches the surface, and this process, like that of an artificial air lift, is effective in bringing the water to the surface. However, to account for any of the phenomena cited in this paper as evidences of compressibility or elasticity, the gas must be liberated before it reaches the well, under the high pressures that prevail within the aquifers.

Sand, Silt, and Clay Removed with the Water.—Large quantities of sand, silt, and clay are in many places brought to the surface by artesian water as it ascends through flowing or pumped wells that end in strata of unconsolidated materials. This has been notably true of some of the wells drilled to the Dakota sandstone many years ago when the pressure was still great. The removal of sand is given by Platt and Johnson as one of the probable causes of the subsidence that has been observed at the surface in the Goose Creek Oil Field, in Texas. However, the removal of the solid substance of a formation will tend to increase its storage capacity whereas the problem that has been outlined in regard to the Dakota sandstone is to dispose of the space which has been vacated by the water that was removed from storage. Moreover, removal of sand or other material through wells would not account for any of the other phenomena cited.

Leakage of Wells.—Many artesian wells that were improperly constructed or in which the casing has become corroded, suffer loss of water by underground leakage. A part of the water that is forced into such a well from the aquifer by artesian pressure escapes through some underground outlet and does not reach the surface. The leakage may occur even in wells that have stopped flowing or have never been flowing wells. Underground leakage can generally be detected in either flowing or non-flowing wells by abnormally low head of the water in them, and the location of the leaks and the rate at which the water is escaping can be definitely determined by the use of a deep-well current meter.[30] None of the phenomena that have been cited can, however, be adequately explained by assuming leakage from wells. Thus, leakage does not account for tidal fluctuations, and in so far as it has occurred in the wells in the Dakota Artesian Basin it only adds to the disparity between discharge and recharge.

The effect of leakage on the rate of recovery of head after a flowing well is closed or the pump on a non-flowing artesian well is stopped is somewhat problematical and depends on the conditions that exist in the stratum into which the water escapes. It

[30] McCombs, John, and Fiedler, A. G., "Methods of Exploring and Repairing Leaky Artesian Wells," U. S. Geol. Survey Water-Supply Paper 596-A. 1927.

would seem that if the head against which the escaping water is discharged remains constant the leakage will not affect the rate but only the extent of recovery, but that if the head in the upper stratum which receives the escaping water is influenced by the rate of leakage into it the effect will be the same as that of the elasticity of the artesian aquifer. It is possible that the lag in the nightly recovery in the Roswell Artesian Basin, shown in Figure 7, is in part a leakage phenomenon.

Leakage of Confining Beds.—The confining beds in many artesian basins are permeable or leaky to the extent that enough artesian water escapes through them so that even when the first well is drilled the artesian water will not rise in the well very far above the water table. If the confining bed is very incompetent to prevent leakage the piezometric surface of the artesian aquifer even before there is any artesian water development may be only a few feet or a few inches above the water table. Under these conditions upward percolation through the confining bed occurs normally. It is accelerated whenever there is any rise in artesian head due to decrease in the rate of withdrawal of artesian water or increase in the rate of recharge of the artesian aquifer. Leakage through a confining bed, like leakage through a well, can account for lag in drawdown or recovery of head only as differences in the rate of leakage produce appreciable effects in lowering or raising the water table. There is a nearly complete gradation from aquifers with essentially impermeable confining beds to those which have no confining beds and whose piezometric surface coincides almost exactly with the water table. Lag in drawdown and recovery are inherent features of water-table wells, in which the cone of depression is formed by the removal of water from storage. On the other hand, where the confining bed is nearly impermeable, fluctuations in rate of leakage will be too slight to affect the water table appreciably and, hence, can not account for lag in drawdown or recovery in wells that tap the aquifer below the confining bed.

The confining bed of the Dakota artesian basin consists of hundreds of feet of dense plastic shale that is impermeable or very nearly so, as is shown by the very high heads that prevailed when the first wells were drilled. Leakage into the sandstone could not have occurred through this dense shale or against the artesian pressure, which is still generally sufficient to lift the water from the Dakota sandstone above the water table. Leakage out of the sandstone was never great or else the pressure would not have been so high when the first wells were drilled, and at any rate it could not account for the disparity between discharge and recharge. Leakage through the confining bed in any basin could not account for such phenomena as fluctuations caused by railroad trains.

SIGNIFICANCE OF COMPRESSIBILITY AND ELASTICITY IN PROBLEMS OF HYDROLOGY.

The amounts of compression and expansion that have been computed are so small when expressed in inches of depth or percentage of volume that the question may arise as to whether, after all, this discussion has any real significance. The answer is, however, near at hand. If the compression of the sandstone accounts for most of the artesian water that has been discharged in the last four decades from 15,000 wells in one of the largest artesian basins in the world, it is obviously of major importance. In the past the principle of elasticity has not been recognized in attempts that have been made to estimate the artesian water supplies from the Dakota sandstone or other artesian aquifers, but estimates have generally been based on computations of transmission capacity, recharge at the intake area, or lowering of the water table at the intake—all on the assumption that the artesian aquifer is a perfectly incompressible and inelastic reservoir or conduit. In fact, artesian aquifers are apparently all more or less compressible and elastic though they differ widely in the degree and relative importance of these properties. In general the properties of compressibility and elasticity are of the most consequence in aquifers that have low permeability, slow recharge, and high head. In many aquifers these properties are evidently important in supplying water not only by permanent reduction of storage but also by temporary reduction that is replenished when the wells are shut down or during the season of minimum use. If it were not for the elasticity, the supplies of water that could be recovered through artesian wells for intermittent or fluctuating needs would be much smaller.

In recent years the United States Geological Survey has conducted intensive investigations of four of the largest and most productive artesian basins in the country. The investigation of the Dakota artesian basin in North Dakota has been conducted in cooperation with the State Geological Survey and has been in charge of Howard E. Simpson, of that Survey, but much intensive work has also been done on the project by Herbert A. Hard; the investigation of the artesian aquifers underlying the Coastal Plain of New Jersey has been conducted in cooperation with the New Jersey Department of Conservation and Development by David G. Thompson; the investigation of the Roswell artesian basin, in the Pecos Valley of New Mexico, has been conducted, in cooperation with the State Engineer, by Albert G. Fiedler and S. Spencer Nye; and the investigation of the Honolulu artesian basin, in Hawaii, has been conducted, in coöperation with the Territorial Commissioner of Public Lands and the Honolulu Sewer and Water Commission, by Max H. Carson, John McCombs, and Harold S. Palmer.

In all of these three States and in the Territory of Hawaii, laws have been enacted regulating to some extent the use of the artesian water, thus making headway against the time-honored but disastrous legal fiction that nothing can be known about the source, quantity, or destination of artesian or other ground water and that therefore anyone has the right to use it or waste it as he pleases without regard to the damage that he may do to others. If these water supplies are to be fully developed and utilized, more progress must be made in giving legal protection to those who develop them and put them to beneficial use. This can, however, be accomplished only as ground-water geologists and engineers establish a broad, secure, scientific foundation for determining ground-water conditions and estimating ground-water supplies. Most of the quantitative work on ground water has been done on the non-artesian aquifers which have water tables, and consequently methods for intensive work on artesian aquifers have been relatively meager and undeveloped. In the investigations that have been mentioned, however, progress has been made on fundamental principles that will lead to effective methods, and it is believed that the theory of compressibility and elasticity of artesian aquifers expresses one of these fundamental principles.

Copyright ©1940 by the American Geophysical Union
Reprinted from *Am. Geophys. Union Trans.* **21**:574-586 (1940)

ON THE FLOW OF WATER IN AN ELASTIC ARTESIAN AQUIFER

C. E. Jacob

Introduction--Slichter showed in 1898 that a solution may be obtained for a given problem in the steady motion of ground-water by solving the familiar Laplace equation and that therefore in steady-state conditions a problem in the motion of ground-water is mathematically analogous to a problem in the steady flow of heat or electricity [see 1 of "References" at end of paper]. More recently it has been recognized that the analogy holds also for the non-steady-state flow of compressible liquids, in elastic systems as well as in rigid systems.

In studying the effect of the discharge of flowing wells on the head in the Dakota sandstone, Meinzer [2, 3] concluded that the water discharged by the wells had largely been derived locally from storage. It was found that the amount of water withdrawn from storage could not be accounted for on the basis of the compressibility of water alone but that it might be accounted for on the basis of the probable compressibility of the sandstone itself. Prior to that time, estimates of water-supplies from artesian aquifers had been based upon the assumption that artesian aquifers are perfectly incompressible and inelastic. However, as Meinzer states [2, p. 289], "artesian aquifers are apparently all more or less compressible and elastic though they differ widely in the degree and relative importance of these properties. In general, the properties of compressibility and elasticity are of the most consequence in aquifers that have low permeability, slow recharge, and high head."

The concept of compressibility and elasticity of artesian aquifers fostered by Meinzer has been extended by Theis [4]. By recourse to the analogy that exists between the laminar flow of water through the interstices of porous materials and the flow of heat through solid media, Theis [5] derived "the relation between the lowering of the piezometric surface and the rate and duration of discharge of a well using ground-water storage." This analogy has as its basis the assumption that owing to elastic compression of the water-bearing beds and the expansion of water confined therein "a specific amount of water is discharged instantaneously from storage as the pressure falls." However, for reasons to be considered presently, it appears that the chief source of the water derived from storage "within" an artesian aquifer is probably the contiguous and intercalated clay beds (or shale-beds, in a sandstone-aquifer) and that because of the low permeability of the clays (or shales) there is a time-lag between the lowering of pressure within the aquifer and the appearance of that part of the water which is derived from storage in those clays (or shales). Such lag-effects are necessarily difficult to handle mathematically. Therefore, in the work which follows it is assumed that the release of the stored water is immediate following the decline of pressure and that the rate at which it is released is directly propor-

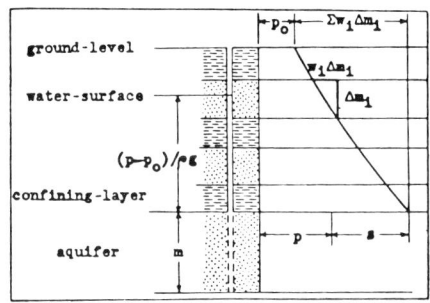

Fig. 1--Distribution of stress in artesian aquifer and overlying beds

tional to the rate of decline of the pressure. The problem is thus reduced to that of flow in an homogeneous elastic artesian aquifer confined between two impermeable planes. These simplifying assumptions are justified because the mathematical solution based thereon appears consistently to yield results that are confirmed by observation.

The writer proposes to derive "from scratch" the fundamental differential equation governing the flow of water in an elastic artesian aquifer, considering in turn each of the assumptions that are necessary to the derivation of the equation. He proposes further to solve the flow-equation for one of the more important cases, that of radial flow in an homogeneous aquifer of uniform thickness and infinite areal extent, thus obtaining the same solution as that derived from the heat-analogy.

Compression of aquifer--The combined weight of the overlying beds and the atmosphere above produces in the aquifer a stress equal to $(p_0 + \sum w\Delta m)$, where p_0 is the atmospheric pressure and where w is the weight of a unit-volume of material in a stratum of thickness Δm (see Fig. 1). This load may be considered as being borne in part by the solid "skeleton" of the aquifer and in part by the water confined therein. By equilibrium of the forces acting on the plane of contact between the aquifer and the confining layer

$$(p_0 + \sum w\Delta m) = (s + bp) \tag{1}$$

where s is the compressive stress in the solid part of the aquifer, where p is the hydrostatic pressure within the aquifer, and where b may be considered ideally to represent the proportion of the plane of contact between the aquifer and the confining layer over which the hydrostatic pressure is effective [6]. Actually, of course, there is no aquifer which is bounded sharply by an absolutely impervious confining bed. The factor b merely represents a concept introduced to facilitate mathematical treatment of the problem.

Neglecting for the moment any changes in barometric pressure and delay in subsidence of the overlying beds, the load on the aquifer may be considered to remain constant. Thus $(ds + bdp) = 0$, or $-(ds/dp) = b$. The factor b is thus defined as the negative ratio of a given change in compressive stress to the change in hydrostatic pressure producing it. In an aquifer composed of uncemented granular material the value of b is unity. In a solid aquifer, as a limestone, having tubular channels, b is apparently equal to the porosity. The value of b for a sandstone doubtless ranges between these limits, the exact value depending on the degree of cementation and probably also on the nature of the contact between the sandstone and contiguous shales.

A volume, V, of a given aquifer contains a mass of water, $M = \rho \theta V$, where ρ is the density of the water and θ is the porosity of the material of which the aquifer is composed. Consider a vertical column of the aquifer of cross-section ΔA_z and height m, equal to the thickness of the aquifer. The mass of water contained therein is $\Delta M = \rho \theta m \Delta A_z$. As the hydrostatic pressure is decreased by Δp, the fluid density is decreased by $\Delta \rho$, and the aquifer is compressed an amount Δm, thus decreasing the porosity by $\Delta \theta$. If the pressure declines at a rate $(\partial p/\partial t)$, the mass of water per unit-surface area contained within the elemental volume of aquifer decreases at a rate

$$\partial^2 M/\partial A_z \partial t = \partial(\rho \theta m)/\partial t \tag{2}$$

The variation of each of the variables ρ, θ, and m may be expressed in terms of the variation of p. As the hydrostatic pressure is reduced, the water contained in the aquifer undergoes elastic expansion, its density being reduced in accordance with

$$d\rho = (\rho/E_w)dp \tag{3}$$

E_w being the bulk modulus of elasticity of water (approximately 300,000 lbs/in^2 at average ground water temperatures).

The increase in compressive stress occasioned by the decline in hydrostatic pressure is accompanied by compression of the aquifer, its thickness being decreased by

$$dm = -(m/E_s)ds = (m/E_s)bdp \tag{4}$$

115

where E_s is the modulus of compression of the solid skeleton of the aquifer as confined in situ.

The volume of solid material in a given column of the aquifer is $V_s = (1 - \theta/m A_z)$. Neglecting the change in volume of the solid material in the aquifer due to compression or deformation of the individual particles, and also neglecting lateral displacement due to inequalities in load

$$dV_s = 0 = A_z[(1 - \theta)dm - md\theta]$$

from which

$$d\theta/(1 - \theta) = dm/m$$

But, from equation (4), $dm/m = bdp/E_s$, therefore

$$d\theta = (1 - \theta)bdp/E_s \tag{5}$$

The pressure at a point of elevation z within the aquifer is $p = [p_0 + \rho g(h - z)]$, where p_0 is the atmospheric pressure, as before, and where h is the "head", or elevation of the free water-surface in a well penetrating the aquifer to the point in question. The elevation of the water-level in a well is ordinarily determined by tape-measurement of the depth to water below a "measuring point" of known elevation and is referred to mean sea-level or some other arbitrary datum, as is z also. In considering purely radial flow in an horizontal aquifer, z may be taken as constant. Then, neglecting fluctuations in atmospheric pressure, $dp = \rho gdh$, and equations (3), (4), and (5) become, respectively,

$$d\rho = (\rho/E_w) \rho gdh \tag{6}$$

$$dm = (bm/E_s) \rho gdh \tag{7}$$

$$d\theta = [(1 - \theta)b/E_s] \rho gdh \tag{8}$$

Substituting in equation (2) the values of $d\rho$, dm, and $d\theta$ given by equations (6), (7), and (8), and reducing

$$\partial^2 M/\partial A_z \partial t = \rho[\rho g \theta m(1/E_w + b/\theta E_s)](\partial h/\partial t) \tag{9}$$

In equation (9) the quantities ρ, θ, and m are considered to represent initial values. It is assumed also that E_w and E_s remain constant over the range of pressure-changes involved. As z is one of the independent variables, $(\partial z/\partial t) = 0$, and $(\partial p/\partial t) = \rho g(\partial h/\partial t)$. Thus the previous restriction that the aquifer be horizontal is removed.

Coefficient of storage--The quantity enclosed in brackets in equation (9) is found by inspection to be dimensionless. This factor is identical to the "coefficient of storage" of Theis [5, p. 894]. It is designated by the letter S. That is,

$$S = \rho g \theta m(1/E_w + b/\theta E_s) \tag{10}$$

Equation (9) thus becomes

$$\partial^2 M/\partial A_z \partial t = \rho S(\partial h/\partial t)$$

by which the coefficient of storage is given as

$$S = (1/\rho A_z)(dM/dh) \tag{11}$$

that is, assuming there is no lag in the release of stored water, or in other words that S is independent of the variable t.

The coefficient of storage is defined by equation (11) as the volume of water of a certain density released from storage within the column of aquifer underlying a unit-surface area during a decline in head of unity. This definition conforms to that given by Theis. The coefficient of storage is presumably a constant which may be determined empirically for a given aquifer from data on the decline of head produced by pumping from wells penetrating the aquifer.

In equation (10) the storage derived from expansion of the confined water is given by $(g\theta m/E_w)$, corresponding to the $(1/E_w)$ in parentheses. The storage derived from compression of the aquifer is given by the remainder, which is represented by the other term in parentheses, $(b/\theta E_s)$. These two terms account for all the water derived from storage within an hypothetical elastic artesian aquifer bounded by impermeable planes. In the actual case, however, a third term is required to account for the water derived from storage in the adjacent and included clay-beds. This term would in general be a function of the rate and perhaps also the magnitude of the decline of pressure, of the thickness and distribution of the intercalated clay-beds, and of the permeability and modulus of compression of the clay; but under ordinary conditions, of course, equation (9) would not hold true. However, if there are a sufficient number of clay-laminae interbedded with the sand so that the release of stored water from the clay is virtually instantaneous, the required additional term may, for all practical purposes, be considered to be independent of both the rate and the magnitude of the decline of pressure and independent of the permeability of the clay. Under such conditions the coefficient of storage may be given as

$$S = \rho g \theta m (1/E_w + b/\theta E_s + c/E_c) \qquad (12)$$

where E_c is the modulus of compression of the clay and where c is a dimensionless quantity that depends largely on the thickness, configuration, and distribution of the intercalated clay-beds.

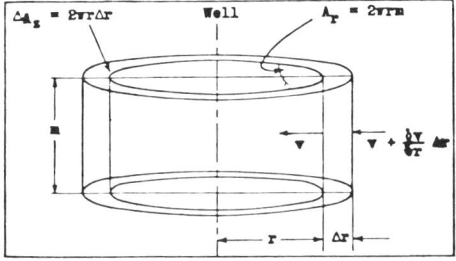

Fig. 2--Radial flow through concentric cylindrical surfaces

Equation of continuity--The time-rate of change of the mass of liquid per unit-volume contained in the elemental volume, $\Delta V = m\Delta A_z$, is

$$\partial^2 M/\partial V \partial t = (1/m)\partial^2 M/\partial A_z \partial t = (\rho S/m)(\partial h/\partial t) \qquad (13)$$

According to the law of conservation of matter this must equal the net flux per unit-volume across the boundaries of the same volume-element. In the case of radially symmetrical flow, as occurs in the vicinity of a single well pumping from an homogeneous aquifer of constant thickness, and with which we are primarily concerned here, the volume-element to be considered is that bounded by two cylindrical surfaces, of radii r and $(r + \Delta r)$, respectively, and by the upper and lower bounding planes of the aquifer, as indicated in Figure 2.

The flux through the inner cylindrical surface is $\rho A_r v$, where $A_r = 2\pi r m$ is the area of the surface, and where v is the radial velocity, considered positive when the motion is away from the origin. Similarly, the flux through the outer cylindrical surface is $(\rho + \Delta\rho)(A_r + \Delta A_r)(v + \Delta v)$. Neglecting differentials of higher order, the net inward flux through the bounding surfaces of the volume considered is

$$\Delta M_r/\Delta t = -\rho A_r v (\Delta\rho/\rho + \Delta A_r/A_r + \Delta v/v)$$

[The leakage from or into the overlying and underlying beds has been neglected. Change in the rate of leakage accompanying the decline in head is accounted for approximately by the third term in parentheses in equation (12).]

Considering in turn each of the terms of the right-hand member of the foregoing equation

$$\begin{cases} \Delta\rho/\rho = (\delta p/\delta r)\Delta r/E_w \\ \Delta A_r/A_r = \Delta r/r + (\delta m/\delta r)\Delta r/m \\ \Delta v/v = (\delta v/\delta r)\Delta r/v \end{cases}$$

For the range of pressure-gradients ordinarily occurring in ground-water bodies, $(\delta p/\delta r)\Delta r/E_w$ is certainly negligible in comparison with $\Delta r/r$. And, since $dm/m = bdp/E_s$, the quantity $[(\delta m/\delta r)\Delta r/m]$ also may be considered negligible, leaving

$$\Delta M_r/\Delta t = -\rho A_r v [\Delta r/r + (\delta v/\delta r)\Delta r/v]$$

Passing to differentials, the net flux per unit-volume is

$$(1/2\pi r m)\partial^2 M_r/\partial r \partial t = -\rho [v/r + (\partial v/\partial r)] \qquad (14)$$

From equation (13) the time-rate of change of the mass of liquid per unit-volume contained in the element whose volume is $\Delta V = m\Delta A_z = 2\pi r m \Delta r$ is

$$(1/2\pi rm)\partial^2 M/\partial r\partial t = (\rho S/m)(\partial h/\partial t) \tag{15}$$

Equating the right-hand members of equations (14) and (15) it is found that in accordance with the law of conservation of matter

$$(v/r + \partial v/\partial r) = -(S/m)(\partial h/\partial t) \tag{16}$$

Equation (16) is a combination of two equations: The "equation of continuity," which alone expresses the law of conservation of matter, and the "equation of state," which expresses the relation between the fluid density and the pressure, the temperature being considered constant. Equation (3) is the equation of state in differential form.

Darcy's law and permeability--The variable v may be eliminated from equation (16) by introducing a further relation known as Darcy's law. This is so named in honor of H. Darcy, who in 1856 found that the rate at which water flows vertically downward through a sand-filter (discharging at atmospheric pressure) is directly proportional to the sum of the thickness of the bed of sand and the depth of water covering it and inversely proportional to the thickness of the bed, the constant of proportionality being a coefficient which depends on the nature of the sand [7]. The principle enunciated by Darcy is but an expression, for a special case, of a more generalized law which may be stated as follows: The macroscopic velocity of movement of a liquid through a permeable medium, that is, the discharge per unit-area, is equal to the negative gradient of a potential which is defined by $\varphi = (k/\mu)\rho gh = (k/\mu)(\rho gz + p - p_0)$. That is, in vector form, $\bar{v} = -\text{grad } \varphi$. Here ρ and μ are the density and viscosity, respectively, of the fluid in question; h is the head, expressed in terms of a column of liquid of density ρ. The constant k is independent of the properties of the fluid and has the dimensions of length squared. It characterizes the flow-medium, being a function of porosity and of size and shape of grain, etc.

It follows from the foregoing equation that the component of velocity in any direction is equal to the negative component of the potential-gradient in that direction. Thus the radial component of the velocity-vector is

$$v = -(\partial \varphi/\partial r) = -(\rho gk/\mu)(\partial h/\partial r) \tag{17}$$

Flow away from the origin is considered positive, hence the negative sign in equation (17). Combining the constants k and g and the quantities ρ and μ, the latter two also being assumed constant,

$$v = -K(\partial h/\partial r) \tag{18}$$

where

$$K = \rho gk/\mu \tag{19}$$

The factor K, known as the "permeability", has the dimensions of length divided by time, or velocity, and is here expressed in the usual velocity-units, that is, in ft/sec. However, it is not to be confused with the macroscopic velocity of movement of the fluid. Actually the permeability is equal to the discharge per unit-area which would obtain under unity hydraulic gradient. The cgs equivalent of K is Muskat's $\bar{k}$, which he terms "the 'effective' permeability for the composite system of porous medium and liquid" [8, p. 72].

For all practical purposes, K may be considered constant for a given homogeneous isotropic aquifer transmitting water at normal ground-water temperature. If the aquifer is not homogeneous and isotropic, the value of the permeability, K, will in general be affected more by the variation of K from place to place within the aquifer than by the variation of any one of the other three factors ρ, μ, and g. Under gradients normally encountered in ground-water bodies the variation of density is negligible, as previously shown. The variation of viscosity with pressure is likewise negligible. And, although both the density and the viscosity vary with the temperature, the variation of the temperature of the water in a given aquifer, both in time and in space, is in most cases so slight that these two factors may be considered constant. However, it should be pointed out that in attempting to effect a direct comparison between values of average permeability determined for different aquifers in the field and values of permeability determined in the laboratory, corrections should be applied for the variation of viscosity with temperature. (Temperature-corrections for density as well as pressure-corrections for both density and viscosity may be omitted. Also, variation in the acceleration of gravity from locality to locality may be neglected without serious consequence.) If, following the convention of Meinzer, 60° F is taken as the standard temperature, the value of permeability corrected for temperature is $K_{60} = (\mu/\mu_{60})K$, where μ_{60} is the viscosity of water at 60° F, and where μ is the viscosity of water at the temperature at which the value of K was determined, whether in the laboratory or in the field.

The coefficient of permeability adopted by the Geological Survey, denoted here by P_{60}, is defined as the rate of flow of water at 60° F, in gallons a day, through a square foot of cross-section, under unity hydraulic gradient [9]. Thus, $P_{60} = 646,000 K_{60}$, one cubic foot per second being approximately equivalent to 646,000 gallons per day.

Integration of fundamental differential equation--Substituting the value of v given by equation (18) in equation (16) and multiplying both members of the equation by $(-r^2/K)$,

$$r(\partial h/\partial r) + r^2(\partial^2 h/\partial r^2) = (r^2 S/Km)(\partial h/\partial t) \tag{20}$$

This is the fundamental differential equation for the flow of a compressible viscous liquid in an elastic artesian aquifer. The particular solution of this equation which satisfies the necessary boundary-conditions is the equation of the piezometric surface, its elevation, h, being expressed as a function of the independent variables r and t, and the constants K, m, and S.

The first integration may be performed by inspection. For, assume

$$r(\partial h/\partial r) = -2c \, e^{-r^2 S/4Kmt} \tag{21}$$

Then

$$r^2(\partial^2 h/\partial r^2) = 2c \, e^{-r^2 S/4Kmt} + c(r^2 S/Kmt)e^{-r^2 S/4Kmt}$$

and

$$r(\partial h/\partial r) + r^2(\partial^2 h/\partial r^2) = (r^2 S/Km)(c/t)e^{-r^2 S/4Kmt}$$

Thus, according to equation (20)

$$(\partial h/\partial t) = (c/t)e^{-r^2 S/4Kmt} \tag{22}$$

By differentiating both sides of equation (22) with respect to r and integrating with respect to t, properly evaluating the constant of integration, the same value is obtained for $(\partial h/\partial r)$ as that given by equation (21). Thus equation (21) is shown to be a particular (partial) solution of the differential equation. The differentiation is performed as follows:

$$\partial^2 h/\partial t \partial r = (c/t)(-2rS/4Kmt)e^{-r^2 S/4Kmt} = (-2c/r)(r^2 S/4Kmt^2)e^{-r^2 S/4Kmt}$$

Integrating the foregoing equation with respect to t

$$(\partial h/\partial r) = (-2c/r)e^{-r^2 S/4Kmt} + c_1$$

The constant c_1 may be evaluated by a consideration of the boundary-conditions. Assuming the aquifer to be infinite in areal extent, the limit of $(\partial h/\partial r)$ is zero as r approaches infinity, and as the limit of $(1/r)e^{-r^2 S/4Kmt}$ is also zero as r approaches infinity, c_1 must equal zero. Thus

$$(\partial h/\partial r) = (-2c/r)e^{-r^2 S/4Kmt} \tag{23}$$

which, on multiplying through by r, becomes identical to equation (21).

The other boundary-condition, namely, that the initial distribution of head is uniform throughout, was satisfied by the original assumption [equation (21)]. For, when $t = 0$ $e^{-r^2 S/4Kmt} = 0$, and hence $(\partial h/\partial r) = 0$.

The value of c may be determined by substituting the proper values of $(\partial h/\partial r)$ and $(\partial h/\partial t)$ in the equation

$$Q = -2\pi rm K (\partial h/\partial r) + \int_0^r S (\partial h/\partial t) 2\pi r \, dr \tag{24}$$

This equation is a statement of the fact that the discharge of the well is equal to the sum of the flow through a cylinder of variable radius, r, and the increase in rate of flow proceeding toward the well caused by compression of the volume of aquifer enclosed by the cylinder. The discharge through the cylindrical surface is $A_r v = -2\pi r m K (\partial h/\partial r)$, and the gain in rate of flow due to compression is $\int S (\partial h/\partial t) dA_z = \int_0^r S (\partial h/\partial t) 2\pi r dr$. The significance of the lower limit zero is that the radius of the well is considered negligible.

Making the proper substitutions in equation (24) and performing the integration indicated, it is found that $Q = 4\pi c K m$, from which

$$c = Q/4\pi K m \qquad (25)$$

Dimensional analysis applied to the factor c yields the following: $[L^3/T]/[L/T][L] = [L]$. In other words, c has the dimensions of length. [This c is not to be confused with that in equation (12).] Inasmuch as the aquifer is assumed to be homogeneous and uniform in thickness, K and m are constant. Therefore, since c is constant, Q must also be constant.

Let $r^2 S/4Km = b$. This quantity [not to be confused with $b = -(ds/dp)$] has the dimensions of time. That is, $[L^2]/[L/T][L] = [T]$. Equation (22) may now be rewritten as

$$(\partial h/\partial t) = (c/t) e^{-b/t} \qquad (26)$$

This equation may be integrated to give h in terms of b, c, and t. In dimensionless form,

$$h/c = \int_0^t (1/t) e^{-b/t} dt$$

where the upper limit of integration is the time elapsed since beginning pumping at a constant rate. By arbitrarily choosing the initial piezometric surface as the plane of reference--that is, the plane to which h is referred--a constant factor, which would ordinarily appear in the left-hand member of the equation, is thereby eliminated.

Let $b/t = u$. Then $t = b/u$, and $dt = (-b/u^2)du$. To determine the new limits of integration: when $t = 0$, $u = \infty$; and when $t = t$, $u = b/t$. Making these substitutions

$$h/c = \int_{\infty}^{b/t} (u/b) e^{-u} (-b/u^2) du$$

That is,

$$h/c = \int_{b/t}^{\infty} (e^{-u}/u) du \qquad (27)$$

This integration may be performed by parts as follows:

$$\int (e^{-u}/u) du = \int (1/u) du + \int [(e^{-u} - 1)/u] du$$

The first integral yields $\log_e u$. The quantity under the second integral sign may be expanded as

$$(e^{-u} - 1)/u = -1 + u/2! - u^2/3! + u^3/4! - \ldots + (-1)^n u^{n-1}/n!$$

As this series converges uniformly for all values of u, it may be integrated term by term, giving

$$\int [(e^{-u} - 1)/u] du = -u + u^2/2 \cdot 2! - u^3/3 \cdot 3! + u^4/4 \cdot 4! - \ldots + (-u)^n/n \cdot n!$$

By properly evaluating the integral for the upper and lower limits of integration the following solution is obtained:

$$h/c = -0.5772 - \log_e u + u - u^2/2 \cdot 2! + u^3/3 \cdot 3! - u^4/4 \cdot 4! + \ldots + (-u)^n/n \cdot n! \qquad (28)$$

in which b/t has been replaced by u corresponding to the fixed value of t, which, however, may now be considered variable. The constant $0.5772 \cdots$ is known as Euler's constant.

It should be noted that equation (28) is in dimensionless form. The solution given by this equation is the same as that derived by Theis [5] and by Muskat [8, p. 667] by recourse to the heat analogy. The integral

$$Ei(b/t) = \int_{\infty}^{-b/t} (e^{-u}/u) du \qquad (29)$$

Flow of Water in an Elastic Artesian Aquifer

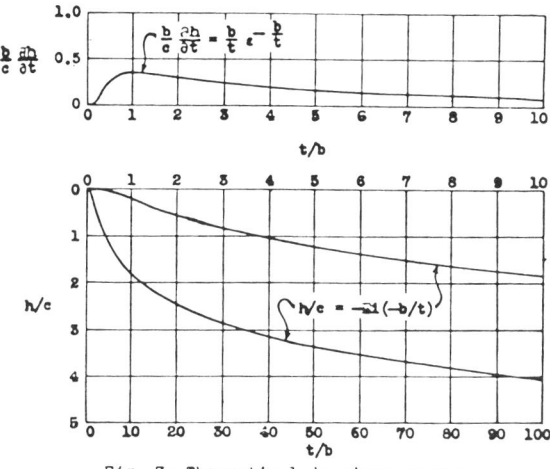

Fig. 3--Theoretical drawdown curves

is known as the "exponential integral." Values of this function have been published in tabular form [10] for a limited range of values of the argument and are reproduced in part by Theis [4].

In terms of the exponential integral, then, the drawdown is given in the following dimensionless form

$$(h/c) = -Ei(-b/t) \quad (30)$$

Figure 3 is a graph of this equation. The abscissas are (t/b), the scale along the upper margin of the graph applying to the upper curve and that along the lower margin to the lower curve. The ordinates are (h/c). The two curves shown in Figure 3 are typical "drawdown curves." Inasmuch as they are in dimensionless form, they are typical of all such curves that might be obtained under the postulated conditions.

Initially the slope of the drawdown curve is zero, and up to a certain point as (t/b) increases the slope also increases. Thereafter, however, the slope decreases continuously as (t/b) increases. It is noted that the maximum slope of the drawdown curve occurs at (t/b) = 1, that is, when $t = r^2S/4Km$. This is also seen from the graph of the equation $(b/c)(\partial h/\partial t) = (b/t)e^{-b/t}$, which also is given in Figure 3. At the point (t/b) = 1, (h/c) = -Ei(-1) = 0.219, and $(b/c)(\partial h/\partial t) = e^{-1} = 0.368$. It is interesting to note also that the value of the drawdown at the time of occurrence of the maximum rate of decline of the head is the same for all values of r. For, according to the above, at this point $h = 0.219 c = 0.219 Q/4\pi Km$.

The drawdown may also be expressed in dimensionless form in terms of the distance, r, and an additional parameter, a, which has the dimensions of length. That is,

$$(h/c) = -Ei(-r^2/a^2) \quad (31)$$

where $a^2 = (4Kmt/S)$, or where $(r^2/a^2) = (r^2S/4Kmt) = (b/t)$.

Figure 4 is a graph of equation (31). The abscissas of the upper curve are given on the upper margin of the graph. Similarly, the abscissas of the lower curve are given on the lower margin. These two curves are typical half-sections of the so-called "cone of depression." As (r/a) becomes infinite, (h/c) approaches zero, satisfying the postulated infinite areal extent of the aquifer. As (r/a) approaches zero, (h/c) theoretically becomes infinite. Practically, however, h is limited to the drawdown at the well-face, corresponding to the theoretical value of drawdown at a distance $r = r_w$, where r_w is the "effective radius" of the well.

Determination of coefficients of storage and transmissibility by graphical method--Equation (30), or equation (31), may be rewritten as

$$h = -(114.6 \, Q/T)[Ei(-1.87r^2S/Tt)] \quad (32)$$

where Q is the pumping rate in gallons a minute; t is the time elapsed since the beginning of pumping, in days; and T is the "coefficient of transmissibility" of Theis, which he defines [4, p. 894] as "the number of gallons of water which will move in one day through a vertical strip of the aquifer one foot wide and having the height of the aquifer when the hydraulic gradient is unity. It is the average coefficient of permeability, adjusted for the temperature, multiplied by the thickness of the aquifer in feet."

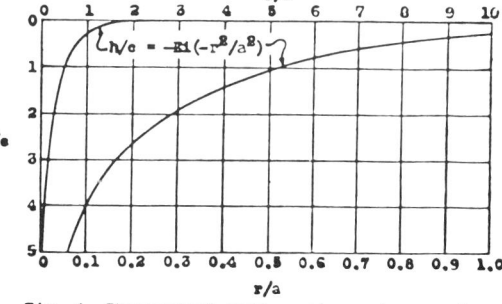

Fig. 4--Theoretical half-sections of cone of depression

It is seen from equation (32) that if S and T are known and if the discharge can be measured, it is then possible to determine the drawdown, h, for any values of distance, r, and time, t. Conversely, if h can be measured for one or more values of r and for several values of t and if the discharge is known, S and T may be determined. However, because there are two unknowns and because of the nature of the exponential integral, an exact analytical solution is not possible. And inasmuch as one of the unknowns, T, occurs twice in the equation, once in the argument of the exponential integral and again in reciprocal form as a multiplier of Ei(-u), solution by trial is rendered very laborious. However, a graphical method has been devised by Theis [11] whereby it is possible to effect a solution and thus determine the value of the coefficient of storage, S, and the coefficient of transmissibility, T. Briefly the method is as follows: (1) Values of -Ei(-u) taken from tables of that function are plotted on logarithmic coordinates against values of u. The curve thus obtained is known as the "type-curve." (2) Values of drawdown, h, obtained from observations of the decline of head in the vicinity of a discharging well are plotted against (r^2/t) on logarithmic tracing paper to the same scale as that used for the type-curve. A convenient scale is five inches to the cycle. (3) The graph of the data plotted on tracing paper is superimposed on the graph of the type-curve. With the axes parallel, a position is found by trial for which the plotted points fall most nearly on the type curve. (4) Then, from the coordinates of an arbitrarily chosen point on one graph and the coordinates of the corresponding point on the other graph, with the graphs in that particular position, the parameters S and T are determined. The value of T is first determined from $T = 114.6 \, q[-Ei(-u_1)]/h_1$. Then, using this value of T the value of S is determined from $S = Tu_1/1.87(r^2/t)_1$. Here u_1 and $-Ei(-u_1)$ are the coordinates of the arbitrarily chosen point on the graph of the type-curve and $(r^2/t)_1$ and h_1 are the coordinates of the corresponding point on the graph of the observational data.

In Table 1 are given values of S and T obtained by application of this method to data from a pumping test conducted at Rockaway Park, Long Island, New York, May 18 to 22, 1939. In the course of construction of well Q 1,030 for the City of New York, a four-inch screen was set in the Lloyd sand between depths of 734 feet and 798 feet below the rotary-table. According to the driller's log there is an 85-foot bed of water-bearing sand between depths of 715 and 800 feet. Well Q 1,030 was pumped for four days at an average rate of 310 gallons a minute. Continuous records were obtained of the decline of water-level in wells Q 288, Q 543, Q 544, and Q 546 nearby. These wells are located on a barrier beach about 2,000 feet wide which separates Jamaica Bay from the Atlantic Ocean. A recording tide-gage was maintained in operation on Jamaica Bay in order to correct the water levels for tidal fluctuations. The ratio between the magnitude of the fluctuations of the water-level in the wells and the actual fluctuations in tide producing them was found to be about 44 per cent for wells Q 543, Q 544, and Q 546, and about 42 per cent for well Q 288. This factor is termed the "tidal efficiency."

Table 1--Coefficients of storage and transmissibility obtained by application of graphical method to pumping test-data

Well	Distance from well Q 1,030, r (feet)	Approximate range of r^2/t (ft^2/sec)	Coefficient of storage, S	Coefficient of transmissibility, T (gpd/ft)
Q 546	67	0.01 to 0.4	1.90 × 10^{-4}	106,000
Q 543	262	0.2 to 6		
Q 544	313	0.3 to 8		
Q 288	4,100	60 to 1,000	8.0 × 10^{-4}	114,000

The values of transmissibility given in Table 1 check within reason values of transmissibility determined from the same test by the Thiem method [12]. Using the combination of wells Q 543 and Q 546 a value of 103,000 was obtained for T, and using wells Q 544 and Q 546 a value of 95,000 was obtained.

Determination of coefficient of storage from tidal efficiency or barometric efficiency--
Theoretically, an artesian well that exhibits tidal fluctuations, which are due to the alternate loading and unloading (or variation in load) on the confining layer occasioned by the rise and fall of the tide, also responds to fluctuations in atmospheric pressure. That is, the same mechanism in the aquifer which produces tidal fluctuations also produces "barometric fluctuations." Again, the barometric fluctuations observed in a well are proportional to the variations in atmospheric pressure which produce them, the constant of proportionality being termed the "barometric efficiency." It appears that barometric and tidal fluctuations may be taken as an index of the elasticity of the aquifer, and further that in the hypothetical case there is a direct relation between the tidal efficiency or the barometric efficiency and the coefficient of storage.

Consider an homogeneous elastic artesian aquifer which is uniform in thickness and infinite in areal extent. Assume that the aquifer is overlain by beds of uniform thickness which are covered by an equally extensive body of tide-water in which the variations in level occur simultaneously and uniformly throughout. Assume further that variations in atmospheric pressure are likewise uniform over the entire aquifer. Under these conditions, tidal and barometric fluctuations of the head in the aquifer will be uniform and there will be no lateral movement of water. (Non-uniform conditions of loading produce inequalities in head which in turn cause lateral movement of the water confined in the aquifer. And, as a result of the inevitable loss of head accompanying the movement of water, a reduction in the magnitude of the fluctuations occurs.)

Equation (1) is a statement of the condition of equilibrium of pressures on the plane of contact. The term $\sum w \Delta m$ may now be considered to include the tide-water in addition to the beds overlying the aquifer. Neglecting for the moment any variation in the surface-level of the body of tide-water, that is, putting $\sum w \Delta m$ = constant

$$\Delta p_0 = (\Delta s + b \Delta p) \tag{33}$$

As the atmospheric pressure changes by an amount Δp_0 the water-level in a well penetrating the aquifer fluctuates $-(B \Delta p_0 / \rho g)$ feet, where B is the barometric efficiency of the well. Thus the net change in hydrostatic pressure within the aquifer is

$$\Delta p = \Delta p_0 - B \Delta p_0 = (1 - B) \Delta p_0 \tag{34}$$

Combining equations (33) and (34)

$$\Delta p = (1 - B)(\Delta s + b \Delta p)$$

from which

$$\Delta p / \Delta s = (1 - B)/[1 - b(1 - B)] \tag{35}$$

When the aquifer and the contained water are compressed by an extensive uniform load, such as represented by an increase in the weight of the atmosphere postulated above, the decrease in volume of water, ΔV_w, is equal to the decrease in volume of aquifer, ΔV. Thus $(\Delta V_w / V_w) = (\Delta V / \theta V)$. It is assumed here that the bulk compression of the individual particles composing the flow-medium is negligible and that there is no leakage from or into contiguous beds. Then, as $(\Delta V_w / V_w) = (-\Delta p / E_w)$ and $(\Delta V / V) = (-\Delta s / E_s)$, it follows that $(\Delta p / E_w) = (\Delta s / \theta E_s)$, from which

$$(\Delta p / \Delta s) = (E_w / \theta E_s) \tag{36}$$

Thus, from equation (35)

$$E_w / \theta E_s = (1 - B)/[1 - b(1 - B)] \tag{37}$$

from which

$$B = 1 - (E_w / \theta E_s)/(1 + b E_w / \theta E_s) \tag{38}$$

Similarly, it may be shown that the tidal efficiency is given by

$$C = (E_w / \theta E_s)/(1 + b E_w / \theta E_s) \tag{39}$$

In other words

$$(B + C) = 1 \tag{40}$$

When b = 1, as for an aquifer composed of an uncemented granular material, such as a quartz-sand,

$$B = 1/(1 + E_w / \theta E_s) \tag{41}$$

and from equation (10)

$$S = (\rho g \theta m / E_w)(1 + E_w / \theta E_s) \tag{42}$$

Thus, for b = 1,

$$S = (\rho g \theta m / E_w)(1/B) \tag{43}$$

The average value of the tidal efficiency, C, for wells Q 543, Q 544, and Q 546 is 0.44. Thus $(1/E) = (1 + E_w/\theta E_s) = [1/(1 - C)] = 1.79$. The average temperature of water in the Lloyd sand is about 65° F. At that temperature the value of $(E_w/\rho g)$ is approximately $300,000 \times 144/62.4 = 690,000$ feet. According to the log of well Q 1,030 the thickness of the aquifer, m, is 85 feet. Assuming the porosity, θ, to be 35 per cent, $S = (0.35 \times 85/690,000)$ $1.79 = 4.3 \times 10^{-5} \times 1.79 = 7.7 \times 10^{-5}$. Similarly, for well Q 288, $[1/(1 - C)] = 1.72$, and $S = 7.4 \times 10^{-5}$.

The tidal efficiency was used to compute the theoretical coefficient of storage because it was more easily determined than the barometric efficiency. In fact, it was not found possible accurately to determine B directly from the data at hand. However, the water-level graphs corrected for tidal fluctuations in the ratio of C and for barometric fluctuations in the ratio of (1 - C) showed no residuals that might be correlated with fluctuations in atmospheric pressure, thus confirming in a way the supposed complementary relation between B and C.

The values of coefficient of storage determined from the tidal efficiency are smaller than the corresponding values of coefficient of storage obtained from the pumping test, in one case by more than one order of magnitude. In order to effect an agreement between a value of S given in column (4) of Table 2 and the corresponding value given in Table 1, the value of (cE_w/E_c) given in column (5) is to be added to the value of $(1 + E_w/\theta E_s)$ given in column (3) and multiplied by $(\rho g \theta m/E_w)$, equal to 4.3×10^{-5}, in accordance with equation (12). For example, from the data for well Q 288, $S = (\rho g \theta m/E_w)[(1 + E_w/\theta E_s) + cE_w/E_c] = 4.3 \times 10^{-5}(1.72 + 16.9) = 4.3 \times 10^{-5} \times 18.6 = 8.0 \times 10^{-4}$.

Table 2--Coefficient of storage computed from tidal efficiency

Well (1)	Tidal efficiency, C (2)	$(1 + E_w/\theta E_s)$ (3)	Coefficient of storage, S (4)	(cE_w/E_c) (5)
Q 546, Q 543, Q 544	0.44	1.79	7.7×10^{-5}	2.63
Q 288	0.42	1.72	7.4×10^{-5}	16.9

According to the data for well Q 288, the amount of "storage" derived from compression of the aquifer and the adjacent and included clay-beds is about 17.6 (that is, 0.72 + 16.9) times as great as the amount derived from expansion of the water; and, according to the data for the other three wells, the amount derived from compression of the aquifer and the adjacent and included clay-beds is about 3.4 (that is, 0.79 + 2.63) times as great as the amount derived from expansion of the water. The determination of the theoretical value of the coefficient of storage from the tidal efficiency (see Table 2) is based on the assumption that "there is no leakage from or into contiguous beds." However, leakage actually does occur, not only from (or into) contiguous beds, but also from (or into) intercalated beds of clay. Therefore, it is not possible to distinguish between that part of the storage derived from compression of the sands and that part derived from compression of the clays. Nevertheless, assuming that all the water derived from storage in the Lloyd sand has been adequately and correctly accounted for, the ratios between the amounts derived from compression of the sands and clays combined and the amounts derived from expansion of the water are as given above.

<u>Flow of compressible fluid in rigid system</u>--The non-steady-state flow of a compressible fluid in a permeable medium has been analyzed by Muskat [8, pp. 621-676] apparently on the assumption that the medium is rigid. He expresses the argument of the exponential integral, used to describe the "transient" effect of a "permanent source" as $-(f\beta\mu r^2/4kt)$; where f is the porosity, identical to θ; where β is the compressibility of the fluid, equivalent to the reciprocal of the bulk modulus, E_w, in atmospheres; and where μ, k, r, and t (in cgs units) have the same meaning as they have here, except that the pressure-factor in k is expressed in atmospheres. Changing the notation to conform to that adopted herein, equating the above to $-u = -(r^2S/4Kmt)$, and substituting for K the value given by equation (19)

$$-\theta \mu r^2/4E_w kt = -\mu r^2 S/4\ gkmt$$

By cancelling like quantities in both members of the above equation and solving for S there is obtained the relation $S = (\rho g \theta m/E_w)$, whereas, according to equation (12),

$$S = (\rho g \theta m/E_w)(1 + bE_w/\theta E_s + cE_w/E_c)$$

In other words, it would appear that consideration has been given only to the compressibility of the fluid, and that the compressibility of the fluid medium has tacitly been neglected. Perhaps in some instances the nature, depth, and configuration of oil-producing structures is such that the amount of compaction accompanying decline of pressure is negligible, in which case the compressibility of the fluid media need not be considered. However, it is believed that oil-bearing structures generally are compressible. The subsidence which occurs in oil-fields may be taken as evidence of the compressibility. Where subsidence has not become visually perceptible its occurrence might be definitely established by differential leveling. Incidentally, it seems that in selecting and establishing bench-marks for precise leveling in areas likely to be affected by the reduction of pressure in oil-reservoirs or in artesian aquifers consideration ought to be given to the possibility of subsidence occurring in those areas.

The production of 500 million barrels of oil from an East Texas oil-field having an estimated total fluid content of 7×10^9 barrels, with a drop in reservoir-pressure of only about 375 pounds per square inch, is attributed by Muskat [8, pp. 644-653] to the "drive" exerted on the oil by the water in the Woodbine sand adjacent to the field, which expands as the pressure is lowered. If the production of oil from this field were to be attributed entirely to expansion of the oil itself, there would be required a compressibility 20 times that actually determined for the oil with its dissolved gas. As the oil is known to be undersaturated at the prevailing reservoir-pressure, such a high value of compressibility could not be justified. In accounting for the production on the basis of the expansion of the water alone, the effective compressibility of the water had to be taken as 12 times that of gas-free water. This abnormally high compressibility "may be explained," Muskat says [8, p. 652] "on the assumption that there are gas pockets dispersed through the water-horizon to the extent of 4.9 per cent of the total pore-volume of the sand." However, it appears from the foregoing discussion that the high effective compressibility assumed for the water in the Woodbine sand might more easily be accounted for as representing not only the compressibility of the water, but also the compressibility of the sand, and, more important perhaps than that, the compressibility, in part, of the adjacent and included clay-beds.

According to the few data available, the Lloyd sand, with the water that it contains, appears to be from about 3.4 to 17.6 times as compressible as water alone. The latter figure is probably more nearly the correct value, because the relative amount of compression of the aquifer accompanying a given decline in pressure, as indicated by the value of the coefficient of storage, is small for small values of the ratio between the lateral distance (r, in Table 1) and the depth to the aquifer (715 feet). This is because the virtual load on the aquifer is decreased in the immediate vicinity of the pumped well owing to the lateral strength of the overlying beds, and as a consequence the value of the compressive stress in the skeleton of the aquifer for a given value of hydrostatic pressure is less than it would be if the full weight of the overlying beds rested continually on the aquifer. It is believed that in the case of the Lloyd sand a greater part of the high apparent compressibility is to be attributed to the clays than to any gas that might be trapped in the aquifer. It seems likely that this might apply also to the Woodbine sand, which, with the water (and gas) that it contains, is apparently about 12 times as compressible as water alone.

Actually the procedure followed by Muskat, wherein the "apparent" compressibility of the fluid is adjusted empirically on the basis of the observed pressure-distributions, is in effect equivalent to that of the graphical method outlined on a previous page, although explanations given for the abnormally high compressibility differ. Of course, there remains the possibility that pockets of gas are present in some artesian aquifers, and that therefore they contribute to the high "effective" compressibility of those composite systems of solid and fluid. However, there is no direct evidence of the presence of undissolved gas in any great proportion in the Lloyd sand. It seems likely that an amount of included gas as great as 4.9 per cent of the total pore-volume of the sand would appreciably diminish the tidal fluctuations. As previously indicated, tidal fluctuations of the head in the Lloyd sand range up to 44 per cent of the fluctuations in the level of the overlying tide-water.

<u>Rate of transmission of artesian pressure</u>--A word might be said here in regard to the rate of transmission of artesian pressure. As pointed out by Muskat [8, p. 669], the "velocity of propagation of disturbances within a fluid medium" is "equal to the velocity of sound in the fluid" and "may, therefore, be considered as effectively instantaneous" (this because of the comparatively low velocity of the fluid particles themselves). What is generally referred to as the "rate of transmission of artesian pressure" is a value of velocity determined by measuring the time required for the effect of pumping, or some such disturbance, to become appreciable or perceptible at a given distance from the source of the disturbance. As a result, rates of transmission determined in this manner are dependent on the means employed to measure the effect

of the disturbance. The more accurate the method of measurements employed, the higher will be the apparent rate of transmission. This is evident from a consideration of equation (30). Assume that $h_1 = -c\,Ei(-u_1)$ is the limit of accuracy of the means employed to measure the effect of a given continuous disturbance of constant strength. (If the depth to water were measured periodically by means of a graduated tape, h_1 would probably be 0.01 foot; if an automatic water-stage recorder were employed, h_1 might be as small as 0.001 foot.) As the exponential integral, $Ei(-u)$, is a single-valued function of the argument, $-u = -(r^2 S/4Kmt)$, there is thus defined a value of u equal to u_1, corresponding to the limiting value of drawdown, $h = h_1$. If $(r^2 S/4Kmt) = u_1$, then $t = (r^2 S/4Kmu_1)$. In other words, the time required for a given continuous disturbance to produce a perceptible effect, that is to say, of magnitude $h = h_1$ at a distance r from the source of the disturbance, is directly proportional to the square of the distance. The apparent average rate of transmission between the source and a point at a distance r from the source is, therefore,

$$(r/t) = (4Kmu_1/rS) \qquad (44)$$

The nature of the exponential integral is such that as the argument of the function increases beyond bound the numerical value of the function approaches zero. Therefore, the smaller the value of h_1, the greater will be the value of u_1, and the greater also will be the value of (r/t) as given by equation (44). From this it might erroneously be concluded that if the limit of accuracy, h_1, could in some way be made to approach zero, the apparent rate of transmission would become infinite. Indeed, such a conclusion logically follows the foregoing statement. However, it should be borne in mind that the above equations are based upon the assumption that the transmission of pressure- effects is instantaneous, whereas actually the velocity of propagation of pressure-disturbances in an artesian aquifer is equal to the velocity of sound in the aquifer. Thus, if the means of measuring changes of pressure in artesian aquifers were sufficiently refined, it should be found that the rate of transmission of such pressure-changes is equal to the velocity of sound.

Acknowledgment--The writer is greatly indebted to his colleagues in the Geological Survey for timely suggestions and criticism, and especially to C. V. Theis, as many of the ideas expressed herein have been taken from his personal communications to the writer.

References

[1] C. S. Slichter, Theoretical investigation of the motion of ground waters, U. S. Geol. Surv., 19th Annual Report, pp. 295-384, 1898.
[2] O. E. Meinzer, Compressibility and elasticity of artesian aquifers, Econ. Geol., v. 23, pp. 263-291, 1928.
[3] O. E. Meinzer and H. A. Hard, U. S. Geol. Surv. W.-S. Paper 520, pp. 90-93, 1925.
[4] C. V. Theis, The significance and nature of the cone of depression in ground-water bodies, Econ. Geol., v. 33, pp. 869-902, 1938.
[5] C. V. Theis, The relation between the lowering of the piezometric surface and the rate and duration of discharge of a well using ground-water storage, Trans. Amer. Geophys. Union, 16th Annual Meeting, pp. 519-524, 1935.
[6] C. V. Theis, Earth tides expressed in fluctuations of water level in artesian wells in New Mexico, paper read at International Union of Geodesy and Geophysics, Seventh Assembly, Washington, D. C., September 1939 (not yet published).
[7] H. Darcy, Les fontaines publiques de la ville de Dijon, Paris, 1856.
[8] Morris Muskat, The flow of homogeneous fluids through porous media, McGraw-Hill, 1937.
[9] N. D. Stearns, Laboratory tests on physical properties of water-bearing materials, U. S. Geol. Surv. W.-S. Paper 596, pp. 148, 149, 1927.
[10] Smithsonian Physical Tables, 8th rev. ed., table 32, 1933.
[11] C. V. Theis, personal communication to the writer, March 1938.
[12] L. K. Wenzel, The Thiem method for determining permeability of water-bearing materials, U. S. Geol. Surv. W.-S. Paper 679-A, 1936.

8

Copyright ©1966 by the American Geophysical Union
Reprinted from Jour. Geophys. Research **71**:4785-4790 (1966)

The Equation of Groundwater Flow in Fixed and Deforming Coordinates[1]

HILTON H. COOPER, JR.

Two forms of the equation of flow of a compressible liquid in an elastic porous medium are derived by considering mass conservation in (1) a control volume whose boundaries are fixed in space and (2) a control volume that deforms and moves through space when the material deforms. The first method yields a form of the equation that necessarily involves the velocity of the grains of the medium. The second yields a form that does not involve the grain velocity. Jacob's equation is found to be correct when negligible terms are omitted.

NOTATION

b, thickness of aquifer (L).
g, gravity field strength (L/T^2).

$$h = z + \frac{1}{g}\int_{p_0}^{p}\frac{dp}{\rho(p)},$$

head above common datum (L).
i, **j**, **k**, unit orthogonal vectors (dimensionless).
$J = \partial z/\partial \xi$, Jacobian (dimensionless).
K, fluid conductivity (L/T).
k, intrinsic permeability of medium (L^2).
n, porosity of medium (dimensionless).
p, fluid pressure (M/LT^2).
q, vector flow per unit area relative to grains of medium (L/T).
S, storage coefficient (dimensionless).
S_s, S^*, specific storage $(1/L)$.
s', surface of deforming volume $v'(L^2)$.
T, transmissibility (L^2/T).
t, time (T).
$v = \Delta x \Delta y \Delta z$, fixed volume element (L^3).
$v' = \Delta x \Delta y \Delta z'$, deforming volume element (L^3).
$v_\xi = \Delta x \Delta y \Delta \xi$, material volume element (L^3).
w, vertical component of flow per unit area relative to fixed point (L/T).
w', vertical component of flow per unit area relative to grains of medium (L/T).
w_g, vertical component of velocity of grains of medium (L/T).
z, fixed spatial coordinate (L).
z', deforming spatial coordinate (L).
α, compressibility of medium (LT^2/M).

[1] Publication authorized by the Director, U. S. Geological Survey.

β, compressibility of liquid (LT^2/M).
ρ, fluid density (M/L^3).
ξ, material coordinate (L).
σ_z, vertical compressive stress on medium (M/LT^2).

INTRODUCTION

The problem of nonsteady flow in an elastic artesian aquifer was first analyzed by *Theis* [1935] when, by recourse to the analogy between groundwater flow and heat flow, he derived the equation which gives 'the relation between the lowering of the piezometric surface and the rate and duration of discharge of a well using ground-water storage.' In a later paper *Theis* [1938] introduced the concepts of the storage coefficient and the transmissibility of an aquifer. As originally defined by Theis, the coefficient of storage is 'the volume of water, measured in cubic feet, released from storage in each column of the aquifer having a base 1 foot square and a height equal to the thickness of the aquifer, when the water table or other piezometric surface is lowered 1 foot,' and the transmissibility is 'the number of gallons of water which will move in one day through a vertical strip of the aquifer 1 foot wide and having the height of the aquifer when the hydraulic gradient is unity.' (Since *Theis*'s [1938] paper was published it has, of course, become customary among most hydrologists, including Theis, to define these terms without reference to units.)

Jacob [1940], in the process of developing the Theis equation from hydrologic concepts, derived a differential equation for nonsteady flow in a uniformly thick, horizontal, compressible sand as

$$\frac{\partial^2 h}{\partial x^2} + \frac{\partial^2 h}{\partial y^2} = \frac{S}{T}\frac{\partial h}{\partial t} \quad (1)$$

where the storage coefficient is

$$S = \rho g b(\alpha + n\beta) \quad (2)$$

Later *Jacob* [1950] derived an equation for three-dimensional flow which can be written

$$\frac{\partial^2 h}{\partial x^2} + \frac{\partial^2 h}{\partial y^2} + \frac{\partial^2 h}{\partial z^2} = \frac{S_s}{K}\frac{\partial h}{\partial t} \quad (3)$$

where

$$S_s = \rho g(\alpha + n\beta) \quad (4)$$

The coefficient S_s was given the name 'specific storage' by *Hantush* [1960] and defined by him as 'the volume of water a unit volume of the aquifer releases from storage under a unit head decline.' From (2) and (4)

$$S = b S_s \quad (5)$$

so that Theis's storage coefficient has been generally accepted as being the product of the specific storage and the thickness of the aquifer (assuming that the aquifer is homogeneous).

De Wiest [1966], however, questioned the validity of (4) and (5). According to him, Jacob's derivation of (3) became the subject of criticism among some of the participants of the 1965 Hydrology Institute, sponsored by the National Science Foundation at Princeton University. He indicated that 'the criticism concerned the fact that in one side of the equation the net inward mass flux was calculated for a volume element without deformation while in the other side of the equation, to compute the rate of change of the mass inside the volume element, the element itself was deformed.' De Wiest [1966, equation 31] derived the equation

$$\nabla^2 h - 2\rho\beta g\frac{\partial h}{\partial z} = \frac{S^*}{K}\frac{\partial h}{\partial t} \quad (6)$$

where

$$S^* = \rho g[(1-n)\alpha + n\beta] \quad (7)$$

The discrepancy between (4) and (7) led De Wiest to redefine the specific storage 'whereby the relation between specific storage and storage coefficient is no longer the thickness of the aquifer....'

De Wiest obtained his result (6) by considering the conservation of mass in a control volume whose boundaries are fixed in space. He did not, however, distinguish between the rate of flow relative to the moving grains of the medium and the rate of flow across the fixed boundaries of his control volume while the material is deforming. The first of these flows obeys Darcy's law; the second does not. My purpose is to show, among other things, that the term $-\rho g\alpha n$, which occurs in De Wiest's expression for specific storage but not in Jacob's, does not arise in a derivation that properly distinguishes between these two flows.

In the analysis that follows two forms of the equation of flow will be derived by considering mass conservation in (1) a volume whose boundaries are fixed in space and (2) a volume that deforms and moves through space when the material deforms.

Analysis

Basic equations. Let us consider an elastic homogeneous porous medium containing a compressible liquid and suppose that this medium is deforming in response to a change in the fluid pressure. Let $v = \Delta x \Delta y \Delta z$ be a small element of the medium which always contains the same grains and which initially occupies the volume $v_\xi = \Delta x \Delta y \Delta \xi$ whose boundaries are fixed in space. As the medium deforms, the medium element deforms and moves through space with some grain velocity $\mathbf{q}_g$. We assume that the horizontal components of the deformation are negligible, so that only the height Δz of the element changes and the element moves vertically with a grain velocity w_g.

Now let us consider the changes in fluid pressure p and vertical compressive stress σ_z on the medium at a point near the center of the moving element. We observe that as the pressure changes the weight of the solid material above this point remains constant. The density and volume of the fluid above this point change, but these changes are so small that the rate of change in compressive stress σ_z is very nearly equal and opposite to that of the fluid. Accordingly, we can write

$$dp/dt = -d\sigma_z/dt$$

where $d/dt = \partial/\partial t + w_g\, \partial/\partial z$ is the material derivative.

The height Δz varies with the stress σ_z in accordance with

$$\frac{\delta \Delta z}{\Delta z} = -\alpha \delta \sigma_z \qquad (8)$$

so that the rate of change of the height is related to the rate of change of σ_z and p by

$$\frac{1}{\Delta z}\frac{d\Delta z}{dt} = -\alpha \frac{d\sigma_z}{dt} = \alpha \frac{dp}{dt} \qquad (9)$$

The grain velocity can be related to the pressure by use of the relation [*Aris*, 1962, p. 84]

$$\frac{1}{J}\frac{dJ}{dt} = \nabla \cdot \mathbf{q}_g = \frac{dw_g}{dz} \qquad (10)$$

Since

$$\frac{1}{J}\frac{dJ}{dt} = \frac{\Delta \xi}{\Delta z}\frac{d(\Delta z/\Delta \xi)}{dt} = \frac{1}{\Delta z}\frac{d(\Delta z)}{dt}$$

$$= \nabla \cdot \mathbf{q}_g = \partial w_g/\partial z \qquad (10)$$

We obtain from (9) and (10)

$$\frac{\partial w_g}{\partial z} = \alpha \frac{dp}{dt} = \rho g \alpha\left(\frac{dh}{dt} - w_g\right) \qquad (11)$$

A grain of the medium that is initially at a height ξ above the bottom of the medium moves in time t to the new height

$$z = \xi + \int_0^t w_g\, dt \qquad (12)$$

The height ξ of the initial position of the grain is the material or Lagrangian coordinate, and the height z of the position to which the grain has moved in time t is the spatial or Eulerian coordinate [*Aris*, 1962, p. 77]. Either z or ξ can be considered as remaining fixed while the other varies with time. Hence we may write $\xi = \xi(z, t)$ or $z = z(\xi, t)$, depending upon whether we wish to consider a property of the fluid opposite a point that remains fixed in space or opposite a point that moves with the grains. If we hold z constant, ξ varies in the sense that the grains which move past the fixed point have started at different initial heights $\xi = \xi(z, t)$. If we hold ξ constant, z is the height of a particular grain as it moves through space.

When z is held constant while ξ varies, z will be referred to as the *fixed* spatial coordinate. When ξ is held constant, the varying height z will be designated hereafter by z' and will be referred to as the *deforming* spatial coordinate.

Equation of flow in fixed coordinates. If we assume that the fluid conductivity K remains constant during the deformation of the material, the vertical component of the Darcy flow (the rate of flow per unit area relative to the grains of the material) is given in terms of the derivative with respect to fixed coordinates by

$$w' = -K\frac{\partial h}{\partial z} = -\frac{\rho g k}{\mu}\frac{\partial h}{\partial z} \qquad (13)$$

The vertical component w of the rate of flow past a point that is fixed in space is

$$w = w' + nw_g \qquad (14)$$

If $\mathbf{q}$ is the vector flow relative to the grains, the vector flow relative to a point fixed in space is $\mathbf{q} + \mathbf{k}nw_g$. The conservation of mass of fluid in a volume element whose boundaries are fixed in space is therefore described by

$$-\nabla \cdot \rho(\mathbf{q} + \mathbf{k}nw_g)\, \Delta x\, \Delta y\, \Delta z$$
$$= \frac{\partial(\rho n)}{\partial t}\, \Delta x\, \Delta y\, \Delta z$$

which reduces to

$$-\nabla \cdot \rho \mathbf{q} - \partial(\rho n w_g)/\partial z = \partial(\rho n)/\partial t$$

or

$$-\nabla \cdot \rho \mathbf{q} - n w_g \frac{\partial \rho}{\partial z} - \rho w_g \frac{\partial n}{\partial z} - \rho n \frac{\partial w_g}{\partial z}$$
$$= n \frac{\partial \rho}{\partial t} + \rho \frac{\partial n}{\partial t}$$

or

$$-\nabla \cdot \rho \mathbf{q} = n \frac{d\rho}{dt} + \rho \frac{dn}{dt} + \rho n \frac{\partial w_g}{\partial z} \qquad (15)$$

The derivative dn/dt can be evaluated by considering continuity on the volume of the solid material. If we accept Jacob's [1950, p. 329] assumption that the volume of each individual grain of the solids remains constant during deformation of the medium, conservation of this volume in the element $\Delta x \Delta y \Delta z$ requires that

$$-\frac{\partial}{\partial z}[(1-n)w_g]\,\Delta x\,\Delta y\,\Delta z$$

$$=\frac{\partial(1-n)}{\partial t}\,\Delta x\,\Delta y\,\Delta z \quad (16)$$

which leads to

$$\frac{dn}{dt}=(1-n)\frac{\partial w_g}{\partial z} \quad (17)$$

Substituting from (11) into (17) we have

$$\frac{dn}{dt}=\rho g\alpha(1-n)\left(\frac{dh}{dt}-w_g\right) \quad (18)$$

The compressibility of the liquid is defined by $\beta = -(1/V)dV/dp$, where V is the volume. Substituting $V = m/p$, where m is the mass of the liquid in V, we can write $d\rho = \rho\beta dp$, which leads to

$$\frac{d\rho}{dt}=\rho\beta\frac{dp}{dt}=\rho^2 g\beta\left(\frac{dh}{dt}-w_g\right) \quad (19)$$

A substitution of (18) and (19) into (15) yields

$$-\nabla\cdot\rho\mathbf{q}=\rho^2 g(\alpha+n\beta)\left(\frac{dh}{dt}-w_g\right) \quad (20)$$

De Wiest [1966, equation 27] showed that when $\mathbf{q} = K\nabla h$ is substitiuted into the first term on the left side of (20) if the dependency of K on the compressibility of the fluid (see equation 13) is not neglected, this term becomes

$$-\nabla\cdot\rho\mathbf{q}=\rho K(\nabla^2 h - 2\rho\beta g\,\partial h/\partial z) \quad (21)$$

Substituting from (21) into (20) we find the equation of flow in terms of fixed coordinates to be

$$\nabla^2 h - 2\rho g\beta\frac{\partial h}{\partial z}$$

$$=\frac{\rho g}{K}(\alpha+n\beta)\left(\frac{dh}{dt}-w_g\right) \quad (22)$$

where $dh/dt = \partial h/\partial t + w_g\,\partial h/\partial z$.

It will be shown in the following section that the term grain velocity is eliminated from the flow equation if the problem is formulated in terms of deforming coordinates.

Equation of flow in deforming coordinates. Holding ξ constant we now consider the conservation of mass of fluid in the material volume element $v'(t) = \Delta x\Delta y\Delta z'$ which deforms and moves vertically through space as the medium deforms and which always contains the same grains. Since there is no movement of the grains of the medium across the boundaries of this element, conservation of the mass of the fluid in it is satisfied by

$$\frac{\partial}{\partial t}\int_{v'(t)}\rho n\,dv + \int_{s'(t)}\rho(\mathbf{q}\cdot\mathbf{v})\,ds = 0 \quad (23)$$

where $\mathbf{v}$ is the unit outward normal of the surface $s'(t)$ of the volume $v'(t)$. (It is important here and with reference to the operations that follow to observe that when ξ is held constant the material derivative d/dt becomes the partial derivative $\partial/\partial t$. Because there is no grain velocity relative to the deforming coordinate z', the term $w_g\,\partial/\partial z'$ vanishes.) Since the integral in the first term is over the varying volume $v'(t)$, we cannot move the time derivative through the integral sign. However, if we first transform the integral over v' into an integral over the fixed volume v_ξ, we can then carry the time derivative through the integral sign [*Aris*, 1962, p. 85]. To transform the integral, we observe that $dv' = J dv_\xi$. Therefore, we can write

$$\frac{\partial}{\partial t}\int_{v'(t)}\rho(z',t)n(z',t)\,dv$$

$$=\frac{\partial}{\partial t}\int_{v_\xi}\rho(\xi,t)n(\xi,t)J\,dv$$

$$=\int_{v_\xi}\left(J\frac{\partial\rho n}{\partial t}+\rho n\frac{\partial J}{\partial t}\right)dv$$

$$=\int_{v_\xi}\left(\frac{\partial\rho n}{\partial t}+\rho n\frac{1}{J}\frac{\partial J}{\partial t}\right)J\,dv \quad (24)$$

Since $dJ/dt = \partial J/\partial t$ in the deforming coordinate system, we obtain from (10)

$$\frac{1}{J}\frac{\partial J}{\partial t}=\frac{\partial w_g}{\partial z}$$

Substituting this result into (24) we have

$$\frac{\partial}{\partial t}\int_{v'(t)}\rho n\,dv$$

$$=\int_{v_\xi}\left(\frac{\partial\rho n}{\partial t}+\rho n\frac{\partial w_g}{\partial z'}\right)J\,dv$$

$$=\int_{v'(t)}\left(\frac{\partial\rho n}{\partial t}+\rho n\frac{\partial w_g}{\partial z'}\right)dv \quad (25)$$

Substituting from (25) into (23) and transforming the second term of (22) into a volume integral by Green's theorem, we can write

$$\int_{v'(t)} \left(\frac{\partial \rho n}{\partial t} + \rho n \frac{\partial w_g}{\partial z'} + \nabla' \cdot \rho \mathbf{q} \right) dv = 0$$

where $\nabla' = \mathbf{i}\partial/\partial x + \mathbf{j}\partial/\partial y + \mathbf{k}\partial/\partial z'$. Since this result is true for an arbitrary volume, the integrand must vanish everywhere. Therefore, the requirement for mass conservation in v' becomes

$$-\nabla' \cdot \rho \mathbf{q} = n \frac{\partial \rho}{\partial t} + \rho \frac{\partial n}{\partial t} + \rho n \frac{\partial w_g}{\partial z'} \quad (26)$$

We next evaluate $\partial n/\partial t$ by considering the conservation of the volume of solid material in v'. Using again the assumption that the volume of each individual grain of the solids remains constant and recognizing that there is no movement of grains across the surface of v', we can write

$$\frac{\partial}{\partial t} \int_{v'(t)} (1 - n) \, dv = 0$$

Employing the procedure used in obtaining (25) we obtain

$$\int_{v'(t)} \left[\frac{\partial (1 - n)}{\partial t} + (1 - n) \frac{\partial w_g}{\partial z'} \right] dv = 0$$

which leads to

$$\frac{\partial n}{\partial t} = \frac{\partial w_g}{\partial z'} - n \frac{\partial w_g}{\partial z'} = \rho g \alpha \frac{\partial h}{\partial t} - n \frac{\partial w_g}{\partial z'} \quad (27)$$

From (19)

$$\frac{\partial \rho}{\partial t} = \rho^2 g \beta \frac{\partial h}{\partial t} \quad (28)$$

Substituting from (27) and (28) into (26) we obtain

$$-\nabla' \cdot \rho \mathbf{q} = \rho^2 g (\alpha + n\beta) \, \partial h/\partial t \quad (29)$$

To proceed from (29), we assume that Darcy's law is approximated closely in the deforming medium by

$$\mathbf{q} = -K \nabla' h \quad (30)$$

which requires that the velocity in the vertical direction be approximated closely by

$$w' = -K \, \partial h/\partial z' \quad (31)$$

Using (30), we find by *De Wiest's* [1966, equation 27] procedure that the left side of (29) can be written

$$-\nabla' \cdot \rho \mathbf{q} = \rho K (\nabla'^2 h - 2\rho \beta g \, \partial h/\partial z') \quad (32)$$

By substituting from (32) into (29) we obtain the equation of flow in terms of deforming coordinates:

$$\nabla'^2 h - 2\rho \beta g \frac{\partial h}{\partial z'} = \frac{\rho g}{K} (\alpha + n\beta) \frac{\partial h}{\partial t} \quad (33)$$

In almost all problems the second term on the left is negligible, and the equation is very closely approximated by

$$\nabla'^2 h = \frac{S_s}{K} \frac{\partial h}{\partial t} \quad (34)$$

which is identical to Jacob's equation 3 except for the use of deforming coordinates in lieu of fixed coordinates.

The result (33) might have been anticipated from (22) by observing that when the problem is considered in terms of deforming coordinates the material derivative becomes equal to the partial derivative.

Conclusions

Equations 22 and 33 both describe the conditions for the nonsteady flow of a compressible liquid through an elastic porous medium. The first is expressed in terms of the fixed coordinate z and involves the grain velocity. The second is expressed in terms of the deforming coordinate x' and does not involve the grain velocity.

Aside from the advantage of eliminating the grain velocity from the flow equation, several factors weigh slightly in favor of deforming coordinates in problems of nonsteady flow in elastic mediums:

1. The use of deforming coordinates resolves a long-standing contradiction in the existing theory of groundwater flow (H. E. Skibitzke, 1952, personal communication). In the customary fixed coordinates two-dimensional nonsteady radial flow cannot occur in an elastic confined aquifer because the top of the aquifer constitutes a moving boundary along which vertical components of flow must occur. On the other hand, it is commonly assumed, as, for example, in the familiar *Theis* [1935] equation,

that two-dimensional radial flow does occur. This contradiction disappears when the problem is conceived in terms of deforming coordinates because the flow relative to the grains has a vertical component only if there is partial penetration of wells or other such boundary conditions.

2. In field and laboratory investigations the piezometers or other instruments for measuring heads do not ordinarily remain fixed in space as the material deforms; more often than not they move with the grains of the medium. Observed gradients are therefore more likely to be in terms of deforming coordinates than fixed coordinates. The difference between the two is negligible in most aquifer materials and can be appreciable only in clays and similar materials that compact substantially when the pressure of the fluid in them is reduced.

3. For a deforming medium the assumption that $w' = K \partial h / \partial z'$, which was used in the development of the equation in deforming coordinates, is a somewhat closer approximation of Darcy's law than $w' = K \partial h / \partial z$. This can be shown most readily by considering unidirectional flow in the z direction. Statistically, a vertical streamline through the deforming element $\Delta x \Delta y \Delta z'$ would traverse the same number of pores regardless of the deformation, whereas one through the fixed element traverses a greater or lesser number when the material deforms. Therefore, w' is more nearly proportional to the head differential across $\Delta z'$ than to one across a fixed element Δz. It follows that K is more nearly constant when Darcy's law is approximated in terms of the deforming coordinate z'. A similar argument could be developed for three-dimension flow, but this argument would be complicated by the fact that the direction of a streamline in three dimensions changes with the deformation.

Jacob's equation 3 for three-dimensional flow, from which many existing solutions have been derived, appears to be exact if one considers his z coordinate to be the deforming coordinate z'.

References

Aris, R., *Vectors, Tensors, and the Basic Equations of Fluid Flow*, 286 pp., Prentice-Hall, Inc., Englewood Cliffs, N. J., 1962.

De Wiest, R. J. M., On the storage coefficient and the equations of groundwater flow, *J. Geophys. Res., 71*, 1117–1122, 1966.

Hantush, M. S., Modification of theory of leaky aquifers, *J. Geophys. Res., 65*, 3713–3725, 1960.

Jacob, C. E., The flow of water in an elastic artesian aquifer, *Trans. Am. Geophys. Union, II*, 574–586, 1940.

Jacob, C. E., chapter 5, pp. 321–386, in *Engineering Hydraulics*, edited by H. Rouse, John Wiley & Sons, New York, 1950.

Theis, C. V., The relation between the lowering of the piezometric surface and the rate and duration of discharge of a well using groundwater storage, *Trans. Am. Geophys. Union, II*, 519–524, 1935.

Theis, C. V., The significance and nature of the cone of depression in ground-water bodies, *Econ. Geol., 33*, 889–920, 1938.

(Manuscript received April 22, 1966; revised July 1, 1966.)

ERRATA

Page 4787, equation (10) should read: $"\frac{1}{J}\frac{dJ}{dt} = \nabla \cdot q_g = \frac{\partial w_g}{\partial z}"$. The second equation (10) should be deleted.

Part II
WELL AND AQUIFER HYDRAULICS

Editors' Comments
on Papers 9 Through 18

9 **THEIS**
   The Relation Between the Lowering of the Piezometric Surface and the Rate and Duration of Discharge of a Well Using Ground-Water Storage

10 **BOULTON**
   The Drawdown of the Water-Table under Non-Steady Conditions near a Pumped Well in an Unconfined Formation

11 **HANTUSH and JACOB**
   Non-Steady Radial Flow in an Infinite Leaky Aquifer

12 **NEUMAN and WITHERSPOON**
   Applicability of Current Theories of Flow in Leaky Aquifers

13 **FERRIS et al.**
   Excerpts from *Theory of Aquifer Tests*

14 **STALLMAN**
   Use of Numerical Methods for Analyzing Data on Ground Water Levels

15 **TYSON and WEBER**
   Ground-Water Management for the Nation's Future—Computer Simulation of Ground-Water Basins

16 **PINDER and BREDEHOEFT**
   Application of the Digital Computer for Aquifer Evaluation

17 **WALTON and PRICKETT**
   Hydrogeologic Electric Analog Computers

18 **TODD**
   Ground-Water Flow in Relation to a Flooding Stream

In the period between 1900 and 1940, the center of excellence in North American hydrogeology clearly lay in the U.S. Geological Survey under the leadership of O. E. Meinzer. During this period, significant theoretical and regional studies were undertaken, but the most fertile field of research at the U.S.G.S. was in well hydraulics.

The steady-state Thiem equation (the so-called equilibrium approach), was by this time well known in America, and its potential application in both a direct and inverse mode recognized. The most common approach was to first apply the formula in inverse mode in the form of a pumping test, whereby steady state drawdowns in observation wells could be used to determine the transmissivity of aquifers; and then in direct mode, whereby the obtained transmissivities could be used to predict the steady-state drawdown cone for any proposed production well. Major uncertainties developed in the interpretation of pumping tests using the Thiem formula, however. The radius of influence of the well at steady state was needed a priori, and it was often doubtful whether steady state had actually been attained during the test. In 1931, L. K. Wenzel ran two exhaustive pumping tests near Grand Island, Nebraska, to test the applicability of the Thiem equation and some ancillary methods suggested by other authors. In the report of his results, Wenzel (1936) concluded that permeability values could successfully be obtained from the Thiem equation but only from observation wells within that portion of the cone of depression where steady flow could be established within the time available for the test.

Despite this relatively favorable finding, the death knell had been sounded for the steady-state method with the publication of Theis's transient analysis (Paper 9) in 1935. Theis recognized the need for a time-dependent analysis and, like Forchheimer and Slichter before him, turned to a heat flow analogy for the solution. With the help of University of Cincinnati mathematician C. I. Lubin, who provided the "mathematical keystone of the paper," Theis developed the transient well hydraulics formula for both the direct and inverse application, for both drawdown and recovery. Theis proved the validity of his theory by comparing his predictions with Wenzel's Grand Island test data.

Wenzel (1942) brought Theis's theory into practical use by publishing a table of exponential function values that appeared in the Theis solution. These tables were the source for the logarithmic *type curves* that appeared with Wenzel's report. Wenzel's *Water-Supply Paper* was the first comprehensive manual of pumping test technology to appear in North America. In an associated development of lasting impact, Cooper and Jacob (1946) recognized an approximation to the Theis solution that allows the use of a semilogarithmic, straight-line, graphical method of interpretation even simpler to apply than the type-curve approach.

Theis was careful in his paper to spell out the assumptions on which his method is based: it applies to an idealized aquifer configuration wherein the aquifer is horizontal, confined between impermeable formations above and below, infinite in horizontal extent, of constant thickness, and homogeneous and isotropic with respect to its hydrogeological parameters. In addition, the system is further constrained to a single well, a constant pumping rate, an infinitesimally small well-diameter, a fully penetrating well, and a uniform hydraulic head throughout the aquifer prior to pumping. It is further assumed that water is released from storage instantaneously with each change in head. The history of development of the methodology of aquifer hydraulics is largely a history of the systematic removal of these assumptions one by one.

Boulton (Paper 10) pioneered the analysis of unconfined aquifers. By 1954 it had been recognized on the basis of field tests that the response of unconfined aquifers was significantly different from the response predicted by the Theis equation. Observation wells usually showed responses with three distinct segments in their time-drawdown curves. His 1954 analysis produced a solution reflecting the second and third segments but failing to account for the first, so to address this problem, Boulton (1955) introduced an empirical constant that accounted for the delayed yield from storage. It was almost twenty years before Neuman (1972) derived a physically based solution that produced all three limbs of the time-drawdown curves without recourse to an empirical constant.

The assumption inherent in the Theis solution that formations overlying and underlying an aquifer are impermeable is seldom satisfied. The development of leaky-aquifer theory, which was designed to address this reality, was presented in two distinct sets of papers. The first—by Jacob (1946), Hantush and Jacob (Paper 11), and Hantush (1960)—provided the original differentiation between the Thiem and Theis responses and that for leaky aquifers. The second, by Neuman and Witherspoon (1969, Paper 12), evaluated the assumptions inherent in the earlier work and presented more general solutions. The original work by Hantush and Jacob, although widely applied in practice, suffers from two rather restrictive assumptions. First, it is assumed that an unpumped aquifer overlying both the leaky aquitard and the pumped aquifer possesses an unlimited capacity for delivering water through the aquitard to the pumped aquifer. Second, the storativity of the aquitard itself is ignored. In a later paper, Hantush (1960) modified his solution to consider the effects of aquitard storage. Neuman and Witherspoon (1969) present a complete solution that considers both water release from storage in the aquitard and head drawdowns in the unpumped aquifer. In Paper 12 they review the history of ideas associated with the hydraulics of leaky aquifers and summarize the applicability of the various theories.

During the 1950s and 1960s a large number of papers appeared in the literature presenting solutions to nonideal well-hydraulics problems. Among the cases analyzed were those involving multiple-well systems, stepped pumping rates, partially penetrating wells, and bounded aquifers. Hantush (1964) provides a comprehensive summary of these solutions.

The fact that this wealth of mathematical solutions led to practical tools for field application is largely due to the appearance of excellent manuals following in the tradition of Wenzel (1942). The most widely used were by Ferris et al. (Paper 13), Bentall (1963a, 1963b), Lohman (1972) of the U.S. Geological Survey, and Walton (Paper 28, 1962) of the Illinois State Water Survey. Paper 13 is a short excerpt from Ferris et al. in which the application of the method of images is examined for analysis of well hydraulics in bounded aquifers.

Hydraulic conductivity values may also be determined by means of recovery tests carried out in single wells or piezometers. The methods of Hvorslev (1951) for a point piezometer, and Cooper et al. (1967) for a confined aquifer, are still widely used.

The well-hydraulics formulas presented in Papers 9 through 13 all arise from analytical solutions to the pertinent boundary-value problems. While these formulas have proved extremely valuable in the design of individual wells, they are inadequate for predicting drawdown in areally extensive, heterogeneous aquifers of irregular shape with irregularly spaced, time-dependent pumping and recharge rates. In these situations, hydrogeologists turned first to electric analog simulation with resistance-capacitance networks and then to numerical simulation with a digital computer. With the growth of computer capacity in recent years, numerical simulation has now become the standard technique for the prediction of aquifer performance.

The early numerical simulations were based on finite-difference mathematics and the method of relaxation, both of which were known long before the advent of computers. The earliest application of numerical simulation to a subsurface flow problem that we have seen is by Shaw and Southwell (1941). They used the method of relaxation in a hand calculation mode to assess seepage through an earth dam. During the early 1950s, numerical methods were applied by soil scientists to agricultural drainage problems, and in one such paper, Luthin and Scott (1952) analyzed radial flow to a well. Most of the

Editors' Comments on Papers 9 Through 18

numerical methodology used subsequently by groundwater hydrologists was adopted from the petroleum industry where it was developed for petroleum reservoir analysis.

Finite-difference concepts were introduced into the hydrogeological literature by Stallman in 1956 (Paper 14). He did not present a numerical simulation per se, but rather showed how a finite-difference analog to the transient flow equation can be used with water-level data to estimate aquifer properties. Much later, Nelson (1968) expanded this approach in a paper that became the foundation for modern research on the inverse problem.

A well-known early aquifer simulation was carried out by Tyson and Weber (Paper 15) in 1964 for the California Department of Water Resources in the coastal plain aquifer of Los Angeles County. They utilized a hybrid approach in which an electric analog was used in the calibration phase and a digital computer in the prediction phase. The Tyson-Weber numerical scheme (a precursor of what is now known as the integrated finite-difference method) did not find favor with subsequent modelers, who turned instead to the methodology of petroleum reservoir analysts. Nevertheless, the paper was significant in demonstrating the practical value of computers in the assessment of aquifer performance. Another paper from this same period that helped popularize the computer approach was Remson et al. (1965), in which the authors reported on a steady-state computer model used to predict the effects of a proposed surface-water reservoir on the heads in an unconfined regional aquifer.

By the early 1970s, computer simulation of aquifers was widely used in water resource evaluation. This advance resulted largely from the development, documentation, and availability of two sophisticated aquifer simulation programs, the first by Pinder and Bredehoeft (Paper 16) of the U.S. Geological Survey, and the second by Prickett and Lonnquist (1971) of the Illinois State Water Survey. The U.S.G.S. model has been continually updated over the years, and it has been applied in a wide variety of hydrogeological environments.

In recent years, research into numerical simulation of groundwater systems has accelerated. Among the most important developments have been the introduction of finite-element methods, the simulation of multilayer aquifer-aquitard systems, consideration of saturated-unsaturated regimes, and the analysis of mass transport in groundwater systems which is treated in *Chemical Hydrogeology* (Back and Freeze, 1983).

Skibitzke (1961) and Walton and Prickett (Paper 17) pioneered electric analog simulation of transient flow in aquifers. Walton and Prickett's lucid explanation of the theory together with the many practical suggestions of Patten (1965) did much to popularize the application of analog simulation during the period preceding the explosion in digital computer technology. Prickett (1975) provided a comparative review of electric analog simulation and digital computer simulation with respect to both theory and practice.

Over the years there has been considerable interest in the interaction between surface-water and groundwater regimes. It has long been recognized that groundwater contributions are the source of baseflow in streams, and there is a long historical record of papers dealing with stream-aquifer systems. Among the earliest to provide a quantitative analysis of a stream-aquifer system is Todd (Paper 18), who used a Hele-Shaw model to simulate the effects of a flood wave on bank storage in an adjacent aquifer. Cooper and Rorabaugh (1963) and Rorabaugh (1964) produced analytical solutions for this same problem. More recently Pinder and Sauer (1971) used numerical simulation to provide the first fully coupled stream-aquifer analysis.

REFERENCES

Back, W., and R. A. Freeze, eds., 1983, *Chemical Hydrogeology*, Benchmark Papers in Geology, vol. 73, Hutchinson Ross Publishing, Stroudsburg, Pa.

Bentall, R., 1963a, Methods of Determining Permeability, Transmissibility and Drawdown, *U.S. Geol. Survey Water-Supply Paper 1536-I*, pp. 243-341.

Bentall, R., 1963b, Shortcuts and Special Problems in Aquifer Tests, *U.S. Geol. Survey Water-Supply Paper 1545-C*, 117p.

Boulton, N. S., 1955, Unsteady Radial Flow to a Pumped Well Allowing for Delayed Yield from Storage, *Internat. Assoc. Sci. Hydrology Pub.* **37**:472-477.

Cooper, H. H., and C. E. Jacob, 1946, A Generalized Graphical Method for Evaluating Formation Constants and Summarizing Well Field History, *Am. Geophys. Union Trans.* **27**:526-534.

Cooper, H. H., and M. I. Rorabaugh, 1963, Ground-Water Movements and Bank Storage Due to Flood Stages in Surface Streams, *U.S. Geol. Survey Water-Supply Paper 1536-J*, pp. 343-366.

Cooper, H. H., J. D. Bredehoeft, and I. S. Papadopoulos, 1967, Response of a Finite-Diameter Well to an Instantaneous Charge of Water, *Water Resources Research* **3**:263-269.

Hantush, M. S., 1960, Modification of the Theory of Leaky Aquifers, *Jour. Geophys. Research* **65**:3713-3725.

Hantush, M. S., 1964, Hydraulics of Wells, *Advances in Hydrosci.* **1**:281-432.

Hvorslev, M. J., 1951, Time Lag and Soil Permeability in Ground Water Observations, *U.S. Army Corps Engineers, Waterways Expt. Sta. Bull.* 36, 50p.

Jacob, C. E., 1946, Radial Flow in a Leaky Artesian Aquifer, *Am. Geophys. Union Trans.* **27**:198-205.

Lohman, S. W., 1972, Ground Water Hydraulics, *U.S. Geol. Survey Prof. Paper 708*, 70p.

Luthin, J. N., and V. H. Scott, 1952, Numerical Analysis of Flow Through Aquifers Towards Wells, *Agric. Engineering* **33:**279-282.

Nelson, R. W., 1968, In-Place Determination of Permeability Distribution for Heterogeneous Porous Media Through Analysis of Energy Dissipation, *Soc. Petroleum Engineers Jour.* **8:**33-42.

Neuman, S. P., 1972, Theory of Flow in Unconfined Aquifers Considering Delayed Response of the Water Table, *Water Resources Research* **8:**1031-1045.

Neuman, S. P., and P. A. Witherspoon, 1969, Theory of Flow in a Confined Two-Aquifer System, *Water Resources Research* **5:**803-816.

Patten, E. P., 1965, Design, Construction and Use of Electric Analog Model, in *Analog Model Study of Ground Water in Houston District*, L. A. Wood and R. K. Gabrysch, eds., Texas Water Commission Bulletin 6508.

Pinder, G. F., and S. P. Sauer, 1971, Numerical Simulation of Flood-Wave Modification Due to Bank Storage Effects, *Water Resources Research* **7:**63-70.

Prickett, T. A., 1975, Modeling Techniques for Groundwater Evaluation, *Advances in Hydrosci.* **10:**1-143.

Prickett, T. A., and C. G. Lonnquist, 1971, Selected Digital Computer Techniques for Groundwater Resource Evaluation, *Illinois Water Survey Bull. 55,* 62p.

Remson, I., C. A. Appel, and R. A. Webster, 1965, Ground-Water Models Solved by Digital Computer, *Am. Soc. Civil Engineers Proc., Jour. Hydraulics Div.* **91:**133-147.

Rorabaugh, M. I., 1964, Estimating Changes in Bank Storage and Ground-Water Contribution to Streamflow, *Internat. Assoc. Sci. Hydrology Pub.* **63:**432-441.

Shaw, F. S., and R. V. Southwell, 1941, Relaxation Methods Applied to Engineering Problems, VII. Problems Relating to the Percolation of Fluids Through Porous Materials, *Royal Soc. London Proc.* **178A:**1-17.

Skibitzke, H. E., 1961, Electronic Computers as an Aid to the Analysis of Hydrologic Problems, *Internat. Assoc. Sci. Hydrology Pub.* **52:**347-358.

Walton, W. C., 1962, Selected Analytical Methods for Well and Aquifer Evaluation, *Illinois Water Survey Bull. 49,* 81p.

Wenzel, L. K., 1936, The Thiem Method for Determining Permeability of Water-Bearing Materials and Its Application to the Determination of Specific Yield, Results of Investigations in the Platte River Valley, Nebraska, *U.S. Geol. Survey Water-Supply Paper 679-A,* 57p.

Wenzel, L. K., 1942, Methods of Determining Permeability of Water-Bearing Materials, with Special Reference to Discharging-Well Methods, *U.S. Geol. Survey Water-Supply Paper 887,* 192p.

THE RELATION BETWEEN THE LOWERING OF THE PIEZOMETRIC SURFACE AND THE RATE
AND DURATION OF DISCHARGE OF A WELL USING GROUND-WATER STORAGE

Charles V. Theis

When a well is pumped or otherwise discharged, water-levels in its neighborhood are lowered. Unless this lowering occurs instantaneously it represents a loss of storage, either by the unwatering of a portion of the previously saturated sediments if the aquifer is nonartesian or by release of stored water by the compaction of the aquifer due to the lowered pressure if the aquifer is artesian. The mathematical theory of ground-water hydraulics has been based, apparently entirely, on a postulate that equilibrium has been attained and therefore that water-levels are no longer falling. In a great number of hydrologic problems, involving a well or pumping district near or in which water-levels are falling, the current theory is therefore not strictly applicable. This paper investigates in part the nature and consequences of a mathematical theory that considers the motion of ground-water before equilibrium is reached and, as a consequence, involves time as a variable.

To the extent that Darcy's law governs the motion of ground-water under natural conditions and under the artificial conditions set up by pumping, an analogy exists between the hydrologic conditions in an aquifer and the thermal conditions in a similar thermal system. Darcy's law is analogous to the law of the flow of heat by conduction, hydraulic pressure being analogous to temperature, pressure-gradient to thermal gradient, permeability to thermal conductivity, and specific yield to specific heat. Therefore, the mathematical theory of heat-conduction developed by Fourier and subsequent writers is largely applicable to hydraulic theory. This analogy has been recognized, at least since the work of Slichter, but apparently no attempt has been made to introduce the function of time into the mathematics of ground-water hydrology. Among the many problems in heat-conduction analogous to those in ground-water hydraulics are those concerning sources and sinks, sources being analogous to recharging wells and sinks to ordinary discharging wells.

C. I. Lubin, of the University of Cincinnati, has with great kindness prepared for me the following derivation of the equation expressing changes in temperature due to the type of source or sink that is analogous to a recharging or discharging well under certain ideal conditions, to be discussed below.

The equation given by H. S. Carslaw (Introduction to the mathematical theory of the conduction of heat in solids, 2nd ed., p. 152, 1921) for the temperature at any point in an infinite plane with initial temperature zero at any time due to an "instantaneous line-source coinciding with the axis of z of strength Q" (involving two-dimensional flow of heat) is

$$v = (Q/4\pi\kappa t)\ e^{-(x^2+y^2)/4\kappa t} \tag{1}$$

where v = change in temperature at the point x, y at the time t; Q = the strength of the source or sink--in other words, the amount of heat added or taken out instantaneously divided by the specific heat per unit-volume; κ = Kelvin's coefficient of diffusivity, which is equal to the coefficient of conductivity divided by the specific heat per unit-volume; and t = time.

The effect of a continuous source or sink of constant strength is derived from equation (1) as follows: Let $Q = \varphi(t')dt'$; then $v_{(x,y,t)} = \int_0^t [\varphi(t')/4\pi\kappa(t-t')]\ e^{-(x^2+y^2)/4\kappa(t-t')}\ dt'$. Let $\varphi(t') = \lambda$, a constant; then $v_{(t)} = (\lambda/4\pi\kappa)\int_0^t [e^{-(x^2+y^2)/4\kappa(t-t')}/(t-t')]\ dt'$. Let $u = (x^2+y^2)/4\kappa(t-t')$; then

$$v_{(t)} = (\lambda/4\pi\kappa) \int_{(x^2+y^2)/4\kappa t}^{\infty} [e^{-u}/(t-t')]\ [(x^2+y^2)/4\kappa]\ [du/u^2]$$

$$= (\lambda/4\pi\kappa) \int_{(x^2+y^2)/4\kappa t}^{\infty} (e^{-u}/u)du \tag{2}$$

The definite integral, $\int_{(x^2+y^2)/4\kappa t}^{\infty} (e^{-u}/u)du$, is a form of the exponential integral, tables of which are available (Smithsonian Physical Tables, 8th rev. ed., table 32, 1933; the values to be used are those given for Ei $(-x)$, with the sign changed). The value of the integral is given by the series

$$\int_x^\infty (e^{-u}/u)du = -0.577216 - \log_e x + x - x^2/2\cdot 2! + x^3/3\cdot 3! - x^4/4\cdot 4! + \ldots \tag{3}$$

Equation (2) can be immediately adapted to ground-water hydraulics to express the draw-down at any point at any time due to pumping a well. The coefficient of diffusivity, , is analogous to the coefficient of transmissibility of the aquifer divided by the specific yield. (The term "coefficient of transmissibility" is here used to denote the product of Meinzer's coefficient of permeability and the thickness of the saturated portion of the aquifer; it quantitatively describes the ability of the aquifer to transmit water. Meinzer's coefficient of permeability denotes a characteristic of the material; the coefficient of transmissibility denotes the analogous characteristic of the aquifer as a whole.) The continuous strength of the sink is analogous to the pumping rate divided by the specific yield. Making these substitutions, we have

$$v = (F/4\pi\tau) \int_{r^2 s/4\tau t}^{\infty} (e^{-u}/u)du \tag{4}$$

in which the symbols have the meanings given with equation (5). In equation (4) the same units must of course be used throughout. Equation (4) may be adapted to units commonly used

$$v = (114.6F/\tau) \int_{1.87 r^2 s/\tau t}^{\infty} (e^{-u}/u)du \tag{5}$$

where v = the draw-down, in feet, at any point in the vicinity of a well pumped at a uniform rate; F = the discharge of the well, in gallons a minute; τ = the coefficient of transmissibility of aquifer, in gallons a day, through each 1-foot strip extending the height of the aquifer, under a unit-gradient--this is the average coefficient of permeability (Meinzer) multiplied by the thickness of the aquifer; r = the distance from pumped well to point of observation, in feet; s = the specific yield, as a decimal fraction; and t = the time the well has been pumped, in days.

Equation (5) gives the draw-down at any point around a well being pumped uniformly (and continuously) from a homogeneous aquifer of constant thickness and infinite areal extent at any time. The introduction of the function, time, is the unique and valuable feature of the equation. Equation (5) reduces to Thiem's or Slichter's equation for artesian conditions when the time of pumping is large.

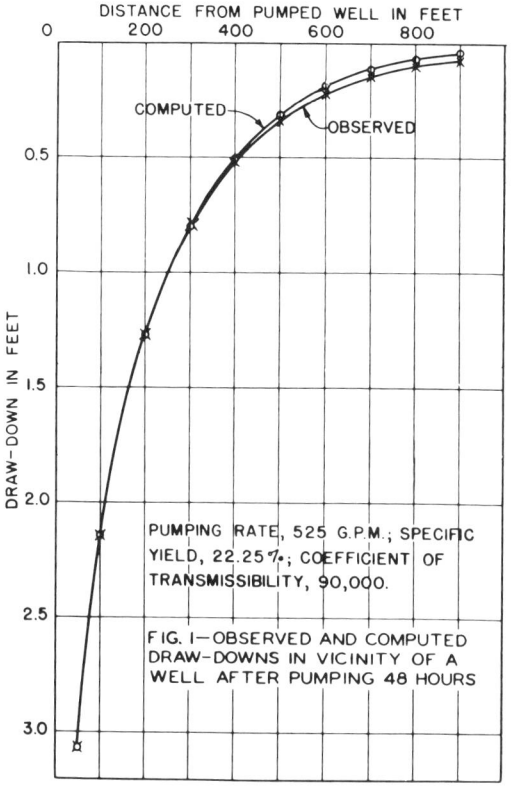

FIG. 1—OBSERVED AND COMPUTED DRAW-DOWNS IN VICINITY OF A WELL AFTER PUMPING 48 HOURS

PUMPING RATE, 525 G.P.M.; SPECIFIC YIELD, 22.25%; COEFFICIENT OF TRANSMISSIBILITY, 90,000.

Empirical tests of the equation are best made with the data obtained by L. K. Wenzel (Recent investigations of Thiem's method for determining permeability of water-bearing sediments, Trans. Amer. Geophys. Union, 13th annual meeting, pp. 313-317, 1932; also Specific yield determined from a Thiem's pumping test, Trans. Amer. Geophys. Union, 14th annual meeting, pp. 475-477, 1933) from pumping tests in the Platte Valley in Nebraska. Figure 1 presents the comparison of the computed and observed draw-downs after two days of pumping. The observed values are those of the generalized depression of the water-table as previously determined by Mr. Wenzel. The computed values are obtained by equation (5), using values of permeability and specific yield that are within one per cent of those determined by Mr. Wenzel by other methods. The agreement represented may be regarded as showing either that the draw-downs have been computed from known values of transmissibility and specific yield or that these factors have been computed from the known draw-downs.

Theoretically, the equation applies rigidly only to water-bodies (1) which are contained in entirely homogeneous sediments, (2) which have infinite areal extent, (3) in which the well penetrates the entire thickness of the water-body, (4) in which the coefficient of transmissibility is constant at all times and in all places, (5) in which the pumped well has an infinitesimal diameter, and (6) - applicable only to unconfined water-bodies - in which the water in the volume of sediments through which the water-table has fallen is discharged instantaneously with the fall of the water-table.

These theoretical restrictions have varying degrees of importance in practice. The effect of heterogeneity in the aquifer can hardly be foretold. The effect of boundaries can be considered by more elaborate analyses, once they are located. The effect of the well failing to penetrate the entire aquifer is apparently negligible in many cases. The pumped well used in the set-up that yielded the data for Figure 1 penetrated only 30 feet into a 90-foot aquifer. The coefficient of transmissibility must decrease during the process of pumping under water-table conditions, because of the diminution in the cross-section of the area of flow due to the fall of the water-table; however, it appears from Figure 1 that if the water-table falls through a distance equal only to a small percentage of the total thickness of the aquifer the errors are not large enough to be observed. In artesian aquifers the coefficient of transmissibility probably decreases because of the compaction of the aquifer, but data on this point are lacking. The error due to the finite diameter of the well is apparently always insignificant.

In heat-conduction a specific amount of heat is lost concomitantly and instantaneously with fall in temperature. It appears probable, analogously, that in elastic artesian aquifers a specific amount of water is discharged instantaneously from storage as the pressure falls. In nonartesian aquifers, however, the water from the sediments through which the water-table has fallen drains comparatively slowly. This time-lag in the discharge of the water made available from storage is neglected in the mathematical treatment here given. Hence an error is always present in the equation when it is applied to water-table conditions. However, inasmuch as the rate of fall of the water-table decreases progressively after a short initial period, it seems probable that as pumping continues the rate of drainage of the sediments tends to catch up with the rate of fall of the water-table, and hence that the error in the equation becomes progressively smaller.

For instance, although the draw-downs computed for a 24-hour period of pumping in Mr. Wenzel's test showed a definite lack of agreement with the observations, similar computations for a

48-hour period gave the excellent agreement shown in Figure 1. Unfortunately data for periods of pumping longer than 48 hours have not been available.

The equation implies that any two observations of draw-down, whether at different places or at the same place at different times, are sufficient to allow the computation of specific yield and transmissibility. However, more observations are always necessary in order to guard against the possibility that the computations will be vitiated by the heterogeneity of the aquifer. Moreover, it appears that the time-lag in the drainage of the unwatered sediments makes it impossible at present to compute transmissibility and specific yield from observations on water-levels in only one observation-well during short periods of pumping. Good data from artesian wells have not been available, but such data as we have hold out the hope that transmissibility and specific yield may be determined from data from only one observation-well.

A useful corollary to equation (5) may be derived from an analysis of the recovery of a pumped well. If a well is pumped for a known period and then left to recover, the residual draw-down at any instant will be the same as if pumping of the well had been continued but a recharge well with the same flow had been introduced at the same point at the instant pumping stopped. The residual draw-down at any instant will then be

$$v' = (114.6F/\tau) \left[\int_{1.87r^2s/\tau t}^{\infty} (e^{-u}/u)du - \int_{1.87r^2s/\tau t'}^{\infty} (e^{-u}/u)du \right] \quad (6)$$

where t is the time since pumping started and t' is the time since pumping stopped.

In and very close to the well the quantity $(1.87r^2s/\tau t')$ will be very small as soon as t' ceases to be small, because r is very small. In many problems ordinarily met in ground-water hydraulics, all but the first two terms of the series of equation (3) may be neglected, so that, if $Z = (1.87r^2s/\tau t)$ and $Z' = (1.87r^2s/\tau t')$ equation (6) may be approximately rewritten

$$v' = (114.6F/\tau) [-0.577 - \log_e Z + 0.577 + \log_e Z(t/t')] = (114.6F/\tau) \log_e(t/t')$$

Transposing and converting to common logarithms, we have

$$\tau = (264F/v') \log_{10}(t/t') \quad (7)$$

This equation permits the computation of the coefficient of transmissibility of an aquifer from an observation of the rate of recovery of a pumped well.

Figure 2 shows a plot of observed recovery-curves. The ordinates are log (t/t'); the abscissas are the distances the water-table lies below its equilibrium-position. According to equation (7) the points should fall on a straight line passing through the origin. Curve A is a plot of the recovery of a well within 3 feet of the well pumped for Mr. Wenzel's test, previously mentioned. Most of the points lie on a straight line, but the line passes below the origin. This discrepancy is probably due to the fact that the water-table rises faster than the surrounding pores are filled. The coefficient of permeability computed from the equation is about 1200, against a probably correct figure of 1000. Curve B is plotted from data obtained from an artesian well near Salt Lake City. The points all fall according to theory.

Curve C shows the recovery of a well penetrating only the upper part of a nonartesian aquifer of comparatively low transmissibility. It departs markedly from a straight line. This curve probably follows equation (6), but it does not follow equation (7), for in this case $(1.87r^2s/\tau t')$ is not small. Equation (6), involving r and s, neither of which may be known in practice, is not of practical value for the present purpose. Further empirical tests may show that it is feasible to project the curve to the origin, in the neighborhood of which $(1.87r^2s/\tau t')$ becomes small, owing to the increase in t and t', and apply equation (7) to the extrapolated values so obtained in order to determine at least an approximate value of the transmissibility.

The paramount value of equation (5) apparently lies in the fact that it gives part of the theoretical background for predicting the future effects of a given pumping regimen upon the water-levels in a district that is primarily dependent on ground-water storage. Such districts may include many of those tapping extensive nonartesian bodies of ground-water. Figure 3 shows the vertical rate of fall of the water-level in an infinite aquifer, the water being all taken from storage. The curves are plotted for certain definite values of pumping rate, transmissibility, and specific yield, but by changing the scales either curve could be made applicable to any values set up.

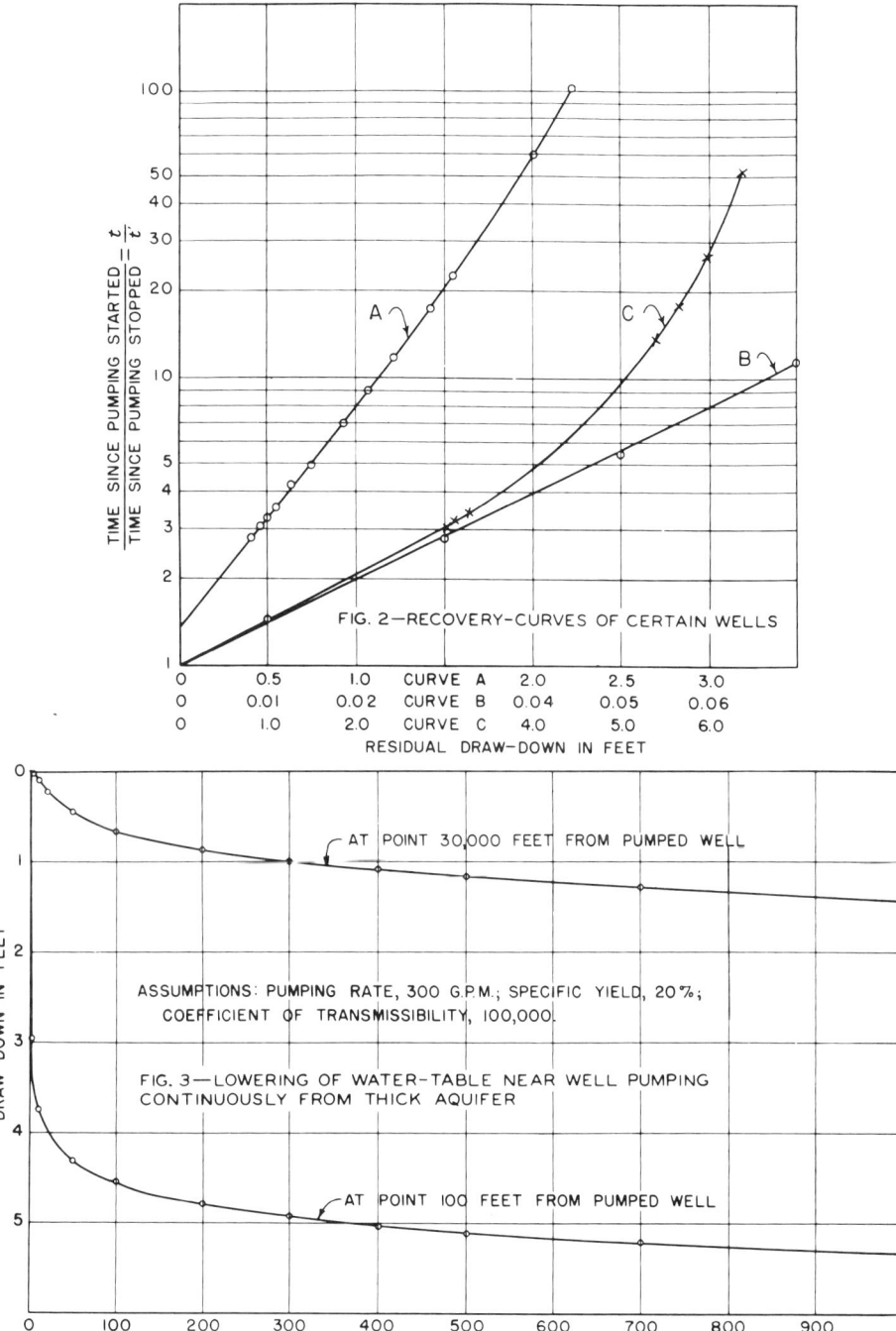

FIG. 2—RECOVERY-CURVES OF CERTAIN WELLS

FIG. 3—LOWERING OF WATER-TABLE NEAR WELL PUMPING CONTINUOUSLY FROM THICK AQUIFER

These theoretical curves agree qualitatively with the facts generally observed when a well is pumped. The water-level close to the well at first falls very rapidly, but the rate of fall soon slackens. In the particular case considered in Figure 3 the water-level at a point 100 feet from the pumped well would fall during the first year of pumping more than half the distance it would fall in 1000 years. A delayed effect of the pumping is shown at distant points.

The water-level at a point about 6 miles from the pumped well of Figure 3 would fall only minutely for about five years but would then begin to fall perceptibly, although at a much less rate than the water-level close to the well. Incidentally the rate of fall after considerable pumping is so small that it might easily lead to a false assumption of equilibrium. The danger in a pumping district using ground-water storage lies in the delayed interference of the wells. For instance, although in 50 years one well would cause a draw-down of only 6 inches in another well 6 miles away, yet the 100 wells that might lie within 6 miles of a given well would cause in it a total draw-down of more than 50 feet.

In the preparation of this paper I have had the indispensable help not only of Dr. Lubin, who furnished the mathematical keystone of the paper, but also of Dr. C. E. Van Orstrand, of the United States Geological Survey, and of my colleagues of the Ground Water Division of the Survey, who cordially furnished data and criticism.

"The Drawdown of the Water-Table under Non-Steady Conditions near a Pumped Well in an Unconfined Formation"

by

Norman Savage Boulton, M.Sc., M.I.C.E.

(Ordered by the Council to be published with written discussion)†

SYNOPSIS

A new equation is obtained for the drawdown of the water-table near a pumped well in an aquifer of any thickness and permeability, under non-equilibrium conditions. Tables of the mathematical functions involved make the equation simple to use in practice. Approximations in deriving the equation are unavoidable, but it is shown that allowance for them can be made in certain cases and, in other cases, their effect on the drawdown is unimportant provided that the variables involved are restricted to a specified range of values. For a thin and very permeable aquifer, the well-known formula involving the exponential integral, originally deduced for artesian conditions, is shown to be accurate provided that a correction is applied to it when necessary.

For an aquifer in which the conditions approximate to those assumed in the Paper (p. 568), it should be possible to use the equation to obtain reliable values of the coefficients of permeability and storage from the results of field pumping tests. The application of the new equation is illustrated by numerical examples, but its application to the results of actual field pumping tests is left for a subsequent Paper.

Simple modifications of the equation are indicated, for a case of anisotropic permeability and for the recovery of the free surface after pumping has ceased. The case when the aquifer is of unlimited depth is considered in Appendix I.

INTRODUCTION

THE main purpose of the Paper is to introduce a new equation for the drawdown of the water-table (free surface) near a gravity well before the flow has attained a state of equilibrium. The exponential integral [1] (Theis's formula) has been used for some time to represent the piezometric surface under artesian flow conditions, and its usefulness in appropriate cases for determining formation constants from the results of field pumping tests has been demonstrated by Wenzel [2] and others in the United States of America and, recently, by Ineson [3, 4] in the United Kingdom.

When equilibrium exists, the drawdown of the water-table in an unconfined formation can be calculated with sufficient accuracy for practical

---

† Correspondence on this Paper should be received at the Institution by the 15th December, 1954, and will be published in Part III of the Proceedings. Contributions should be limited to about 1,200 words.—SEC. I.C.E.

[1] The references are given on p. 578.

purposes, as the Author has shown in a recent Paper.[5] For non-equilibrium conditions, however, no satisfactory theory has yet been published, and Ineson [6] has recently drawn attention to the need for such a theory.

As applied to water-table conditions, the exponential integral is theoretically unsound, since it takes no account of the vertical velocity-component of the pore-water approaching the well. This criticism applies particularly to the early stages of a pumping test in a deep well, when the motion of the water in the upper part of the aquifer is more vertical than horizontal. However, a time factor $\tau = \dfrac{kt}{Sh_e}$,* and not merely the duration of pumping t, determines whether the exponential integral gives a good approximation to the drawdown of the free surface. This time factor involves the coefficients of permeability and storage, and the depth of the saturated part of the aquifer, as well as the duration of pumping.

It is shown on p. 571 that the exponential integral gives a good approximation to the drawdown only when $\tau > 5$. From the definition of τ it is evident that, for a thin and very permeable aquifer, this condition may be satisfied after quite a short duration of pumping. In fact, for a well penetrating coarse sand and gravel, 27 feet thick, described by Wenzel,[7] the time factor at the end of 48 hours pumping is found to be 122. On the other hand, for a typical pumped well in the Bunter sandstone of the Midlands, about 900 feet thick, the time factor for the same duration of pumping is of the order of 0·04. Whereas the exponential integral can justifiably be applied to a 48-hour pumping test in the first case, it would be very inaccurate in the second case. It follows from these two examples that, under practical conditions, the time factor may vary between a small fraction and many hundreds.

The question as to which water-bearing formations the theory described in the Paper is applicable can only be answered by comparing predictions from the theory with the results of actual pumping tests. The agreement obtained would depend upon the extent to which the assumptions (p. 567) underlying the theoretical analysis are satisfied under field conditions.

By means of the equation, average values of the formation constants can be determined from a pumping test of any duration from a few hours upwards, in a formation of any thickness and permeability.

Theoretically, the solution of the flow problem is difficult, particularly because of the changing position of the free surface. A large sand model or an electro-analogy model might provide the most satisfactory solution, especially when the water level in the well is lowered to a considerable depth; although, owing to the very small ratio of diameter to length of well in many pumped wells, it might be impossible to reproduce accurately the actual boundary conditions at the well surface.

The method adopted is to obtain a solution by potential theory, which

---

* The notation is defined on pp. 566 and 567.

only strictly satisfies the given boundary conditions at the free surface and at entrance to the well when the water level in the well is lowered by a small amount as compared with the thickness of the aquifer. Corrections are then applied to this solution, where possible, in cases when it is known to be seriously in error.

For the case $\tau > 5$, when the exponential integral is applicable, the Author has obtained a simple method, which is described later in the Paper, for correcting the drawdown, based on his previous free-surface curves for steady flow. This correction is necessary because the exponential integral does not satisfy the correct boundary conditions at the free surface and at the well surface. The correction enables the integral to be used with readings from observation wells, which are quite close to the pumped well, and for pumping levels near the base of the aquifer.

NOTATION

h denotes height of free surface above impermeable stratum during pumping.

h_e ,, height of undisturbed free surface above impermeable stratum.

h_s ,, value of h when $r = r_w$.

h_w ,, height of water level in well above impermeable stratum during pumping.

k ,, coefficient of permeability of aquifer.

k_r ,, coefficient of permeability in horizontal direction.

k_z ,, coefficient of permeability in vertical direction.

l ,, length of well below water-table for an aquifer of unlimited depth.

m ,, symbol defined by equation (18).

p ,, pressure intensity in pore-water (atmospheric pressure being taken as zero).

Q ,, constant rate of discharge of well during pumping.

r ,, horizontal distance from well-axis to any point.

r_w ,, radius of well.

s ,, drawdown of free surface at distance r.

s_c ,, drawdown given by equation (15).

S ,, coefficient of storage.

t ,, time since pumping commenced.

T ,, coefficient of transmissibility.

$u = \dfrac{\rho^2}{4\tau}$

v_r denotes " discharge velocity " component in horizontal direction away from well.

v_z ,, discharge velocity component in upward vertical direction.

V ,, integral in equation (10).

X_0 ,, correction term defined by equation (13).

X_1 denotes correction term defined by equation (12).
y ,, depth of any point below initial water-table.
z ,, height of any point above impermeable stratum.
γ_w ,, density of water.

$$\mu = \sqrt{\frac{k_z}{k_r}}$$

$$\rho = \frac{r}{h_e} \text{ and } \rho_w = \frac{r_w}{h_e}$$

σ denotes a correction factor.

$\tau = \dfrac{kt}{Sh_e}$, a dimensionless time factor.

ϕ denotes pressure head plus potential head at any point of saturated aquifer : $\phi = \dfrac{p}{g\gamma_w} + z$

Statement of Problem and Assumptions

The conditions assumed in the main problem investigated are :—

(1) The aquifer is homogeneous and, except where otherwise stated, isotropic ; it extends to the ground surface, is of infinite lateral extent, and is underlain by a horizontal impermeable bed.

(2) The well completely penetrates the aquifer and is unlined.

(3) The coefficient of storage (S) is constant.

(4) The flow in the aquifer obeys Darcy's law (k = constant).

(5) The water-table is initially horizontal. The replenishment of the water-table near the well by the rainfall is not considered, since the flow system due to the rainfall may be regarded as an independent system in equilibrium ; its effect on the drawdown of the free surface due to pumping is likely to be small.

(6) The well is pumped at a constant rate from the instant $t = 0$.

Basic Equations

Darcy's law is expressed by the equations :

$$\left. \begin{array}{l} v_r = -k \dfrac{\partial \phi}{\partial r} \\ v_z = -k \dfrac{\partial \phi}{\partial z} \end{array} \right\} \qquad \ldots \ldots \ldots (1)$$

where ϕ denotes the sum of the pressure and potential heads at the point in the flow considered.

The well-known equation which results from Darcy's law and the equation of continuity of flow of incompressible fluids is:

$$\nabla^2(\phi) = \frac{\partial^2 \phi}{\partial r^2} + \frac{1}{r} \cdot \frac{\partial \phi}{\partial r} + \frac{\partial^2 \phi}{\partial z^2} = 0 \quad \ldots \ldots (2)$$

which must be satisfied at all points of the saturated aquifer.

The differential equation to be satisfied at the variable free-surface boundary may be obtained in the following way. Since the pressure on this surface is atmospheric (assumed zero), the equation to the surface is:

$$\phi(r, z, t) - z = 0.$$

Also, since a particle of fluid, which is once in the free surface, never leaves it,

$$\frac{D}{Dt}(\phi - z) = 0$$

where D denotes differentiation following the motion of the particle. Therefore, since,

$$\frac{D}{Dt} = \frac{\partial}{\partial t} + \frac{v_r}{S}\frac{\partial}{\partial r} + \frac{v_z}{S}\frac{\partial}{\partial z}$$

it follows that
$$\frac{\partial \phi}{\partial t} + \frac{v_r}{S} \cdot \frac{\partial \phi}{\partial r} + \frac{v_z}{S}\left(\frac{\partial \phi}{\partial z} - 1\right) = 0$$

where $\frac{v_r}{S}$ and $\frac{v_z}{S}$ are "seepage velocity" components.

Substituting for v_r and v_z from equations (1), the free-surface boundary equation obtained is:

$$\frac{\partial \phi}{\partial t} = \frac{k}{S}\left\{\left(\frac{\partial \phi}{\partial r}\right)^2 + \left(\frac{\partial \phi}{\partial z}\right)^2 - \frac{\partial \phi}{\partial z}\right\} \quad \ldots \ldots (3)$$

Since equation (3) is non-linear, a solution satisfying it cannot easily be found. However, if the ϕ-gradients are small, their squares may be neglected and equation (3) then reduces to the linear equation:

$$\frac{\partial \phi}{\partial t} + \frac{k}{S}\frac{\partial \phi}{\partial z} = 0 \quad \ldots \ldots \ldots (4)$$

which is to be satisfied when $z = h$

Since the undisturbed water-table is assumed to be a horizontal plane, $z = h_e$, it follows that:

$$\phi = h_e \text{ when } t = 0 \quad \ldots \ldots (5)$$

At the surface of the well:

$$\left.\begin{array}{l}\phi = h_w; \quad 0 \leqslant z \leqslant h_w \\ \phi = z; \quad h_w \leqslant z \leqslant h_s\end{array}\right\} \quad \ldots \ldots (6)$$

At the impermeable layer

$$\frac{\partial \phi}{\partial z} = 0; \quad z = 0 \quad \ldots \ldots \quad (7)$$

Also, as $r \to \infty$, $\phi \to h_e$; $0 \leqslant z \leqslant h_e$.

Equation (4) should, strictly speaking, be satisfied at all points of the free surface, but this is impossible except by using a laborious step-by-step method of calculation. Advantage may, however, be taken of the fact that the drawdown of the free surface is usually a small fraction of h_e.* Thus, subject to an error of the order already neglected, equation (4) may be written:

$$\frac{\partial \phi}{\partial t} + \frac{k}{S}\frac{\partial \phi}{\partial z} = 0; \quad z = h_e \quad \ldots \ldots \quad (4a)$$

A solution can be obtained, satisfying equations (2), (5), (6), (7), and (4a), assuming that the water level in the well is constant. The solution, however, is difficult to evaluate numerically. Also, a constant rate of pumping instead of a constant water level is the usual requirement in practice. To satisfy the former requirement, a simplification of the boundary condition at the well surface is necessary. In place of equations (6), it is assumed that the discharge, per unit length of well of vanishingly small radius, is constant. That is:

$$Q = 2\pi k h_e r \frac{\partial \phi}{\partial r}; \quad r \to 0; \quad 0 \leqslant z \leqslant h_e. \quad \ldots \quad (8)$$

The correction to be applied, when the error due to this approximation is important, is considered later in the Paper.

The Drawdown Equation

The required solution is:

$$h_e - \phi = \frac{Q}{2\pi k h_e} \int_0^\infty \frac{J_0(\beta r)}{\beta}\left\{1 - \frac{\cosh \beta z}{\cosh \beta h_e} \exp\left(-\frac{k}{S} t\beta \tanh \beta h_e\right)\right\}d\beta \quad (9)$$

where J_0 denotes the Bessel function of the first kind of zero order.

It can be verified that equation (9) satisfies equations (2), (5), (7), and (4a), using the properties of the Bessel function. It is also easy to show that the exponential term, involving the time, makes no contribution to the discharge into a well of vanishingly small radius, and hence that equation (8) is satisfied.

Assuming, as previously stated, that the drawdown of the water-table is small, a first approximation for the drawdown is obtained by putting $z = h_e$ in equation (9). Since the effect of this approximation is to make the calculated drawdown too small, whereas the effect of using equation (4a)

---

* This fact is illustrated for the case when $h_w < 0\cdot 5 h_e$ in reference 5, p. 538 (*Fig. 2*).

instead of equation (3) is to make the drawdown too large, the errors due to these two approximations tend to cancel each other.

For numerical computation, it is convenient to introduce the non-dimensional quantities $\rho = \dfrac{r}{h_e}$ and $\tau = \dfrac{kt}{Sh_e}$. Then, making these substitutions and also changing the variable of integration by writing $\lambda = \beta h_e$, equation (9) gives for the drawdown:

$$s = \frac{Q}{2\pi k h_e} \int_0^\infty \frac{J_0(\lambda \rho)}{\lambda} \left\{ 1 - \exp(-\tau \lambda \tanh \lambda) \right\} d\lambda \quad . \quad . \quad (10)$$

Denoting the definite integral by $V(\rho, \tau)$ and introducing the "transmissibility coefficient," $T = kh_e$, equation (10) becomes:

$$s = \frac{Q}{2\pi T} V(\rho, \tau) \quad . \quad . \quad . \quad . \quad . \quad . \quad (11)$$

which is the proposed general equation for the drawdown of the water-table.

When the time factor τ is sufficiently large, $\tanh \lambda$ may be replaced by λ in equation (10) without appreciably changing the value of the integral. By using Weber's first integral,[8] it can then be shown that:

$$V(\rho, \tau) = -\tfrac{1}{2} \mathrm{Ei}\left(-\frac{\rho^2}{4\tau}\right) - X_1 \quad . \quad . \quad . \quad (12)$$

where the correction term X_1 is small when τ is large. In equation (12), Ei denotes the exponential integral[*] used by Theis in his non-equilibrium formula for pumping under artesian conditions. Thus, the exponential integral is shown to apply to water-table conditions in an unconfined aquifer in the limiting case when the duration of pumping is sufficiently great.

The function V in equation (11), unlike the exponential integral, has not previously been available in tabular form. A few values which have been computed for this Paper are given in Table 3. Sufficient tabulation for all practical purposes would not be difficult, making equation (11) as simple to use as any other drawdown formula.

For small values of the time factor ($\tau < 0.05$), the function V is closely given by the simple equation (22), derived in Appendix I, for the case when the permeable medium extends downwards indefinitely and there is no underlying impermeable stratum. Introducing the correction X_0, to be added to the right-hand side of equation (22) to obtain V:

$$V(\rho, \tau) = \sinh^{-1} \frac{1}{\rho} + \sinh^{-1} \frac{\tau}{\rho} - \sinh^{-1} \frac{1+\tau}{\rho} + X_0 \quad . \quad . \quad (13)$$

where X_0 is small when τ is small.

---

[*] C. E. Jacob and others have used the notation, $W(u)$, for $-\mathrm{Ei}(-u)$.

Numerical Evaluation of V

To determine numerical values of V, it is convenient to find by quadrature, values of the correction X_0 for the smaller values of τ, and values of X_1 for the larger values of τ. Values of V can then be found from equations (13) and (12), using the published tables of $\sinh^{-1} x$ and $\text{Ei}(-x)$.

Tables 1 and 2 give some values of X_0 and X_1. Intermediate values may be found with sufficient accuracy by linear interpolation. The corresponding values of V are given in Table 3, which, however, is not so suitable for interpolation.

TABLE 1.—VALUES OF $X_0(\rho, \tau)$

τ	$\rho = 0.0$	$\rho = 0.2$	$\rho = 0.4$	$\rho = 0.6$	$\rho = 0.8$	$\rho = 1.0$	$\rho = 1.5$
0.05	0.0150	0.0143	0.0125	0.0101	0.0077	0.0056	0.0021
0.20	0.0564	0.0541	0.0480	0.0398	0.0312	0.0234	0.0099

TABLE 2.—VALUES OF $X_1(\rho, \tau)$

τ	$\rho = 0.0$	$\rho = 0.2$	$\rho = 0.4$	$\rho = 0.6$	$\rho = 0.8$	$\rho = 1.0$	$\rho = 1.5$
1.00	0.1810	0.1747	0.1575	0.1331	0.1057	0.0787	0.0250
5.00	0.0344	0.0343	0.0338	0.0330	0.0320	0.0306	0.0264

TABLE 3.—VALUES OF $V(\rho, \tau)$

τ	$\rho = 0.2$	$\rho = 0.4$	$\rho = 0.6$	$\rho = 0.8$	$\rho = 1.0$	$\rho = 1.5$
0.05	0.214	0.092	0.051	0.032	0.021	0.008
0.20	0.756	0.358	0.207	0.132	0.088	0.035
1.00	1.844	1.183	0.826	0.599	0.443	0.220
5.00	2.785	2.098	1.696	1.416	1.203	0.832

From the values of X_0 for $\tau = 0.05$, the error in drawdown, due to neglecting X_0, does not exceed 0.2 foot when the constant $\dfrac{Q}{2\pi T} = 10$.

Comparing the values of X_1 and V for $\tau = 5$, it is found that the error involved in using the exponential integral instead of V does not exceed about 3 per cent. For $\tau = 1$, the corresponding error is about 18 per cent. Hence, when $\tau > 5$, the function V is given with sufficient accuracy for practical purposes by the exponential integral.

Accuracy and Limitations of the Drawdown Equation. Method of Correction

Three cases arise when considering the error in drawdown due to the approximations involved in deriving equation (11).

Case I ($\tau < 0.05$).—For this small range of τ, consideration of a numerical example (not given here) indicates that the flow pattern only varies slightly from its initial form. Consequently, if s' denotes the drawdown at any radius r, using equations (6) at the well surface, and s the drawdown using the assumed equation (8), the ratio $\dfrac{s'}{s}$ may be replaced by the ratio of the initial rates of drawdown :

$$\sigma = \frac{[\partial \phi/\partial t]'}{\partial \phi/\partial t} \ ; \quad t = 0 \ ; \quad z = h_e$$

the dash indicating the true value. From equation (4a), this ratio is also equal to :

$$\sigma = \frac{[\partial \phi/\partial z]'}{\partial \phi/\partial z} \ ; \quad t = 0 \ ; \quad z = h_e$$

The ratio σ has been calculated from the Fourier–Bessel series * which are obtainable for $\dfrac{\partial \phi}{\partial z}$, for both boundary conditions under consideration, and values of $100(\sigma - 1)$ are shown plotted on a base of ρ, to a logarithmic scale, in *Fig. 1*.

Curves are shown for two values of the constant $\dfrac{\pi r_w}{2h_e}$ and four values of $\dfrac{Q}{kh_e{}^2}$. The corresponding values of $\dfrac{h_w}{h_e}$ are also given, h_w here denoting the initial depth of water in the well immediately after pumping is started. The vertical ordinate gives the percentage correction to be added to the drawdown, the correction being obtained by interpolating between the curves, if necessary. The variation in the correction is fairly small as $\dfrac{r_w}{h_e}$ varies over the range considered.

Case II ($0.05 \leqslant \tau \leqslant 5$).—Considering the lower end of this range of τ, it is seen from curves A, B, and E (*Fig. 1*) that if :

$$\frac{Q}{kh_e{}^2} \leqslant 0.18 \quad \text{when} \quad \frac{r_w}{h_e} = 0.0013 \ ;$$

---

* The series are given in Appendix II.

$$\frac{Q}{kh_e^2} \leqslant 0{\cdot}25 \text{ when } \frac{r_w}{h_e} = 0{\cdot}0064\,;$$

and $$\rho \geqslant 0{\cdot}2,$$

then the drawdown correction does not exceed 6 per cent. Also, considering the upper end of the range of τ, it is seen from *Fig. 2* (Case III) that the drawdown correction does not exceed 6 per cent if $\rho \geqslant 0{\cdot}2$. Now it is reasonable to suppose that, for intermediate values of τ, the drawdown correction will lie between its values at the ends of the range

Fig. 1

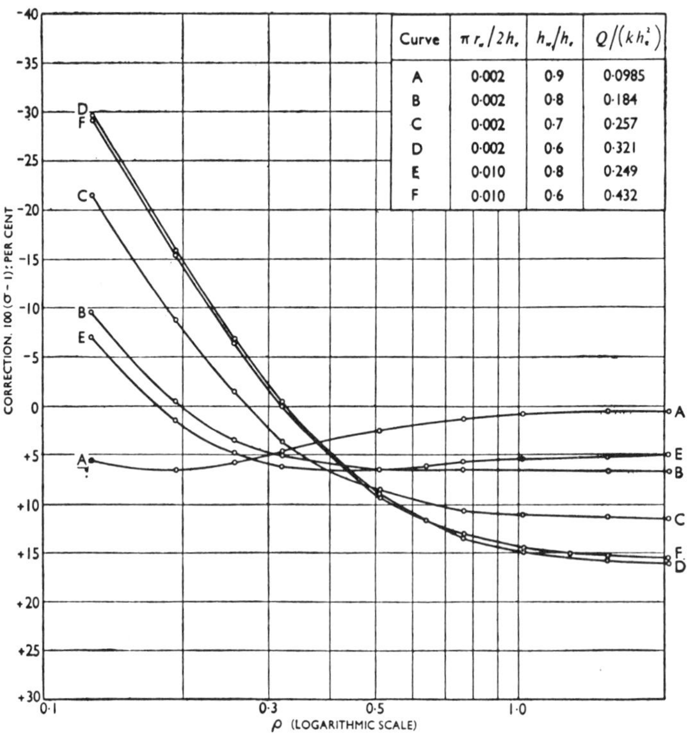
Curve
πr_w/2h_e
h_w/h_e
Q/(kh_e²)
A 0·002 0·9 0·0985
B 0·002 0·8 0·184
C 0·002 0·7 0·257
D 0·002 0·6 0·321
E 0·010 0·8 0·249
F 0·010 0·6 0·432
-40
-35
-30
-25
-20
-15
-10
-5
0
+5
+10
+15
+20
+25
+30
CORRECTION. 100(σ − 1): PER CENT
0·1
0·3
0·5
1·0
ρ (LOGARITHMIC SCALE)
A
B
C
D
E
F

CURVES FOR CORRECTING DRAWDOWN OF FREE SURFACE WHEN $\tau < 0{\cdot}05$

of τ, and consequently will be less than 6 per cent. Since corrections to the drawdown are not available when $0{\cdot}05 < \tau < 5$, drawdown calculations should in this case be restricted to the above ranges of Q and ρ.

Case III ($\tau > 5$).—The restrictions imposed on the magnitudes of Q and ρ in Case II are unnecessary in this case and the following method of correction enables the drawdown to be accurately obtained, even for a low pumping level and at a point quite near to the pumped well.

The expansion of the exponential integral, used by Jacob [10] and others, gives the expression for the drawdown:

$$s = \frac{Q}{4\pi T}\left(-0{\cdot}5772 - \log_e u + u - \frac{u^2}{4} \ldots\right) \quad . \quad . \quad (14)$$

Fig. 2

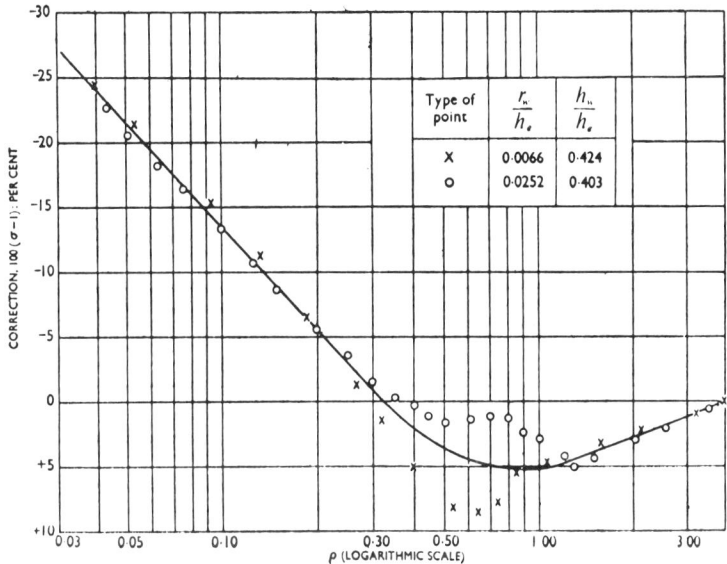

Curve for correcting Drawdown of Free Surface when $\tau > 5$

where $u = \dfrac{\rho^2}{4\tau}$, using the present notation. If u is small so that the power series on the right-hand side may be neglected,

$$s_c \simeq \frac{Q}{2\pi T} \log_e \frac{\rho_e}{\rho} \quad . \quad . \quad . \quad . \quad . \quad . \quad (15)$$

where $\rho_e = 1{\cdot}5\sqrt{\tau}$.

Equation (15), which approximates to equation (14) with increasing accuracy as the well is approached, is the equation (except for the changed notation) proposed by Cambefort [11] to represent the drawdown of the free surface under conditions of steady flow. In a previous Paper,[12] the Author has shown that equation (15) accurately represents the drawdown only when $\dfrac{r_w}{h_e} = 0\cdot 11$.* For the much smaller values of this ratio which are usual, equation (15) requires a correction, the magnitude of which is proportional to the vertical distance of the appropriate curve in *Fig. 5* of reference 5 from the straight line representing equation (15).

Writing $\sigma = \dfrac{s}{s_c}$, where s_c denotes Cambefort's value of s, the required correction to the drawdown, as given by equation (12) or equation (14), is $100(\sigma - 1)$ per cent. This correction is shown by the ordinates in *Fig. 2*, on a base of ρ (to a logarithmic scale). It must be added to the calculated drawdown. The correction strictly applies only when $\rho_e = 4$. However, it is found that considerable departure from this value has little practical effect on the value of the correction. Also, the correction is practically independent of the ratio $\dfrac{h_w}{h_e}$

Different Horizontal and Vertical Permeability

If the coefficients of permeability in horizontal and vertical directions have different but constant values, k_r and k_z respectively, k is replaced by k_z in boundary equation (4a) and in the expression for τ. Equations (9) to (13), previously derived for the isotropic case, will then apply provided r (or ρ) in them is replaced by μr (or $\mu\rho$), the quantity μ denoting $\sqrt{\dfrac{k_z}{k_r}}$. This is the simple transformation [13] applicable to the case of steady flow, which also applies to the case of non-steady flow since the time does not appear in the differential equation (2). The boundary equation (4a) is clearly satisfied when μ is introduced. Also, boundary equation (8) is satisfied since $r\,\dfrac{\partial\phi}{\partial r}$ is unaltered by the transformation.

Equation for Recovery

If τ and τ' are the time factors, reckoned from the commencement of pumping and from the end, respectively, the drawdown of the free surface during recovery is found from equation (11) to be:

$$s = \dfrac{Q}{2\pi T}\left\{V(\rho, \tau) - V(\rho, \tau')\right\} \quad \ldots \quad (16)$$

---

* Inaccurately shown as 0·13 in reference 5, p. 544.

CALCULATION OF PUMPING LEVEL

Owing to the approximate nature of boundary equation (8), assumed in obtaining the drawdown equation, the depth h_w of water in the well does not appear explicitly and there is some uncertainty as to how it should be calculated. As an approximation it may be assumed that $h_w = \phi_0$, where ϕ_0 denotes the value of ϕ at the bottom of the well. Then, putting $z = 0$ and $\phi = h_w$ in equation (9), and changing the variables r, t and β as before:

$$h_e - h_w = \frac{Q}{2\pi T} \int_0^\infty \frac{J_0(\lambda \rho_w)}{\lambda} \left\{1 - \operatorname{sech} \lambda \exp(-\tau \lambda \tanh \lambda)\right\} d\lambda \quad . \quad . \quad (17)$$

which gives the depth of pumping level below rest level.

It has been estimated that h_w, as given by equation (17), is about 1 per cent too great, for small values of τ and for $\dfrac{h_w}{h_e} = 0.8$.

The integral in equation (17) has been computed for values of the time factor assumed in Table 3. Writing:

$$h_e - h_w = (-\log_e \rho_w + m) \frac{Q}{2\pi T} \quad . \quad . \quad . \quad (18)$$

the values of m, defined by equation (18), are given in Table 4.

TABLE 4

τ	m
0·05	− 0·043
0·20	+ 0·087
1·00	+ 0·512
5·00	+ 1·228

When the time factor $\tau > 5$, equation (14) is applicable and there is then a simple method for calculating h_w, which is accurate even when the drawdown of the pumping level is large. This method is based on the Dupuit equation:[14]

$$h_w^2 = h_e^2 - \frac{Q}{\pi k} \log_e \frac{\rho_e}{\rho_w} \quad . \quad . \quad . \quad . \quad (19)$$

which is known to give an accurate relationship between h_w and Q for steady flow conditions. The depth h_w of water in the well is easily calculated from equation (19), using the previously determined relation:

$$\rho_e = 1.5\sqrt{\tau} \quad . \quad . \quad . \quad . \quad . \quad . \quad (20)$$

Examples of Drawdown Calculation

The following examples demonstrate the use of the foregoing equations:—

Example 1.—Data for thick and moderately permeable aquifer

Assume: $r_w = 1$ foot, $h_e = 800$ feet, $k = 2 \times 10^{-5}$ foot per second, $S = 0.15$, $Q = 1$ cubic foot per second, $t = 3.47$ days $= 3 \times 10^5$ seconds. Then time factor $\tau = \dfrac{kt}{Sh_e} = 0.05$; $\dfrac{Q}{2\pi T} = 9.95$.

From Table 3, $V(0.2, 0.05) = 0.214$. Hence, from equation (11), for $\rho = 0.2$, drawdown $s = 9.95 \times 0.214 = 2.13$ feet.

The constant $\dfrac{\pi r_w}{2h_e} = 0.002$ and $\dfrac{Q}{kh_e^2} = 0.078$. Hence from curve A, *Fig. 1*, the correction to the drawdown is about 6 per cent, making the corrected drawdown 2.26 feet.

The drawdown of the water level in the well is found from equation (18) to be: $h_e - h_w = 9.95(\log_e 800 - 0.043) = 66.1$ feet.

Example 2

Consider the well in **Example 1**, after prolonged pumping for 69.4 days. Then $\tau = 1$. From Table 3, for $\rho = 0.2$, $V = 1.844$. Therefore $s = 18.35$ feet.

From equation (18), $h_e - h_w = 71.6$ feet.

Example 3.—Data for thin and very permeable aquifer

Assume: $r_w = 1$ foot, $h_e = 100$ feet, $k = 10^{-3}$ foot per second, $S = 0.2$, $Q = 4$ cubic feet per second, $t = 27.8$ hours $= 10^5$ seconds. Then $\tau = 5.0$, $T = kh_e = 0.1$, $\dfrac{Q}{2\pi T} = 6.366$.

From Table 3, for $\rho = 0.2$, $V = 2.785$.
Therefore $s = 6.366 \times 2.785 = 17.73$ feet.

From *Fig. 2*, the correction to be subtracted from s, for $\rho = 0.2$, is 5.5 per cent. Thus, the corrected value of s is 16.76 feet. Also, from equation (20), $\rho_e = 1.5 \times \sqrt{5} = 3.354$; and since $\dfrac{Q}{\pi k} = 1{,}273.2$ square feet, equation (19) gives $h_w = 51$ feet.

Conclusions

Equation (11) gives the drawdown of the water-table for all values of the time factor τ.

For $\tau < 0.05$ the drawdown should be corrected using the curves in *Fig. 1*.

For $0.05 \leqslant \tau \leqslant 5$, ρ and Q in equation (11) are limited to the ranges stated under Case II (p. 572). Some values of the function V are given in Table 3.

For $\tau > 5$, the exponential integral (equation (12)) gives the drawdown for any value of Q (that is for all pumping levels) provided a correction is applied using the curve in *Fig. 2*.

REFERENCES

1. C. V. Theis, "The Relation between the Lowering of the Piezometric Surface and the Rate and Duration of Discharge of a Well using Ground-Water Storage." Trans Amer. geophys. Un., vol. 16 (1935), p. 519.
2. L. K. Wenzel, "Methods for Determining Permeability of Water-Bearing Materials." U.S. Geol. Survey, Water Supply Paper 887 (1942).
3. J. Ineson, "Notes on the Theory and Formulae associated with Pumping Tests for the Determination of Formation Constants." J. Instn Wat. Engrs, vol. 6, p. 443 (Oct. 1952).
4. J. Ineson, "Some Observations on Pumping Tests carried out on Chalk Wells." J. Instn Wat. Engrs, vol. 7. p. 215 (May 1953).
5. N. S. Boulton, "The Flow Pattern near a Gravity Well in a Uniform Water-Bearing Medium." J. Instn Civ. Engrs, vol. 36, p. 534 (Dec. 1951).
6. See reference 4, p. 220.
7. See reference 2, pp. 142 to 146.
8. Gray, Mathews, and MacRobert, "Treatise on Bessel Functions." MacMillan, 1922, p. 68.
9. See reference 1, p. 519.
10. C. E. Jacob, "Drawdown Test to Determine Effective Radius of Artesian Well." Trans Amer. Soc. Civ. Engrs, vol. 112 (1947), p. 1052.
11. H. Cambefort, "*Les puits filtrants et la formule de Dupuit*." ("Filter Wells and Dupuit's Formula.") *Travaux*, vol. 164, p. 392 (June 1948).
12. See reference 5, p. 544.
13. Morris Muskat, "The Flow of Homogeneous Fluids through Porous Media." McGraw-Hill, 1937, p. 226.
14. See reference 5, p. 536.

The Paper is accompanied by two sheets of diagrams, from which the Figures in the text have been prepared, and by the following two Appendices.

APPENDIX I

The formula corresponding to equation (11), when the permeable medium extends downwards indefinitely below the bottom of the well, is derived as follows:—

Consider water being removed at a constant rate q from a small spherical cavity on the well-axis, at a depth η below the rest level of the water-table. It is convenient to employ the symbol y for the depth below the initial water-table of any point. Then after an interval of time t from the instant when water is first removed:

$$\phi = \frac{q}{4\pi k}\left[\frac{1}{\sqrt{\{r^2 + (y-\eta)^2\}}} + \frac{1}{\sqrt{\{r^2 + (y+\eta)^2\}}} - \frac{2}{\sqrt{\{r^2 + (y+\eta+kt/S)^2\}}}\right] \quad \ldots \quad (21)$$

It can easily be verified that equation (21) satisfies boundary equation (4a) when $y = 0$.

The solution for a well of length l below the initial water-table may now be obtained by integration. Thus replacing q by $\frac{Q}{l} \cdot d\eta$ and integrating (21) between the limits of 0 and l gives:

$$\phi = \frac{Q}{4\pi kl}\left(\sinh^{-1}\frac{l-y}{r} + \sinh^{-1}\frac{l+y}{r} - 2\sinh^{-1}\frac{l+y+kt/S}{r} + 2\sinh^{-1}\frac{y+kt/S}{r}\right)$$

Putting $y = 0$ and writing $\rho = \frac{r}{l}$, $\tau = \frac{kt}{Sl}$, the drawdown of the free surface is given approximately by:

$$s = \frac{Q}{2\pi kl}\left(\sinh^{-1}\frac{1}{\rho} - \sinh^{-1}\frac{1+\tau}{\rho} + \sinh^{-1}\frac{\tau}{\rho}\right) \quad \ldots \quad (22)$$

Appendix II

The series referred to in this Paper, satisfying the correct boundary equations (6) at the well, are:

$$\frac{\partial \phi}{\partial z}_{z=h_e} = \frac{4}{\pi}\sum_{n=1,3,5\ldots}^{\infty}(-1)^{\frac{1}{2}(n-1)}\frac{1}{n}\cos\frac{n\pi h_w}{2h_e}\frac{K_0\left(\frac{n\pi r}{2h_e}\right)}{K_0\left(\frac{n\pi r_w}{2h_e}\right)} \ldots (n \text{ odd})$$

$$Q = \frac{16}{\pi}kr_w h_e \sum_{n=1,3,5\ldots}^{\infty}(-1)^{\frac{1}{2}(n-1)}\frac{1}{n^2}\cos\frac{n\pi h_w}{2h_e}\frac{K_1\left(\frac{n\pi r_w}{2h_e}\right)}{K_0\left(\frac{n\pi r_w}{2h_e}\right)} \ldots (n \text{ odd})$$

The series satisfying the approximate boundary equation (8) at the well is

$$\frac{\partial \phi}{\partial z}_{z=h_e} = \frac{Q}{\pi k h_e^2}\sum_{n=1,3,5\ldots}^{\infty} K_0\left(\frac{n\pi r}{2h_e}\right) \ldots (n \text{ odd})$$

11

Copyright ©1955 by the American Geophysical Union
Reprinted from *Am. Geophys. Union Trans.* 36:95-100 (1955)

NON-STEADY RADIAL FLOW IN AN INFINITE LEAKY AQUIFER

M. S. Hantush and C. E. Jacob

Abstract--The non-steady drawdown distribution near a well discharging from an infinite leaky aquifer is presented. Variation of drawdown with time and distance caused by a well of constant discharge in confined sand of uniform thickness and uniform permeability is obtained. The discharge is supplied by the reduction of storage through expansion of the water and the concomitant compression of the sand, and also by leakage through the confining bed. The leakage is assumed to be at a rate proportional to the drawdown at any point. Storage of water in the confining bed is neglected. Two forms of the solution are developed. One is suitable for computation for large values of time and the other suitable for small values of time. This solution is compared with earlier solutions for slightly different boundary conditions.

Introduction--The differential equation for radial flow in an elastic artesian aquifer with linear leakage has been given by JACOB [1946]. He also obtained the non-steady drawdown distribution produced by a well of constant discharge situated in the center of a circular region whose outer boundary is maintained at constant head. The head distribution in his problem is initially uniform.

In this paper the solution is obtained for the problem in which the outer boundary is removed to infinity.

Statement of the problem--The problem is to determine the variation with time of the drawdown induced by a well steadily discharging from an infinite leaky aquifer in which the initial head is uniform. Leakage into the aquifer is assumed vertical and proportional to the drawdown. Stated mathematically the boundary-value problem is

$$\partial^2 s/\partial r^2 + (1/r)\, \partial s/\partial r - s/B^2 = (S/T)\, \partial s/\partial t \quad \ldots \ldots \ldots \ldots (1)$$

$$s(r, 0) = 0 \qquad r \geq 0 \quad \ldots \ldots \ldots \ldots \ldots (2a)$$

$$s(\omega, t) = 0 \qquad t \geq 0 \quad \ldots \ldots \ldots \ldots \ldots (2b)$$

$$\lim_{r \to 0} r\, \partial s(r, t)/\partial r = -Q/2\pi T \qquad t > 0 \quad \ldots \ldots \ldots \ldots (2c)$$

where

$s(r, t)$ is the drawdown at any time and any distance from the well.
r is the distance to any point measured from the axis of the well.
S is the storage coefficient of the artesian aquifer (a non-dimensional constant) defined as 'the product of the thickness of the artesian bed and the relative volume of water released from storage by a unit decline of head' [JACOB, 1946].
K and K' are the hydraulic conductivities (or 'permeabilities') of the artesian sand and confining bed, respectively. They have the dimension L/t.
b and b' are the thicknesses of the artesian sand and confining bed, respectively.
$T = Kb$ is the transmissibility (of dimension L^2/t) of the artesian sand. The ratio K'/b' may be termed 'specific leakage' or leakance [HANTUSH, 1949, p. 8]. It has the dimension $1/t$.
The transmissibility divided by the leakance (of dimension L^2) is symbolized by B^2.
Q is the discharge of the well.

Solution of the problem--After separating the variables, it can be shown that

$$J_0(\alpha r/B)\, \exp[-(\alpha^2 + 1)\, Tt/SB^2] \quad \text{and} \quad K_0(r/B)$$

are particular solutions of (1), where J_0 and K_0 are respectively the Bessel function of the first kind of zero order and the modified Bessel function of the second kind of zero order, and where α is any real constant.

Due to the linearity and homogeneity of (1), the above particular solutions can be combined linearly to obtain the following solution

$$s = c\left[\int_0^\infty A(\alpha) J_0(\alpha r/B) \exp[-(\alpha^2 + 1) Tt/SB^2] d\alpha + K_0(r/B)\right] \quad \dots \dots \dots (3)$$

in which c is an arbitrary constant, and in which $A(\alpha)$ is a function depending upon α alone and hence constant or independent of the variables of (1).

That (3) satisfies condition (2b) is seen immediately. Recalling that

$$\lim_{r \to 0} K_0(r/B) = -0.5772 - \ln(r/2B)$$

and applying condition (2c), the value of c is found to be $Q/2\pi T$. The function $A(\alpha)$ may be found by using condition (2a) as follows. When $t = 0$, $s = 0$ and (3) becomes

$$0 = K_0(r/B) + \int_0^\infty A(\alpha) J_0(\alpha r/B) d\alpha \quad \dots \dots \dots \dots \dots (4)$$

Using the integral relation [WATSON, 1944, p. 425]

$$K_0(r/B) = \int_0^\infty [\alpha/(\alpha^2 + 1)] J_0(\alpha r/B) d\alpha$$

Eq. (4) reduces to

$$\int_0^\infty J_0(\alpha r/B) [A(\alpha) + \alpha/(\alpha^2 + 1)] d\alpha = 0 \quad \dots \dots \dots \dots (5)$$

whence

$$A(\alpha) = -\alpha/(\alpha^2 + 1)$$

Substituting the values of c and $A(\alpha)$ thus obtained, the formal solution of the problem is

$$s/(Q/2\pi T) = K_0(r/B) - \int_0^\infty [\alpha/(\alpha^2 + 1)] J_0(\alpha r/B) \exp[-p(\alpha^2 + 1)] d\alpha \quad \dots \dots (6)$$

where

$$p = Tt/SB^2$$

As pumping continues indefinitely, that is, as $t \to \infty$, the integral in (6) becomes zero and the drawdown is represented by $s/(Q/2\pi T) = K_0(r/B)$, which is the steady-state solution of the problem [JACOB, 1946].

To evaluate the infinite integral of (6), hereafter represented by I, one proceeds as follows. The integral representation of $1/(\alpha^2 + 1)$ is

$$\int_0^\infty \exp[-(\alpha^2 + 1)x] dx$$

Thus, replacing $1/(\alpha^2 + 1)$ by its integral form and changing the order of integration, one obtains

$$I = \int_0^\infty \exp[-(p + x)] dx \int_0^\infty J_0(\alpha r/B) \exp[-(p + x)\alpha^2] \alpha \, d\alpha \quad \dots \dots \dots (7)$$

Reversing the order of integration is justified, since the integral I is convergent [BROMWICH, 1947, p. 504]. Integration with respect to α gives [WATSON, 1944, p. 394]

$$\int_0^\infty J_0(\alpha r/B) \exp[-(p+x)\alpha^2]\alpha \, d\alpha = [1/2(p+x)] \exp[-(r^2/B^2)/4(p+x)]$$

Hence, (7) becomes

$$I = (1/2) \int_0^\infty [1/(p+x)] \exp[-p - x - (r^2/B^2)/4(p+x)] \, dx$$

and upon the substitution of y for $(p+x)$, it becomes

$$I = (1/2) \int_p^\infty (1/y) \exp[-y - (r^2/B^2)/4y] \, dy \quad \dots \dots \dots \dots \dots (8)$$

If, in the integral of (8), $\exp[-r^2/4B^2 y]$ is replaced by the absolutely and uniformly convergent series

$$\sum_{n=0}^\infty \frac{(-1)^n (r^2/4B^2)^n}{n! \, y^n}$$

and if the order of integration and summation are changed, integration by parts will result in

$$I = (1/2) \exp(-p) \sum_{n=1}^\infty \sum_{m=0}^{n-1} \frac{(-1)^{n+m}(n-m-1)! \, (r^2/4B^2)^n}{(n!)^2 \, p^{n-m}}$$

$$+ (1/2) \int_p^\infty \left\{ [\exp(-y)]/y \right\} dy \left[\sum_{n=0}^\infty \frac{(r/2B)^{2n}}{(n!)^2} \right]$$

$$I = -(1/2) \exp[-r^2/4B^2 u] \sum_{n=0}^\infty \sum_{m=0}^n \frac{(-1)^{n+m}(n-m)! \, (r^2/4B^2)^m \, u^{n-m+1}}{(n+1)!^2}$$

$$+ (1/2) [-\text{Ei}(-r^2/4B^2 u)] I_0(r/B) \quad \dots \dots \dots \dots \dots \dots \dots \dots \dots \dots (9)$$

where $I_0(r/B)$ is the modified Bessel function of the first kind and zero order, and where

$$u = r^2/4B^2 p = r^2 S/4Tt$$

That the series in (9) is uniformly and absolutely convergent can be seen by applying the ratio test for any given value of m. Because of the absolute and uniform convergence of this double series, then, summation in any manner is justified. Separating the first column $m = 0$ and the diagonal $m = n$ from the double series, the series of (9) becomes

$$\sum_{n=0}^\infty \frac{(-1)^n u^{n+1}}{(n+1)(n+1)!} + \sum_{n=1}^\infty \frac{(r^2/4B^2)^n u}{(n+1)!^2} + \sum_{n=2}^\infty \sum_{m=1}^{n-1} \frac{(-1)^{n+m}(n-m)!}{(n+1)!^2} (r^2/4B^2)^m u^{n-m+1}$$

$$\dots \dots \dots \dots \dots \dots \dots (10)$$

Recognizing the first series in (10) as $0.5772 + \ln u + [-\text{Ei}(-u)]$, and the second series as $-u + u[I_0(r/B) - 1]/(r^2/4B^2)$, and by changing the index of summation from $n = 2$ to $n = 1$ in the third series; and then substituting in (9), and then in (6), one obtains

$$s/(Q/4\pi T) = 2K_0(r/B) - I_0(r/B)[-\text{Ei}(-r^2/4B^2u)]$$
$$+ [\exp(-r^2/4B^2u)]\left\{0.5772 + \ln u + [-\text{Ei}(-u)] - u + u[I_0(r/B) - 1]/(r^2/4B^2)\right.$$
$$\left. - u^2 \sum_{n=1}^{\infty} \sum_{m=1}^{n} \frac{(-1)^{n+m}(n-m+1)!}{(n+2)!^2}(r^2/4B^2)^m u^{n-m}\right\} \quad \ldots \ldots \ldots \ldots (11)$$

The series of (11) is rapidly convergent for the values $r/2B \leq 1$ and $u \leq 1$. Only a few terms are required to obtain results accurate to four decimal places. When $r/2B \leq 0.10$ and $u \leq 1$, the solution can be reduced to

$$s/(Q/4\pi T) = 2K_0(r/B) - I_0(r/B)[-\text{Ei}(-r^2/4B^2u)]$$
$$+ [\exp(-r^2/4B^2u)][0.5772 + \ln u + [-\text{Ei}(-u)] + (u/4)(r^2/4B^2)(1-u/9)] \quad \ldots (12)$$

which gives values of the drawdown accurate to four decimal places.

For the range of values $r/2B \leq 0.10$ and $u \leq 0.10$, the solution may be approximated by

$$s/(Q/4\pi T) = 2K_0(r/B) - I_0(r/B)[-\text{Ei}(-r^2/4B^2u)]$$
$$+ [\exp(-r^2/4B^2u)][(u/4)(r^2/4B^2)(1-u/9) + u - u^2/2.2! + u^3/3.3!] \quad \ldots (13)$$

<u>Solution for short times</u>--A solution that is more convenient for computations when t is small can be developed as follows. Making use of the integral relation [WATSON, 1944, p. 183]

$$2K_0(r/B) = \int_0^{\infty} (1/y) \exp[-y - r^2/4B^2 y] \, dy$$

combining (8) with (6) and making the substitutions, $r^2/4B^2 y = Z$, (6) reduces to

$$s/(Q/4\pi T) = \int_u^{\infty} (1/Z) \exp[-Z - r^2/4B^2 Z] \, dZ \quad \ldots \ldots \ldots \ldots (14)$$

where

$$u = r^2 S/4Tt$$

It is of interest to note that if $B \to \infty$, that is, either when leakage is very small or when the time is very small so that leakage does not have time to enter the main flow, (14) becomes

$$s/(Q/4\pi T) = \int_u^{\infty} (1/Z) \exp(-Z) \, dZ \quad \ldots \ldots \ldots \ldots \ldots \ldots (15)$$

which is the solution to the problem if leakage is not present [THEIS, 1935].

The integral of (14), being similar to that of (8), can be evaluated in the same way. The final expression is

$$s/(Q/4\pi T) = I_0(r/B)[-\text{Ei}(-u)] - e^{-u}\left\{0.5772 + \ln(r^2/4B^2 u)\right.$$
$$+ [-\text{Ei}(-r^2/4B^2 u)] - (r^2/4B^2 u) + [I_0(r/B) - 1]/u\Big\}$$
$$+ (e^{-u}/u^2) \sum_{n=1}^{\infty} \sum_{m=1}^{n} \frac{(-1)^{n+m}(n-m+1)!(r^2/4B^2)^n}{(n+2)!^2 u^{n-m}} \quad \ldots \ldots \ldots (16)$$

which is in a form convenient for computation with values of $u \geq 1$. Very few terms of the series need be used when $r/2B \leq 1$ and $u \geq 1$. For values of $u \geq 5$ and values of $r/2B \leq 1$, the series may be safely neglected and results obtained accurate to four decimal places. If $r/2B \leq 0.10$ and $u \geq 1$, the solutions can be approximated by

$$s'/(Q/4\pi T) = I_0(r/B)[-Ei(-u)] - e^{-u}[(r^2/4B^2)(1/u - 1/36u^2) + (r^2/4B^2)^2(1/4u - 1/4u^2)]$$
$$\dots \dots \dots \dots \dots \dots (17)$$

<u>Drawdown at the well</u>--The drawdown at the face of the well is found by substituting r_w for r in any of the previous solutions, r_w being the radius of the well. However, if the ratio $r_w/B \leq 0.01$ the drawdown at the face of the well will, for all practical purposes, be given by

$$s_w/(Q/4\pi T) = -Ei(-u_w) \dots \dots \dots \dots \dots \dots (18)$$

which is the drawdown at the well if there is no leakage. Eq. (18) is obtained by making r/B very small in either (12) or (17).

<u>Calculated example</u>--Tables 1 and 2 were prepared to facilitate evaluation of (11) and (16). It gives a few coefficients of the double series that appears in both equations. The numbers in brackets in the body of the table are negative powers of ten by which the other numbers are multiplied.

Table 1--<u>Values of $(n-m+1)!/(n+2)!^2$ found in Equations (11) and (16)</u>

n	m							
	1	2	3	4	5	6	7	8
1	2.7778[-2]							
2	3.4722[-3]	1.7361[-3]						
3	4.1667[-4]	1.3889[-4]	6.9444[-5]					
4	4.6296[-5]	1.1574[-5]	3.8580[-6]	1.9290[-6]				
5	4.7240[-6]	9.4481[-7]	2.3620[-7]	7.8734[-8]	3.9367[-8]			
6	4.4287[-7]	7.3814[-8]	1.4763[-8]	3.6907[-9]	1.2302[-9]	6.1512[-10]		
7	3.8274[-8]	5.4677[-9]	9.1129[-10]	1.8226[-10]	4.5564[-11]	1.5188[-11]	7.5940[-12]	
8	3.0619[-9]	3.8274[-10]	5.4677[-11]	9.1129[-12]	1.8226[-12]	4.5564[-13]	1.5188[-13]	7.5940[-14]

Table 2--<u>Values of $(r^2/4B^2)^m$ for B = 20,000 ft</u>

r	r/2B	m			
		1	2	3	4
ft					
1,000	0.025	6.25[-4]	3.9062[-7]		
2,000	0.05	2.5[-3]	6.25[-6]	1.5625[-8]	
5,000	0.125	1.5625[-2]	2.4414[-4]	3.7903[-6]	5.9223[-8]
10,000	0.25	6.25[-2]	3.9062[-3]	2.4414[-4]	1.5259[-5]
20,000	0.50	2.5[-1]	6.25[-2]	1.5625[-2]	3.9062[-3]
50,000	1.25	1.5625	2.4414	3.7903	5.9223

Drawdown values for one set of assumed values of the parameters are given in Table 3 and are represented graphically in Figure 1. The factor B is taken to be 20,000 ft. The abscissa is $4Tt/S = r^2/u$ and represents the time multiplied by a constant. The straight line marked 'zero leakage' shows the trend that the drawdown would take at a distance of 1000 ft if the leakage were zero ($B \to \infty$). The effect of leakage is shown by the departure of the drawdown curve from this straight line as time increases. The effect of leakage is shown also by the drawdown curves for the more remote distances.

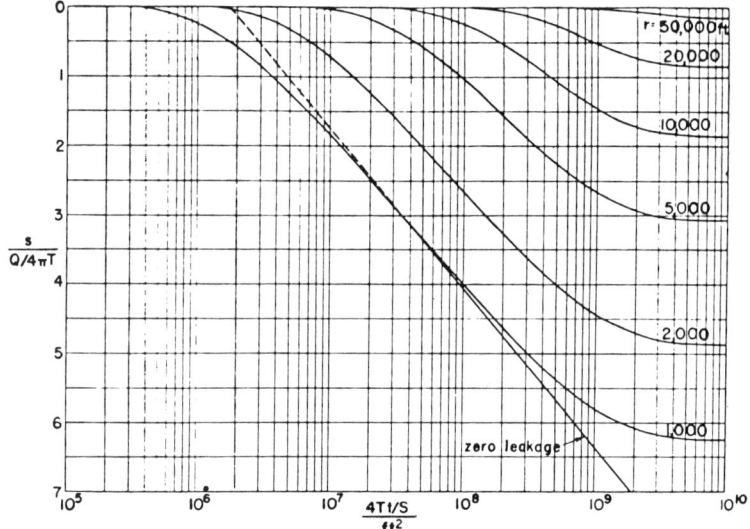

Fig. 1--Variation of drawdown with time at different distances in an infinite leaky aquifer with B = 20,000 ft

Table 3--Values of $s/(Q/4\pi T)$ for B = 20,000 ft

u	r (in feet)					
	1000	2000	5000	10,000	20,000	50,000
0.0002	6.217	....	....	....	....	....
0.0005	6.082	....	....	....	....	....
0.001	5.796	4.829	....	....	....	....
0.002	5.354	4.708	....	....	....	....
0.005	4.608	4.296	3.072	....	....	....
0.01	3.979	3.815	2.993	....	....	....
0.02	3.326	3.244	2.766	1.838	....	....
0.05	2.457	2.427	2.230	1.708	....	....
0.10	1.818	1.805	1.715	1.443	0.819	....
0.20	1.221	1.215	1.187	1.060	0.715	....
0.50	0.559	0.558	0.550	0.521	0.420	0.117
1.0	0.219	0.218	0.218	0.210	0.185	0.080

References

BROMWICH, T. J., Theory of infinite series, MacMillan and Co., 1947.
HANTUSH, M. S., Plain potential flow of ground water with linear leakage, doctoral dissertation, Univ. of Utah, 1949.
JACOB, C. E., Radial flow in a leaky artesian aquifer, Trans. Amer. Geophys. Union, v. 27, pp. 198-205, 1946.
THEIS, C. V., The relation between the lowering of the piezometric surface and the rate and duration of discharge of a well using ground-water storage, Trans. Amer. Geophys. Union, v. 16, pp. 519-524, 1935.
WATSON, G. N., Theory of Bessel functions, MacMillan and Co., 1944.

College of Engineering (M. S. H.),
   Baghdad, Iraq

Brigham Young University (C. E. J.),
   Provo, Utah

(Manuscript received January 25, 1954; presented at the Thirty-First Annual Meeting, Washington, D. C., May 1, 1950; open for formal discussion until July 1, 1955.)

12

Copyright ©1969 by the American Geophysical Union
Reprinted from *Water Resources Research* **5**:817-829 (1969)

Applicability of Current Theories of Flow in Leaky Aquifers

SHLOMO P. NEUMAN AND PAUL A. WITHERSPOON

University of California, Berkeley, California 94720

NOTATIONS

b_i, thickness of ith aquifer, L;
b_j', thickness of jth aquitard, L;
$J_0(x)$, zero order Bessel function of the first kind;
K_i, permeability of ith aquifer, L/T;
K_j', permeability of jth aquitard, L/T;
Q_i, pumping rate from ith aquifer, L^3/T;
r, radial distance from pumping well, L;
r/B_{ij}, $r\sqrt{K_j'/K_i b_i b_j'}$;
S_{s_i}, specific storage of ith aquifer, L^{-1};
S_{s_j}', specific storage of jth aquitard, L^{-1};
s_i, drawdown in ith aquifer, L;
s_j', drawdown in jth aquitard, L;
s_D, dimensionless drawdown $(4\pi T_i s/Q_i)$;
T_i, transmissibility of ith aquifer $(K_i b_i)$, L^2/T;
t, pumping time, T;
t_{D_i}, dimensionless time for pumped (ith) aquifer $(K_i t/S_{s_i} r^2)$;
z, vertical distance above bottom of aquitard 1, L;
β_{ij}, $r/4b_i \sqrt{K_j' S_{s_j}'/K_i S_{s_i}}$.

INTRODUCTION

When water is being withdrawn from the subsurface, the question often arises as to how much of the production comes from the aquifer in which the well is completed and how much comes from beds lying above or below. An aquifer that receives a significant inflow from adjacent beds while being pumped is called a 'leaky' aquifer. In reality a leaky aquifer is often just one part of a multiple aquifer system, i.e., a succession of aquifers separated by varying thicknesses of relatively low permeability aquitards and/or aquicludes. If these intervening beds are sufficiently low in permeability, they act like no-flow boundaries, and the behavior of any aquifer that is being pumped should follow the Theis solution [*Neuman and Witherspoon*, 1968].

However, as permeability increases, leakage becomes appreciable and flow is not restricted to the pumped aquifer alone. The additional water is derived primarily from storage in the aquitards and the adjoining unpumped aquifers. Since flow is not restricted to the pumped aquifer alone, the leaky aquifer must be viewed as part of a complex multiple aquifer system where the drawdown in each layer depends on the flow behavior of the entire system. Therefore one cannot develop a complete understanding of flow in the pumped aquifer without having analyzed the system as a whole.

Since a rigorous approach to flow in multiple aquifer systems involves boundary conditions that make the problem intractable analytically, it has been customary to simplify the mathematics by assuming that flow is essentially horizontal in the aquifers and vertical in the aquitards. The validity of this assumption has recently been investigated by *Neuman and Witherspoon* [1969a] for the case of two aquifers and one aquitard. It was found that when the permeabilities of the aquifers are more than two orders of magnitude greater than that of the aquitard the errors introduced by this assumption are usually less than 5%. These errors increase with time, decrease with radial distance from the pumping well, and are smallest in the pumped aquifer and greatest in the aquitard. Since the permeability contrast between aquifer and aquitard is often greater than three orders of magnitude, it would appear that the above assumptions can be used safely.

In dealing with steady state problems the assumption of vertical flow in the aquitards implies that the rate of leakage into the pumped aquifer is proportional to the potential drop across the leaky aquitard. An approach to steady flow in leaky aquifers based on this assumption was first introduced into the literature by *De Glee* [1930] and later was used by *Steggeventz and Van Ness* [1939]. Additional work along these lines by *Glebov* [1940], *Myatiev* [1946, 1947], and *Girinsky* [1947] has been summarized by *Polubarinova-Kochina* [1962] and *Aravin and Numerov* [1965]. Only in one case described by *Polubarinova-Kochina* [1962, p. 385] has the drawdown been allowed to vary in more than one aquifer at a time.

Jacob [1946] used the same approach to develop a partial differential equation for nonsteady flow in a leaky aquifer. However, it is important to recognize that under transient conditions the assumption that leakage is proportional to the potential drop across the aquitard is tantamount to ignoring the storage capacity of the aquitard. In addition Jacob assumed that head in the unpumped aquifer remained constant. A similar approach has been used by *Polubarinova-Kochina* [1962, p. 546] to investigate the problem of seepage from a ditch into an aquifer underlain by an aquitard.

The above assumptions have greatly facilitated mathematical treatment of the problem. As a result between the years 1949-1960 *Hantush* [1949, 1956, 1957, 1959] and *Hantush and Jacob* [1954, 1955a, 1955b, 1960] have used this approach to develop a large number of solutions to various problems involving flow in aquifers with vertical leakage. One of these solutions [*Hantush and Jacob*, 1955b] describes nonsteady radial flow to a well that completely penetrates an infinite leaky aquifer and discharges at a constant rate. This solution is of particular interest in the present work because it has been extensively tabulated [*Hantush*, 1956], and the resulting type curves are being widely used by ground-water hydrologists to evaluate the properties and potential yield of leaky aquifers [*Ferris et al.*, 1962; *Walton*, 1960, 1962; *Slater*, 1963; *De Wiest*, 1965; *Narasimhan*, 1968]. The results of this solution are usually presented in terms of a dimensionless parameter

$$r/B = r\sqrt{\frac{K'}{Kbb'}} \quad (1)$$

which we note is a function of the permeability contrast between the pumped aquifer and an adjacent aquitard. We shall refer to this as the 'r/B solution.'

Subsequent to the r/B solution *Hantush* [1960a] published a modified version of the differential equations for nonsteady flow in a leaky system in which for the first time consideration was given to the effect of storage in the aquitards. Assuming that drawdown in the unpumped aquifers remains zero, Hantush was able to develop asymptotic solutions for the pumped aquifer that apply only at small and large values of time. We shall demonstrate later that Hantush's solutions are quite good over a broader time span than he has indicated. His solution for small values of time has been extensively tabulated [*Hantush*, 1960b] in terms of a dimensionless parameter

$$\beta = \frac{r}{4b}\sqrt{\frac{K'S_s'}{KS_s}} \quad (2)$$

which we see differs from r/B in that the effects of storage are now included. We shall refer to this as the 'β solution.'

To our knowledge Hantush's modified approach has not been used in subsequent work on leaky aquifers. Between 1961-1967 *Hantush* [1962, 1964, 1967a, 1967b] and *De Wiest* [1961, 1963] analyzed various problems involving flow in leaky aquifers, but in all of these investigations storage in the aquitard was consistently neglected. Only in one analysis [*Hantush*, 1967b] has drawdown in the unpumped aquifer been allowed to vary with time.

Recently we have investigated the problem of flow in a system consisting of two aquifers separated by an aquitard [*Neuman and Wither-

spoon, 1969*a*, 1969*b*]. In contrast to previous work we considered the effects of storage in the aquitard and allowed drawdown in the unpumped aquifer to vary with time. Our solutions have been evaluated numerically [*Neuman and Witherspoon*, 1969*a*] and verified by the finite element method [*Javandel and Witherspoon*, 1968*b*; *Witherspoon et al.*, 1968].

The β solution corresponds to a special case of this two aquifer problem wherein drawdown in the unpumped aquifer is assumed to remain zero, and the r/B solution corresponds to another special case where it is further assumed that storage in the aquitard can be neglected. To investigate the applicability of the β and r/B solutions as well as the underlying assumptions that were used in obtaining these solutions we must first introduce a complete solution for a two aquifer system with zero drawdown in the unpumped aquifer and consider the effects of storage in the aquitard.

SOLUTION FOR TWO AQUIFER SYSTEM WITH ZERO DRAWDOWN IN UNPUMPED AQUIFER

Consider a system composed of two aquifers and one aquitard (Figure 1). A well of infinitesimal diameter completely penetrates one of the aquifers and discharges at constant rate Q_1. Each layer is homogeneous, isotropic, horizontal, and of infinite radial extent. No-flow boundaries are assumed for the upper and lower limits of the complete system, and it is further assumed that the aquifers and aquitard remain water-saturated at all times.

In the particular case where drawdown in the unpumped aquifer remains zero ($s_2 = 0$), the above problem becomes equivalent to Case 1 in *Hantush*'s [1960*a*] 'Modification of the Theory of Leaky Aquifers.' His asymptotic solution for small values of time, which is the β solution, can be written

$$s_1(r, t) = \frac{Q_1}{4\pi T_1} \int_{1/(4t_{D_1})}^{\infty} \exp(-y)$$
$$\cdot \operatorname{erfc}\left(\frac{\beta_{11}}{\sqrt{y(4t_{D_1}y - 1)}}\right) \frac{dy}{y} \quad (3)$$

where Hantush's criterion for the validity of this solution can be expressed as

$$t_{D_1} \leq \frac{1.6\beta_{11}^2}{(r/B_{11})^4} \quad (4)$$

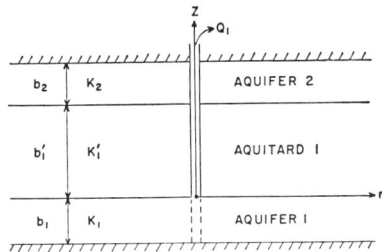

Fig. 1. Schematic diagram of two aquifer system.

Hantush's asymptotic solution for large values of time can be written

$$s_1(r, t) = \frac{Q_1}{4\pi T_1}$$
$$\cdot \int_{\delta_1/(4t_{D_1})}^{\infty} \exp\left[-y - \frac{(r/B_{11})^2}{4y}\right] \frac{dy}{y} \quad (5)$$

where

$$\delta_1 = 1 + \frac{16\beta_{11}^2}{3(r/B_{11})^2} \quad (6)$$

and his criterion for the validity of this solution can be expressed as

$$t_{D_1} \geq \frac{80\beta_{11}^2}{(r/B_{11})^4} \quad (7)$$

It is seen by virtue of the expressions given for δ_1 and t_{D_1} in equations 4, 6, and 7 that (3) and (5) are functions of both β_{11} and r/B_{11}. In addition it is evident from (4) and (7) that for given values of β_{11} and r/B_{11} these solutions cover the entire time domain except for an interval whose span is less than two log cycles.

As mentioned above, a complete solution for this problem for all values of time has recently been developed by *Neuman and Witherspoon* [1969*a*, 1969*b*]. This solution may be written in terms of β_{11} and r/B_{11} as

$$s_1(r, t) = \frac{Q_1}{4\pi T_1}$$
$$\cdot 2 \int_0^{\infty} (1 - e^{-y^2 \bar{t}_{D_1}}) J_0[\omega(y)] \frac{dy}{y} \quad (8)$$

where

$$\bar{t}_{D_1} = t_{D_1}(r/B_{11})^4/(16\beta_{11}^2)$$

$$\omega^2(y) = \frac{(r/B_{11})^4}{16\beta_{11}^2} y^2 - (r/B_{11})^2 y \cot y$$

and $J_0[\omega(y)]$ must be set to zero when $\omega^2(y) < 0$. A solution for the more general case when the pumped aquifer is enclosed between two leaky aquitards has also been developed [*Neuman and Witherspoon*, 1969a].

Equation 8 has been evaluated numerically, and the results for $\beta_{11} = 0.01$, 0.1, and 1.0 are shown in Figures 2, 3, and 4, respectively. In each figure we see an envelope from which a family of r/B_{11} curves extend. The position of the envelope depends on the magnitude of β_{11}, and as this parameter increases in magnitude one sees that deviations from the Theis solution also increase. As the r/B_{11} parameter increases, the curves diverge from the envelope at earlier and earlier values of dimensionless time. These curves reach steady state values for dimensionless drawdown that are identical with those obtained by *Hantush* [1956].

Each curve on Figures 2, 3, and 4 has been divided by a set of parentheses into three sections. The section to the left of the parentheses corresponds to *Hantush's* [1960a] definition of small values of time (equation 4) from his β solution. Our results as obtained from equation 8 are identical with his β solution (given by equation 3) over these same time intervals.

The section to the right of the parentheses on each curve corresponds to Hantush's solution for large values of time (equation 7). Again our results as obtained from equation 8 are identical with those evaluated by *Hantush* [1960a] from equation 5.

The sections on each curve of Figures 2, 3, and 4 that are enclosed by parentheses therefore represent the time intervals for which *Hantush* [1960a] stated his asymptotic solutions should not be used. Our results indicate that his criteria for the validity of his solutions are on the conservative side and could be relaxed somewhat.

EVALUATING THE ASSUMPTION THAT STORAGE
IN THE AQUITARD MAY BE NEGLECTED

As indicated earlier *Hantush and Jacob* [1955b] have solved the above problem of flow in a leaky aquifer by assuming that the storage capacity of the aquitard can be neglected. Their solution for all values of time can be expressed as

$$s_1(r, t) = \frac{Q_1}{4\pi T_1} \cdot \int_{1/(4t_{D_1})}^{\infty} \exp\left[-y - \frac{(r/B_{11})^2}{4y}\right] \frac{dy}{y} \quad (9)$$

and has been referred to here as the r/B solution. One may note that if storage is to be neglected, then $\beta_{11} = 0$ in (6), and equations 5 and 9 are identical.

Anyone familiar with the literature on leaky aquifers will immediately recognize that the family of curves shown in Figure 2 for $\beta = 0.01$ is almost identical with that of the r/B solution. The only difference is that the envelope of curves instead of coinciding with the Theis

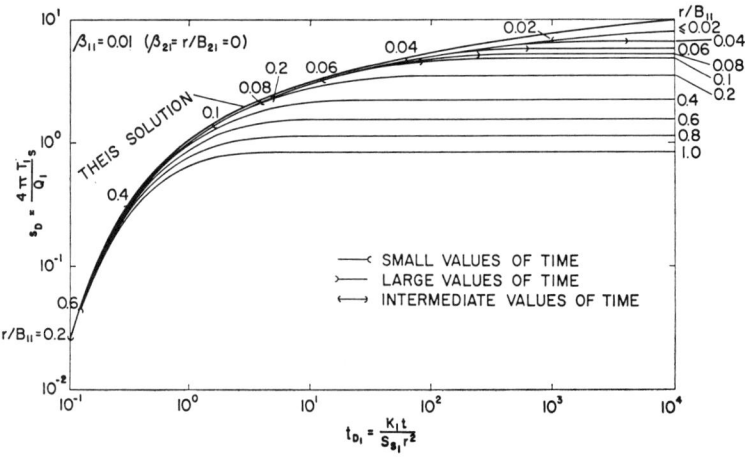

Fig. 2. Dimensionless drawdown in pumped aquifer when drawdown in unpumped aquifer is zero for $\beta_{11} = 0.01$.

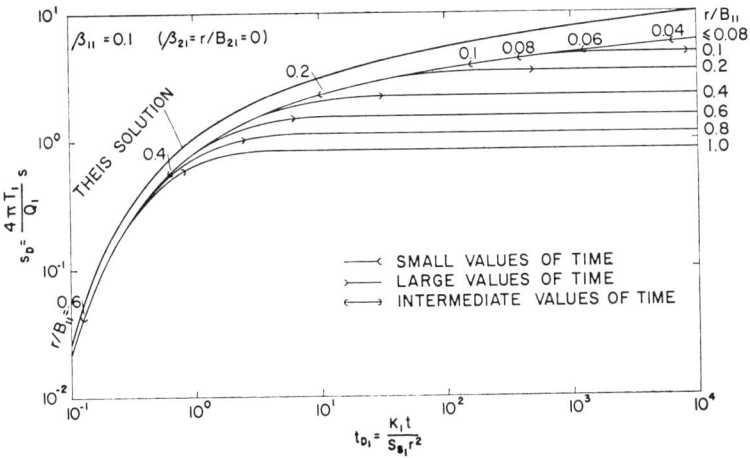

Fig. 3. Dimensionless drawdown in pumped aquifer when drawdown in unpumped aquifer is zero for $\beta_{11} = 0.1$.

solution slowly diverges from it as time increases.

Thus if one is analyzing field data at large values of time for a system with $\beta_{11} = 0.01$, ignoring the effects of storage in the aquiclude will only introduce a slight error as long as r/B_{11} is also small (i.e., $\lesssim 0.01$). From a practical standpoint, however, it would appear that neglecting storage in the aquitard should not affect the solution for the pumped aquifer as long as $\beta_{11} \lesssim 0.01$.

An examination of Figures 3 and 4 shows that as β_{11} increases, the r/B solution becomes less and less representative of the actual behavior in the pumped aquifer. The errors involved in the r/B solution become significant when β_{11} reaches 0.1 (Figure 3), and they are large when $\beta_{11} = 1.0$ (Figure 4).

The nature of these errors for large values of β_{11} can be better understood by superposing the r/B solution on the $\beta_{11} = 1.0$ solution of Figure 4 as shown in Figure 5. One sees immediately that the errors involved in the r/B solution increase as the magnitude of r/B_{11} decreases. However, from the convergence of the curves for the two solutions at large values of time, it is also apparent that these errors decrease with time and disappear altogether for those values of t_{D_1} given by (7).

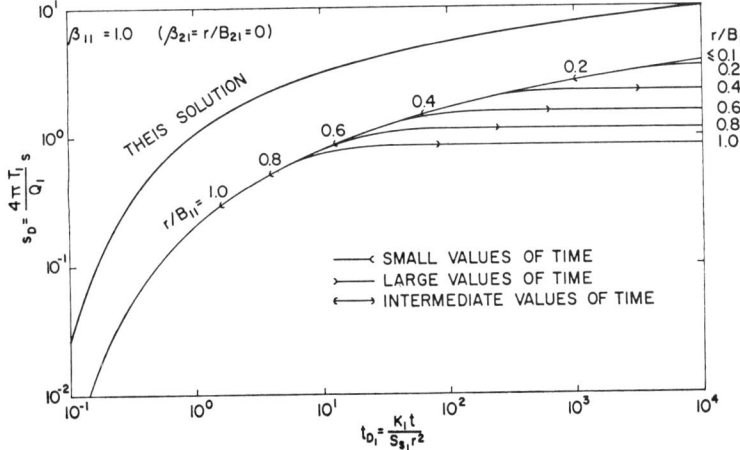

Fig. 4. Dimensionless drawdown in pumped aquifer when drawdown in unpumped aquifer is zero for $\beta_{11} = 1.0$.

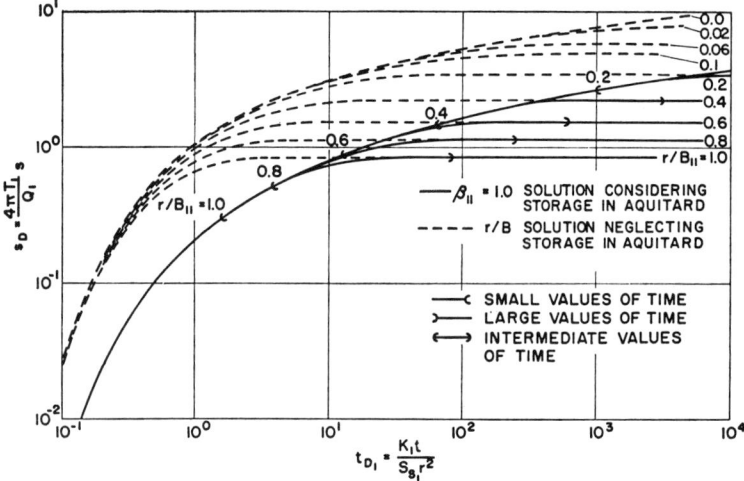

Fig. 5. Comparison of solution for pumped aquifer when $\beta_{11} = 1.0$ with r/B solution.

We therefore conclude that with large values of β_{11} and zero drawdown in the unpumped aquifer, the r/B solution is subject to significant errors whenever $t_{D_1} \leq 80\beta_{11}^2/(r/B_{11})^4$.

For the particular case of $\beta_{11} = 1.0$ as shown on Figure 5, this means that the entire non-steady state period of pumping cannot be analyzed using the r/B solution. In other words the r/B solution would only be applicable to the steady state after drawdowns have become constant. It should be kept in mind that the steady state regions on all the curves of Figures 2 through 5 are the result of the initial assumption that drawdown remains zero in the unpumped aquifer.

There is a simple physical explanation for the errors in the r/B solution when β_{11} is large. At small values of time the disturbance created by withdrawing water has not yet significantly affected the unpumped aquifer. Most of the early leakage is derived from the aquitard, and the amount depends on the specific storage of this part of the system, i.e., on the magnitude of β_{11}. At this stage, disregarding storage in the aquitard is equivalent to neglecting leakage altogether. Thus the resulting curves for the r/B solution fail to show the true effects of leakage at early time.

As time increases, more and more leakage is being contributed by the unpumped aquifer, and the relative amount of water that comes from storage in the aquitard diminishes. By the time steady state is reached and drawdowns are constant, all of the leakage is supplied by the unpumped aquifer. The aquitard merely acts as a conduit for flow from one aquifer to another. The storage capacity of the aquitard has no influence on the behavior of the system, and therefore the r/B solution is applicable.

The assumption that storage in the aquitard may be neglected obviously fails completely if one is interested in the transient behavior of the aquitard itself. When drawdown in the unpumped aquifer remains zero, a solution has been obtained for the aquitard [*Neuman and Witherspoon*, 1969a, 1969b] which can be written

$$s_1'(r, z, t) = \frac{Q_1}{4\pi T_1} \frac{4}{\pi} \sum_{n=1}^{\infty} \frac{1}{n} \sin \frac{n\pi z}{b_1'}$$

$$\cdot \int_0^\infty \left[e^{-n^2\pi^2 \bar{t}_{D_1}} - 1 - \frac{e^{-n^2\pi^2 \bar{t}_{D_1}} - e^{-y^2 \bar{t}_{D_1}}}{1 - y^2/(n^2\pi^2)} \right]$$

$$\cdot J_0[\omega(y)] \frac{dy}{y} \qquad (10)$$

where $\bar{t}_{D_1}$ and $\omega(y)$ are as defined in (8) and $J_0[\omega(y)]$ must be set to zero when $\omega^2(y) < 0$. Equation 10 has been evaluated for selected values of β_{11} and r/B_{11}, and the results demonstrate that s_1' does not become linear in z until the system has reached steady state (Figures 4-2 through 4-9, *Neuman and Witherspoon* [1969a]).

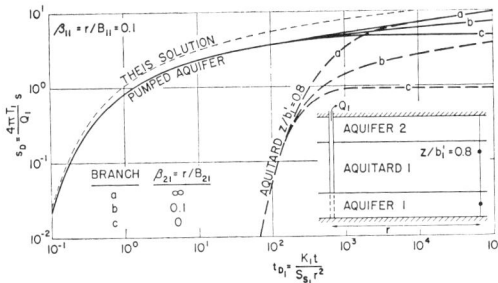

Fig. 6. Effect of unpumped aquifer on dimensionless drawdown in pumped aquifer and in aquitard at $z/b_1' = 0.8$.

EVALUATING THE ASSUMPTION THAT DRAWDOWN IN THE UNPUMPED AQUIFER MAY BE NEGLECTED

As mentioned earlier we have recently obtained a complete solution to the problem of flow in a two aquifer system for the case when drawdown in the unpumped aquifer is not necessarily zero [*Neuman and Witherspoon*, 1969a, 1969b]. The results must be expressed in terms of four parameters: (a) β_{11} and r/B_{11} with reference to the pumped aquifer and (b) β_{21} and r/B_{21} with reference to the unpumped aquifer. For the pumped aquifer the solution may be expressed as

$$s_1(r, t) = \frac{Q_1}{4\pi T_1} \int_0^\infty (1 - e^{-y^2 t_{D_1}})$$
$$\cdot \{[1 + G(y)]J_0[\omega_1(y)]$$
$$+ [1 - G(y)]J_0[\omega_2(y)]\} \frac{dy}{y} \quad (11)$$

For the aquitard the solution is

$$s_1'(r, z, t) = \frac{Q_1}{4\pi T_1} \frac{2}{\pi} \sum_{n=1}^\infty \frac{1}{n} \sin \frac{n\pi z}{b_1'}$$
$$\cdot \int_0^\infty \left[1 - e^{-n^2\pi^2 t_{D_1}} + \frac{e^{-n^2\pi^2 t_{D_1}} - e^{-y^2 t_{D_1}}}{1 - y^2/(n^2\pi^2)}\right]$$
$$\cdot \left\{\left[\frac{2(r/B_{21})^2(-1)^n y}{F(y) \sin y} - G(y) - 1\right]J_0[\omega_1(y)]\right.$$
$$- \left[\frac{2(r/B_{21})^2(-1)^n y}{F(y) \sin y} - G(y) + 1\right]$$
$$\left. \cdot J_0[\omega_2(y)]\right\} \frac{dy}{y} \quad (12)$$

and for the unpumped aquifer

$$s_2(r, t) = \frac{Q_1}{4\pi T_1} \int_0^\infty (1 - e^{-y^2 t_{D_1}}) \frac{2(r/B_{21})^2}{F(y)}$$
$$\cdot \{J_0[\omega_1(y)] - J_0[\omega_2(y)]\} \frac{dy}{\sin y} \quad (13)$$

where

$$G(y) = M(y)/F(y)$$
$$\omega_1^2(y) = \tfrac{1}{2}[N(y) + F(y)]$$
$$\omega_2^2(y) = \tfrac{1}{2}[N(y) - F(y)]$$
$$F^2(y) = M^2(y) + \left[\frac{2(r/B_{11})(r/B_{21})y}{\sin y}\right]^2$$
$$M(y) = \left[\frac{(r/B_{11})^4}{\beta_{11}^2} - \frac{(r/B_{21})^4}{\beta_{21}^2}\right] \frac{y^2}{16}$$
$$- \left[\left(\frac{r}{B_{11}}\right)^2 - \left(\frac{r}{B_{21}}\right)^2\right] y \cot y$$
$$N(y) = \left[\frac{(r/B_{11})^4}{\beta_{11}^2} + \frac{(r/B_{21})^4}{\beta_{21}^2}\right] \frac{y^2}{16}$$
$$- \left[\left(\frac{r}{B_{11}}\right)^2 + \left(\frac{r}{B_{21}}\right)^2\right] y \cot y$$

As in the case of (8), $J_0[\omega_1(y)]$ must be set to zero when $\omega_1^2(y) < 0$, and the same is true of $J_0[\omega_2(y)]$.

These solutions have been evaluated for selected values of the four controlling parameters, and some of the results for the particular case when $\beta_{11} = r/B_{11} = 0.1$ are shown in Figure 6. These results have also been checked by using the finite element method [*Javandel and Witherspoon*, 1968b; *Witherspoon et al.*, 1968].

As might be expected the results in the pumped aquifer are independent of β_{21} and r/B_{21} at small values of time. In general we found that the transient behavior of the pumped aquifer and the aquitard was not affected by conditions in the unpumped aquifer as long as equation 4 is satisfied. However, transient effects in the unpumped aquifer are dependent on β_{21} and r/B_{21} as well as on β_{11} and r/B_{11} at all values of time. Thus the assumption that drawdown in the unpumped aquifer may be neglected is valid at small values of time everywhere in the system except in the unpumped aquifer itself.

At larger values of time when $t_{D_1} > 1.6\beta_{11}^2/(r/B_{11})^4$, the behavior of the unpumped aquifer

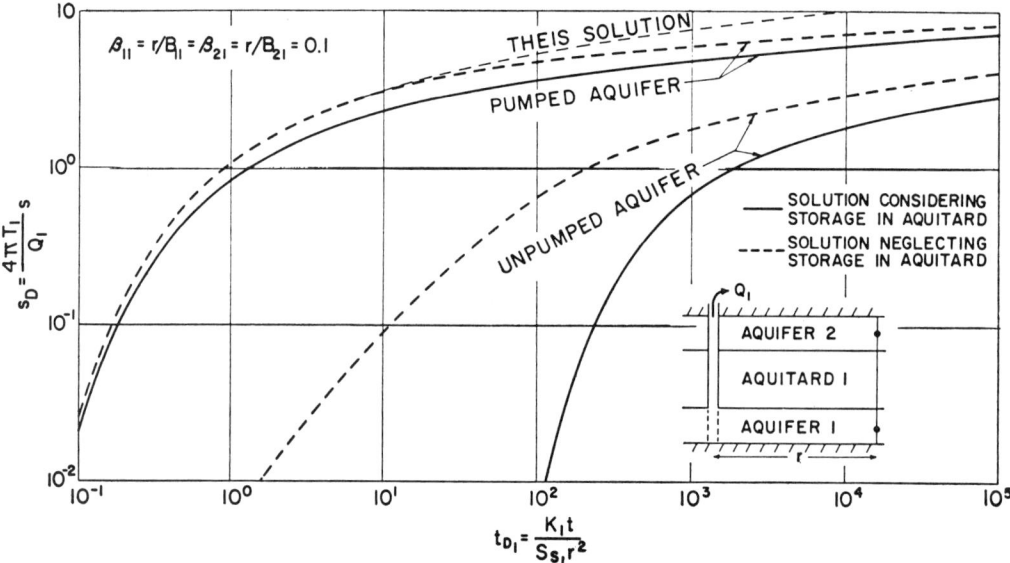

Fig. 7. Effect of neglecting storage in aquitard on dimensionless drawdown in pumped and unpumped aquifers (evaluated from equations 11, 13, 14, and 15).

may have a significant effect on drawdown in other parts of the system. For example, reference to Figure 6 will show that the results for the pumped aquifer that are uniquely defined for a given β_{11} and r/B_{11} at small values of time become a family of curves at large values of time, depending on the values of β_{21} and r/B_{21}.

A family with three branches is shown for the drawdown curve of the pumped aquifer in Figure 6. The lower branch corresponds to the special case previously discussed in connection with Figures 2-4, where it was assumed that there is no drawdown in the unpumped aquifer. A necessary condition for zero drawdown in the unpumped aquifer is that its transmissibility be infinitely large. Thus the lower branch of the curves for the pumped aquifer on Figure 6 corresponds to the special case when $T_2 = \infty$, which means $\beta_{21} = r/B_{21} = 0$.

When β_{21} and r/B_{21} are greater than zero, the assumption of no drawdown in the unpumped aquifer will no longer be valid. For example the middle branch of the drawdown curve for the pumped aquifer on Figure 6 represents another special case where the hydraulic properties of both aquifers are identical, i.e. $\beta_{11} = \beta_{21}$ and $r/B_{11} = r/B_{21}$. The upper branch represents the limiting case when $\beta_{21} = r/B_{21} = \infty$. This latter case implies that the permeability of the unpumped aquifer is zero ($K_2 = 0$) and means that a no-flow boundary exists at the interface between the aquiclude and the unpumped aquifer. This corresponds to *Hantush*'s [1960a] case 2 in his modified treatment of leaky aquifers. It is clear from Figure 6 that when the transmissibility of the unpumped aquifer is not infinitely large, as would usually be the case in the field, drawdown in the pumped aquifer does not reach steady state.

The assumption of zero drawdown in the unpumped aquifer can therefore lead to errors that will depend primarily on the magnitude of β_{11} and r/β_{11}. As is indicated on Figure 6, where $\beta_{11} = r/B_{11} = 0.1$, these errors increase with time as the ratios β_{21}/β_{11} and $(r/B_{21})/(r/B_{11})$ increase, i.e., as the transmissibility and storage capacity of the unpumped aquifer become less in relation to those of the pumped aquifer.

However, as the values of β_{11} and r/B_{11} decrease, the errors introduced by assuming zero drawdown in the unpumped aquifer will decrease for given ratios of T_2/T_1 and S_{s_2}/S_{s_1}. Theoretically these errors will not disappear completely unless the unpumped aquifer is replaced by a ponded body of water whose head remains constant. From a practical standpoint these errors can probably be neglected in analyzing drawdowns in the pumped aquifer when both β_{11} and $r/B_{11} < 0.01$ or when the ratios T_2/T_1 and S_{s_2}/S_{s_1} are sufficiently large. Our present data suggest that β_{21} and r/B_{21} should be about ten times smaller than β_{11} and r/B_{11}, respectively, in order for the errors to be negligible. This theory needs further investigation.

If we now turn our attention briefly to the aquitard, one would anticipate that the properties of the unpumped aquifer will have a much more profound effect on drawdown in the aquitard than we have just seen for the pumped aquifer. To illustrate this point we have included results on Figure 6 determined at only one location in the aquitard ($z/b_1' = 0.8$). Radial distance from the pumping well also becomes important, and these aquitard results are for a distance of more than twice the combined thicknesses of pumped aquifer and aquitard ($b_1 + b_1'$).

The aquitard results shown on Figure 6 for the limiting case of $\beta_{21} = r/B_{21} = \infty$ reveal an interesting result. It will be noted that dimensionless drawdown in the aquitard coincides with that of the pumped aquifer at large values of time. This means that in this limiting case of a no-flow boundary on one side of the aquitard, flow in all parts of the system away from the immediate vicinity of the pumping well becomes radial at large values of time. *Javandel and Witherspoon* [1968a] report the same effect has been noted for a well that partially penetrates a two-layered aquifer regardless of the permeability contrast.

Thus since it has been customary to assume vertical flow in the aquitard in developing solutions for leaky aquifers, the special case of a no-flow boundary presents a problem because the direction of flow is essentially vertical only at early time and then slowly changes to become radial at large values of time, especially as distance from the pumping well increases. *Hantush's* [1960a] asymptotic solution at large values of time for case 2 in his modified approach to leaky aquifers does not take this into consideration. However, in checking his results for the pumped aquifer, we found that our solution, which in this case was obtained by the finite element method, is in good agreement with his, indicating that direction of flow

in the aquitard at this stage does not affect the result in the pumped aquifer.

As mentioned earlier, *Hantush* [1967b] has also analyzed this two aquifer case and developed solutions for drawdown in both the pumped and the unpumped aquifer. However, he assumed that storage in the intervening aquitard could be neglected. It is therefore of interest to see how our results, which include the effects of storage in the aquitard, compare with his. If we choose $T_1 = T_2$(e.g., $r/B_{11} = r/B_{21}$), then Hantush's equations 17 and 18 for the special case of equal diffusivities in both aquifers ($K_1/S_{s_1} = K_2/S_{s_2}$) may be written

$$s_1(r, t) = \frac{Q_1}{8\pi T_1} \int_{1/(4t_{D_1})}^{\infty} \left\{ \exp(-y) + \exp\left[-y - \frac{2(r/B_{11})^2}{4y}\right] \right\} \frac{dy}{y} \quad (14)$$

$$s_2(r, t) = \frac{Q_1}{8\pi T_1} \int_{1/(4t_{D_1})}^{\infty} \left\{ \exp(-y) - \exp\left[-y - \frac{2(r/B_{11})^2}{4y}\right] \right\} \frac{dy}{y} \quad (15)$$

Figure 7 shows a comparison of results for the pumped and unpumped aquifers as obtained using our solutions (equations 11 and 13) and as obtained using Hantush's solutions (equations 14 and 15). In both cases we arbitrarily chose $\beta_{11} = \beta_{21} = r/B_{11} = r/B_{21} = 0.1$. One may note that in the pumped aquifer our results differ somewhat from those of Hantush. His solution lies along the Theis solution at early time and then diverges from it as time increases so as to lie between our solution and the Theis curve. If we had chosen the case for $\beta_{11} = \beta_{21} = 1.0$, these differences would be considerably greater.

It should be recalled that when we had zero drawdown in the unpumped aquifer, errors due to neglecting storage in the aquitard did not disappear until steady state was reached. Now we see from Figure 7 that when there is no steady state these errors occur at all values of time.

For the unpumped aquifer, however, it is seen that there is a large difference between our results and those of Hantush. His solution gives an earlier response and a much greater drawdown than ours. This of course is to be

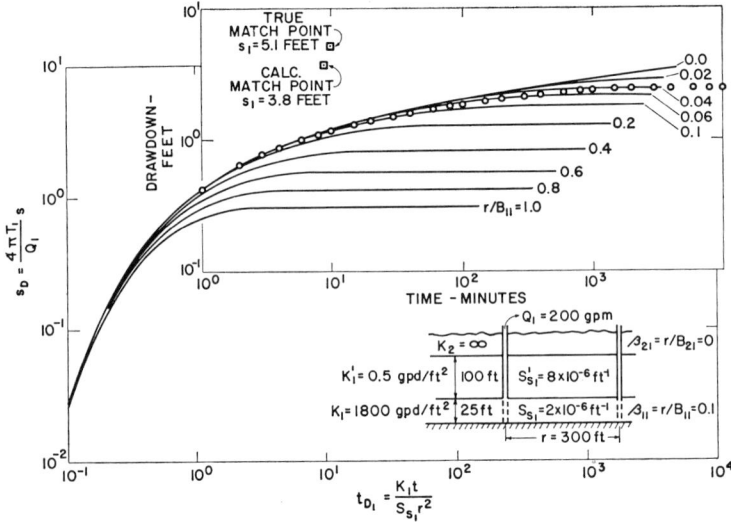

Fig. 8. Comparison of hypothetic field data with r/B solution when drawdown in unpumped aquifer is zero.

expected because in our system the aquitard contributes water from storage and therefore acts as a buffer between the two aquifers. As might be expected the differences between our two solutions for the unpumped aquifer will be even greater if $\beta_{11} = \beta_{21} = 1.0$.

This naturally raises the question whether the magnitude of the storage coefficient for an aquitard is so small that it can be neglected. This question is of course difficult to answer because of the general lack of data on the hydrologic characteristics of aquitards. We have found one comprehensive report containing a number of laboratory measurements on core samples from several different aquitards of Central California [*Johnson et al.*, 1965]. The materials range from sandy silts to clays. Permeabilities range from 10^{-6} to over 10^{-3} gpd/ft^2, and specific storage ranges from 3×10^{-6} to 5×10^{-4} ft^{-1}. In other words these data indicate that the specific storage of unconsolidated aquitard materials is at least as large as that of most confined aquifer sands and gravels, if not significantly greater. Under these circumstances it is not difficult to show that β may easily reach 0.1 and at times may exceed 1.0. If this is the case, our analysis indicates that, in general, storage in the aquitard must be taken into consideration when evaluating leaky systems.

ERRORS INTRODUCED BY USING r/B SOLUTION TO ANALYZE FIELD DATA

The following two examples of hypothetic field data will be used to illustrate the kind of errors that can arise when one uses the r/B solution to analyze the results of a pumping test.

For the first example we shall assume a two aquifer system (Figure 8) where $b_1 = b_2 = 25$ feet, $b_1' = 100$ feet, $K_1 = 1800$ gpd/ft^2 and $S_{s_1} = 2 \times 10^{-6}$ ft^{-1} for the aquifer being pumped, and $K_1' = 0.5$ gpd/ft^2 and $S_{s_1}' = 8 \times 10^{-6}$ ft^{-1} for the overlying aquitard. We shall further assume that $T_2 \gg T_1$ such that $\beta_{21} \approx r/B_{21} \approx 0$, which means that drawdown in the unpumped aquifer is negligible.

If an observation well is completed in aquifer 1 at a distance of $r = 300$ feet from the pumping well, the values of β_{11} and r/B_{11} become

$$\beta_{11} = \frac{r}{4b_1} \sqrt{\frac{K_1' S_{s_1}'}{K_1 S_{s_1}}}$$

$$= \frac{300}{(4)(25)} \sqrt{\frac{(0.5)(8 \times 10^{-6})}{(1800)(2 \times 10^{-6})}} = 0.1$$

$$\frac{r}{B_{11}} = r \sqrt{\frac{K_1'}{K_1 b_1 b_1'}}$$

$$= 300 \sqrt{\frac{0.5}{(1800)(25)(100)}} = 0.1$$

Under these conditions a pumping test is performed with $Q_1 = 200$ gpm, and the drawdown data at the observation well would of course follow the curve for $\beta_{21} = r/B_{21} = 0.0$ as shown on Figure 6. This fact, however, is unknown to the analyst, and in using the r/B solution he would obtain the best match with $r/B = 0.04$ (Figure 8). Using the indicated match point at $s_D = 10$, where $s_1 = 3.8$ feet, he would then obtain

$$K_1 = \frac{114.6 Q_1}{b_1} \frac{s_D}{s_1}$$

$$= \frac{(114.6)(200)(10)}{(25)(3.8)} = 2410 \text{ gpd/ft}^2$$

$$K_1' = \frac{(r/B_{11})^2 K_1 b_1 b_1'}{r^2}$$

$$= \frac{(0.04)^2(2410)(25)(100)}{(300)^2} = 0.11 \text{ gpd/ft}^2$$

Thus we see that in using the r/B solution for this situation the calculated permeability is too high for the aquifer ($\sim 33\%$) and almost five times too low for the aquitard. The calculated permeability contrast would be $K_1/K_1' = 2410/0.11 = 21{,}900$, whereas the true value is only 3600.

If a second observation well were available at $r = 3000$ feet, then drawdown data would follow the curve for $\beta_{11} = r/B_{11} = 1.0$ shown on Figure 4. If the analyst again used the r/B solution in the same manner as above, he would probably obtain the best match with $r/B = 0.3$. At the same match point of $s_D = 10$, where $s_1 = 1.5$ feet, he would then obtain

$$K_1 = \frac{(114.6)(200)(10)}{(25)(1.5)} = 6110 \text{ gpd/ft}^2$$

$$K_1' = \frac{(0.3)^2(6110)(25)(100)}{(3000)^2} = 0.15 \text{ gpd/ft}^2$$

From this second set of results we see that the calculated permeability is three times too high for the aquifer and three times too low for the aquitard. The calculated permeability contrast is worse than before, $K_1/K_1' = 6110/0.15 = 40{,}700$, or over ten times the true value. This illustrates how the errors increase as β_{11} and r/B_{11} increase. The analyst could be badly misled from the results of this analysis by (a) overpredicting the productivity of the aquifer, (b) underestimating the leakage contribution from the aquitard, and (c) concluding that the aquifer is radially inhomogeneous when the apparent increase in permeability with distance is due to a misinterpretation of the effects of leakage.

For the second example we shall use the same two aquifer system as before, except that now we shall assume that the hydraulic properties of both aquifers are identical, i.e., $K_1 = K_2 = 1800$ gpd/ft^2 and $S_{s_1} = S_{s_2} = 2 \times 10^{-6}$ft^{-1}. If the aquitard properties are the same as before, then an observation well at $r = 300$ feet would mean $\beta_{11} = \beta_{21} = r/B_{11} = r/B_{21} = 0.1$.

Under these conditions a pumping test is performed with $Q_1 = 200$ gpm, and the drawdown data would now follow the curve for $\beta_{21} = r/B_{21} = 0.1$ shown on Figure 6. This fact is again unknown to the analyst, and he might choose to match the field results with the r/B solution as indicated on Figure 9. If he does so, then at $s_D = 10$, $s_1 = 3.6$, and the permeability result for the aquifer would be $K_1 = 2540$ gpd/ft^2, or about 40% too high.

This result might be acceptable, but a more serious problem now arises in that one will have difficulty deciding what value of r/B to choose. The analyst who expects the drawdown data to flatten may be tempted to choose an r/B value at the end of the test period where he presumes the flattening is about to appear. However, the longer the test runs, the lower the value of r/B that will be selected, and the more the result will be in error.

Another alternative for the analyst is to ignore the early drawdown data and shift the plot of field data so as to force a match with the Theis solution. This would yield even larger errors in K_1 and also lead to the erroneous conclusion that there is no leakage.

CONCLUSIONS

It is our conclusion that present methods of analyzing field data from leaky aquifers need to be improved. These methods are based on two assumptions that we believe are not generally applicable. The assumption that storage in the aquitard is negligible can lead to significant errors when β of the pumped aquifer is greater than 0.01. It appears that large values

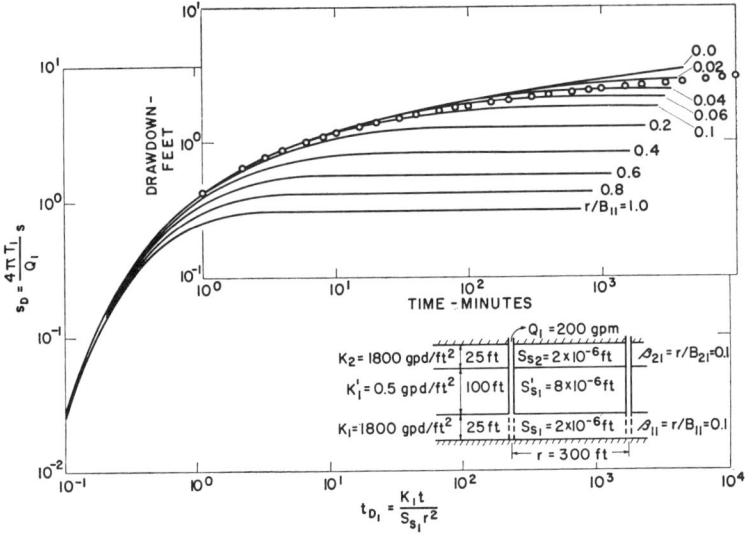

Fig. 9. Comparison of hypothetic field data with r/B solution when properties of aquifers are identical.

of β of the order of 0.1 to 1.0 are certainly possible in many field situations.

The second assumption of zero drawdown in the unpumped aquifer can also lead to significant errors at large values of time. These errors cannot be neglected unless the transmissibility of the unpumped aquifer is significantly greater than that of the pumped aquifer. Since the r/B solution relies on both of these assumptions, one must be cautious in using it to analyze field data.

Our entire discussion has been restricted to the case of a rather simple two aquifer system. Obviously the problems of evaluating field data are further complicated by the fact that leakage may occur not only from above but also from below. This will tend to restrict further the applicability of the above two assumptions.

It seems to us that relying entirely on drawdown data obtained from the pumped aquifer is not sufficient to characterize a leaky system. Our theory [*Neuman and Witherspoon*, 1969a, 1969b] suggests that by using drawdown data from both aquifers and aquitards one should be able to develop improved methods of analysis. We are currently attempting to develop such methods.

Acknowledgments. The authors would like to acknowledge with thanks the research funds provided by the State of California, Department of Water Resources, under standard agreements 756472 and 957669 in support of this work.

REFERENCES

Aravin, V. I., and S. N. Numerov, *Theory of Fluid Flow in Undeformable Porous Media*, 511 pp., Israel Program for Scientific Translations, Jerusalem, 1965.

De Glee, G. J., *Over Grondwaterstroomingen bij Wateronttrekking door Middel van Putten*, J. Waltman, Jr., Delft, 1930.

De Wiest, R. J. M., On the theory of leaky aquifers, *J. Geophys. Res.*, 66, 4257, 1961.

De Wiest, R. J. M., Flow to an eccentric well in a leaky circular aquifer with varied lateral replenishment, *Geofis. Pura Appl.*, 54, 87, 1963.

De Wiest, R. J. M., *Geohydrology*, 366 pp., John Wiley, New York, 1965.

Ferris, J. G., D. B. Knowles, R. H. Brown, and R. W. Stallman, Theory of aquifer, tests; a summary of lectures, *U. S. Geol. Surv. Water Supply Paper 1536-E*, 1962.

Girinsky, N. K., Nekotorye voprosy dinamiki podzemnykh vod, in *Gidrogeologiya i Inzhenernaya Geologiya, Sbornik Statei*, 9, Moscow, 1947.

Glebov, P. D., Pritok infiltratsionnoi vody k kolodtsam pri gorizontalnom zaleganii gruntov razlichnoi vodopronitsaemosti, *Tr. Leningr. Ind. Inst.*, 1, 1940.

Hantush, M. S., Plain potential flow of groundwater with linear leakage, Ph. D. dissertation, University of Utah, Salt Lake City, 1949.

Hantush, M. S., Analysis of data from pumping tests in leaky aquifers, *Trans. Amer. Geophys. Union*, 37, 702, 1956.

Hantush, M. S., Nonsteady flow to a well partially penetrating an infinite leaky aquifer, *Proc. Iraqi Sci. Soc., 1,* 10, 1957.

Hantush, M. S., Nonsteady flow to flowing wells in leaky aquifers, *J. Geophys. Res., 64,* 1043, 1959.

Hantush, M. S., Modification of the theory of leaky aquifers, *J. Geophys. Res., 65,* 3713, 1960a.

Hantush, M. S., *Tables of the Function $H(u,\beta)$*, Document 6427, U. S. Library of Congress, Washington, D. C., 1960b.

Hantush, M. S., Drainage wells in leaky water-table aquifers, *Proc. Amer. Soc. Civil Engrs., 88*(HY2), 123, 1962.

Hantush, M. S., Depletion of storage, leakage, and river flow by gravity wells in sloping sands, *J. Geophys. Res., 69,* 2551, 1964.

Hantush, M. S., Flow of groundwater in relatively thick leaky aquifers, *Water Resour. Res., 3*(2), 583, 1967a.

Hantush, M. S., Flow to wells in aquifers separated by a semipervious layer, *J. Geophys. Res., 72*(6), 1709, 1967b.

Hantush, M. S., and C. E. Jacob, Plane potential flow of ground water with linear leakage, *Trans. Amer. Geophys. Union, 35,* 917, 1954.

Hantush, M. S., and C. E. Jacob, Nonsteady Green's functions for an infinite strip of leaky aquifers, *Trans. Amer. Geophys. Union, 36,* 101, 1955a.

Hantush, M. S., and C. E. Jacob, Nonsteady radial flow in an infinite leaky aquifer, *Trans. Amer. Geophys. Union, 36,* 95, 1955b.

Hantush, M. S., and C. E. Jacob, Flow to an eccentric well in a leaky circular aquifer, *J. Geophys. Res., 65,* 3425, 1960.

Jacob, C. E., Radial flow in a leaky artesian aquifer, *Trans. Amer. Geophys. Union, 27,* 198, 1946.

Javandel, I., and P. A. Witherspoon, Analysis of transient fluid flow in multi-layered systems, *Water Resour. Center Contrib. 124*, University of California, Berkeley, 1968a.

Javandel, I., and P. A. Witherspoon, Application of the finite element method to transient flow in porous media, *Soc. Petrol. Eng. J., 8*(3), 241, September 1968b.

Johnson, A. I., R. P. Moston, and D. A. Morris, *Physical and Hydrologic Properties of Water-Bearing Deposits in Subsiding Areas in Central California*, U. S. Geological Survey Open-File Report, Denver, Colorado, 1965.

Myatiev, A. N., Deistvie kolodtsa v napornom basseine podzemnykh vod, *Izv. Turkm. Filiala AN SSSR, 3–4,* 1946.

Myatiev, A. N., Napornyi kompleks podzemnykh vod i kolodtsy, *Izv. AN SSSR, Otd. Tekhn. Nauk, 9,* 1947.

Narasimhan, T. N., Ratio method for determining characteristics of ideal, leaky and bounded aquifers, *Bull. Int. Ass. Sci. Hydrol., 13*(1), 70, 1968.

Neuman, S. P., and P. A. Witherspoon, Theory of flow in aquicludes adjacent to slightly leaky aquifers, *Water Resour. Res., 4*(1) 103, 1968.

Neuman, S. P., and P. A. Witherspoon, *Transient Flow of Ground Water to Wells in Multiple-Aquifer Systems*, Geotechnical Engineering Report, University of California, Berkeley, January 1969a.

Neuman, S. P., and P. A. Witherspoon, Theory of flow in a two aquifer system, *Water Resour. Res., 5*(4), 1969b.

Polubarinova-Kochina, P. Ya., *The Theory of Ground Water Movement*, 613 pp., Princeton University Press, New Jersey, 1962.

Slater, R. J., Application and limitations of pumping tests, *Inst. Water Engrs. J., 17*(3), 189, 1963.

Steggeventz, J. H., and B. A. Van Ness, Calculating the yield of a well taking account of replenishment of the groundwater from above, *Water & Water Eng., 41,* 561, 1939.

Walton, W. C., Leaky artesian aquifer conditions in Illinois, 27 pp., *Illinois State Water Surv. Rep. Invest. 39*, Urbana, Illinois, 1960.

Walton, W. C., Selected analytical methods for well and aquifer evaluation, 81 pp., *Illinois State Water Surv. Bull. 49*, Urbana, Illinois, 1962.

Witherspoon, P. A., I. Javandel, and S. P. Neuman, Use of the finite element method in solving transient flow problems in aquifer systems, in *The Use of Analog and Digital Computers in Hydrology,* AIHS Publication 81, 2, 687, 1968.

(Manuscript received March 7, 1969.)

Reprinted from pages 69 and 144-153 of *U.S. Geol. Survey Water-Supply Paper 1536-E,* 1962, pp. 69-174.

THEORY OF AQUIFER TESTS

By J. G. Ferris, D. B. Knowles, R. H. Brown, and R. W. Stallman

ABSTRACT

The development of water supplies from wells was placed on a rational basis with Darcy's development of the law governing the movement of fluids through sands and with Dupuit's application of that law to the problem of radial flow toward a pumped well. As field experience increased, confidence in the applicability of quantitative methods was gained and interest in developing solutions for more complex hydrologic problems was stimulated. An important milestone was Theis' development in 1935 of a solution for the nonsteady flow of ground water, which enabled hydrologists for the first time to predict future changes in ground-water levels resulting from pumping or recharging of wells. In the quarter century since, quantitative ground-water hydrology has been enlarging so rapidly as to discourage the preparation of comprehensive textbooks.

This report surveys developments in fluid mechanics that apply to ground-water hydrology. It emphasizes concepts and principles, and the delineation of limits of applicability of mathematical models for analysis of flow systems in the field. It stresses the importance of the geologic variable and its role in governing the flow regimen.

The report discusses the origin, occurrence, and motion of underground water in relation to the development of terminology and analytic expressions for selected flow systems. It describes the underlying assumptions necessary for mathematical treatment of these flow systems, with particular reference to the way in which the assumptions limit the validity of the treatment.

INTRODUCTION

Lectures on ground-water hydraulics by John G. Ferris provide most of the source material for this paper, which was organized by Doyle B. Knowles. Subsequent refinements of concepts and standardization of nomenclature and method of presentation were accomplished by Russell H. Brown and Robert W. Stallman, with the important collaboration of Edwin W. Reed. Appropriate individual authorship is recognized for several sections of the text.

The material presented herewith concerns the theory supporting many hydraulic concepts. Applications of the theory to field problems are to be shown in another report.

[*Editors' Note:* Material has been omitted at this point.]

THEORY OF IMAGES AND HYDROLOGIC BOUNDARY ANALYSIS

The development of the equilibrium and nonequilibrium formulas discussed in the preceding sections was predicated in part on the assumption of infinite areal extent of the aquifer, although it is recognized that few if any aquifers completely satisfy this assumption. In many instances the existence of boundaries serves to limit the continuity of the aquifer, in one or more directions, to distances ranging from a few hundred feet to as much as tens of miles. Thus when an aquifer is recognized as having finite dimensions, direct analysis of the test data by the equations previously given is often precluded. It is often possible, however, to circumvent the analytical difficulties posed by the aquifer boundary. The method of images, widely used in the theory of heat conduction in solids, provides a convenient tool for the solution of boundary problems in ground-water flow. Imaginary wells or streams, usually referred to as images, can sometimes be used at strategic locations to duplicate hydraulically the effects on the flow regime caused by the known physical boundary. Use of the image thus is equivalent to removing a physical entity and substituting a hydraulic entity. The finite flow system is thereby transformed by substitution into one involving an aquifer of infinite areal extent, in which several real and imaginary wells or streams can be studied by means of the formulas already given. Such substitution often results in simplifying the problem of analysis to one of adding effects of imaginary and real hydraulic systems in an infinite aquifer.

An aquifer boundary formed by an impermeable barrier, such as a tight fault or the impermeable wall of a buried stream valley that cuts off or prevents ground-water flow, is sometimes termed a "negative boundary." Use of this term is discouraged, however, in favor of the more meaningful and descriptive term "impermeable barrier." A line at or along which the water levels in the aquifer are controlled by a surface body of water such as a stream, or by an adjacent segment of aquifer having a comparatively large transmissibility or water-storage capacity, is sometimes termed a "positive boundary." Again, however, use of the term is discouraged in favor of the more precise terms line source or line sink, as may be appropriate.

Although most geologic boundaries do not occur as abrupt discontinuities, it is often possible to treat them as such. When conditions permit this practical idealization, it is convenient for the purpose of analysis to substitute a hypothetical image system for the boundary conditions of the real system.

In this section, where the analysis of pumping-test data is considered, several examples are given of image systems required to duplicate, hydraulically, the boundaries of certain types of areally restricted aquifers. It should be apparent that similar methods can be used to analyze flow to streams or drains through areally limited aquifers.

PERENNIAL STREAM—LINE SOURCE AT CONSTANT HEAD

An idealized section through a discharging well in an aquifer hydraulically controlled by a perennial stream is shown in figure 35A. For thin aquifers the effects of vertical-flow components are small at relatively short distances from the stream, and if the stream stage is not lowered by the flow to the real well there is established the boundary condition that there shall be no drawdown along the stream position. Therefore, for most field situations it can be assumed for practical purposes that the stream is fully penetrating and equivalent

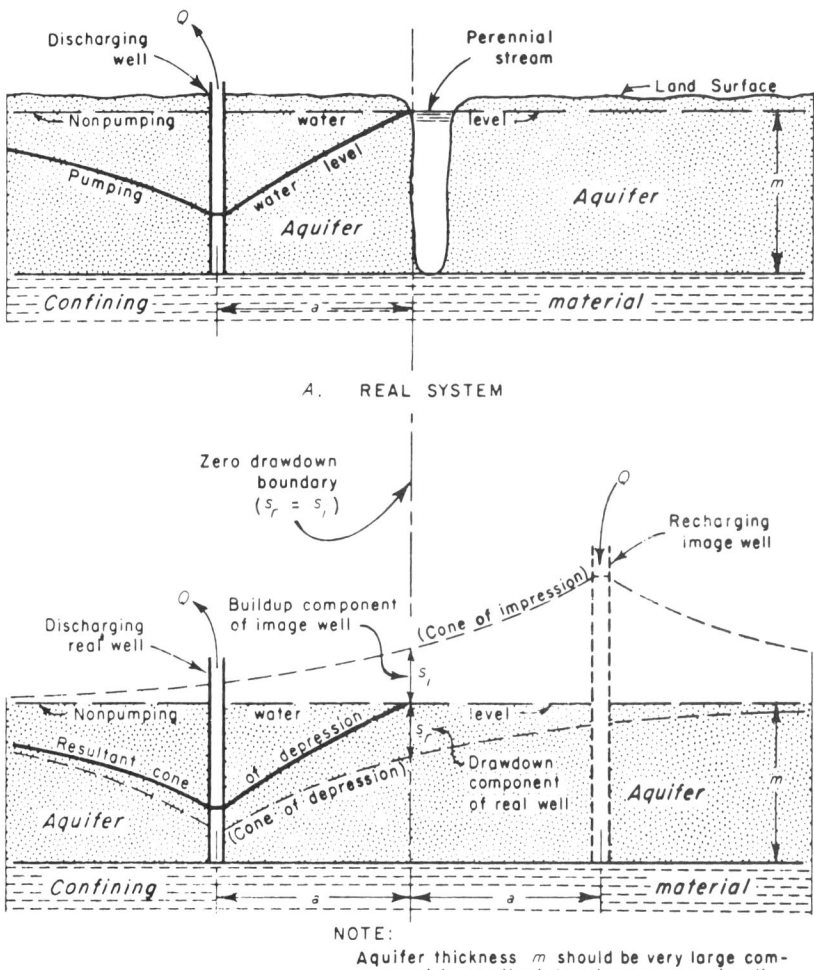

FIGURE 35.—Idealized section views of a discharging well in a semi-infinite aquifer bounded by a perennial stream, and of the equivalent hydraulic system in an infinite aquifer.

to a line source at constant head. An image system that satisfies the foregoing boundary condition, as shown in figure 35B, allows a solution of the real problem through use, in this example, of the Theis nonequilibrium formula. Note in figure 35B that an imaginary recharging well has been placed at the same distance as the real well from the line source but on the opposite side. Both wells are situated on a common line perpendicular to the line source. The imaginary recharge well operates simultaneously with the real well and returns water to the aquifer at the same rate that it is withdrawn by the real well. It can be seen that this image well produces a buildup of head everywhere along the position of the line source that is equal to and cancels the drawdown caused by the real well which satisfies the boundary condition of the problem. The resultant drawdown at any point on the cone of depression in the real region is the algebraic sum of the drawdown caused by the real well and the buildup produced by its image. The resultant profile of the cone of depression, shown in figure 19B, is flatter on the landward side of the well and steeper on the riverward side, as compared with the shape it would have if no boundary were present. Figure 36 is a generalized plan view of a flow net for the situation given in figure 35A. The distribution of stream lines and potential lines about the real discharging well and its recharging image, in an infinite aquifer, is shown. If the image region is omitted, the figure represents the stream lines and potential lines as they might be observed in the vicinity of a discharging well obtaining water from a river by induced infiltration.

IMPERMEABLE BARRIER

An idealized section through a discharging well in an aquifer bounded on one side by an impermeable barrier is shown in figure 37A. It is assumed that the irregularly sloping boundary can, for practical purposes, be replaced by a vertical boundary, occupying the position shown by the vertical dashed line, without sensibly changing the nature of the problem. The hydraulic condition imposed by the veritcal boundary is that there can be no ground-water flow across it, for the impermeable material cannot contribute water to the pumped well. The image system that satisfies this condition and permits a solution of the real problem by the Theis equation is shown in figure 37B. An imaginary discharging well has been placed at the same distance as the real well from the boundary but on the opposite side, and both wells are on a common line perpendicular to the boundary. At the boundary the drawdown produced by the image well is equal to the drawdown caused by the real well. Evidently, therefore, the drawdown cones for the real and the image wells will be symmetrical and will produce a ground-water divide at every point along the boundary line. Because there can be no flow

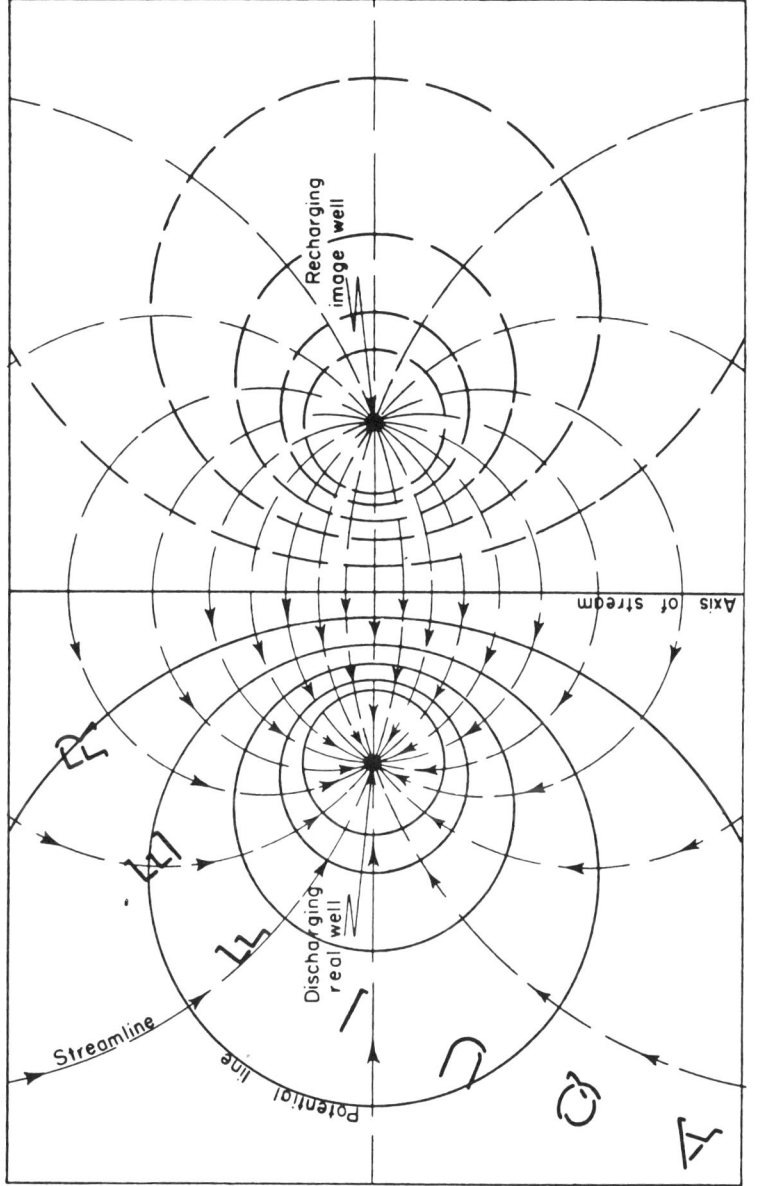

FIGURE 36.—Generalized flow net showing stream lines and potential lines in the vicinity of a discharging well dependent upon induced infiltration from a nearby stream.

THEORY OF AQUIFER TESTS

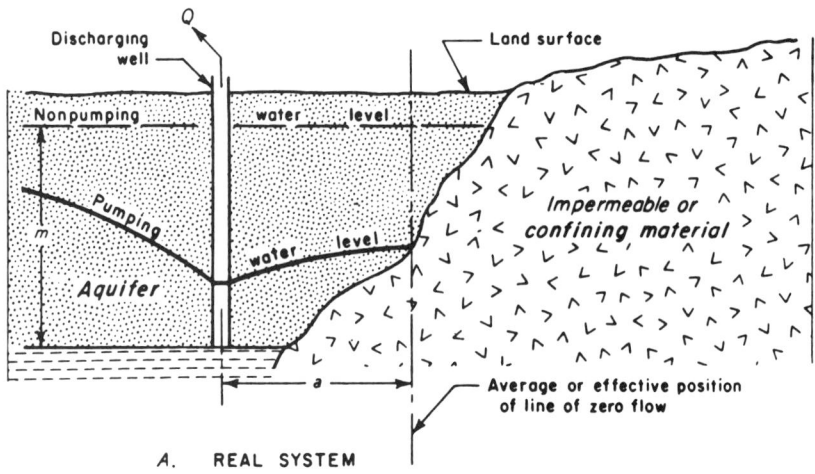

A. REAL SYSTEM

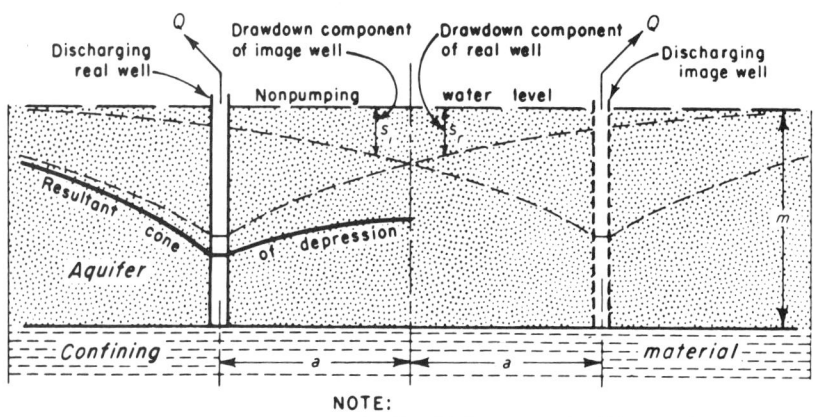

NOTE:
Aquifer thickness m should be very large compared to resultant drawdown near real well

B. HYDRAULIC COUNTERPART OF REAL SYSTEM

FIGURE 37.—Idealized section views of a discharging well in a semi-infinite aquifer bounded by an impermeable formation, and of the equivalent hydraulic system in an infinite aquifer.

across a divide, the image system satisfies the boundary condition of the real problem and analysis is simplified to consideration of two discharging wells in an infinite aquifer. The resultant drawdown at any point on the cone of depression in the real region is the algebraic sum of the drawdowns produced at that point by the real well and its image. The resultant profile of the cone of depression, shown in figure 37B, is flatter on the side of the well toward the boundary and steeper on the opposite side away from the boundary than it would be if no boundary were present. Figure 38 is a general-

150 GROUND-WATER HYDRAULICS

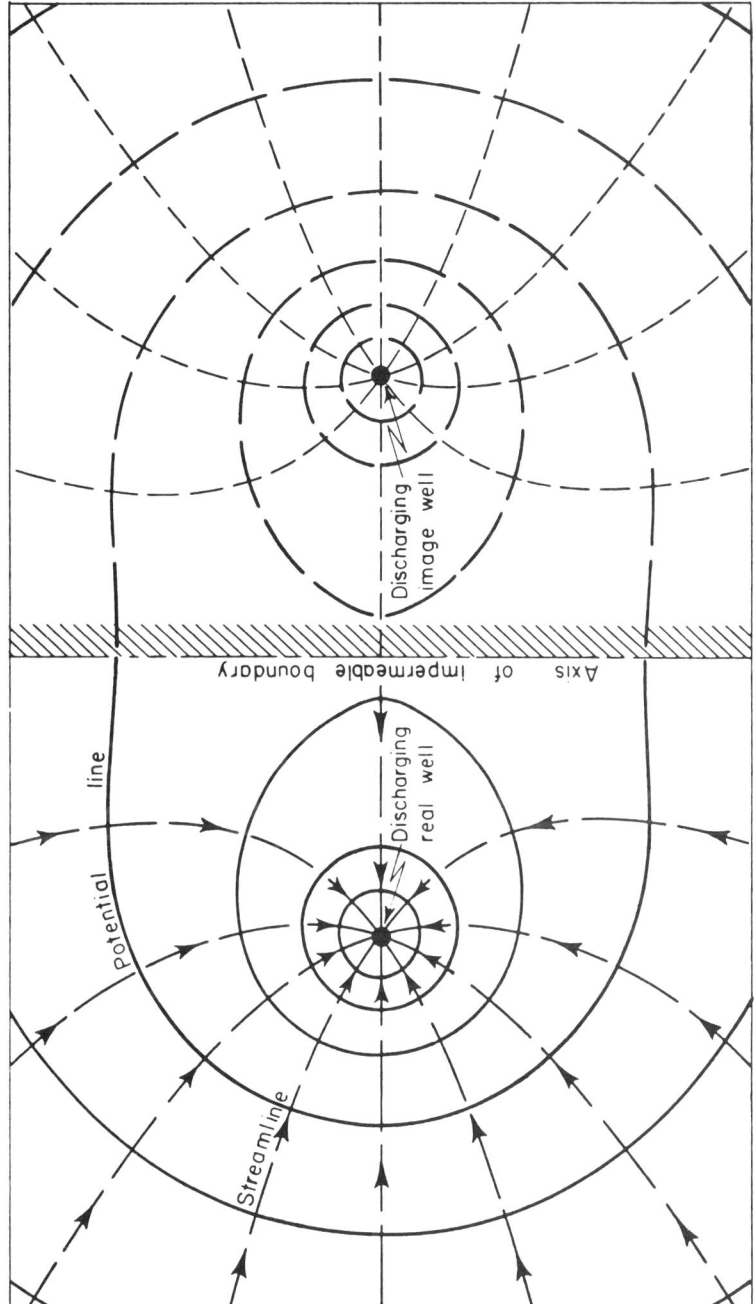

FIGURE 38.—Generalized flow net showing stream lines and potential lines in the vicinity of a discharging well near an impermeable boundary.

THEORY OF AQUIFER TESTS

ized plan view of a flow net for the situation given in figure 37A. The distribution of stream lines and potential lines about the real discharging well and its discharging image, in an infinite aquifer, is shown. If the image region is omitted, the diagram represents the flow net as it might be observed in the vicinity of a discharging well located near an impermeable boundary.

TWO IMPERMEABLE BARRIERS INTERSECTING AT RIGHT ANGLES

The image-well system for a discharging well in an aquifer bounded on two sides by impermeable barriers that intersect at right angles is shown in figure 39. Although the drawdown effects of the primary image wells, I_1 and I_2, combine in the desired manner with the effect

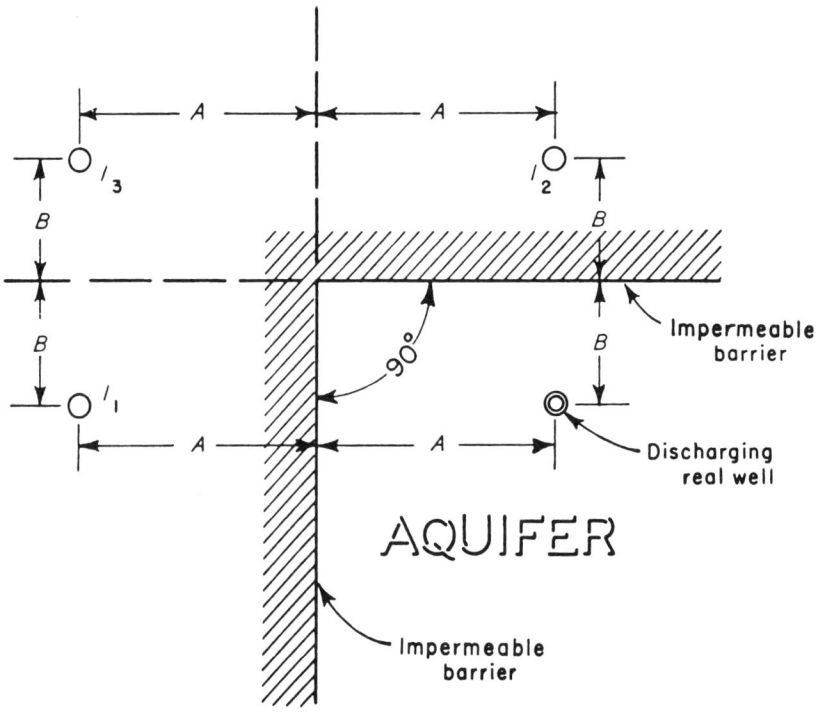

NOTES:

Image wells, I, are numbered in the sequence in which they were considered and located

Open circles signify discharging wells

FIGURE 39.—Plan of image-well system for a discharging well in an aquifer bounded by two impermeable barriers intersecting at right angles.

of the real well at their respective boundaries, each image well produces an unbalanced drawdown at the extension (reflection) of the other boundary. These unbalanced drawdowns at the boundaries produce a hydraulic gradient, with consequent flow across the extension of each boundary, and therefore do not completely satisfy the requirement of no flow across the boundaries of the real system. It is necessary, therefore, to use a secondary image well, I_3, which balances the residual effects of the two primary image wells at the two extensions of the boundaries. The image system is then hydraulically in complete accord with the physical boundary conditions. The problem thereby has been simplified to consideration of four discharging wells in an infinite aquifer.

IMPERMEABLE BARRIER AND PERENNIAL STREAM INTERSECTING AT RIGHT ANGLES

The image-well system for a discharging well in an aquifer bounded on two sides by an impermeable barrier and a perennial stream which intersect at right angles is shown in figure 40. The perennial stream of figure 40 might also represent a canal, drain, lake, sea, or any other line source of recharge sufficient to maintain a constant head at this boundary. As before, the drawdown effects of the primary images, I_1 and I_2, combine in the desired manner with the effects of the real well at their respective boundaries. However, discharging image well I_1 produces a drawdown at the extension of the line source, which is a no-drawdown boundary, and recharging image well I_2 causes flow across the extension of the impermeable barrier, which is a no-flow boundary. By placing a secondary recharging image well, I_3, at the appropriate distance from the extension of each boundary, the system is balanced so that no flow occurs across the impermeable barrier and no drawdown occurs at the perennial stream. Thus again the problem has been simplified to consideration of an infinite aquifer in which there operate simultaneously two discharging and two recharging wells.

The simplest way to analyze any multiple-boundary problem is to consider each boundary separately and determine how best to meet the condition of no flow or no drawdown, as the case may be, at that boundary. After the positions of the primary image wells have been established, the boundary positions should be reexamined to see if the net drawdown effects of the primary image wells satisfy all stipulated conditions of no flow or no drawdown. For each primary image causing an unbalance at a boundary position, or extension thereof, it is necessary to place a secondary image well at the same distance from the boundary but on the opposite side, both wells occupying a common line perpendicular to the boundary. When the combined drawdown (or buildup) effects of all image wells are found to produce the desired effect at this boundary the same procedure is executed with

THEORY OF AQUIFER TESTS

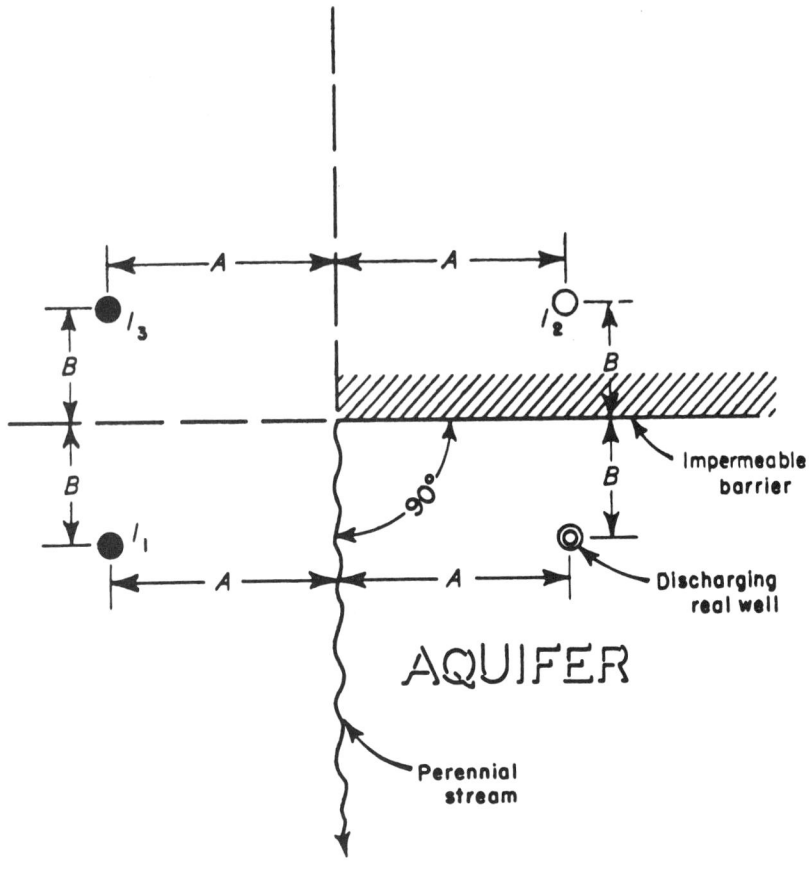

NOTES:

Image wells, I, are numbered in the sequence in which they were considered and located

Open circles signify discharging wells

Filled circles signify recharging wells

FIGURE 40.—Plan of image-well system for a discharging well in an aquifer bounded by an impermeable barrier intersected at right angles by a perennial stream.

respect to the second boundary. Thus, the inspection and balancing process is repeated around the system until everything is in balance and all boundary conditions are satisfied, or until the effects of additional image wells are negligible compared to the total effect.

[*Editors' Note:* Material has been omitted at this point.]

USE OF NUMERICAL METHODS
FOR ANALYZING DATA ON GROUND WATER LEVELS [1]

by

R. W. STALLMAN [2]

Abstract

In analyzing water-level data collected over a large area, particular solutions of the basic differential equations of ground-water flow become impractical if the boundary conditions or aquifer characteristics are complex functions of space and time. The analysis becomes practicable, however, if finite-difference approximations or general solutions of the differential equations are employed. Methods of computing the hydrologic properties of an aquifer, using water-level data collected at widely spaced points, are given for both homogeneous and nonhomogeneous aquifers. The methods are based on simultaneous solution of a set of equations constructed from observed data. Although both direct and least-squares solutions can be used, the least-squares methods result in more reasonable findings, probably because vagaries in the data available are resolved to mean values. The analytical methods described should prove valuable in determining the regional ground-water hydrology from water-level data.

Introduction

The academic foundation of the engineering techniques used in modern studies of ground-water hydrology was developed by Darcy and Boussinesq. Their work was done far in advance of today's comparatively general need for engineering estimates related to the availability of ground water. In the wake of their published work, others have applied their basic concepts to build the wealth of quantitative analytical methods now used by hydrologists around the world. Fundamentally these methods are all founded on a common approach in which selected analytical equations are used either to compute the hydrologic properties of the water-bearing formation or to predict the relationship of head versus discharge. These data are valuable for describing the water-bearing potentialities of a formation in quantitative terms so that it may be compared with another formation, and are also an aid to the location and economical design of well fields or other systems for pumping or recharging ground water.

Most of the analytical equations commonly used for determining hydrologic constants are solutions to relatively simple differential equations and account for equally simple boundary conditions. In applying the analytical equations to field problems, the inevitably more complex boundary conditions evident from studies of the areal geology must be idealized so as to fit the comparatively elementary geometric forms assumed in deriving the analytical expressions. For many problems, this approach yields usable estimates of the hydrologic constants in the vicinity of test facilities, such as pumping wells and surface streams. However, in this approach, quantitative study of the hydrologic properties of the water-bearing formation is limited by the availability of suitable testing facilities. Because testing consists in part of observing water-level changes, the test facilities, to be of use in quantitative studies, must generally be of such size that measurable changes can be induced over a rather large area in a reasonably short time. Unfortunately, in most areas of interest, facilities meeting these qualifications are either concentrated in a small segment of the formation, or do not exist. Construction of the required facilities for testing purposes usually is not economically feasible. In order to obtain quantitative data on the hydrologic constants over an entire formation, it appears evident that this approach must be augmented by others which will (1) sample larger areas; (2) provide more detail on the nonhomogeneous formations; (3) apply to more general conditions than are covered by the analytical equations now used;

---

[1] Publication authorized by Director, U. S. Geological Survey.
[2] Hydraulic Engineer, U. S. Geological Survey, Washington, D. C.

228 R. W. Stallman

(4) utilize data easily obtained at low cost and already abundantly available for many areas and, (5) employ analytical processes that can be reduced to routine computations.

Many of the limitations on present quantitative techniques seem to result directly or indirectly from the difficulty of finding analytical solutions to the differential equations describing the more complex ground-water flow systems. Thus it appears likely that some of the restrictions on solving the more complex problems might be reduced or eliminated by devising methods of analysis that employ the differential equations in such a way that quantitative results can be obtained without recourse to the presently accepted rigorous solutions.

The finite-difference equation as an analytical expression

In ground-water hydrology, differential equations are used for describing the continuity of water movement and for relating head changes to the hydrologic constants of the water-bearing formation. For example, the equation

$$mP\,(\partial^2 h/\partial x^2 + \partial^2 h/\partial y^2) = -W \tag{1}$$

(where m is the saturated thickness of the water-bearing formation, P is the coefficient of permeability, W is the rate of recharge, and h is head) simply states that the net change in the rate at which water moves past point x, y equals the rate of recharge at that point. By inference, equation (1) applies only for the equilibrium state, and only if the thickness m is constant at x, y. Equation (1) can be rewritten as

$$\partial^2 h/\partial x^2 + \partial^2 h/\partial^2 y^2 = -W/mP \tag{2}$$

in which the hydrologic constants are grouped more clearly to show their relation to the configuration of the piezometric surface. It is evident that the W/mP term can be evaluated simply from observations on the piezometric surface, and that, in addition, one measurement of velocity at x, y should suffice for estimating P.

Although the numerical value of the left side of equation (1) cannot be obtained from field measurements at a point on the piezometric surface, the mechanism for estimating its value from observations at several points has been described by Southwell (1946). In order to reconcile Southwell's approach with the mathematical concept of a point, one need only visualize a point as a small area of length and width such that it is insignificant in size compared with the size of the flow field. Thus the mathematical concept of a point might be revised

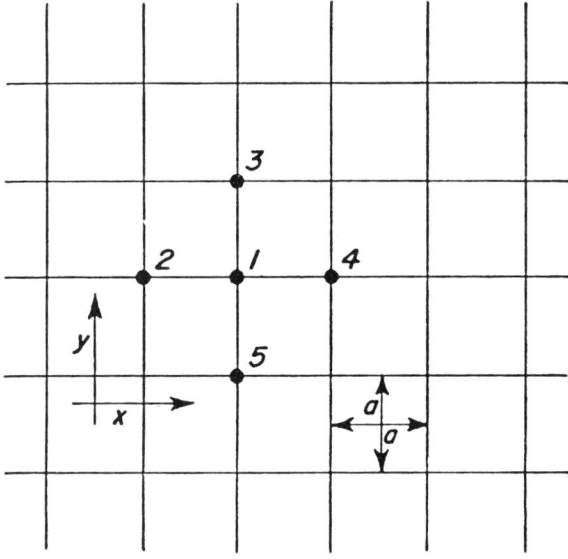

Fig. 1
Finite-difference grid on flow field.

Numerical Methods for Analyzing Data 229

to include finite-dimensional size. Accepting this revision as an approximation leads directly to conversion of the differential equation to a finite-difference equation.

As an example, note the rectangular grid over a portion of the flow field as shown on figure 1. The subdivisions are equally spaced in both directions. It is assumed that the head values are known at nodes 1 to 5. From Southwell (1946, p. 19):

$$\partial^2 h/\partial x^2 \approx \frac{h_2 + h_4 - 2h_1}{a^2} \quad (3)$$

and

$$\partial^2 h/\partial y^2 \approx \frac{h_3 + h_5 - 2h_1}{a^2} \quad (3a)$$

Equations (3) and (3a) may be substituted in equation (2) to form the finite-difference equation:

$$h_2 + h_3 + h_4 + h_5 - 4h_1 \approx \frac{-a^2 W}{mP} \quad (4)$$

Southwell's interest was focused on determining, in this type of problem, the head distribution within a given area for given boundary conditions, where the term a^2W/mP is known. To some extent this approach parallels the procedure of applying analytical expressions to find h as a function of the space coordinates when W/mP is known. Conversely, values of the hydrologic coefficient W/mP can be found by the relaxation or numerical methods described by Southwell, using finite-difference expressions such as equation (4), if the head distribution is known. For complex systems, admittedly long and laborious computations are required to find the relationship among head, space, and the hydraulic characteristics of a formation. However, the aim of most basic hydrologic studies is to find the value of W/mP or similar hydrologic constants. The finite-difference equations are well suited for this purpose, because the only processing of data required is the evaluation of head differentials. If the data are taken from wells spaced on the regular pattern shown on figure 1 the computations involved in using equation (1) require only the simple summation indicated by equation (4). Thus it is apparent that finite-difference approximations of the basic differential equations of flow, exemplified by equation (4), might be utilized for analyzing observed water-level data without the necessity of deriving or finding suitable analytical expressions as had to be done by Theis or Thiem (Wenzel, 1942). Of course, their equations serve a very useful and necessary purpose in ground-water hydrology. However, it is unlikely that convenient expressions of this form can be derived for the more complex flow problems.

Water-level analysis for the nonhomogeneous aquifers

Following the analytical method briefly outlined above, a quantitative description of the nonhomogeneous formation can be obtained from water-level data. Consider the following equation which applies to two-dimensional nonsteady flow in nonhomogeneous formations being recharged at the rate W:

$$mP(\partial^2 h/\partial x^2 + \partial^2 h/\partial y^2) + (\partial mP/\partial x)(\partial h/\partial x) + (\partial mP/\partial y)(\partial h/\partial y) = S\partial h/\partial t - W \quad (5)$$

where S is the coefficient of storage and t is time. As was done in constructing equation (4), appropriate finite-difference approximations may be computed from field data and substituted for the head differentials in equation (5). It is convenient to divide equation (5) by an arbitrarily selected reference value of $(mP)_{0,0}$. Then, by Taylor's theorem,

$$(mP)_{x,y}/(mP)_{0,0} = K_{x,y} \approx 1 + \Delta x \partial k_{x,y}/\partial x + \frac{\Delta x^2}{2!} \partial^2 K_{(x,y)}/\partial x^2 + \ldots \frac{\Delta x^n}{n!} \frac{\partial^n K_{x,y}}{\partial x^n}$$

$$+ \Delta y \partial K_{x,y}/\partial y + \frac{\Delta y^2}{2!} \partial^2 K_{x,y}/\partial y^2 \ldots \frac{\Delta y^n}{n!} \frac{\partial^n K_{x,y}}{\partial y^n} \quad (6)$$

Equation (6) may be substituted in the left side of equation (5). The value of the right side of equation (5) will likely change in space, time, or both, but can be eliminated by using head observations from two areas of different size, both centering about a common reference point. By finite-difference equations, the flow through the perimeter of the area also may be expressed as a function of $(mP)_{0,0}$ and head differencials, which in turn can be set equal to $(S\partial h/\partial t - W)/(mP)_{0,0}$ times the smaller area. In this way a finite-difference equation can be constructed which contains only the numerical approximations of differentials of head and of K. An equation of this type can be constructed for each point on the piezometric surface, using a common reference point for $(mP)_{0,0}$. If the finite areas from which head differentials are estimated partially overlap, the finite-difference equations thus formed can be solved simultaneously for the differentials of $K_{x,y}$. If only the first differentials are sought, only two equations are needed, whereas if the first and second differentials are sought, four equations are needed, and so forth. In each case the higher differentials of $K_{x,y}$ are assumed equal to zero. Once the numerical value of the finite-difference approximation of the left side of equation (5) is obtained, obvious further analysis will yield estimates of S/mP and W/mP, using approximations of $\partial h/\partial t$ as obtained from field observations.

Application of method to field data

Selected water-level data from both homogeneous and nonhomogeneous formations have been used for computing hydraulic constants by the approach outlined above (Stallman, 1956). The results were in good agreement with those obtained by orthodox methods of analysis for the homogeneous aquifer. The accuracy of the results computed for the nonhomogeneous aquifer could not be evaluated because of a lack of control. However, in the latter case qualitative considerations suggest that the results were correct. In both cases, the flow studied was unconfined. The water-level altitudes were observed to the nearest 0.01 foot at intervals (a) of 400 feet in the homogeneous formation, and to the nearest 0.5 foot at intervals of 5,280 feet in the nonhomogeneous formation.

The effects on the computed results caused by inaccuracies in the observed water-level altitude can be quite large, particularly if the sum $(\partial^2 h/\partial x^2 + \partial^2 h/\partial y^2)$ nearly equals zero. Errors arising from this source were minimized by solving the finite-difference equations by least squares. The statistical solution gave reasonable results from data on the nonhomogeneous formation described above, whereas direct solution of a minimum number of equations yielded what appeared to be erratic results. Inadequacy of equation (6) for representing the areal variations of K also was considered a possible source of the erratic results, but this source of error could not be evaluated quantitatively since $f(x, y, K)$ was unknown. In either case, whether erratic results are caused by errors in observed water-level altitudes or by inadequate conditional equations, it is to be expected that a least-squares solution of the finite-difference equations will produce a smoothed result that will be more likely to reflect the regional characteristics of the formation.

In order to perform the analysis as described, water-level altitudes must be known at many points. The number of points required depends on the complexity of the flow system studied. As indicated on figure 1, and by equation (4), water-level altitudes are required at a minimum of 5 points for study of a homogeneous formation. Assuming that mP varies linearly in the nonhomogeneous formation, or that all differentials of mP higher than the first are zero, water-level altitudes must be known at a minimum of 14 points. More complex definition of the hydraulic characteristics can be obtained by including additional water-level data in the analysis. Compared with the data requirements of the quantitative analytical procedures now used in studies of ground-water flow, the amount of data required by the finite-difference methods may appear to be excessive. However, it should be recognized that the finite-difference methods permit analysis of data without the need for strategically located testing facilities, and also permit the quantitative description of $f(x, y, K)$.

In the United States, many of the water-level measurements obtained by the hydrologist are made in selected privately owned wells that are used primarily to supply water for domestic, agricultural, or industrial use. The wells are ordinarily scattered in space and do not conform to any conveniently described geometric pattern. The finite-difference approximations of head differentials as used by most workers in numerical methods are, however, dependent on a more regular spacing or alinement of the data in space. For the case illustrated by figure 1, the head differentials were obtained by a simple summation. Where the observed water-levels are not located in such a regular pattern, Taylor's theorem can be applied to obtain estimates of the head differentials as indicated by Rowe (1955), although in the latter case the computing requirements are greatly increased.

Conclusion

Quantitative methods of studying ground-water levels have been developed largely by following the comparatively narrow approach of finding solutions to the basic differential equations of ground-water motion. This approach, though fruitful, is not adequate for all the needs of the ground-water hydrologist. Difficulties with the complex mathematics required, and the limited existence of suitable testing facilities, seriously impair efforts at quantitative description of water-bearing formations by these methods. A great need for alternative and complementary methods of analysis is indicated. Numerical methods of solving partial differential equations may be used for estimating hydrologic coefficients. The latter approach circumvents the necessity for developing the highly complex analytical expressions required for rigorous determination of the flow through nonhomogeneous formations. Also of importance is the comparatively ready availability of water-level data on which such analysis can be founded. Numerical analysis by the writer of two sets of water-level data indicate that such an approach may prove to be a valuable aid in quantitative studies.

BIBLIOGRAPHY

ROWE, P. R., 1955, Difference approximations to partial derivatives for uneven spacings in the network: *Am. Geophys. Union Trans.*, v. 36, no. 6, p. 995-1008.
SOUTHWELL, R. V., 1946, *Relaxation methods in theoretical physics*: London, Oxford Univ. Press, 248 p.
STALLMAN, R. W., 1956, Numerical analysis of regional water levels to define aquifer hydrology: *Am. Geophys. Union Trans.* (in press).
WENZEL, L. K., 1942, Methods for determining permeability of waterbearing materials: *U. S. Geol. Survey Water-Supply Paper* 887, 192 p.

15

Copyright ©1964 by the American Society of Civil Engineers
Reprinted from *Am. Soc. Civil Engineers Proc., Jour. Hydraulics Div.* **90**:59-77 (1964)

GROUND-WATER MANAGEMENT FOR THE NATION'S FUTURE—
COMPUTER SIMULATION OF GROUND-WATER BASINS[a]

By Howell N. Tyson, Jr.[1] and Ernest M. Weber[2]

SYNOPSIS

General purpose analog and digital computers are used in developing and testing a two-dimensional diffusion model of a ground-water basin. The two-part computational procedure is outlined: (1) the development of the model; and (2) the use of the model for predicting dynamic behavior of the basin, under imposed conditions.

INTRODUCTION

The Department of Water Resources of the State of California is presently (1964) conducting an investigation of planned use of the major ground-water basins in southern California. Planned use means formulating and executing the most economical plan for operating these ground-water basins in coordination with surface storage and distribution facilities.

Note.—Discussion open until December 1, 1964. Separate discussions should be submitted for the individual papers in this symposium. To extend the closing date one month, a written request must be filed with the Executive Secretary, ASCE. This paper is part of the copyrighted Journal of the Hydraulics Division, Proceedings of the American Society of Civil Engineers, Vol. 90, No. HY4, July, 1964.
[a] This paper was originally presented May 18, 1963 at the ASCE Water Resources Engineering Conference at Milwaukee, Wis. The four papers comprising this Symposium were all published in the July 1964 Journal of the Hydraulics Division.
[1] Advisory Systems Engr., Data Processing Div., IBM Corp., Los Angeles, Calif.
[2] Senior Engrg. Geologist, California Dept. of Water Resources, Los Angeles, Calif.

The purpose herein is to set forth a general method of analysis of the dynamics of a ground-water basin. For illustrative purposes, a study of the ground-water basin of the Los Angeles Costal Plain is examined. Both an electronic differential analyzer and a general purpose digital computer are used in the study.

Because of its critically overdrawn conditions, the Coastal Plain of Los Angeles County was the first ground-water basin studied by the department. The coastal plain is in the southwestern portion of the county. The watershed area encompasses approximately 600 sq miles, with the ground-water basin comprising approximately 480 sq miles. Approximately half of the water supply for the greater Los Angeles metropolitan complex of four million people is supplied from this ground-water basin.

Currently (1964), as in the past, the extractions from the coastal plain have exceeded its natural and imported replenishment. As a result, water levels continue to recede. In addition, the ground-water quality near the seacoast is impaired by the intrusion of sea water. These difficulties can be surmounted, in part, by importing greater quantities of fresh water for additional recharge and by controlling extractions. These and other remedial measures are evaluated and applied through ground-water basin management. Optimum results will be more quickly and directly achieved when the dynamics of the basin are better understood.

A better understanding of the basin dynamics will also assist in solving some additional problems that arise in the process of artificially recharging the basin. In the first place, imported water may be spread only at certain areas suitable for this purpose. The locations of these spreading grounds are dictated primarily by: (1) suitable surface and subsurface hydrologic characteristics of the land; e.g., ability of the land to absorb large quantities of imported water per unit time; and (2) availability of cheap land. Riverbeds usually satisfy both requirements and, in addition, they can often be made to act as catch basins for mountain rain water runoff. Secondly, recharge of imported water at an excessive rate may result in: (1) flooding of nearby public and private property; (2) water logging of remote property; and (3) saturating the spreading grounds so that surface runoff from local rainfall cannot be absorbed. Further difficulties are often imposed by the removal of excessive quantities of water from the basin by heavy local pumping.

In order to study the dynamic behavior of the ground-water basin, the details of its geometry, physical characteristics, and its associated replenishment and extraction data must be obtained. Unfortunately, the cost of determining the physical characteristics (transmission and storage properties) in the detail required by field measurements is prohibitively large. On the other hand, the geometric details can, in many instances, be readily determined from a variety of records that accrue from geophysical studies, petroleum exploration, water quality analyses, well logs, etc. In addition, water level elevation histories may often be obtained from various local water agencies.

If such records are available for a given basin, and if replenishment and extraction data can be obtained in detail, then the computer can be used to find a representative set of transmission and storage properties or coefficients. This is accomplished by a cut-and-try adjustment of the coefficients until the dynamic response of the model satisfactorily matches the historical water level elevations measured in the field. This method of attack has long been in use by investigators of the behavior of oil reservoirs.[3,4]

The procedure outlined previously constitutes the first of two phases of a typical ground-water basin computer study. The second phase is devoted to

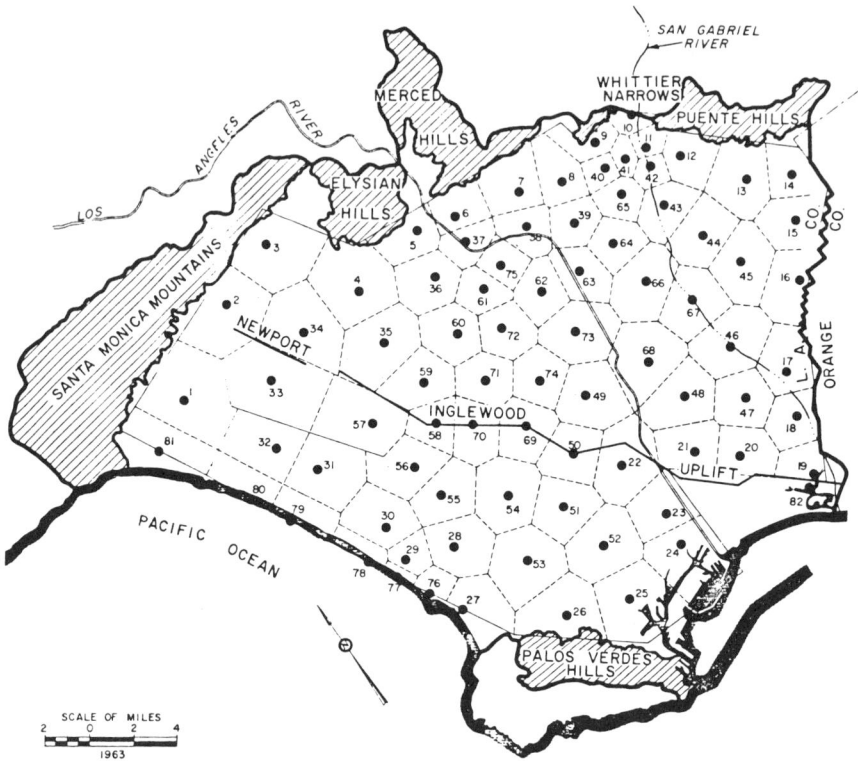

FIG. 1.—POLYGONAL ZONES AND NODE POINTS

the forecasting of the dynamic behavior of a ground-water basin, when subjected to certain "operating conditions." Two different computers are used in the study: an electronic differential analyzer for Phase I, and a general purpose digital computer for Phase II.

---

[3] Bruce, W. A., "An Electrical Device for Analyzing Oil Reservoir Behavior," Transactions, AIME, Vol. 151, 1943.

[4] Patterson, O. L., Montague, K. E., and Weiss, B., Jr., "High Speed Electronic Reservoir Analyzer," Drilling and Production Practice, Amer. Petroleum Inst., Dallas, Tex., 1951, p. 47.

Notation.—The symbols adopted for use in this paper are defined where they first appear and are listed alphabetically in the Appendix.

GEOLOGY AND HYDROLOGY

The Los Angeles Coastal Plain (Fig. 1) is bounded on the north by the Santa Monica Mountains, Elysian Hills, and Merced Hills; on the east by the Puente Hills and the Los Angeles-Orange County line; and on the south and west by the Pacific Ocean and the Palos Verdes Hills. Crossing the basin diagonally from southeast to northwest is the Newport-Inglewood Uplift.

The mountains and hills are nonwater bearing and present impervious barriers to the flow of water. The Newport-Inglewood Uplift acts as a partial barrier whose resistance to the flow varies along its length. The boundary condition along the seacoast is satisfied by holding the ground-water level at sea level.

The San Gabriel River enters the coastal plain at the Whittier Narrows between the Merced and Puente Hills. Ground-water levels in the Narrows area have been relatively constant during the period of record (1947-1962). Therefore, a constant head is assumed as a boundary condition at this point. The eastern boundary of the Los Angeles Coastal Plain ground-water basin is the Los Angeles-Orange County line. This is a political boundary of no geologic or hydrologic significance. Only small quantities of ground water have moved across this boundary, usually as outflow during the early years, and as inflow more recently. Therefore, for this study, the boundary flow along this line is considered to be a known function of time.

Within the boundaries outlined previously, the ground-water basin consists of a system of eleven major water-producing aquifers. The geometry of this system is immensely complicated. The aquifers overlay one another and are separated, for the most part, by fine-grained sediments. However, the aquifers merge together with sufficient frequency to permit large transfers of ground water from one to another.

For the initial computer analysis, the ground-water complex described previously is replaced by a simplified model. This model consists of a single equivalent aquifer whose local properties are composites of the corresponding properties of the several aquifers that make up the actual structure. The thickness of the single aquifer is allowed to vary with position, and yet it is considered to be small compared to its lateral dimensions. The bounding edges of the model are irregular in shape; in plan view, the boundaries are the same as those described previously for the Los Angeles Coastal Plain.

The model is divided into small polygonal zones. The dynamic response of the portion of the model included within each zone is represented by a single water level elevation. The size of the zones is dependent on the variations in replenishment, extraction, transmission, storage, and water level data. It is particularly important that these zones be small in regions of large spatial rates of change of the water level elevation. For purposes of testing the model against historical water level data, provision is made for the extraction (or injection) of time-varying flow rates from each of the zones.

The basis for division of the model into polygonal zones is examined subsequently.

MATHEMATICAL MODEL

The equation of continuity of an unconfined aquifer, in which there is no vertical variation of properties, is given by

$$\nabla \bar{v} + S \frac{\partial h}{\partial t} + Q = 0 \quad \ldots \ldots \ldots \ldots \ldots \ldots (1)$$

in which

$$h = \delta + z \quad \ldots \ldots \ldots \ldots \ldots \ldots (2a)$$

and

$$\delta = \delta_o + \bar{\delta} \quad \ldots \ldots \ldots \ldots \ldots \ldots (2b)$$

δ_o and $\bar{\delta}$ are, respectively, the mean and perturbation components of δ. Darcy's Law provides the equation of motion

$$\bar{v} = -\rho g \frac{k}{\mu} \nabla h \quad \ldots \ldots \ldots \ldots \ldots \ldots (3)$$

In Eqs. 1, 2a, 2b, and 3 z is the reference elevation, h is the head, δ refers to the local thickness of saturated portion of the aquifer, $\bar{v}$ denotes the velocity, S refers to the storage coefficient, Q describes the volumetric flow rate per unit area, ρ is the density, g refers to the acceleration of gravity, k is the permeability, μ is the absolute viscosity, and t denotes the time.

Eqs. 1 and 3 are combined and linearized to yield a single equation that, subject to appropriate boundary conditions, describes the dynamics of flow in the aquifer. The thickness of the aquifer is assumed to be small compared to its lateral dimensions. This equation is

$$\nabla T \nabla h - S \frac{\partial h}{\partial t} - Q = 0 \quad \ldots \ldots \ldots \ldots \ldots \ldots (4)$$

in which

$$T = \frac{\delta_o \rho g k}{\mu} \quad \ldots \ldots \ldots \ldots \ldots \ldots (5)$$

The quantities T and S are, respectively, the local transmissibility and storage coefficient of the aquifer. The source flow rate, Q, is, in most cases, time dependent. This flow rate is the algebraic sum of several component extraction and replenishment flows. The replenishment flows are precipitation, imported water, stream percolation, artificial recharge, and subsurface inflow across boundaries. The extraction flows consist mainly of the water pumped from the aquifer for consumptive use and subsurface outflow across boundaries.

Herein, Eq. 4 is replaced by an equivalent system of difference-differential equations, the simultaneous solution of which gives the wanted function, h, at a finite number of node points lying within the boundaries of the aquifer.

If the node points at which difference-differential equations are defined are regularly spaced, the equations have a particularly simple symmetrical form that is identical for all the node points. However, there are two troublesome problems that often exist if such spacing is to be used. The first of these comes about if the boundaries are irregularly shaped. The node points near such a boundary must be connected to the boundary by grid elements of lengths. This necessitates the use of special equations at these points. The second problem concerns the change of mesh size within the boundaries. In the neighborhood of expected large rates of change of the wanted function, the mesh size must be reduced if an accuracy is to be obtained that is comparable with the accuracy in the rest of the aquifer. Because it would not be economical in terms of computer cost to preserve a minimum spacing of node points throughout the aquifer, a rational approach to the problem of mesh size change is required.

Both of these difficulties are encountered in the case of the thin aquifer, shown in Fig. 1. The boundaries are irregular, and the expected spatial rates of change of the head, h, and the properties of the aquifer (i.e., local transmissibility, etc.) are large. Extensive work has been done by R. H. MacNeal[5] in the development of an asymmetric network of node points for two dimensional second order boundary value problems that overcomes the difficulties outlined previously. Such a network pertaining to the thin aquifer is also shown in Fig. 1. The attendant system of difference-differential equations is

and

$$\sum_i \left(h_i - h_B \right) Y_{i,B} = A_B S_B \frac{dh_B}{dt} + A_B Q_B$$

$$Y_{i,B} = \frac{J_{i,B} T_{i,B}}{L_{i,B}}$$

.. (6)

in which A_B denotes the area associated with node B in acres, $Y_{i,B}$ is the conductance of path between nodes i and B, in acre feet per year feet, S_B is the storage coefficient of polygonal zone associated with node B, $Q_{i,B}$ is the volumetric flow rate per unit area at acre feet per year acre; $T_{i,B}$ denotes the transmissibility at midpoint between nodes i and B, in acre feet per year feet; $L_{i,B}$ refers to the distance between nodes i and B, in feet; and $J_{i,B}$ describes the length of perpendicular bisector associated with nodes i and B, in feet.

A typical node point, its associated polygonal zone, and contiguous node points are shown in Fig. 2.

The terms in the first equation of Eqs. 6 are interpreted physically with the aid of Fig. 2 as follows: The left side represents the sum of the flows within the aquifer to the polygonal zone, A_B, (across the dashed lines, whose lengths are $J_{i,B}$); and the first term on the right side represents the rate of water storage within A_B. The remaining term represents the extraction (or replenishment) flow from A_B.

Eqs. 6 can be represented by a network of electrical elements; e.g., resistors, capacitors, current generators, and arbitrary function generators.

---

[5] MacNeal, R. H., "An Asymmetrical Finite Difference Network," Quarterly of Applied Mathematics, Vol. XI, No. 3, 1953.

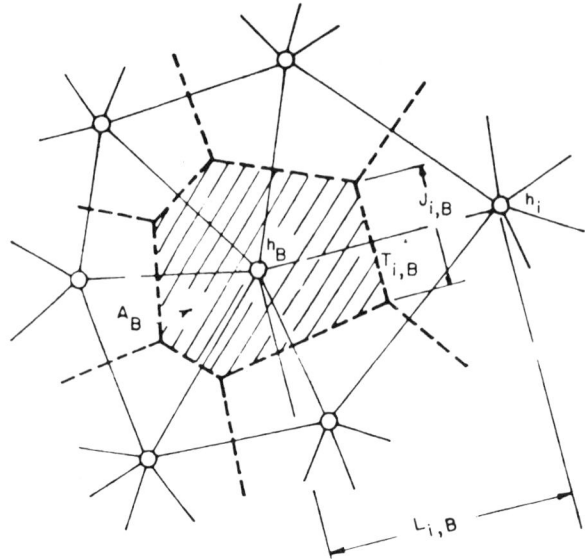

FIG. 2.—POLYGONAL GEOMETRY

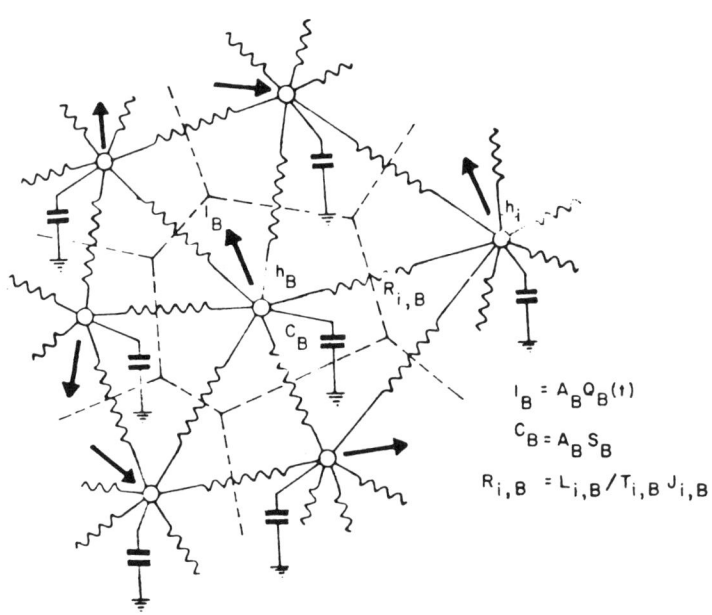

$I_B = A_B Q_B(t)$
$C_B = A_B S_B$
$R_{i,B} = L_{i,B}/T_{i,B} J_{i,B}$

FIG. 3.—TYPICAL RESISTOR-CAPACITOR NETWORK

Fig. 3 illustrates the portion of such a network corresponding to the node point array of Fig. 2. Currents in the various branches of the network are analogous to flow rates of water. Voltages at the nodes correspond to water level elevations. For example, a current flowing in a resistor from one node to a neighbor represents the subsurface flow rate of water across the common boundary of the polygonal zones corresponding to the nodes. A current flowing into a capacitor represents a flow rate of water into storage. Finally, the time dependent currents introduced at each node (indicated by the arrows in Fig. 3) represent the extraction or replenishment flow rates associated with a polygonal zone. Each of these time dependent currents requires a current generator and an arbitrary function generator for its implementation.

In addition to the analogous behavior of each of the electrical elements to those of the physical model, the circuit of Fig. 3 also bears topological similarity to the model. Networks of this type are referred to as direct analogs.[6]

Special direct analog computers containing the electrical elements of Fig. 3 have been constructed in the past to solve diffusion problems.[3,7,8] More recently, the general purpose electronic differential analyzer has been used in the solution of this kind of problem.[9] The general purpose digital computer has also been used, for some time, to solve diffusion problems, particularly those concerned with the transfer of heat.[10,11] However, in view of the cut-and-try nature of Phase I and the desirability of relatively immediate "on-line" graphical output, the analog computer appears, at present, (1964) to be the appropriate tool for this phase of the study. The choice between the electronic differential analyzer and the comparable direct analog computer should probably be made in favor of the machine that is most easily accessible. In general, the electronic differential analyzer is the most accessible, due to its widespread use in the aerospace industry, and is the computer (PACE 231-R) used in the work described herein.

ANALOG COMPUTER SOLUTION

Eqs. 6 can be mechanized on the electronic differential analyzer in the usual way,[12,13] as shown in Fig. 4. However, this approach has two unattractive

---

[6] McCann, G. D., "The Direct Analogy Electric Analog Computer," Journal, ISA, April, 1956.

[7] Wilts, C. H., and McCann, G. D., "Application of Electric Analog Computer to Heat Transfer and Fluid Flow Problems," Journal of Applied Mathematics, Vol. 16, No. 3, September, 1949.

[8] Paschkis, V., and Baker, H. D., "A Method for Determining Unsteady-State Heat Transfer by Means of an Electrical Analogy," Transactions, ASME, 1942.

[9] Darms, D. A., and Tyson, H. N., Jr., "Analog Simulation of Underground Water Flow in the Los Angeles Coastal Plain," Proceedings, WJCC, 1961, p. 535.

[10] Peaceman, D. W., and Rachford, H. H., "The Numerical Solution of Parabolic and Elliptic Differential Equations," Journal of the Society of Industrial and Applied Mathematics, Vol. 3, 1955, pp. 28-41.

[11] Douglas, J., Jr., and Rachford, H. H., "On the Numerical Solution of Heat Conduction Problems in Two and Three Space Variables," Transactions, Amer. Mathematical Soc., Vol. 82, 1956, pp. 421-439.

[12] Korn, G. A., "Electronic Analog Computers," McGraw-Hill Book Co., Inc., New York, N. Y., 1956.

[13] Jackson, A. S., "Analog Computation," McGraw-Hill Book Co., Inc., New York, N. Y., 1960.

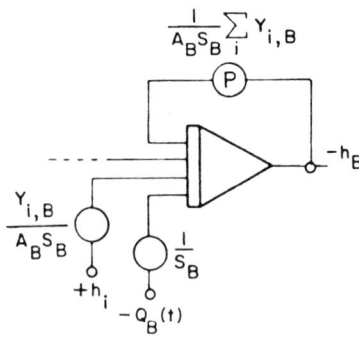

FIG. 4.—NORMAL ELECTRONIC DIFFERENTIAL ANALYZER CIRCUIT

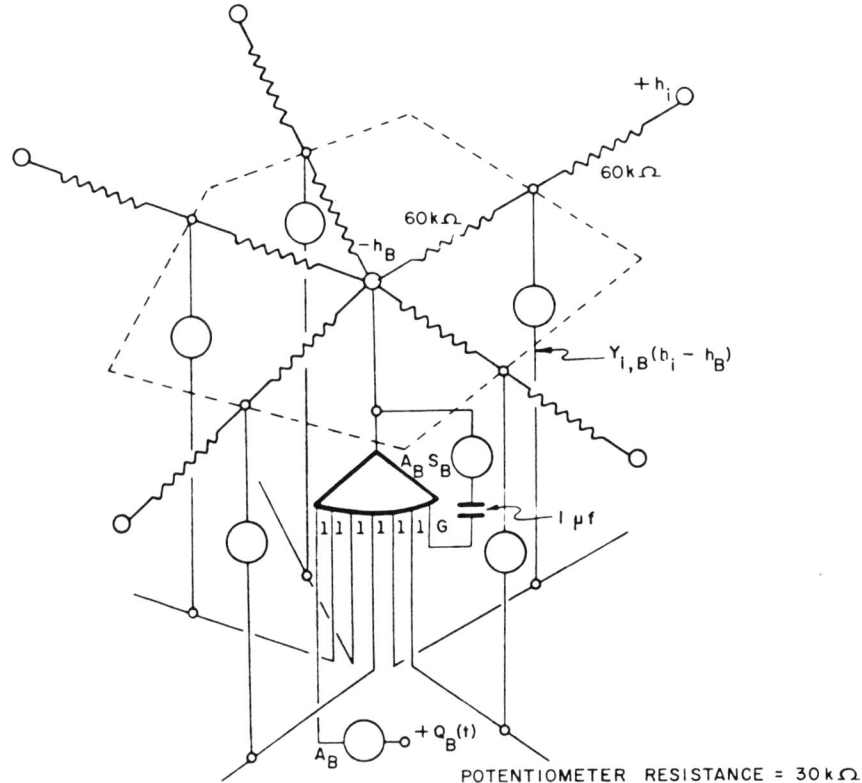

FIG. 5.—MODIFIED ELECTRONIC DIFFERENTIAL ANALYZER CIRCUIT

features. First, the conductance $Y_{i,B}$ is represented by two different potentiometer settings for every node. A slight error in setting one of the two has the effect of introducing an additional flow rate. For example, if the feedback potentiometer, P in Fig. 4, is set slightly high, the resultant error can be considered a "leakage" flow; e.g., a flow from the node in question to a sink at h = 0. This flow rate has, of course, no physical counterpart in the ground-water model, but, unfortunately, it can be disproportionately large compared to the error in the potentiometer setting. Such an error is especially disconcerting in those cases in which all flow rates should actually be small or zero at steady state (uniform head distribution in a ground-water basin).

Additional undesirable features of the circuit (Fig. 4) are the computational labor involved in changing the conductance value and the resetting of at least two potentiometers. These difficulties are made particularly unattractive by the many adjustments in conductance values required by the cut-and-try modeling process.

To avoid these difficulties, and at the same time retain an economy of equipment, two nonstandard circuits are used: a voltage divider circuit and a variable capacitor circuit. These are shown for a typical polygonal zone in Fig. 5.

Each of the terms $(h_i - h_{i,B}) Y_{i,B}$ of Eqs. 6 is mechanized by a simple voltage divider potentiometer network between the two nodes. Pairs of 60,000 ohm resistors, matched to within 1%, are chosen for compatibility with the 30,000 ohm potentiometers. By setting the conductance potentiometers with one node at the computer reference voltage and the other at ground, the loading of the dividers is compensated for. The outputs of these potentiometers now read directly the flow between two adjacent nodes.

To obtain a variable capacitance, a potentiometer is inserted in series with the integrating capacitor (usually a standard value, for example, of 1.0 μ fd). The error introduced by this procedure is negligible, if the potentiometer resistance is small compared to the input resistors of the integrator.[9]

The number and similarity of the analog circuits of the type shown in Fig. 5, which are used in the simulation of a ground-water basin the size of the Los Angeles Coastal Plain, is sufficiently large to require many circuit checking procedures. The procedures should be carefully designed to indicate all errors in programming, patchboard wiring, computer operation, and parameter scaling.

MODEL DEVELOPMENT

When the errors have been completely eliminated from the analog circuit, the task of obtaining the physical constants (transmission and storage coefficients) by a cut-and-try process is begun.

The replenishment and extraction data, in the form of time dependent source flow rates, are inserted at the appropriate node points of the analog circuit. The resulting water level responses (node voltages versus time) are recorded graphically, using an X-Y plotter. When the historical water level elevations (functions of time) are superposed on these plots, the portions of the ground-water basin in which poor agreement exists between the analog model and the actual structure are easily spotted. The local transmission

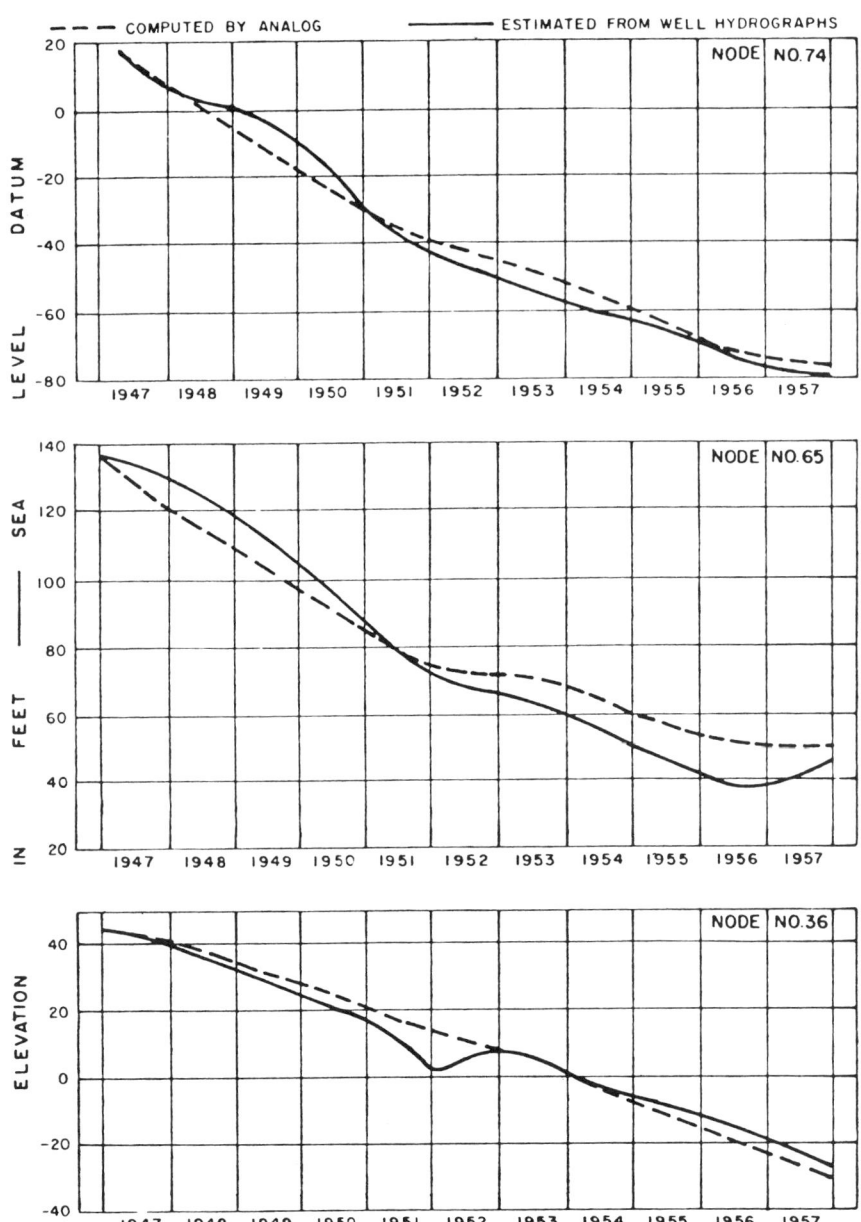

FIG. 6.—COMPARISON OF CHANGES IN GROUND WATER ELEVATIONS AT SELECTED NODES IN THE COASTAL PLAIN OF LOS ANGELES COUNTY

or storage constants, or both, are then suitably adjusted, and new responses are obtained.

The cut-and-try process of adjusting the physical constants is continued in this manner until satisfactory agreement is reached between the computed water level responses and the historical data.

A comparison between the water level elevations simulated by the computer and those of the prototype is shown in Fig. 6. Agreement between the machine produced water level elevations and historic measured water level elevations is on the order of 10%.

OPERATIONAL STUDIES

After the model is perfected to the engineer's satisfaction, a more general study is usually made of the dynamic behavior of the basin. Because the model studied herein is linear, an exhaustive study of it could be made by obtaining influence functions. That is, the response (water level elevation versus time) of the entire network due to the sudden application of a unit flow rate at any node (all other source flow rates set to zero) could be plotted. Similar responses would be obtained for successive applications of the unit flow rate at all other interior nodes. Any desired solution could be constructed from these influence functions by superposition. However, this procedure normally requires a large number of computer runs. Consequently, in general, it may be more economical to make investigations of the dynamic behavior of the aquifer under certain specific conditions of replenishment or extraction that are expected to take place.

The study reported herein involves a compromise between the two approaches. A set of responses is obtained for the simultaneous injection or extraction of large flow rates at selected groups of nodes throughout the basin. Then, by an appropriate superpositioning of these solutions, the water level responses for a large number of proposed additional replenishment and extraction combinations are obtained.

Studies of the type outlined previously are perhaps best performed on a general purpose digital computer. Although a complete set of influence functions is not obtained, the number of solutions desired is still quite large. In addition, it is helpful to group the runs in small batches, spaced far enough apart in time to allow for a review and study of the results.

A program designed to secure information from small batches of computer runs over a period of several months would entail a costly manual set-up and check-out procedure on the electronic differential analyzer for each group of runs to be made. Fortunately, on the general purpose digital computer, this process is automatic and thus relatively inexpensive. Once the digital program to solve the difference-differential equations associated with the ground-water basin model has been checked out, it is permanently recorded on cards or paper tape. The cards or tape can then be reintroduced to the computer at any later time without further manual checking.

The digital program for the ground-water basin model is written using the FORTRAN (FORmula TRANslation) system. FORTRAN is an automatic coding system that enables the engineer to communicate with any digital computer using the system, in a language resembling algebra. By using FORTRAN, or an equivalent automatic coding system, the time required to write and checkout a program for a ground-water basin (or, in fact, diffusion problems in general) can be quite short. The principal reason for this is that the only special checking procedures the user must design for the digital computer program check-out process are those needed to catch logic errors in the FORTRAN program. The user is relieved from having to design additional special checks for the machine setup of the problem. Automatic

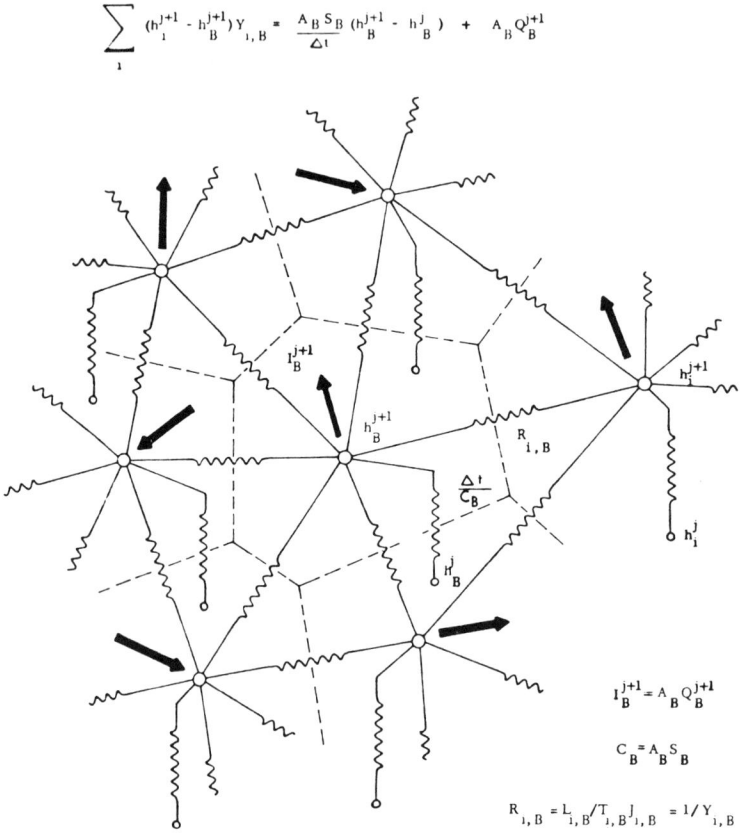

FIG. 7.—RESISTANCE NETWORK EQUIVALENT FOR STEP-BY-STEP IMPLICIT INTEGRATION PROCESS

self-checking features are built into the logic circuitry of the digital computer, and diagnostic routines are incorporated in the FORTRAN compiler for this purpose. This, together with the fact that no parameter scaling is required in the writing of the FORTRAN program, represents a considerable saving in ground-water problem setup time and cost over that of the electronic differential analyzer (or direct analog computer).

DIGITAL COMPUTER SOLUTION

Eqs. 4(a) and 4(b) are solved on the general purpose digital computer by an implicit numerical integration technique.[14] Using this technique, Eqs. 5(a) and 5(b) are replaced by

$$\sum_i \left(h_i^{j+1} - h_B^{j+1} \right) Y_{i,B} = \frac{A_B S_B}{\Delta t} \left(h_B^{j+1} - h_B^j \right) + A_B Q_B^{j+1} \quad \ldots \quad (7)$$

in which the superscripts j denote points along the time coordinate. This method of integration has the advantage that the magnitude of the time step, Δt, does not depend on a stability criterion.

The corresponding electrical network representation of Eq. 7 is shown in Fig. 7, which illustrates the mechanics of the simultaneous solution of Eq. 7 at every node within the aquifer. The procedure is as follows: Initial values $h_B(0)$ are impressed at the terminals labeled h_B^j (B = 1, 2, ... N; N = number of interior nodes). Then, for a given set of coefficients, $R_{i,B}$ and C_B, and currents I_B^{j+1}, the values h_B^{j+1} are implicitly determined at the end of a step in time, Δt. Once determined, these values become the initial water elevations for the next succeeding step in time.

To facilitate the writing of the FORTRAN program, it is found useful to first assign consecutive numbers to all the nodes shown in Fig. 1. Consecutive numbers are also given to the line segments (branches) connecting the nodes. These assignments are illustrated in Fig. 8.

A portion of the FORTRAN program that was used is shown in Fig. 9 and its corresponding flow chart in Fig. 10. Note that for simplicity in writing the program, all the node-to-node subsurface flows (Q) are calculated first. Then, all the storage flows (S) are calculated. Next, all the flows (subsurface, storage, and extraction) are balanced at each node by setting their sum equal to a residual (RES) term. The water level elevation at the node is adjusted by the magnitude of the residual, attenuated by a (relaxation) coefficient. After all the values of h have been adjusted in this manner, a sum is formed of the nodal residuals. This sum is compared with a threshold value (ERROR). The threshold value is a tolerance level; e.g., the maximum acceptable sum of nodal flow residuals at any time step. If the sum of the residuals is less than or equal to the threshold value, the calculation of the values of h is complete for that time step. Otherwise, the calculation is repeated as many times as is required to reduce the sum of residuals to a value less than or equal to the threshold value. It is important to choose this quantity with sound engineering judgment, because it determines the amount of computer time spent on each time step.

Next, a suitable formula must be found for the relaxation coefficient (RELAX). Because the product of the residual term (RES) and the relaxation coefficient results in a change in water level elevation, Δh, and because the residual term is a flow rate, the relaxation coefficient must be an impedance.

---

[14] Richtmyer, R. D., "Difference Methods for Initial Value Problems," Interscience Publishers, Inc., New York, N. Y., 1957, Chapter 1.

Each node is, of course, the junction for one or more branches, so that a logical choice for the relaxation coefficient might be the equivalent impedance of the branches joining a node to its neighbors. Thus,

$$\text{RELAX}_B = \frac{1}{\sum_i Y_{i,B} + \frac{A_B S_B}{\Delta t}} \quad \ldots \ldots \ldots \quad (8)$$

The method outlined previously for the simultaneous solution of Eq. 7 is essentially that of Gauss-Seidel.[15] For linear, symmetric systems of

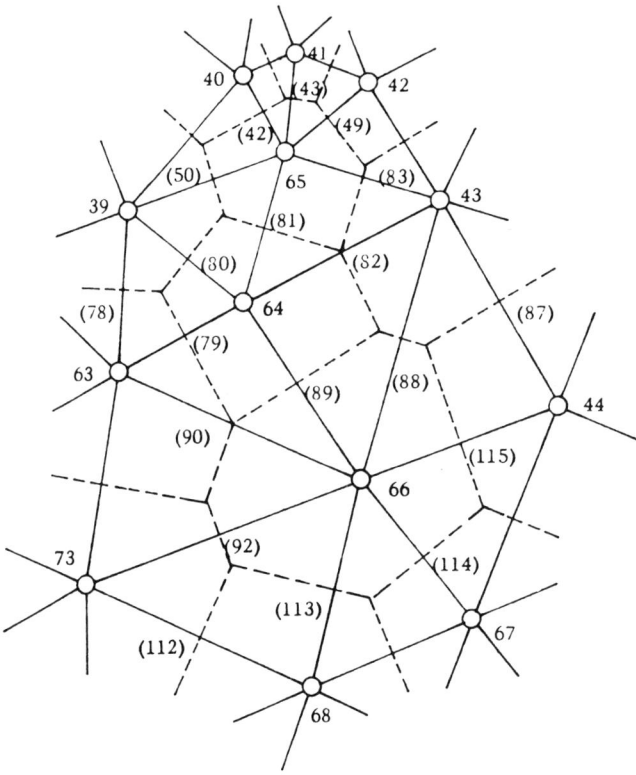

FIG. 8.—PORTION OF LOS ANGELES COASTAL PLAIN ASYMMETRIC NETWORK

equations, such as the system representing the ground-water problem, the Gauss-Seidel method is known to be unconditionally convergent.

The digital computer used in the work reported herein was an IBM 7074. Using this computer, a complete set of transient responses (all nodes) for a

---

[15] Ralston, A., and Wilf, H. S., "Mathematical Methods for Digital Computers," John Wiley & Sons, Inc., New York, N. Y., 1961, Chapter 3.

given set of boundary conditions and over a time period of 30 yr can be obtained in the order of 10 min. Still faster numerical procedures exist for the linear problem,[10,16] but their setup is more difficult. By using the method outlined, engineers totally inexperienced at programming can write and check out a program for a ground-water basin the size of the Los Angeles Coastal Plain in several days.

The agreement between corresponding electronic differential analyzer and digital solutions is on the order of 1%.

SOLUTION OF NONLINEAR PROBLEMS

The numerical procedure outlined previously for the analysis of ground-water basins can be extended to handle certain nonlinearities. For example, if the values of the conductances, $Y_{i,B}$, or the capacitances, C_B, or both, depend on the local water level elevation, it is necessary to amend the procedure to recalculate $Y_{i,B}$ and C_B, and the relaxation coefficients, $RELAX_B$, at the end of every iteration.

This approach has been applied successfully in a study of a primary surface pipeline network for the Los Angeles Coastal Plain. The extended method has also been used in the solution of nonlinear problems in heat transfer and thermodynamic systems.

In the case of the pipeline network study, the flows in the branches of the network are nonlinear, and are described by the relation[17]

$$Q_{i,B} = \alpha_{i,B} \left(h_i - h_B \right) \frac{1}{N} \quad \ldots \ldots \ldots \ldots (9)$$

in which $\alpha_{i,B}$, and N are constants. The value of $\alpha_{i,B}$ is dependent on the characteristics of the fluid and the pipeline. The constant N lies between 1.0 and 2.0. If Eq. 8 is rewritten as

$$\left. \begin{array}{l} Q_{i,B} = Y_{i,B} \left(h_i - h_B \right) \\[6pt] Y_{i,B} = \dfrac{\alpha_{i,B}}{\left| h_i - h_B \right|^{\frac{N-1}{N}} + \sigma} \end{array} \right\} \quad \ldots \ldots \ldots (10)$$

then, as indicated previously, the conductances $Y_{i,B}$ and the relaxation coefficients

$$RELAX_B = \frac{1}{\sum\limits_i Y_{i,B}} \quad \ldots \ldots \ldots \ldots (11)$$

must be recalculated after each iteration. A constant, σ, has been introduced to the formula for $Y_{i,B}$, in order to prevent $Y_{i,B}$ from growing without bound,

---

[16] Young, D., Jr., "Iterative Methods for Solving Partial Difference Equations of the Elliptic Type," Transactions, Amer. Mathematical Soc., Vol. 76, 1954, p. 92.

[17] Hoag, L. N., and Weinberg, G., "Pipeline Network Analysis by Electronic Digital Computer," Journal, Amer. Water Works Assn., Vol. 49, No. 5, May, 1957.

T = 0.0

  DO 80 M=1, MAJOR
  DO 70 J=1, MINOR

T = T + DELTA

  DO 40 K=1, N
40 HO(K) = H(K)

41 Q(2) = Y(2)*(H(81) - H(1))
  Q(3) = Y(3)*(H(11) - H(2))
  Q(4) = Y(4)*(H(2) - H(3))
  Q(5) = Y(5)*(H(11) - H(32))

  .
  .

  Q(79) = Y(79)*(H(63) - H(64))
  Q(80) = Y(80)*(H(39) - H(64))
  Q(81) = Y(81)*(H(64) - H(65))

  .
  .

  Q(187) = Y(187)*(H(69) - H(50))

  DO 50 K=1, N
50 S(K) = AS(K)*(H(K) - HO(K))/DELTA

  RES(1) = Q(2) - Q(3) - Q(5) - S(1) + AQ(1)
  RES(2) = Q(3) + Q(7) - Q(8) - S(2) + AQ(2)

  .
  .

  RES(64) = Q(79) + Q(80) - Q(81) - Q(82) - Q(89) - S(64) + AQ(64)

  .
  .

  RES(75) = Q(55) + Q(39) - Q(40) - Q(53) - S(75) + AQ(75)

  DO 100 K=1, N
100 H(K) = H(K) + RELAX(K)*RES(K)

  SUM = 0.0
  DO 60 K=1, N
60 SUM = SUM + ABSF(RES(K))
  IF(SUM - ERROR) 70, 70, 41

70 CONTINUE

  PRINT, T
  DO 80 K=1, N
80 PRINT, K, H(K)

  PAUSE

Definitions:

DELTA = Time Step (Δt), Years
ERROR = Max. Acceptable Sum of Nodal Flow Residuals at Any Time Step, Acre-Ft./Yr.
AQ(K) = Source Flow Rate at Node (K) at Time $(j+1)\Delta t$, Acre-Ft./Yr.
AS(K) = Capacitance at Node (K), Acres
H(K) = Water Level Elevation at Node (K) at Time $j(\Delta t)$, Ft.
HO(K) = Water Level Elevation at Node (K) at Time $j(\Delta t)$, Ft.
S(K) = Storage Flow Rate at Node (K) at Time $(j+1)\Delta t$, Acre-Ft./Yr.
Q(Bn) = Sub-surface Flow Rate Along Node-to-node Branch (Bn) at Time $(j+1)\Delta t$, Acre-Ft./Yr.
Y(Bn) = Conductance of Node-to-node Branch (Bn), Acre-Ft./Yr.-Ft.
T = $(j+1)\Delta t$, Years
RELAX(K) = Relaxation Coefficient at Node (K), Yr.-Ft./Acre-Ft.
RES(K) = Nodal Flow Residual, Acre-Ft./Yr.

FIG. 9.—FORTRAN IMPLICIT INTEGRATION ROUTINE

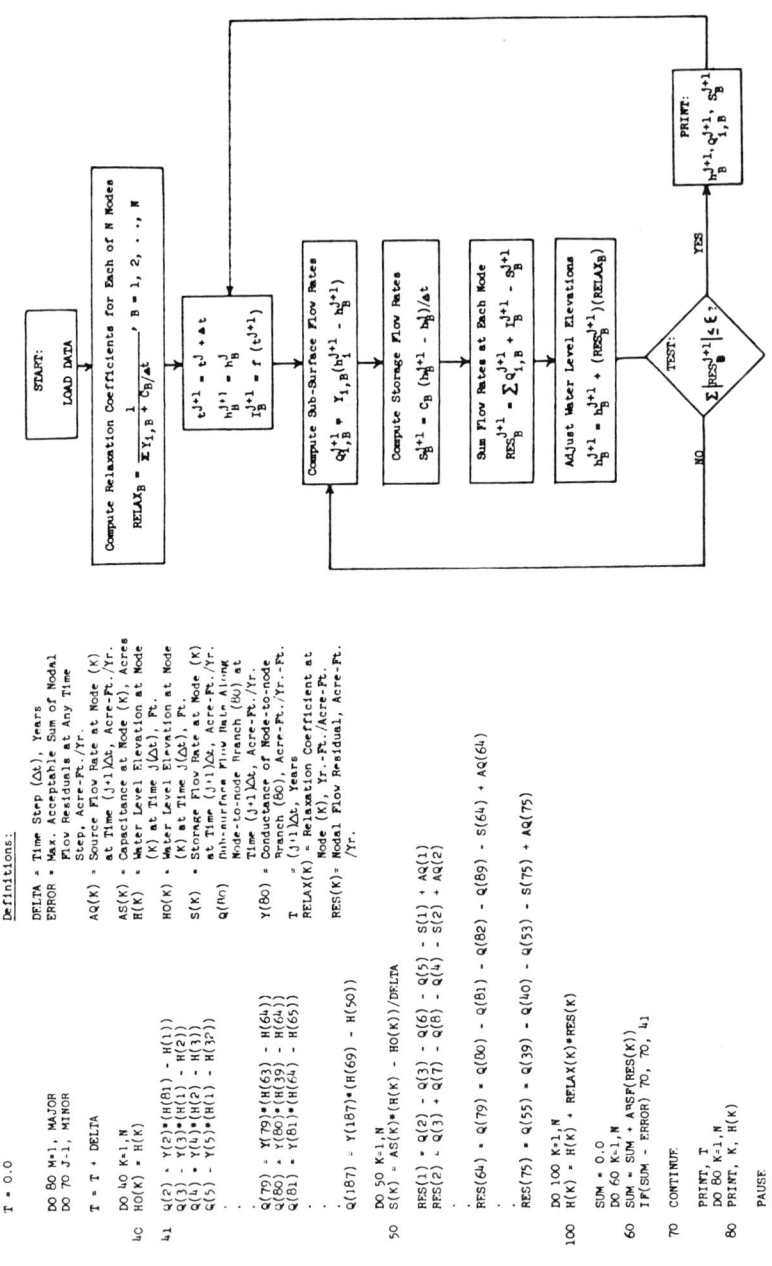

FIG. 10.—SIMPLIFIED FLOW CHART FOR DIGITAL COMPUTER SOLUTION OF GROUND WATER PROBLEM

should the difference in water level elevations, $h_i - h_B$, go to zero or become small during the solution. The choice of σ is arbitrary as long as it is small compared to $h_i - h_B$ when the final solution is reached. Should the solution result in a difference, $h_i - h_B$ comparable to σ in magnitude, then at least two possibilities exist: (1) σ can be made still smaller, and a new solution can be obtained; or (2) on inspection of the solution, it may become evident that the flow in the branch associated with the difference, $h_i - h_B$, is itself small compared to the neighboring flows to or from nodes i and B. If the latter is the case, the error introduced by σ is of little consequence.

The analysis of flow in pipeline networks has long been successfully achieved by the Hardy Cross technique.[17] The principal difference between the method presented herein and that of Hardy Cross lies in the equations to be solved. Nodal equations are written in the first case, and loop equations are written in the latter case. The selection of the best method is largely determined by the ratio of nodes to loops in the network to be analyzed. If the number of nodes is small compared to the number of loops, nodal equations should be written; and vice versa.

CONCLUSIONS

The techniques developed herein have been used successfully in evolving and testing a mathematical model of a ground-water basin.

The ability of the analog computer to produce immediate graphical output, allowing for quick visual observations of the model dynamics, makes this computer the appropriate machine in the modeling phase.

The digital computer is best suited for an operational analysis of the model, because this phase usually requires a large number of solutions to be run in batches over an extended period of time. The ease with which ground-water basin programs are written and checked out for the digital computer means that engineers totally untrained in programming will be able to write and check out similar programs in a matter of days.

Computers will undoubtedly be assigned a greater role in future ground-water basin studies. Improvements in the speed of computer setup and checkout for the model development phase will considerably increase the economy of computer application, as will shorter solution times for the operational phase. Better methods of collecting and evaluating the geologic and hydrologic data will improve the quality of the models developed, and a better understanding of the dynamic behavior of ground-water basins will be achieved.

APPENDIX.—NOTATION

The following symbols have been adopted for use in this paper:

A_B = area associated with node B;
g = acceleration of gravity;
h = head (water level elevation);
J_{iB} = length of perpendicular bisector associated with nodes i and B;
k = permeability;
L_{iB} = distance between nodes i and B;
Q = volumetric flow rate per unit volume;
Q_B = extraction or replenishment flow associated with node B;
Q_{iB} = volumetric flow rate per unit area at node B;
S = storage coefficient;
S_B = storage coefficient of polygonal zone associated with node B;
T = transmissibility coefficient;
T_{iB} = transmissibility of mid-point between nodes i and B;
t = time;
$\bar{v}$ = velocity;
Y_{iB} = conductance of path between nodes i and B;
z = reference elevation;
ρ = density;
μ = absolute viscosity;
$\sum$ = summation; and
σ = constant.

16

Copyright ©1968 by the American Geophysical Union
Reprinted from Water Resources Research **4**:1069-1093 (1968)

Application of the Digital Computer for Aquifer Evaluation

GEORGE F. PINDER

J. D. BREDEHOEFT

Abstract. The unsteady-state flow of a fluid in a confined aquifer can be approximated by linear, parabolic, partial-differential equations. A digital-computer program was written to solve these equations based upon an implicit finite-difference technique. Calculated values compared favorably with analytical solutions for homogeneous, isotropic aquifers of simple geometry. An analysis of an aquifer at Musquodoboit Harbour, Nova Scotia, was made using this technique. Aquifer parameters in the digital model were modified until the results computed fitted the pump-test data. The results indicate that this aquifer would come to equilibrium in about 100 days and would easily provide an adequate supply of water for the village of Musquodoboit Harbour. As a check of the digital model for this problem an electric analog of this aquifer was also constructed; the two solutions obtained compared favorably. (Key words: Ground water; digital computers; Nova Scotia)

SYMBOLS

b, thickness of the aquifer;
$Ei(u)$, the exponential integral;
h, hydraulic head (L);
i, index in the x dimension;
j, index in the y dimension;
k, index in the t dimension;
K_s, hydraulic conductivity of the stream bottom (L/T);
K_z, hydraulic conductivity in the confining layer (L/T);
m, thickness of the confining layer (L);
Q, volume rate of withdrawal or injection at any node (L^3/T);
r, distance from the pumping well to the observation well (L);
S, storage coefficient (dimensionless);
t, time (T);
T, transmissibility (L^2/T);
Δt, increment in the t dimension (T);
$\Delta x, \Delta y$, increments in space (L);
ρ, density of the fluid (M/L^3).

INTRODUCTION

The objective of quantitative methods in ground-water hydrology is to predict accurately the response of an aquifer system to various hydrologic stresses.

Analytical Methods

Before the introduction of the numerical methods, analytical techniques were employed in studying problems in ground-water flow. Initially, solutions dealt primarily with the problems encountered in well hydraulics because of the immediate application in the field of engineering. The equilibrium equation developed by *Thiem* [1906] permitted evaluation of the coefficient of permeability for the aquifer. The nonequilibrium expression derived by *Theis* [1935] allowed determination of both the storage coefficient and the permeability. *Hantush and Jacob* [1955] extended this concept to treat aquifers that received leakage from confining beds. A number of available analytical solutions pertinent to problems of ground-water flow may be found in *Muskat* [1937], *Polubarinova-Kochina* [1962], and *Hantush* [1964]. An extensive compilation of expressions for heat flow is found in *Carslaw and Jaeger* [1959]; these solutions may be applied directly to ground-water flow in most instances. In general, however, the analytical solutions apply only to simple geometry with constant aquifer coefficients, which fact limits their use.

Electric Analogs

Many of the restrictions inherent in the analytical approach are not encountered in electric analog simulators. The electric analog, as employed in ground-water hydrology, consists essentially of a network of resistors and capacitors that can be considered analogous to aquifer transmissibility and storage. The electric analog was initially developed to treat heat flow problems [*Paschkis and Baker*, 1942] and was quickly adopted by *Bruce* [1942] to study oil-reservoir behavior. *Skibitzke* [1961] applied this method to hydrologic problems and developed a theoretical verification of his approach using the relaxation methods of applied mathematics. *Stallman* [1963a] employed the same mathematical technique to determine the source of the error inherent in the electric analog simulation. *Prickett* [1967] considered the error in the immediate vicinity of the well to be due to the finite-difference approximation and introduced methods of simulating partial penetration as well as pumped wells with different radii, screen lengths, gravel pack sizes, and well loss constants. Electric analogs have been developed for many complex aquifer problems [*Brown*, 1962]; *Walton and Prickett* [1963]; *Wood and Gabrysch* [1965].

Digital Models

The finite-difference approximations of the flow equations, which provide a theoretical foundation for the electric analog approach, also act as the basis for the digital model. The mechanics of solving the finite-difference equations have been known for some time, and the various methods may be found in *Southwell* [1946], *Allen* [1954], *Crank* [1956], *Todd* [1962], *Saul'yev* [1964], *Volynskii and Bukhman* [1965], and *Richtmeyer and Morton* [1967]. In practice, however, the finite-difference technique requires such a large number of arithmetic calculations that without a digital computer the time required to solve even the simplest problem is prohibitive. It is not surprising, therefore, that the popularity of this method for solving complex physical systems has been dependent upon the development of sophisticated digital computers.

The first attempt to obtain solutions to a porous flow problem using numerical methods and punched card equipment was made by *Terwilliger et al.* [1951]. He solved the two-phase flow problem encountered in gravity drainage of an oil reservoir, using a finite-difference approximation for the acceleration term in the Beckley-Leverett equation.

The next significant contribution was made by *Bruce et al.* [1953], who investigated the problem of linear and radial unsteady-state gas flow through porous mediums. They employed a technique developed by L. H. Thomas, known as the alternating direction method, for solving the nonlinear partial-differential equations. This procedure had the advantage of minimizing both the number of calculations and computer storage; later this approach was introduced into other branches of science where parabolic partial-differential equations are encountered [*Peaceman and Rachford*, 1955; *Douglas and Peaceman*, 1955; *Douglas and Rachford*, 1956].

In an attempt to reproduce more accurately the response of the oil reservoir to pumping stress, *West et al.* [1954] used two second-order, nonlinear partial-differential equations. They solved these equations simultaneously using an implicit iterative technique, Newton's method [*Willers*, 1947].

It was not until 1956 that numerical methods were introduced into ground-water hydrology by *Stallman* [1956], who outlined a procedure for determining the permeability of an aquifer by numerical analysis of the piezometric surface.

The next important application of the numerical approach was in a study of the hydrodynamics of large scale geologic basins by *Fayers and Sheldon* [1962]. They considered the problem of evaluating quantitatively the magnitude and direction of steady-state flow in aquifer systems within a geologic basin. It was necessary to solve the Laplace equation in three dimensions for a nonhomogeneous porous medium. A successive overrelaxation technique was used, although the authors pointed out that the alternating direction technique of *Peaceman and Rachford* [1955] required potentially fewer computations.

A similar problem was considered by *Freeze* [1966], who applied both the analytical and numerical approach to the two-dimensional steady regional-flow problem.

Since the attempts by Bruce and Peaceman and Rachford in the early 1950's, reservoir engineers have made great strides in the numerical simulation of petroleum reservoirs. Three phase nonsteady flow can now be effectively simulated in large petroleum reservoirs. The equations are identical to those for the flow of ground water, and techniques for solutions can be adapted to the hydrologic problems.

Interesting flow problems have also been considered in soil physics. *Klute et al.* [1965] used an iterative numerical technique to solve the nonlinear form of the flow equation in which the permeability is defined as a function of the dependent variable. A similar problem was studied by *Liakopoulos* [1965], who applied the implicit alternating-direction technique to the problem of one-dimensional simultaneous flow of water and air through a porous medium. He simplified the problem considerably, however, by assuming that the functions relating fluid pressure to water content and permeability to pressure were already determined.

Recently *Rubin* [1968] applied the iterative alternating-direction technique to unsteady-state unsaturated flow in the study of horizontal infiltration and ditch drainage.

The first attempt to evaluate numerically the response of a ground-water reservoir to pumping stresses was made by *Fiering* [1964]. He applied an iterative implicit technique with a set of finite-difference approximations developed from mass balance considerations. His technique, however, required that

$$[(\Delta t)T]/[(\Delta x)^2 S]$$

should not exceed 0.15; this approach requires a prohibitive amount of computer time if it is applied to the long-term response of an aquifer with a large transmissibility and a small storage coefficient.

A more general technique for long-term aquifer evaluation was described by *Eshett and Longenbaugh* [1965]. They proposed using either the Gaussian elimination or the alternating-direction scheme for solving the finite-difference approximations to the basic two-dimensional, nonlinear, second-order partial-differential equation describing flow in a homogeneous isotropic aquifer. The model was designed to handle a water table aquifer with impermeable lateral boundaries or a constant head boundary, such as a fully penetrating hydraulically connected river. The method of solution was verified for a homogeneous isotropic aquifer by comparing the calculated values against the analytical solution from heat-flow theory.

Bittinger et al. [1967] briefly described the mathematical basis for the digital model and various computing schemes that could be employed in solving the resulting algebraic equations. They described the various boundary conditions that could be treated, as well as several problems that they were currently investigating. There was not, however, any rigorous evaluation of the model.

Purpose and Scope of Investigation

This research was designed to provide a digital model capable of describing the response of an aquifer system to a wide range of hydraulic stresses and to apply this digital technique to a selected field problem. It was necessary to design the model to handle (1) nonhomogeneous anisotropic porous mediums; (2) irregular boundary conditions; and (3) vertical leakage to the aquifer. The memory capacity of the computer and the execution time required to solve the necessary differential equations were additional constraints on its design.

The development and application of the model were carried out in several steps. A program was written for the IBM 360 digital computer to solve the flow equations for the prescribed set of boundary conditions and an arbitrary distribution of permeability. The computer program was verified for a confined, homogeneous, isotropic aquifer with simple boundary geometry by comparing the values calculated from the digital model against those obtained from the analytical solutions. Finally, the model was used in an attempt to treat numerically an aquifer under investigation at

Musquodoboit Harbour. Since it is impossible to solve analytically the equations of motion for this complex hydrologic system, an electric-analog simulator was constructed to test the precision of the digital model. The digital and the electric analogs were pumped for selected periods of time, and a comparison was made between the results of the two models. Finally, the transmissibility and recharge boundaries of the digital model were adjusted until pump-test results from the field were reproducible. The digital model was then subjected to a long pumping period in an effort to evaluate the aquifer as a source of supply of ground water for the village of Musquodoboit Harbour.

THEORY OF AQUIFER MODELS

The digital and electric-analog models are essentially techniques for solving the flow equations for specified aquifer parameters and boundary conditions. The theoretical basis for either approach, therefore, is a correct mathematical expression for describing the physics of flow of a homogeneous compressible fluid through a porous medium. The derivation of these equations is well known for a homogeneous, isotropic aquifer, and recently both *Papadopulos* [1965] and *Hantush* [1966] presented solutions for a homogeneous, anisotropic medium. The following discussion considers the two-dimensional flow equations for a homogeneous, slightly compressible fluid in a nonhomogeneous, anisotropic aquifer; from the finite-difference approximations of these equations an analogy is developed between the flow of fluid and electricity.

Equations of Flow

The velocity for two-dimensional laminar flow in an anisotropic, nonhomogeneous porous medium is described by Darcy's law, which may be expressed in matrix notation as

$$\begin{bmatrix} V_x \\ V_y \end{bmatrix} = - \begin{bmatrix} K_{xx} & K_{xy} \\ K_{yx} & K_{yy} \end{bmatrix} \begin{bmatrix} \frac{\partial h}{\partial x} \\ \frac{\partial h}{\partial y} \end{bmatrix} \qquad (1)$$

or in cartesian tensor notation using Einstein's convention as

$$V_i = -K_{ij} \, \partial h / \partial x_j \qquad (2)$$

where

V_x and V_y or V_i are the components of the Darcy velocity of flow vector (L/T);

K_{xx}, K_{xy}, K_{yx} and K_{yy} or K_{ij} are the components of the permeability tensor (L/T) and are a function of the space co-ordinates (x, y); and

h is the hydraulic head (L).

The differential equation describing nonsteady ground-water flow can be stated as [*Jacob*, 1950]

$$\mathrm{div}\left(K_{ij} \frac{\partial h}{\partial x_j}\right) = \frac{S}{b} \frac{\partial h}{\partial t} + \frac{L'}{\rho}(x, y, t) \qquad (3)$$

where L' is a source function.

For convenience the transmissibility tensor may be generated by multiplying the permeability tensor by the thickness of the aquifer and (3) becomes

$$\mathrm{div}\left(T_{ij} \frac{\partial h}{\partial x_j}\right) = S \frac{\partial h}{\partial t} + \frac{bL'}{\rho}(x, y, t) \qquad (4)$$

where $T_{ij} = bK_{ij}$ and T_{ij} is the transmissibility tensor. Equation 4 may also be expressed as

$$\frac{\partial}{\partial x_i}\left(T_{ij} \frac{\partial h}{\partial x_j}\right) = S \frac{\partial h}{\partial t} + \frac{bL'}{\rho}(x, y, t) \qquad (5)$$

Substituting $x_1 = x$ and $x_2 = y$, 5 may be written as

$$\frac{\partial}{\partial x}\left(T_{xx} \frac{\partial h}{\partial x}\right) + \frac{\partial}{\partial x}\left(T_{xy} \frac{\partial h}{\partial y}\right) + \frac{\partial}{\partial y}\left(T_{yx} \frac{\partial h}{\partial x}\right)$$
$$+ \frac{\partial}{\partial y}\left(T_{yy} \frac{\partial h}{\partial y}\right) = S \frac{\partial h}{\partial t} + \frac{bL'}{\rho}(x, y, t) \qquad (6)$$

Equation 6 may be simplified somewhat by selecting a new set of coordinate axes, which we will continue to call x and y, chosen in such a manner that the principal components of the transmissibility T_{xx} and T_{yy} are co-linear with the coordinate directions x and y. The terms T_{xy} and T_{yx} become zero under these conditions, and (6) reduces to

$$T_{xx} \frac{\partial^2 h}{\partial x^2} + \frac{\partial T_{xx}}{\partial x} \frac{\partial h}{\partial x} + T_{yy} \frac{\partial^2 h}{\partial y^2} + \frac{\partial T_{yy}}{\partial y} \frac{\partial h}{\partial y}$$
$$= S \frac{\partial h}{\partial t} + \frac{W'}{\rho}(x, y, t) b \qquad (7)$$

where W' is the function describing the mass flux of the fluid into or out of the system in the new coordinate system (M/L^3T). Let

$$W = W'(x, y, t)/\rho b (L/T)$$

Finite-Difference Approximations

To solve the equations of flow numerically or to see the analogy between Kirchhoff's law for the flow of electricity in a resistor capacitor network and the flow of fluid through a porous medium, it is necessary to develop the finite-difference approximations to equation 7. In the finite-difference approximation the derivative is replaced by values of the difference quotients of the function at separate discrete points. *Volynskii and Bukhman* [1965] show that

$$\frac{\partial h}{\partial x} \approx \frac{h_{i-1jk} - h_{i+1jk}}{2 \Delta x}$$

and

$$\frac{\partial^2 h}{\partial x^2} \approx \frac{h_{i-1jk} - 2h_{ijk} + h_{i+1jk}}{(\Delta x)^2} \qquad (8)$$

and similarly

$$\frac{\partial h}{\partial y} \approx \frac{h_{ij-1k} - h_{ij+1k}}{2 \Delta y}$$

and

$$\frac{\partial^2 h}{\partial y^2} \approx \frac{h_{ij-1k} - 2h_{ijk} + h_{ij+1k}}{(\Delta y)^2}$$

(9)

The approximation for the time derivative (see Figure 1) is expressed as [*Crank*, 1956]

$$\partial h/\partial t \approx (h_{ijk} - h_{ijk-1})/\Delta t \qquad (10)$$

where i, j, and k are the indices in the x, y, and t dimensions respectively, and Δx, Δy, and Δt are increments in the x, y, and t dimensions, respectively.

To obtain the finite-difference approximation for the flow equation, expressions 8, 9, and 10 are substituted into 7 to give

$$T_{xx}\left[\frac{h_{i-1jk} - 2h_{ijk} + h_{i+1jk}}{(\Delta x)^2}\right]$$
$$- \frac{\partial T_{xx}}{\partial x}\left[\frac{h_{i-1jk} - h_{i+1jk}}{2 \Delta x}\right]$$
$$+ T_{yy}\left[\frac{h_{ij-1k} - 2h_{ijk} + h_{ij+1k}}{(\Delta y)^2}\right]$$
$$- \frac{\partial T_{yy}}{\partial y}\left[\frac{h_{ij-1k} - h_{ij+1k}}{2 \Delta y}\right]$$
$$= S\left[\frac{h_{ijk} - h_{ijk-1}}{\Delta t}\right] + W(x, y, t) \qquad (11)$$

To simplify the theoretical development to be considered shortly, 11 may also be written as

$$A_{i-1j}h_{i-1jk} + A_{ij-1}h_{ij-1k} + A_{i+1j}h_{i+1jk}$$
$$+ A_{ij+1}h_{ij+1k} - 4A_{ij}h_{ijk}$$
$$= S\left[\frac{h_{ijk} - h_{ijk-1}}{\Delta t}\right] + W(x, y, t)$$

where

$$A_{i-1j} = T_{xx}/(\Delta x)^2$$
$$- (1/2 \Delta x) \partial T_{xx}/\partial x \qquad (11a)$$
$$A_{ij-1} = T_{yy}/(\Delta y)^2$$
$$- (1/2 \Delta y) \partial T_{yy}/\partial y \qquad (11b)$$
$$A_{i+1j} = T_{xx}/(\Delta x)^2$$
$$+ (1/2 \Delta x) \partial T_{xx}/\partial x \qquad (11c)$$
$$A_{ij+1} = T_{yy}/(\Delta y)^2$$
$$+ (1/2 \Delta y) \partial T_{yy}/\partial y \qquad (11d)$$

and

$$A_{ij} = A_{i-1j} + A_{ij-1}$$
$$+ A_{i+1j} + A_{jj+1} \qquad (11e)$$

Saul'yev [1964] shows that in the case of variable, and even discontinuous coefficients, the transmissibility value lying midway between the nodal points can be used in the

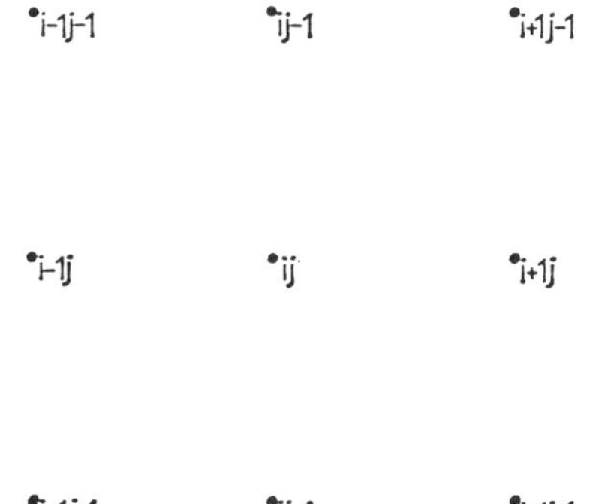

Fig. 1. Nodal array for development of finite-difference expressions.

finite-difference expression. The justification and significance of this technique may be shown in the following manner.

To obtain an expression that utilizes the transmissibility at the midpoint between two nodes, a Taylor series is written about the points (ij) and $(i-1j)$. Consider, for simplicity, the equation in one dimension only. The transmissibility midway between (ij) and $(i-1j)$ is given by

$$T_{xx(i-1/2j)} = T_{xx(ij)} - \frac{\Delta x}{2}\frac{\partial T_{xx(ij)}}{\partial x}$$
$$- \frac{(\Delta x)^2}{8}\frac{\partial^2 T_{xx(ij)}}{\partial x^2} \quad (12)$$

or, rearranging

$$T_{xx(ij)} = T_{xx(i-1/2j)} + \frac{\Delta x}{2}\frac{\partial T_{xx(ij)}}{\partial x}$$
$$+ \frac{(\Delta x)^2}{8}\frac{\partial^2 T_{xx(ij)}}{\partial x^2} \quad (13)$$

similarly

$$T_{xx(i-1j)} = T_{xx(i-1/2j)}$$
$$- \frac{\Delta x}{2}\frac{\partial T_{xx(i-1j)}}{\partial x} - \frac{(\Delta x)^2}{8}\frac{\partial^2 T_{xx(ij)}}{\partial x^2} \quad (14)$$

and finally

$$T_{xx(ij)} = T_{xx(i+1/2j)}$$
$$- \frac{\Delta x}{2}\frac{\partial T_{xx(ij)}}{\partial x} - \frac{(\Delta x)^2}{8}\frac{\partial^2 T_{xx(ij)}}{\partial x^2} \quad (15)$$

Equations 11 and 12 can now be used to calculate the coefficients at each node as follows:

$$\frac{\partial}{\partial x}\left(T_{xx}\frac{\partial h}{\partial x}\right)_{ijk} = A_{i-1j}h_{i-1jk} + A_{i+1j}h_{i+1jk} - A_{i-1j}h_{i-1jk} - A_{i+1j}h_{ijk}$$
$$= A_{i-1j}(h_{i-1jk} - h_{ijk}) + A_{i+1j}(h_{i+1jk} - h_{ijk})$$
$$= \left[\frac{T_{xx(ij)}}{(\Delta x)^2} - \frac{1}{2\Delta x}\frac{\partial T_{xx(ij)}}{\partial x}\right][h_{i-1jk} - h_{ijk}]$$
$$+ \left[\frac{T_{xx(ij)}}{(\Delta x)^2} + \frac{1}{2\Delta x}\frac{\partial T_{xx(ij)}}{\partial x}\right][h_{i+1jk} - h_{ijk}] \quad (16)$$

similarly

$$\frac{\partial}{\partial x}\left(T_{xx}\frac{\partial h}{\partial x}\right)_{i-1jk} = \left[\frac{T_{xx(i-1j)}}{(\Delta x)^2} - \frac{1}{2\Delta x}\frac{\partial T_{xx(i-1j)}}{\partial x}\right][h_{i-2jk} - h_{i-1jk}]$$
$$+ \left[\frac{T_{xx(i-1j)}}{(\Delta x)^2} + \frac{1}{2\Delta x}\frac{\partial T_{xx(i-1j)}}{\partial x}\right][h_{ijk} - h_{i-1jk}] \quad (17)$$

Substituting equation 13 and 15 into equation 16, the finite-difference approximation becomes

$$\frac{\partial}{\partial x}\left(T_{xx}\frac{\partial h}{\partial x}\right)_{ijk}$$
$$= \left\{\frac{1}{(\Delta x)^2}\left[T_{xx(i-1/2j)} + \frac{\Delta x}{2}\frac{\partial T_{xx(ij)}}{\partial x}\right]\right.$$
$$\left. - \frac{1}{2\Delta x}\frac{\partial T_{xx(ij)}}{\partial x}\right\}(h_{i-1jk} - h_{ijk})$$
$$+ \left\{\frac{1}{(\Delta x)^2}\left[T_{xx(i+1/2j)} - \frac{\Delta x}{2}\frac{\partial T_{xx(ij)}}{\partial x}\right]\right.$$
$$\left. + \frac{1}{2\Delta x}\frac{\partial T_{xx(ij)}}{\partial x}\right\}(h_{i+1jk} - h_{ijk})$$

or

$$T_{xx(i-1/2j)}\frac{h_{i-1jk} - h_{ijk}}{(\Delta x)^2}$$
$$+ T_{xx(i+1/2j)}\frac{h_{i+1jk} - h_{ijk}}{(\Delta x)^2} \quad (18)$$

Similarly, considering the finite-difference equation for node h_{i-1jk} and the coefficient of the term $(h_{i-1jk} - h_{ijk})$ only

$$\frac{\partial}{\partial x}\left(T_{xx}\frac{\partial h}{\partial x}\right)_{i-1jk}$$
$$= \left[\frac{1}{(\Delta x)^2}\left[T_{xx(i-1/2j)} - \frac{\Delta x}{2}\frac{\partial T_{xx(i-1j)}}{\partial x}\right]\right.$$
$$\left. + \frac{1}{2\Delta x}\frac{\partial T_{xx(i-1j)}}{\partial x}\right](h_{i-1jk} - h_{ijk})$$
$$= \frac{T_{xx(i-1/2j)}}{(\Delta x)^2}(h_{ijk} - h_{i-1jk}) \quad (19)$$

In equations 18 and 19 the Taylor series substitutions have been truncated after the second term. Expressions of the form $(\Delta x)^2/8 \; [\partial^2 T_{xx(i,j)}]/\partial x^2$ have been neglected; these expressions are negligible in a medium in which the transmissibility is a continuous function and will be undefined at any point where it is discontinuous. The final finite-difference approximation for the flow equation has the form

$$T_{xx(i-1/2 j)} \left(\frac{h_{i-1jk} - h_{ijk}}{(\Delta x)^2} \right)$$
$$+ T_{xx(i+1/2 j)} \left(\frac{h_{i+1jk} - h_{ijk}}{(\Delta x)^2} \right)$$
$$+ T_{yy(ij-1/2)} \left(\frac{h_{ij-1k} - h_{ijk}}{(\Delta y)^2} \right)$$
$$+ T_{yy(ij+1/2)} \left(\frac{h_{ij+1k} - h_{ijk}}{(\Delta y)^2} \right)$$
$$= S \left[\frac{h_{ijk} - h_{ijk-1}}{\Delta t} \right] + W(x, y, t) \quad (20)$$

The approximate expressions for the partial derivatives given by (18) and (19) illustrate two important points: (1) the coefficients for the terms $(h_{i+1jk} - h_{ijk})$ and $(h_{ijk} - h_{i-1jk})$ are identical and are given by the arithmetic mean of $T_{xx(i+1,j)}$ and $T_{xx(i,j)}$: this is an essential property of a finite-difference expression if it is to be used in an electric-analog model [*Volynskii and Bukhman*, 1965]; (2) terms of the form $[\partial T_{xx(i,j)}]/\partial x$ do not appear in the final expression for the finite-difference approximation when the mean values of the transmissibility are used.

Sources and Sinks

The term $W(x, y, t)$ has been used heretofore to represent conveniently the net effect of recharge and discharge from the aquifer system. Included in this term were man-made effects such as pumping and artificial recharge, as well as the flux from vertical leakage. The net effect of sources and sinks may, therefore, be represented in finite-difference form by the expression

$$W(x, y, t) \, \Delta x \, \Delta y$$

$$= Q(x, y, t) - K_z \frac{H_0(x, y) - h(x, y, t)}{m(x, y)} \Delta x \, \Delta y \quad (21)$$

where

$Q(x, y, t)$ is the volume rate of withdrawal or injection at any node (L^3/T);

K_z is the hydraulic conductivity in the confining layer (L/T);

$H_0(x, y)$ is the head at the water table referred to the same datum as the aquifer head (L); and

$m(x, y)$ is the thickness of the confining layer (L).

Boundary Conditions

The two kinds of boundary conditions generally encountered in problems of flow through porous mediums are described when the head (Dirichlet problem) or the derivative of the head (Neumann problem) is specified.

In the treatment of the problems considered here, both the Dirichlet and the Neumann conditions were encountered. An impermeable boundary is a kind of Neumann condition and is treated by making the head on each side of the boundary equal; this makes the derivative of the head normal to the boundary zero, and no flow occurs across the boundary. A barrier boundary may also be treated by making the hydraulic conductivity beyond the boundary zero. Both the Neumann condition and the zero-permeability technique were employed.

A constant-head boundary is encountered in the case of a fully penetrating, hydraulically connected river or lake; at this boundary the head is defined at the appropriate nodes at any point in time.

Application of the Electric-Analog Model

The construction of an electric-analog simulator was thought desirable in this research to verify the technique employed in the digital approach.

The resistor-capacitor network shown in Figure 2 can be used to simulate unsteady-state, two-dimensional flow in a confined aquifer. From Kirchhoff's law for an electrical network the flow of current to any node may be written as

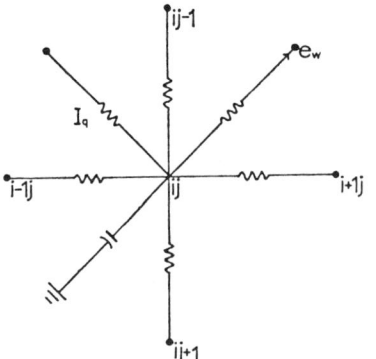

Fig. 2. Resistor-capacitor network for development of the electric analog.

$$\frac{1}{R_{xx(i-1/2j)}} (e_{i-1j} - e_{ij}$$
$$+ \frac{1}{R_{yy(ij-1/2)}} (e_{ij-1} - e_{ij})$$
$$+ \frac{1}{R_{xx(i+1/2j)}} (e_{i+1j} - e_{ij})$$
$$+ \frac{1}{R_{yy(ij+1/2)}} (e_{i+1j} - e_{ij})$$
$$= C \frac{\partial e}{\partial t} + I_q + \frac{1}{R_z} (e_{ij} - e_W) \quad (22)$$

where

e is the electromotive force (volts);
I is the current (coulombs/sec);
R is the resistance (volt-sec/coulomb); and
C is the capacitance (coulombs/volt).

The analogy between equations 20 and 22 is apparent.

To design an electric-analog model for aquifer evaluation the geologic parameters of the aquifer (spatial distribution of transmissibility and storage) and the boundary conditions of the hydrologic system must be determined. This information is usually obtained from geologic maps, well logs, and, where possible, pump test records. Unfortunately, these data are generally difficult to obtain, and the largest error in aquifer analysis is in the interpretation of the available geohydrologic information.

Once the field data have been processed, a contoured transmissibility and storage map of the area of interest is usually prepared. A rectangular grid is superimposed on this map, and a mean value of the transmissibility and storage between contours is assigned to all points within the area bounded by the contours. Although this procedure reduces the number of different resistors and capacitors necessary to construct the model, it generates a discontinuity at the boundary between different values of the lumped aquifer parameters.

Head data that record the response of the hydrologic system to known and described stresses are necessary to check and adjust the original assumptions for the distribution and magnitude of the aquifer parameters and thereby obtain maximum precision for the model. Combinations of resistors and capacitors are adjusted until the response of the model closely reproduces the historical data. Once a historical match is obtained it is possible to apply any desired pumping stress to the system and to extrapolate the response of the aquifer into the future. A review of the mechanics for the design and construction of electric analogs is given in *Karplus* [1958], *Walton and Prickett* [1963], and *Patten* [1965].

Digital Model

The digital model is related to the electric analog, inasmuch as both techniques are designed to solve the equations of flow for the complex hydrologic systems encountered in nature, and both are based upon finite-difference approximations to the flow equations. The digital model, however, evaluates the finite-difference expressions mathematically, whereas the electric analog simulates the flow of ground water, and the response of the system is measured directly. The relative merits of each approach will be discussed later.

Since a very large number of arithmetic calculations are required to solve even the simplest ground water system, the practical application of the finite-difference mathematical scheme depends upon the accessibility of a high-speed digital computer with a large memory capacity. The University of Illinois IBM 360 model 50/75 with a memory of 64 K words was found adequate for a problem with a moderate number of nodes.

The equation of flow as described by (20) must be written for each node of the finite-difference network to completely define the physics of the system. This procedure generates N equations in N unknowns, where N is the number of nodes in the network. There are several efficient methods of solving simultaneous equations; an elementary discussion of the various techniques may be found in *McCracken and Dorn* [1964], and more comprehensive treatments are presented in *Todd* [1962], *Varga* [1962], *Saul'yev* [1964], *Ralston and Wilf* [1965], and *Richtmeyer* [1967].

In this program the alternating-direction technique developed by *Peaceman and Rachford* [1955] and by *Douglas and Peaceman* [1955] was selected for solution of the equations. It has been used extensively by the petroleum industry for predicting oil and gas reservoir behavior. To develop the necessary difference expressions, equation 20 is rewritten for the case of a square mesh of side length Δx as follows:

$$T_{xx(i-1/2j)}(h_{i-1jk} - h_{ijk})$$
$$+ T_{xx(i+1/2j)}(h_{i+1jk} - h_{ijk})$$
$$+ T_{yy(ij-1/2)}(h_{ij-1k} - h_{ijk})$$
$$+ T_{yy(ij+1/2)}(h_{ij+1k} - h_{ijk})$$
$$= S \frac{h_{ijk} - h_{ijk-1}}{\Delta t} (\Delta x)^2 + W(x, y)(\Delta x)^2 \quad (23)$$

The expressions are similar to (23), except that each has unknown values of head in only one co-ordinate direction, the known head values that were calculated during the kth time step

$$T_{xx(i-1/2j)}(h_{i-1jk} - h_{ijk})$$
$$+ T_{xx(i+1/2j)}(h_{i+1jk} - h_{ijk})$$
$$+ T_{yy(ij-1/2)}(h_{ij-1k+1/2} - h_{ijk+1/2})$$
$$+ T_{yy(ij+1/2)}(h_{ij+1k+1/2} - h_{ijk+1/2})$$
$$= S \frac{(h_{ijk+1/2} - h_{ijk})}{\Delta t/2} (\Delta x)^2 + W(x, y)(\Delta x)^2$$
$$(24)$$

and

$$T_{xx(i-1/2j)}(h_{i-1jk+1} - h_{ijk+1})$$
$$+ T_{xx(i+1/2j)}(h_{i+1jk+1} - h_{ijk+1})$$
$$+ T_{yy(ij-1/2)}(h_{ijk+1/2} - h_{ijk+1/2})$$

$$+ T_{yy(ij+1/2)}(h_{ij+1k+1/2} - h_{ijk+1/2})$$
$$= S \frac{(h_{ijk+1} - h_{ijk+1/2})}{\Delta t/2} (\Delta x)^2$$
$$+ W(x, y)(\Delta x)^2 \quad (25)$$

During one complete cycle of $2N$ calculations the entire matrix is solved twice, once using each of the above equations. There is a maximum of three unknowns for each of the N equations generated by (24); in this case equations are implicit in the y direction and explicit in the x direction. The equations obtained for each node using (25) are implicit in the x direction and explicit in the y direction. In the alternating-direction technique, therefore, values are calculated first by sweeping the matrix column by column (using equation 24), and then the entire matrix is recalculated sweeping row by row (using equation 25). The main advantage of this scheme over an entirely explicit technique is that it is unconditionally stable, and there are no stability restrictions on the size of Δt as in the explicit technique. The advantage of this method over Gaussian elimination is that the computer storage necessary is greatly reduced. *Eshett and Longenbaugh* [1965] report that computing time for a 50-grid model is approximately the same for both the alternating direction and Gaussian elimination techniques.

For any column the finite-difference expression may be rearranged for ease of calculation to read

$$T_{yy(ij-1/2)}h_{ij-1k+1/2}$$
$$- [T_{yy(ij+1/2)} + T_{yy(ij-1/2)} + \gamma]h_{ijk+1/2}$$
$$+ T_{yy(ij+1/2)} h_{ij+1k+1/2}$$
$$= T_{xx(i-1/2j)} h_{i-1jk}$$
$$+ [T_{xx(i-1/2j)} + T_{xx(i+1/2j)} - \gamma]h_{ijk}$$
$$- T_{xx(i+1/2j)} h_{i+1jk} + W(x, y)(\Delta x)^2$$
$$(26)$$

where

$$\gamma = [2S(\Delta x)^2]/\Delta t$$

(An analogous expression may be developed for calculating along rows.) These equations are of the form

$$B_1 h_1 + C_1 h_2 = D_1 \quad (27)$$

$$A_i h_i - 1 + B_i h_i + C_i h_i + 1 = D_i$$
$$2 \leq i \leq \eta - 1 \quad (28)$$

$$A_\eta h_\eta - 1 + B_\eta h_\eta \quad (29)$$

where

η is the length of the row or column under consideration;

A, B, and C are the coefficients of the unknown head values;

D is the sum of all known parameters (terms occurring on the right hand side of equation 26).

Equations 27 and 29 are modified forms of equation 28. Equation 27 is written for the first node of either a row or a column, since at this point in the network the head at $(j-1)$ is defined. If the edge of the network is a barrier boundary and the reflection technique is employed

$$h_{j-1} = h_{j+1}$$

and

$$A_1 = C_1$$

whereby A_1 is dropped from the expression, and C_1 is replaced by $2C_1$. When the transmissibility is given a zero value, equation 28 immediately reduces to 29, although the location of the boundary is shifted $\Delta \times 12$ outside the edge of the network. When the head is specified at the boundary, the value of h_{j-1} is given, and the first term of equation 28 is transferred to the right-hand side, again reducing (28) to the form of (27).

Equation 28 becomes of the form of (29) when the last node in the row or column is reached. At this point the $(j+1)$ node will be defined according to the boundary condition encountered at the edge of the network, and the last term of the expression will be dropped as in the case of the first node.

Recalling from equation 21 that $W(x, y, t)$ actually encompasses several source and sink functions, we can write

$$W_{ijk} = Q_{(ijk)}$$
$$- K_z \frac{[H_{0(ij)} - h_{ijk} - h_{ijk-1}]}{2m_{(ij)}} \Delta x \Delta y \quad (30)$$

In the digital model the term $Q(x, y, t)$, representing a pumping or recharge well, is written as a known value in cfs (cubic feet per second) at the specified nodes. The network can accommodate up to N different wells pumping at various rates, which themselves may be time-dependent functions. The leakage, however, is not as easily handled; since the term h_{ijk} is an unknown value it is necessary to transfer the expression $(K_z h_{ijk})/2m_{(ij)} \Delta x \Delta y$ to the left-hand side of equation 26, where it is incorporated into the Bj coefficient. The expression $-K_z[H_{0(ij)} - h_{ijk-1}]/[2m_{(ij)}]$ contains only known values and remains on the right-hand side of equation 26 to become part of D_j.

To solve equations 27, 28, and 29 we make the following substitutions: let

$$w_i = (B_i - A_i C_{i-1})/w_{i-1}$$
$$g_i = (D_i - A_i g_{i-1})/w_i$$

where wj and gj are dummy variables. The values of wj and gj for any row or column are computed in order of increasing j. When the last node in the row or column is reached, the head is calculated in order of decreasing j as follows:

$$h_i = g_i - (C_i h_{i+1})/w_i$$

It is apparent that at the ηth node the parameter C_j will not exist, and this expression will reduce to

$$h_i = g_i$$

A complete discussion of the application of this technique to boundary-value problems encountered in heat flow is presented in *Douglas and Peaceman* [1955]. The techniques were developed for heat-flow problems and generally are directly applicable to flow through porous mediums.

The geohydrologic information required to solve an aquifer problem is the same whether the digital computer and the finite-difference technique or the electric analog is used. The transmissibility and storage coefficient values at each node representing the aquifer must be known. These values are fed into the computer as two-dimensional arrays. Other input parameters are as follows:

1. size of the grid intervals Δx and Δy;
2. function defining the length of the time steps for each cycle of calculations Δt;
3. the location of the boundaries, including lakes and streams;
4. the constant head values for Dirichlet problems;
5. the recharge or discharge from wells;
6. the thickness of the leaky confining layer m;
7. the hydraulic conductivity of the leaky confining layer K_s;
8. the initial areal distribution of head in the aquifer.

Analytical Checks

To check the program, problems were initially selected that were described by known analytic functions. The approach permitted a theoretical check on the validity of the model as differing boundary conditions were considered. Problems involving nonhomogeneous, anisotropic porous mediums and irregular aquifer geometry could not be verified using analytical methods. To study the precision of the digital technique in problems of this kind an electric-analog, resistor-capacitor network was constructed.

Step-Function Boundary: Semi-Infinite Aquifer

The hydrologic system consists of an aquifer A, confined above and below by impermeable stratigraphic units as shown in Figure 3. The aquifer is bounded at x equal to zero by a fully penetrating hydraulically connected stream or lake; the aquifer is assumed to extend to infinity in the positive x direction. Initially, $t < 0$, the hydrologic system is at equilibrium and the head in the aquifer is assumed to be everywhere equal to $h_0(x) = 0$. At time $t = 0$ the river level is raised to a new elevation. The migration of the influence of the new head (H'_0) through the aquifer is similar to the problem encountered in bank storage.

The boundary value problem is described for the homogeneous isotropic aquifer by

$$\partial^2 h/\partial x^2 = S/T \, \partial h/\partial t$$
$$h(x, 0) = 0$$
$$h(0, t) = H_0'$$
$$h(\infty, t) = 0$$

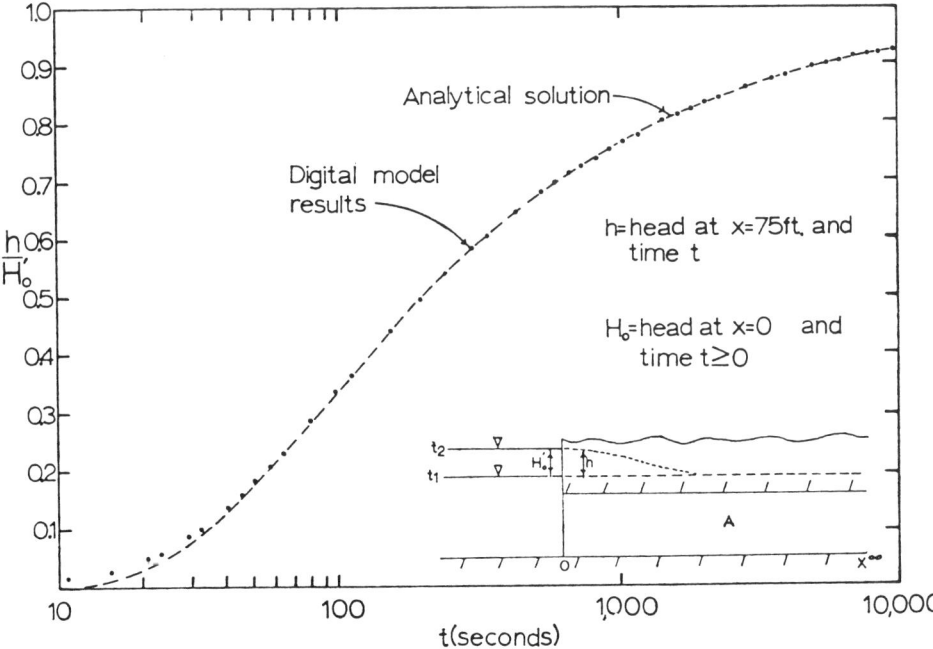

Fig. 3. Ratio of head in aquifer at $x = 75$ to head in stream versus time for a step function boundary, semi-infinite aquifer. Inset is a diagrammatic cross-section for the hydrologic problem.

when the initial head in the system is assumed to be zero. An analytical solution for the problem is given by *Carslaw and Jaeger* [1959]

$$h/H_0' = \text{erfc } x/2\sqrt{Tt/S} \quad (31)$$

where erfc is the complementary error function.

Values calculated using the alternating-direction technique are compared with those obtained by evaluating (31) in Figure 3. It is apparent that at very early times the calculated values deviate from the theoretical solution; after approximately 60 seconds, however, the two methods provide essentially identical values.

Well in an Infinite Aquifer

The aquifer system consists of a pumping well located near the center of a confined, homogeneous, isotropic aquifer which is, by necessity, bounded in the digital program. The system is shown in Figure 4. The cone of depression, however, is considered only until the boundary effects become significant.

The boundary-value problem describing the flow of water in this system is the Theis condition, which may be stated

$$\nabla^2 h = (S/T) \, \partial h/\partial t$$

$$h(\infty, 0) = 0 \quad t = 0$$

$$h(\infty, t) = 0 \quad t > 0$$

$$-K \frac{\partial h(0, t)}{\partial r} = V_r \quad t > 0$$

The analytical solution for this equation was developed by Theis in radial co-ordinates for an infinite aquifer, and the head is given by *Theis* [1935]

$$h = H_0 - \left[\frac{Q}{4\pi T}[-Ei(-u)]\right]$$

where

$-Ei(-u)$ is the exponential integral;

$$u = (r^2 S)/4Tt;$$

and

r is the distance from the pumping well to the observation well.

The Theis solution is compared with results computed by the digital program in Figure 4. The crosses on this diagram represent the calculated values using an initial time step of 0.1 second; the dots indicate values obtained

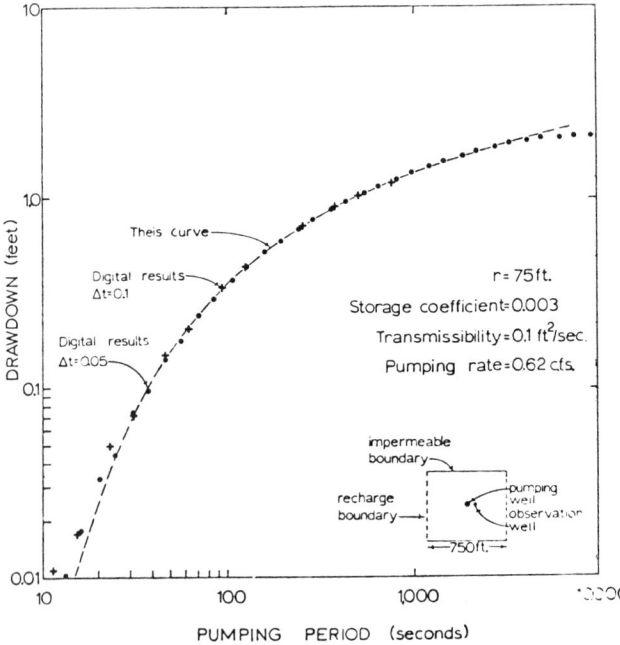

Fig. 4. Time-drawdown values; inset is a plan view of hydrologic problem.

with an initial time step of 0.05 second. Comparison of the two sets of values suggests that for early periods of pumping, with a pumping duration of less than 1 minute, more precise solutions may be obtained using smaller time steps. The calculated values deviate from the theoretical solution when the boundary effects of the aquifer become significant.

Well in a Bounded Aquifer

The presence of a constant-head recharge boundary in an otherwise infinite aquifer system with one pumping well was also considered. A solution to this problem can be obtained from the theory of images and is [*Stallman*, 1963c, among others]

$$h = H_0 - \left[\frac{Q}{4\pi T} \left[-Ei(u)_p + Ei(u)_i \right] \right]$$

where $-Ei(u)_p$ and $-Ei(u)_i$ are the values of $-Ei(u)$ for the pumped and image well, respectively

$$u_p = (r_p^2 S)/4Tt,$$
$$u_i = (r_i^2 S)/4Tt;$$

r_p is the distance from the pumped well to the observation well; and

r_i is the distance from the image well to the observation well.

The theoretical solution is compared with the digital results in Figure 5. In the early periods of pumping the calculated values deviate from the true solution, but after approximately 2 minutes the analytical and digital solutions are essentially the same. As in the case considered earlier, a more precise solution for early stages of pumping could have been obtained by selecting smaller initial values for Δt.

Error Analysis

As indicated earlier, the precision of the digital technique for the nonhomogeneous, anisotropic case with irregular boundary conditions could be checked with an electric analog. Both models of a field problem were pumped at the same rate for a specified duration to compare their responses to identical stresses. The potential at predetermined nodes of the

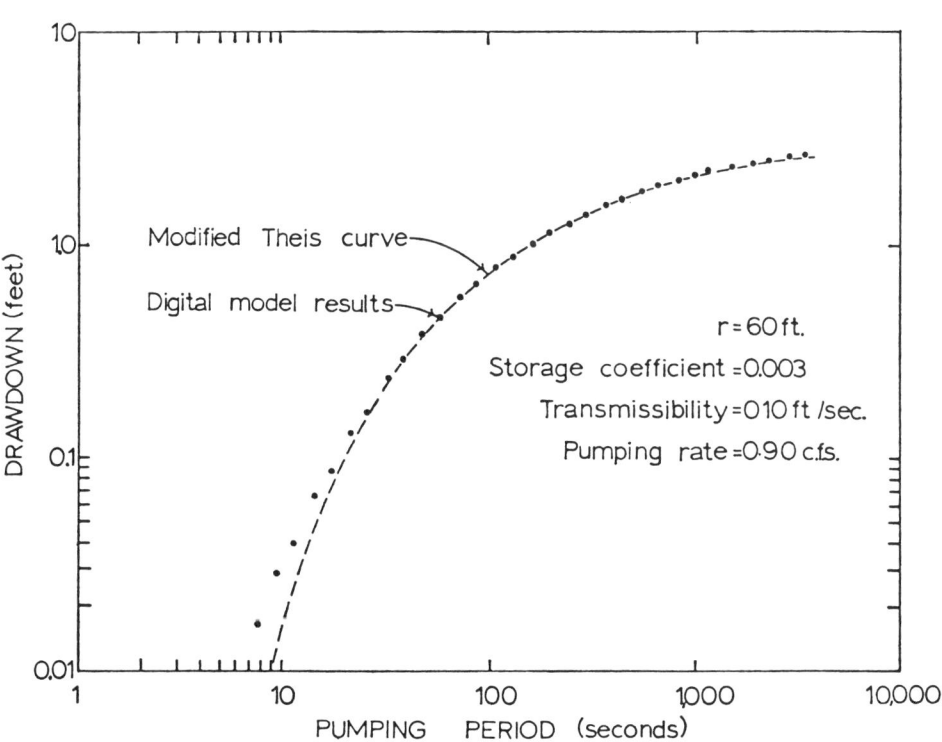

Fig. 5. Time-drawdown curve; homogeneous isotropic aquifer with a constant head boundary.

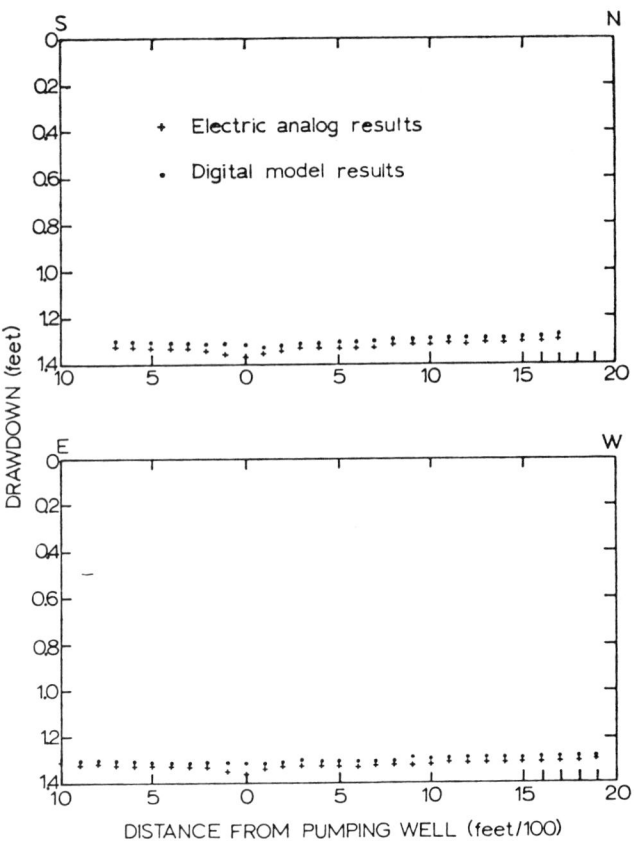

Fig. 6. Distance-drawdown values obtained from the digital and electric analog models after 5.6 days of pumping at a rate of 0.078 cfs.

electric analog was recorded and converted to head values; these results, along with calculated values from corresponding nodes in the digital model, are compared in Figure 6. Owing to the high transmissibility of the aquifer the cone of depression is relatively flat after 5.6 days of pumping; the difference between the values calculated at analogous points using the two techniques is, on the average, within 2% of the total drawdown.

Since the digital model was designed initially to obtain water only from storage, summation of water released from storage could be used as a measure of the model's precision. At any time the quantity released from storage is given by

$$Q = -S \int_0^{360} \int_0^R \mathcal{S} \, dr \, d\theta$$

or numerically

$$Q = -S \sum \sum (h_{ij0} - h_{ijk})$$

where

$\mathcal{S}(r, \theta)$ is the drawdown due to the well, and R is the radial distance to the aquifer boundary.

The error defined here as the arithmetic difference between the pumped volume calculated directly from the discharge rate and the volume determined from numerical integration of the cone of depression was observed to be a linear function dependent upon the duration of pumping during early time periods; after long pumping periods, however, the function apparently becomes more complex. The magnitude of the error for the same time depends upon the values of the transmissibility (see Figure 7).

It was noted above that the technique of using large areas of homogeneous transmissibility leads to a discontinuity in the space dependent transmissibility function. The use of a discontinuous function violates the mathe-

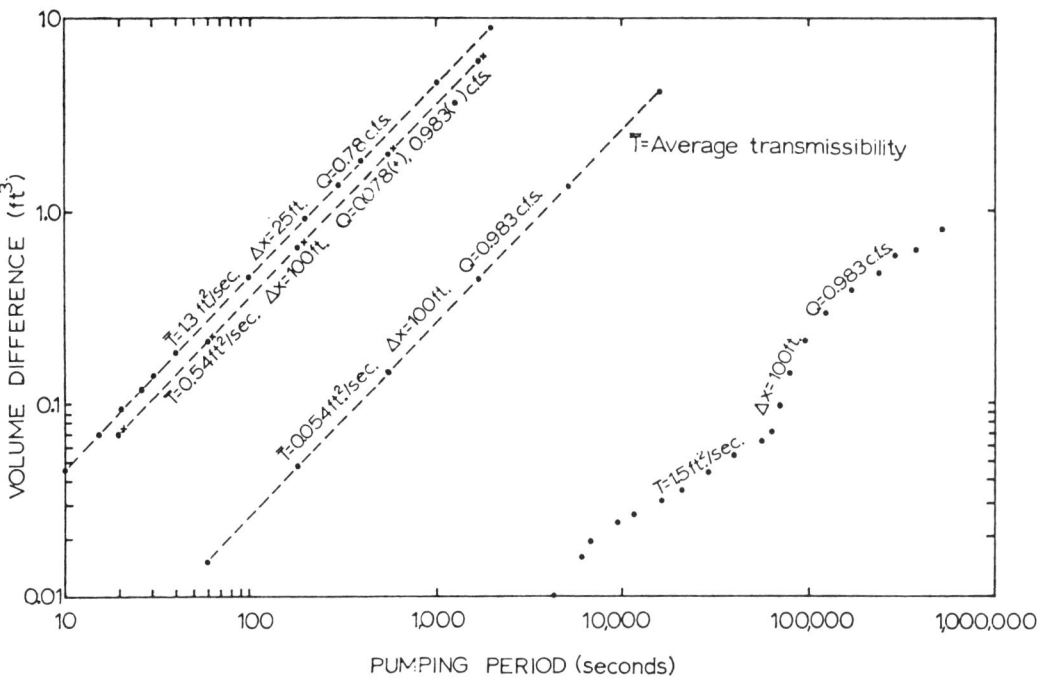

Fig. 7. Difference between two methods of calculating total discharge from the aquifer versus time since pumping started.

matical assumption of continuous function for which derivatives can be defined, and an error may be associated with equations 13, 14, and 15; this problem can be minimized by employing a reasonably continuous transmissibility function.

Varga, as quoted in *Spanier* [1967], suggested that the alternating-direction technique may not be entirely valid when applied to the nonhomogeneous, anisotropic case. He indicated that the assumption of the commutativity of the row and column solutions may not hold for the case where transmissibility is a space-dependent variable. To ensure that the theory of convergence may be applied rigorously, Spanier suggests a transformation matrix for his case that effectively relaxes the restrictions on the transmissibility functions. Since a transformation was not applied to equation 26, the use of variable coefficients may not be entirely justified, and we must proceed formally.

The most precise solution was obtained using an initial time step of 0.001 sec, which was increased in a geometric progression after each cycle of calculations according to the expression

$$\Delta t_{k+1} = \Delta t_k + 0.25(\Delta t_k)$$

The discrepancy between the computed quantity of water released from storage and that computed from the discharge rate varied from 0.001 to 5.0%.

HYDROLOGIC ANALYSIS OF AN AQUIFER AT MUSQUODOBOIT HARBOUR

The senior author has been engaged in a reconnaissance ground-water investigation in the Musquodoboit Valley (see Figure 8), under the auspices of the Nova Scotia Department of Mines. The village of Musquodoboit Harbour, located in the lower segment of the valley, depends entirely on domestic wells for a limited supply of poor quality water. Bedrock wells, on which most of the village now depends, provide water with iron concentrations in excess of 10 parts per million; shallow drift wells will not supply household demands during the summer months, owing to the thin saturated thickness and low permeability of the drift.

Well logs, surficial geologic mapping, and

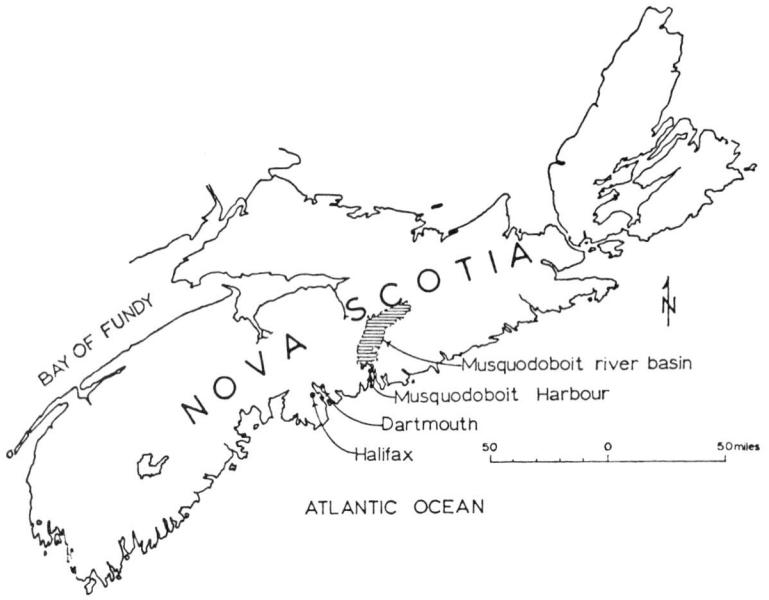

Fig. 8. Location map of the Musquodoboit river basin.

chemical analyses of selected ground water samples indicate that a local unconsolidated deposit should provide an adequate supply of very good quality water.

This aquifer is located ¼ mile NW of the village and is surrounded on all sides by exposed bedrock. Upstream, just west of the aquifer area, the river flows through a narrow rugged valley incised in a granite upland. Downstream, below the aquifer area, the river is confined for several hundred feet by a steep, narrow gorge before it empties into the head

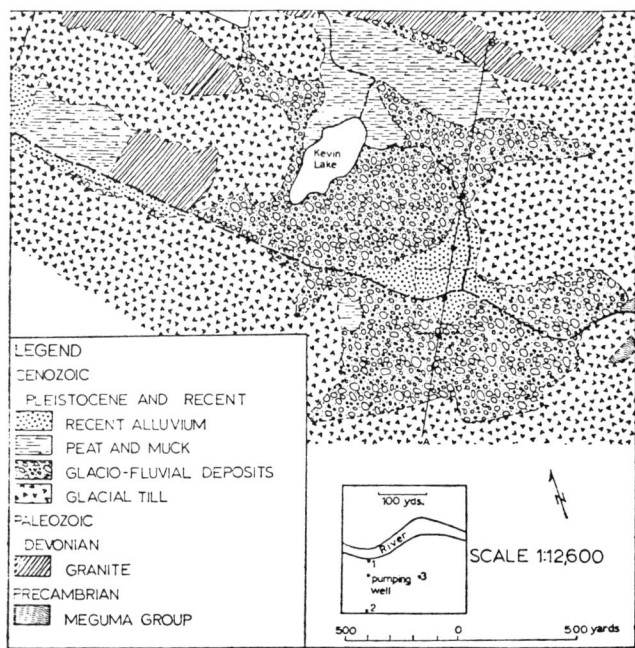

Fig. 9. Geologic map of Musquodoboit Harbour area, Nova Scotia. Inset is the well configuration for the pump test conducted on this aquifer (numbered wells are observation wells).

of a drowned estuary 7 miles from the Atlantic Coast.

The areal distribution of the unconsolidated deposits was outlined using air photo interpretation and detailed surficial mapping (Figure 9). A test site was selected based upon the geologic mapping. A pumping test was conducted in an effort to evaluate the aquifer transmissibility and storage and to determine whether the river was a significant source of recharge. Because of the close proximity of boundaries the pumping test is difficult, if not impossible, to analyze by the usual methods.

Hydrologic Units

Bedrock. The oldest rocks in the area are the slates and quartzites of the Meguma group, which are of questionable Precambrian age (see Figure 9). These rocks have been folded into a series of parallel, northeast-striking folds which are tightly compressed. The strata in the area dip at angles ranging from 60 to 90 degrees, and schistosity is particularly well developed in the more competent quartzitic beds. An elongate granite body intrudes the Meguma group about one mile north of Musquodoboit Harbour (Figure 9).

The porosity of both the granite and the metasediments is low owing to consolidation, cementation, and recrystallization of these rocks. The permeability of these units must be attributed to interconnected joints or local porous zones associated with fault planes. The contrast in permeability between the granite and metamorphic rocks and the valley fill is so great (approximately 10^6) that the bedrock was considered as impermeable in the aquifer analysis.

Glacial till. The oldest surficial material in the valley is the glacial till that was deposited as the ice sheet advanced southeastward towards the continental shelf and the sea. A sandy till overlies the resistant slates and quartzites of the Meguma group and generally occurs as a thin veneer on the upland areas. The permeability of the till unit suggests that it would yield satisfactory domestic supplies if it occurred in sufficiently thick and saturated deposits.

Glaciofluvial deposits. As the ice sheet began to retreat northward during late Pleistocene time, large quantities of meltwater discharging down the Musquodoboit River deposited coarse outwash sand, gravel, cobbles, and boulders in the central and lower segments of the valley. In the area of the aquifer the underfit Musquodoboit River flows in a typical U-shaped glacial valley filled with outwash.

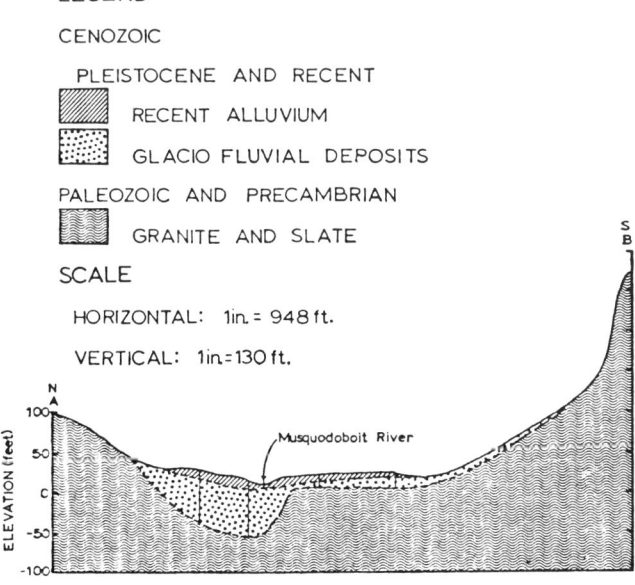

Fig. 10. Geologic cross-section, Musquodoboit Harbour area, Nova Scotia.

The glaciofluvial deposits at this location are composed of angular medium to coarse sand, gravel, cobbles, and boulders. These deposits, which are up to 65 feet in thickness, form a highly permeable aquifer (see Figure 10). In the axial position of the valley the deposits are coarsest and saturated to within 2 to 3 feet of the surface; toward the periphery the material becomes finer grained, and the unit thins, particularly north of the river.

Alluvium. Recent alluvium, in the form of floodplain deposits of sand, silt, and clay, covers large areas of the aquifer. Both the pump test and the well logs suggest that this less permeable material confines to a degree the more permeable outwash deposits.

Aquifer analysis. At the pump test site the aquifer, although irregularly shaped, is approximately 4800 feet wide and extends along the river about 5700 feet (see Figure 9). Granite bedrock contains the aquifer on three sides; slate and glacial till bound the deposit to the south. At the pumping well (see inset to Figure 9 for well configuration) 75 feet of interbedded medium sand to fine gravel overlies granite bedrock (see Figure 10). Near the river, at observation well number 1, 36 feet of glaciofluvial gravel and sand is overlain by 18 feet of alluvial silt and clay with sand and gravel lenses. The aquifer does not thin appreciably south of the river; approximately 450 feet southwest of the pumping well there is 62 feet of sand and gravel. On the northeast side of the river the aquifer thins rapidly; approximately 450 feet north from the river there is only 18 feet of unconsolidated material overlying slate bedrock. The saturated thickness of the aquifer at the pumping well at the time of the test was approximately 71 feet.

A 6-inch pumping well and three 4-inch observation wells were drilled for the test. A pumping test was conducted for 36 hours in which a constant discharge of 0.963 cubic foot per second (432 gallons per minute) was maintained. The test was discontinued when the water level in the pumping well became stable; as can be seen from Figure 11, the water levels in the observation wells had not stabilized at this time.

The aquifer parameters were initially calculated using the early segment of the draw-

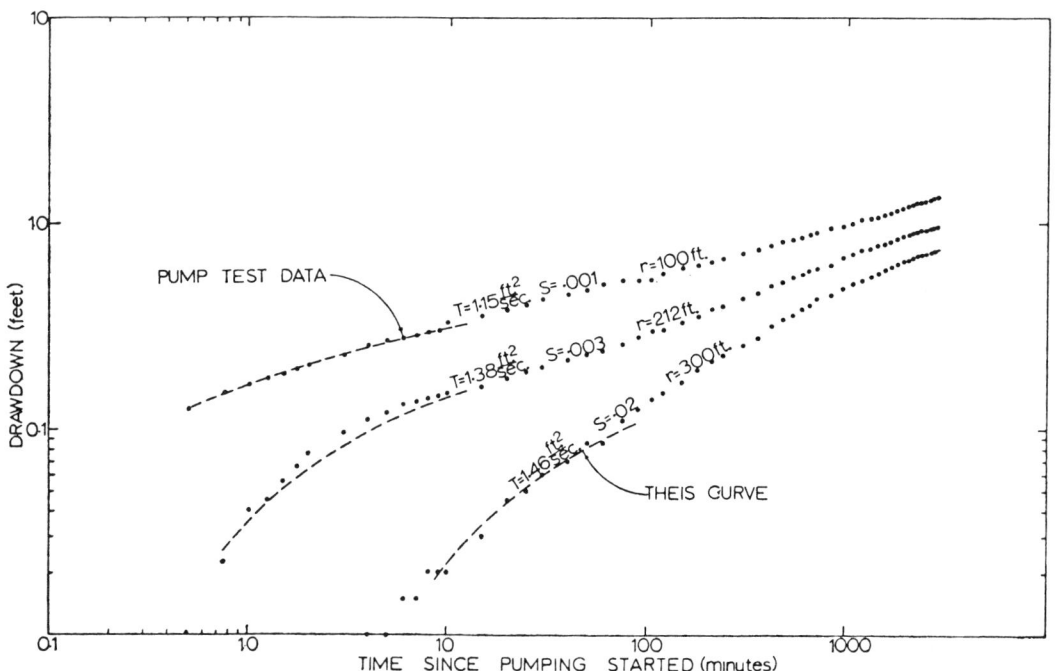

Fig. 11. Time-drawdown curves for a pump test conducted at the Musquodoboit Harbour aquifer.

down curve before the effects of the impermeable boundaries became significant. The fit with the Theis curve as well as the values of transmissibility and storage computed from these fits are indicated on Figure 11.

It is apparent that this aquifer is not fulfilling the assumptions of the Theis derivation, and that a number of T and S values could be calculated, depending upon the segment of the data thought to best fit the type curve. Moreover, it is obvious from these curves that a meaningful understanding of this aquifer system could not be obtained from the usual sort of pumping-test analysis.

Aquifer models. The geologic and hydrologic information were combined in the digital model of the system. An electric-analog simulator was also constructed to verify the results of the digital approach. Both models were constructed on a scale of 1 inch to 100 feet with 1145 nodes within the aquifer.

The electric-analog network consisted of 1145 capacitors and approximately 2300 resistors. The transmissibility distribution was initially estimated from the geologic information and from the initial analysis of the pumping-test data (Figure 12). Resistors were chosen using the average value between transmissibility contours, which resulted in five resistor values. An average storage coefficient of 0.003 was assumed everywhere. No sources of recharge were built into this model of the aquifer system.

Design and construction of the electric network were completed in approximately 38 hours.

To analyze this aquifer problem digitally, a 45 by 57 rectangular matrix was used in which approximately half the nodes were outside the aquifer area and were assigned a zero transmissibility. This approach provides a simplified technique for generalizing the model to any boundary configuration. Since the complex aquifer system at Musquodoboit Harbour undoubtedly receives recharge from the river, the ability to simulate this inflow was included in the digital model. Well logs suggest that the river acts as a source of recharge through a bed of low permeability beneath the stream.

Vertical inflow proportional to the drawdown, 'leakage', was permitted along the length of the river. The necessary flux of water was obtained by adjusting equation 30 at the ap-

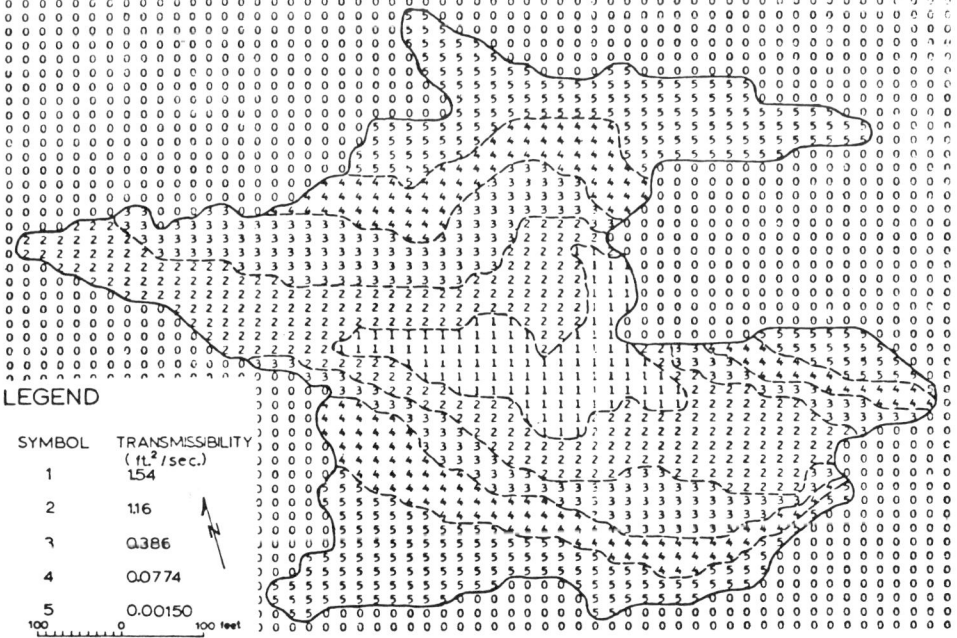

Fig. 12. Initial estimate of the transmissibility matrix determined from hydrogeologic information.

propriate nodes as follows:

$$W_{ijk} \Delta x \Delta y - Q_{ijk} - K_s \frac{H_{r(ij)} - h_{ijk} - h_{ijk-1}}{2m_{r(ijk)}} \Delta x \Delta y$$

where

K_s is the permeability of the stream bottom;
H_r is the head in the river; and
m_r is the thickness of the river bottom.

Results of analysis. Initially, transmissibility and storage-coefficient values as interpreted from the geology and the pumping-test results were introduced into the model. The aquifer parameters were then adjusted until a reasonable approximation for the time-drawdown curves for the three observation wells was obtained from the digital model. Three additional test wells drilled during the winter of 1967 in the area provided additional geologic data that modified our original ideas about the distribution of aquifer materials and suggested that the deposits thinned rapidly north of the river.

Approximately 37 computer runs involving various adjustments and combinations of the aquifer parameters were made before a satisfactory fit was obtained. This involved approximately 3 hours of IBM 360/75 computer time and does not include the initial investment in time necessary to develop and perfect the program. The final set of graphs in which the digital results are compared to the pumping-test results are shown in Figure 13. The transmissibility distribution as modified from test drilling and digital results is shown in Figure 14.

A decrease in transmissibility results in a greater drawdown after a given period of pumping. The storage coefficient affects the shape of the time-drawdown curve before equilibrium is reached in the aquifer system. The most pronounced effect of an increase in the storage coefficient was a decrease in the drawdown during the early periods of pumping. Although the storage coefficient value calculated from the pumping test data resulted in a reasonable response from the digital model during the first hour of pumping, later drawdown values were too large. The most accurate model response was obtained using a time-dependent storage coefficient. A value of

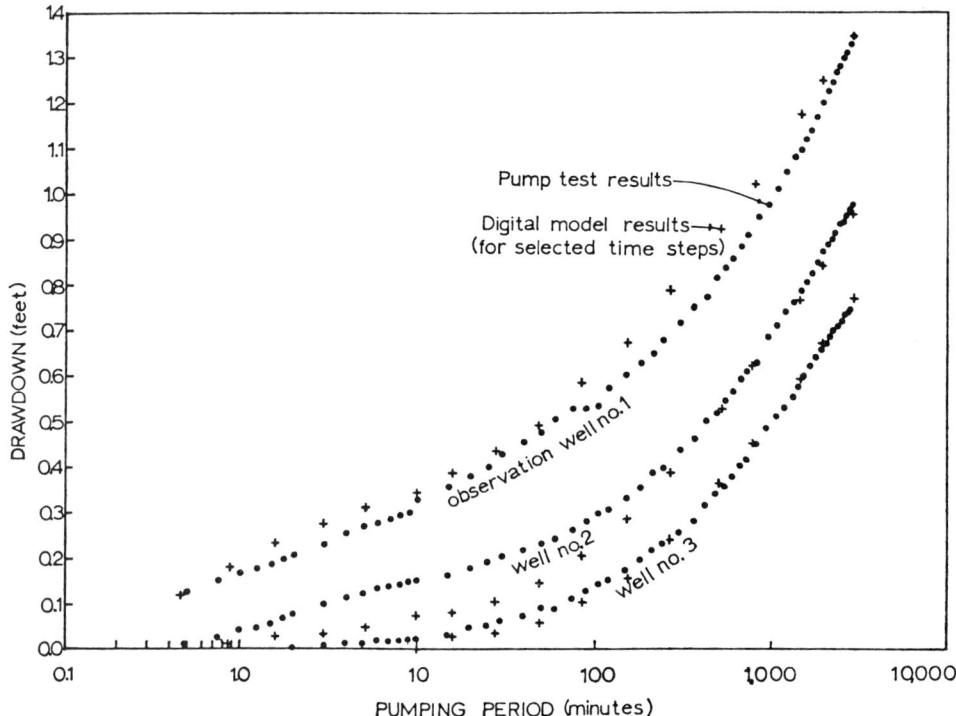

Fig. 13. Time-drawdown values observed in the field and calculated from the digital model.

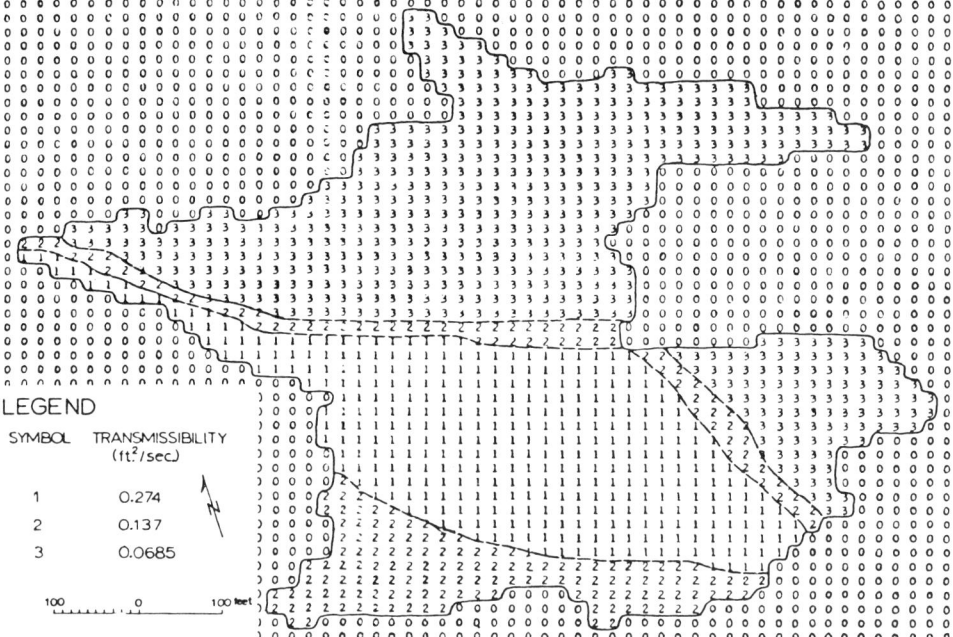

Fig. 14. Modified transmissibility matrix adjusted on the basis of three additional test-well logs and digital model results.

0.003 was introduced at the beginning of pumping and increased linearly to a maximum of 0.06 after 10 minutes. The time dependency of the storage coefficient implies delayed drainage in the aquifer system. *Boulton* [1954] suggested that two components of the storage coefficient be assumed to account for delayed drainage. One component that determines the instantaneous release of water with decline in head is independent of time, whereas the second term allows for delayed yield from storage and is a function of time. In our model, the storage coefficient is assumed to vary with time over the entire aquifer, which is at best only a crude approximation for the field situation.

Although the theoretical development of the digital model above assumes a confined aquifer, the drawdown is small, compared with the saturated thickness, and, therefore, equation 4 would describe the flow quite closely [*Jacob*, 1950].

Steady-flow conditions in the aquifer depend upon the quantity of water entering the system through the river bed. The closest approximation to the pump test results was obtained using a permeability value of 0.00002 foot/sec for 10-foot thickness of river bottom. The shape of the time-drawdown curve was changed markedly by adjusting this value as little as 0.000005 foot/sec.

The areal head distribution in the aquifer after selected pumping periods is indicated in Figures 15, 16, and 17. Owing to the high transmissibility of the aquifer, a rapidly expanding, flat cone of depression develops. The influence of the Musquodoboit River is felt within a minute after pumping begins, and after 300 minutes the drawdown at the closest impermeable boundary is greater than 0.1 foot. Pumping-test results for this system would be exceedingly difficult to interpret, since the complicated boundary conditions quickly influence the drawdown at the pumping well.

The drawdown for long pumping periods was computed for the three observation wells used in the pumping test (Figure 18). It is interesting to note that the steeply rising time-drawdown curve levels off rapidly after approximately 273 days of pumping and has essentially attained steady state after 5000 days. This reservoir could easily supply the village of Musquodoboit Harbour indefinitely at a rate of 0.963 cfs (approximately 0.6 mgd).

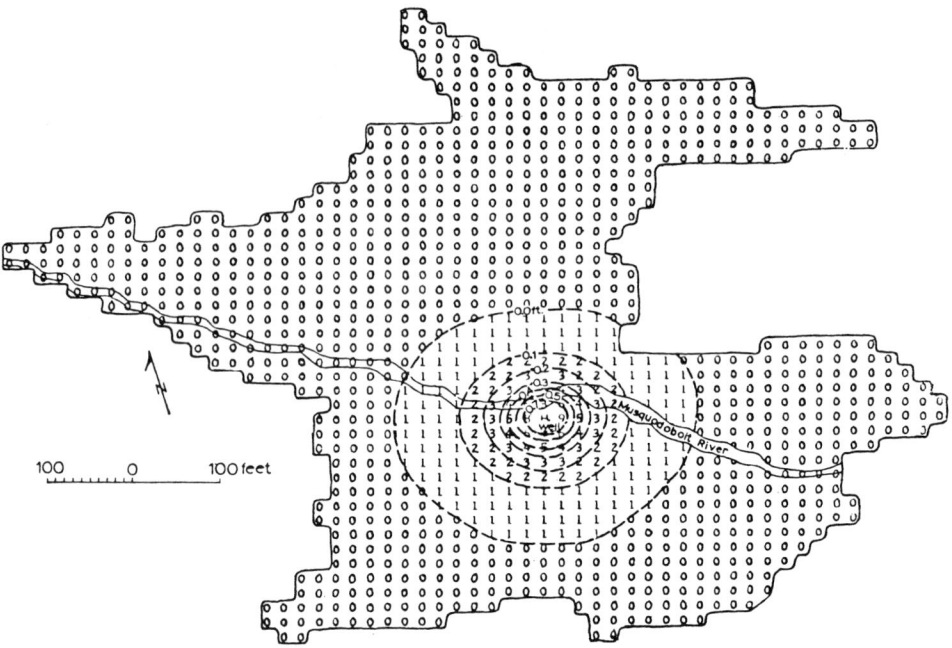

Fig. 15. Contoured computer output showing the piezometric surface determined from the digital model after 0.36 days of pumping at a rate of 0.963 cfs.

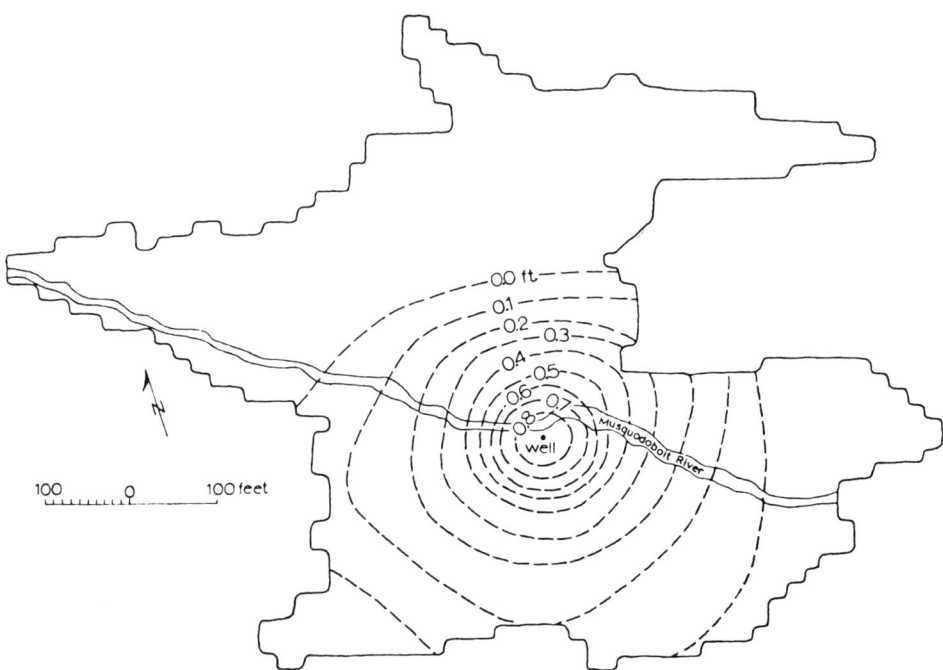

Fig. 16. Piezometric surface from the digital model after 1.58 days of pumping at a rate of 0.963 cfs.

Aquifer Evaluation

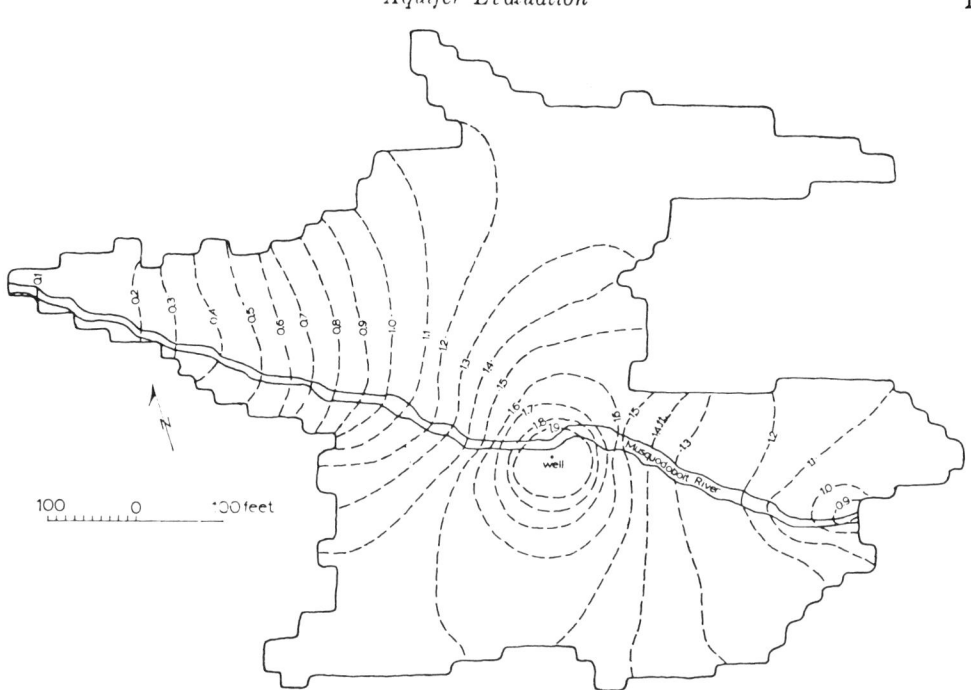

Fig. 17. Piezometric surface determined from the digital model after 206.65 days of pumping at a rate of 0.963 cfs.

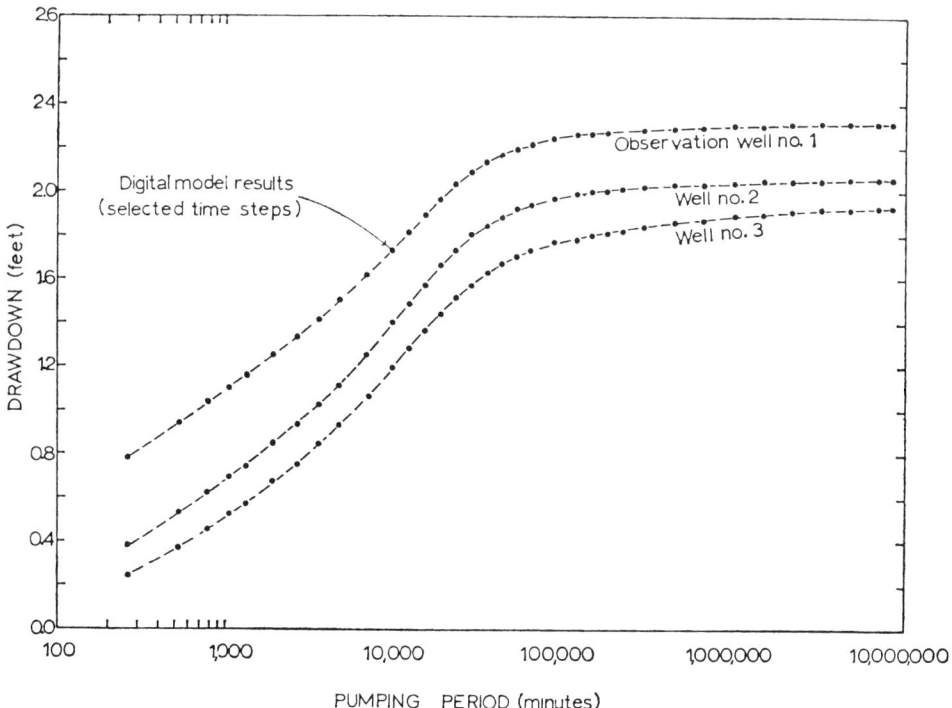

Fig. 18. Time-drawdown curves obtained from the digital model.

This quantity of water is more than adequate to supply the village needs for the immediate future.

SUMMARY AND CONCLUSIONS

A mathematical model was developed in which the finite-difference approximations to the flow equations were solved using an alternating direction scheme; the approach is flexible and efficient.

Model results were compared with analytical solutions for problems in homogeneous media of simple geometry and with electric-analog results for a realistic aquifer. A third check was made by comparing the volume of water pumped with the amount released from storage as computed by numerical integration of the cone of depression. All of these checks indicate that the model gives good results. Irregular boundaries as well as differing boundary conditions provide little difficulty, and the model is readily adjusted to reproduce historical potential data. The principal advantages of the mathematical approach are that

1. the basic program, once developed, can be reused, with some modification, for a variety of hydrologic problems;
2. the technique requires only the accessibility of a digital computer; the treatment of multiple wells and complex hydrologic boundaries does not demand elaborate electronic equipment as in the case of the electric analog;
3. the computer provides a rapid, comprehensive picture of the potential distribution in the aquifer after any desired period of pumping;
4. the digital program may be called from the computer tape library to provide a rapid and accurate picture of the potential distribution in the aquifer for use in other aspects of water resource system analysis.

Acknowledgments. The field investigation was sponsored by the Province of Nova Scotia under authority of the Agricultural and Rural Development Act, under the direction of J. Jones of the Groundwater Section, Geological Division, Nova Scotia Department of Mines. The authors are indebted to Professor C. L. Chen of the University of Illinois for his helpful suggestions concerning the finite-difference technique and to T. Maddock, III, of the U. S. Geological Survey for his guidance in the mathematical development of the flow equations.

REFERENCES

Allen, D. N., de G., *Relaxation Methods*, 257 pp., McGraw-Hill Book Company, New York, 1954.

Bittinger, M. W., H. R. Duke, and R. A. Longenbaugh, Mathematical simulations for better aquifer management, *IASH Hydrol. Symp. on Artificial Recharge and Management of Aquifers*, 15 pp., Haifa, Israel, 1967.

Boulton, N. S., Unsteady radial flow to a pumped well allowing for delayed yield from storage, *Intern. Assoc. Sci. Hydrol., Publ. 37*, 472–477, 1954.

Boulton, N. S., Analysis of data from non-equilibrium pumping tests allowing for delayed yield from storage, *Proc. Inst. Civil Engrs., 26*, 469–482, 1963.

Bredehoeft, J. D., H. H. Cooper, Jr., and I. S. Papadopulos, Inertial and storage effects in well-aquifer systems: An analog investigation, *Water Resources Res., 2*(4), 697–707, 1966.

Brown, R. H., Progress in ground water studies with the electric-analog model, *J. Am. Waterworks Assn., 54*(8), 943–958, 1962.

Bruce, W. A., An electrical device for analyzing oil-reservoir behavior, *Trans. Am. Inst. Mech. Engrs., 151*, 112–124, 1942.

Bruce, G. H., D. W. Peaceman, H. H. Rachford, and J. P. Rice, Calculations of unsteady-state gas flow through porous media, *Trans. Am. Inst. Mech. Engrs., 198*, 79–92, 1953.

Carslaw, H. S., and J. C. Jaeger, *Conduction of Heat in Solids*, 510 pp., Oxford University Press, London, 1959.

Crank, J., *The Mathematics of Diffusion*, 347 pp., Oxford University Press, London, 1956.

De Wiest, R. J. M., *Geohydrology*, 366 pp., John Wiley & Sons, New York, 1965.

Douglas, J., Jr., and D. W. Peaceman, Numerical solution of two dimensional heat flow problems, *J. Am. Inst. Chem. Eng., 1*(4), 505–512, 1955.

Douglas, J., Jr., and H. H. Rachford, Jr., On the numerical solution of heat conduction problems in two and three space variables, *Trans. Am. Math. Soc., 82*, 421–439, 1956.

Eshett, A., and R. A. Longenbaugh, Mathematical model for transient flow in porous media, *Progress Rept., Civil Eng. Sect., Colorado State Univ., Fort Collins, Colorado*, 14 pp., 1965.

Fayers, F. J., and J. W. Sheldon, The use of a high-speed digital computer in the study of the hydrodynamics of geologic basins, *J. Geophys. Res., 67*(6), 2421–2431, 1962.

Fiering, M. B., A digital computer solution for well field drawdown, *Bull. Intern. Assoc. Sci. Hydrol., 9*(4), 16–23, 1964.

Freeze, R. A., Theoretical analysis of regional ground water flow, Ph.D. dissertation, 304 pp., Univ. of California, Berkeley, 1966.

Hantush, M. S., Hydraulics of wells *in* Chow, V. T., (ed.), *Advances in Hydroscience*, Vol. 1, 282–432, Academic Press, New York, 1964.

Hantush, M. S., Wells in homogeneous anisotropic aquifers, *Water Resources Res.*, 2(2), 273-279, 1966.

Hantush, M. S., and C. E. Jacob, Non-steady radial flow in an infinite leaky aquifer, *Trans. Am. Geophys. Union*, 36, 95-100, 1955.

Jacob, C. E., Flow of groundwater *in* Rouse, H., (ed.), *Engineering Hydraulics*, John Wiley & Sons, New York, 321-386, 1950.

Karplus, W. J., *Analog Simulation*, 434 pp., McGraw-Hill Book Company, New York, 1958.

Klute, A., F. D. Whisler, and E. J. Scott, Numerical solution of the nonlinear diffusion equation for water flow in a horizontal soil column of finite length, *Soil Sci.*, 29(4), 353-358, 1965.

Liakopoulos, A. C., Theoretical solution of the unsteady unsaturated flow problems in soils, *Bull. Intern. Assoc. Sci. Hydrol.*, 10(1), 5-39, 1965.

McCracken, D. D., and W. S. Dorn, Numerical methods and fortran programming, 457 pp., John Wiley & Sons, Inc., New York, 1964.

Muskat, M., *The Flow of Homogeneous Fluids through Porous Media*, 763 pp., McGraw-Hill Books Company, New York, 1937.

Papadopulos, I. S., Nonsteady flow to a well in an infinite anisotropic aquifer, *IASH Symp. Dubrovnik*, 21-31, 1965.

Paschkis, V., and H. D. Baker, A method for determining unsteady-state heat transfer by means of an electric analogy, *Trans. Am. Soc. Mech. Engrs.*, 64, 105-112, 1942.

Patten, E. D., Jr., Design, construction, and use of electric analog models *in* Wood, L. A., Analog model study of ground water in Houston district, *Texas Water Comm. Bull. 6508*, 41-103, 1965.

Peaceman, D. W., and H. H. Rachford, Jr., The numerical solution of parabolic and elliptical difference equations, *J. Soc. Indust. Appl. Math.*, 3(11), 28-41, 1955.

Polubarinova-Kochina, P. Y., *Theory of Ground-Water Movement*, 613 pp., Princeton University Press, Princeton, New Jersey, 1962.

Prickett, T. A., Designing pumped well characteristics into electric analog models, *Ground Water*, 5(4), 38-46, 1967.

Ralston, A., and H. S. Wilf (ed.), *Mathematical Methods for Digital Computers*, 293 pp., John Wiley & Sons, New York, 1965.

Richtmeyer, R. D., and K. W. Morton, *Difference Methods for Initial-Value Problems* (second edition), 405 pp., Interscience Publishers, New York, 1967.

Rubin, J., Theoretical analysis of two dimensional transient flow of water in unsaturated and partly unsaturated soils, *Proc. Soil Sci. Soc. Am.* (in press), 1968.

Saul'yev, V. K., *Integration of Equations of Parabolic Type by the Method of Nets*, 346 pp., Pergamon Press, Oxford, England, 1964.

Skibitzke, H. E., Electronic computers as an aid to the analysis of hydrologic problems, *Bull. Intern. Assoc. Sci. Hydrol., Publ. 52*, 347-358, 1961.

Southwell, R. V., *Relaxation Methods in Theoretical Physics*, 248 pp., Oxford University Press, London, 1946.

Spanier, J., Alternating direction methods applied to heat conduction problems *in* Ralston, A., and H. S. Wilf (ed.), *Mathematical Methods for Digital Computers*, 11, John Wiley & Sons, New York, 1967.

Stallman, R. W., Numerical analysis of regional water levels to define aquifer hydrology, *Trans. Am. Geophys. Union*, 37(4), 451-460, 1956.

Stallman, R. W., Calculation of resistance and error in an electric analog of steady flow through nonhomogeneous aquifers, *U. S. Geol. Surv., Water-Supply Paper 1544-G*, G1-G20, 1963a.

Stallman, R. W., Electric analog of three-dimensional flow to wells and its application to unconfined aquifers, *U. S. Geol. Surv., Water-Supply Paper 1536-H*, 205-242, 1963b.

Stallman, R. W., Type curves for the solution of single boundary problems *in* Bentall, R., Short cuts and special problems in aquifer tests, *U. S. Geol. Surv., Water-Supply Paper 1545-C*, C45-C47, 1963c.

Terwilliger, P. L., L. E. Wilsey, H. N. Hall, P. M. Bridges, and R. A Morse, An experimental and theoretical investigation of gravity drainage performance, *Trans Am. Inst. Mech. Engrs.*, 192, 285-296, 1951.

Theis, C. V., The relation between the lowering of the piezometric surface and the rate and duration of discharge of a well using groundwater storage, *Trans. Am. Geophys. Union*, 16, 519-524, 1935.

Thiem, A., *Hydrologische Methoden*, 56 pp., Gebhardt, Leipzig, 1906.

Todd, J. (ed.), *Survey of Numerical Analysis*, 589 pp., McGraw-Hill Book Company, New York, 1962.

Varga, R. S., *Matrix Iterative Analysis*, 322 pp., Prentice-Hall, Inc., Englewood Cliffs, N. J., 1962.

Volynskii, B. A., and V. Ye Bukhman, *Analogues for the Solution of Boundary-Value Problems*, 460 pp., Pergamon Press, New York, 1965.

Walton, W. C., and T. A. Prickett, Hydrogeologic electric analog computer, *J. Hydraul. Div., Am. Soc. Civil Engrs.*, 89, HY6, 67-91, 1963.

West, W. J., W. W. Garvin, and J. W. Sheldon, Solution of the equations of unsteady state two-phase flow in oil reservoirs, *Trans. Am. Inst. Mech. Engrs.*, 201, 217-229, 1954.

Willers, F. A., *Practical Analysis*, 218 pp., Dover Publications, Inc., New York, 1947.

Wood, L. A., and R. K. Gabrysch, Analog model study of groundwater in the Houston district, Texas, *Texas Water Comm. Bull. 6508*, 103 pp., 1965.

(Manuscript received May 14, 1968; revised June 25, 1968.)

ERRATA

Page 1073, (10) should read: $\dfrac{\partial h}{\partial t} \approx \dfrac{(h_{ijk} - h_{ijk-1})}{\Delta t}$

Page 1078, (28) should read: $A_j h_{j-1} + B_j h_j + C_j h_{j+1} = D_j$
$$2 \leq j \leq \eta - 1$$

Page 1078, (29) should read: $A_\eta h_{\eta-1} + B_\eta h_\eta = D_\eta$

Page 1078, line 27 should read: ". . . location of the boundary is shifted ½Δx outside . . . ."

Page 1088, line 2 should read: $W_{ijk} \Delta x \, \Delta y = Q_{ijk}$

HYDROGEOLOGIC ELECTRIC ANALOG COMPUTERS

By Willia C. Walton,[1] M. ASCE, and Thomas A. Prickett[2]

SYNOPSIS

Electric analog computers can play an important role in the forecast of the consequences of developing nonhomogeneous aquifers having irregular shapes and boundaries and a variety of head and discharge controls. Analog computers are versatile and simple and of low to moderate cost. The use of analog computers enables ground-water development schemes to be tested rapidly and accurately, thus permitting the appraisal of the relative merits of alternate choices of development.

The electric analog computer consists of an analog model and excitation-response apparatus, i.e., waveform generator, pulse generator, and oscilloscope. The analog model is a regular array of resistors and capacitors, and is a scaled-down version of the aquifer. Resistors are inversely proportional to the hydraulic conductivity of the aquifer, and capacitors store electrostatic energy in a manner analogous to the storage of water in an aquifer. The behavior of the electrical network is described by an equation that has the same form as the finite-difference equation for nonsteady state, three-dimensional flow of ground water. Electrical units (voltage, coulombs, amperes, and seconds) and corresponding hydraulic units (feet, gallons, gallons per day and days) are connected by four scale factors.

Excitation-response equipment forces electrical energy in the proper time phase into the analog model and measure energy levels within the energy-dissipative resistor-capacitor network. Oscilloscope traces, i.e., time-voltage graphs, are analogous to time-drawdown graphs that would result

---

Note.—Discussion open until April 1, 1964. To extend the closing date one month, a written request must be filed with the Executive Secretary, ASCE. This paper is part of the copyrighted Journal of the Hydraulics Division, Proceedings of the American Society of Civil Engineers, Vol. 89, No. HY6, November 1963.

[1] Engr. in charge of ground-water research, Illinois State Water Survey, Urbana, Ill.
[2] Asst. Engr., Illinois State Water Survey, Urbana, Ill.

after a step function-type change in discharge or head. A catalog of time-voltage graphs provides data for construction of a series of water-level change maps.

Close agreement between analog computer and exact analytical solutions for three selected idealized aquifer situations is noted. A comparison of the analog computer and simplified analytical solutions for a selected complex aquifer situation indicates that evaluations based on incomplete data and analog or idealized mathematical models can be meaningful and useful.

INTRODUCTION

The ground-water hydrologist is concerned chiefly with the quantitative description of aquifers and their physical parameters and with the response of aquifers to development. It is his responsibility to evaluate ground-water resources and to forecast the consequences of the utilization of aquifers. Proper planning of ground-water developments requires testing of all possible schemes and appraising of the relative merits of various alternates. Thus, the ground-water hydrologist must consider many choices of development and describe their effects. Ground-water resource development and management personnel are concerned with more than the qualitative, abstract description of aquifers; they are interested in the sustained yields of wells and aquifers, the interference between wells and well fields, and the interrelation between surface water and ground water.

Questions pertaining to the use of ground-water resources require that pumping be related to water-level change with reference to time and space. The ground-water hydrologist must then determine the change in water levels due to the withdrawal of water from aquifers. The two factors to be considered are the cause and effect—pumpage and changes—in water level.[3] The hydraulic characteristics and dimensions of the aquifer, and existing confining beds and the boundaries of the aquifer are of utmost importance in relating cause and effect.

The hydraulic characteristics to be considered in calculating aquifer response are the coefficients of permeability and/or transmissibility and storage of the aquifer and the coefficients of vertical permeability of existing confining beds. Changes in water levels can be evaluated only if the widths, lengths, and thicknesses of the aquifer and the confining beds are known. Boundaries, such as folds, faults, or relatively impervious layers of shale or clay, and beds of rivers, lakes, and other bodies of surface water, influence the response of an aquifer to pumping. Thus, cause can not be related to effect until hydrogeologic maps are available, which describe the following factors: 1. Coefficient of transmissibility and/or permeability of the aquifer; 2. coefficient of storage of the aquifer; 3. area of the aquifer; 4. sat-

---

[3] "Electronic Computers as an Aid to the Analysis of Hydrologic Problems," by H. E. Skibitzke, Publication 52, Internatl. Assn. of Scientific Hydrology, 1961.

urated thickness of the aquifer; 5. coefficients of vertical permeability of existing confining beds; 6. saturated thicknesses of existing confining beds; and 7. location, extent, and nature of the aquifer and the confining bed boundaries. These prescribed maps must encompass all nonhomogeneous and irregular hydrogeologic conditions.

The response of an aquifer to pumping cannot be predicted with great precision unless the basic definition of the hydrogeologic conditions of the aquifer is precise. However, the task of defining the hydrogeologic conditions of the aquifer is difficult, because available basic data are seldom sufficient to permit rigorous descriptions of aquifers, and economic limitations often prohibit the collection of extensive and detailed information concerning the complexities of these aquifers. Insufficient basic data require much interpretation, extrapolation, and application of geologic principles together with intuition in the preparation of requisite quantitative hydrogeologic maps. Unless geologic methods and techniques are refined so that aquifers may be more adequately defined in quantitative terms despite the incompleteness of data, further developments in the field of hydrogeologic system analysis will be severely limited.

Electric analog computer techniques[3] that allow the study of cause and effect relationships involving complex aquifer conditions have recently been developed. The computer consists of an analog model and excitation-response apparatus. The theory and construction features of the analog computer are described herein, and the accuracy and reliability of the computer are assessed by comparing analog and exact analytical results for selected simple aquifer conditions. The analog solution for a complex aquifer situation and the analytical solution for a corresponding highly idealized model aquifer are compared. From these studies it is concluded that the analog computer will assist greatly the Illinois State Water Survey in making quantitative appraisals of developed and undeveloped aquifers in Illinois.

Notation.—The symbols adopted for use in this paper are defined where they first appear and are arranged alphabetically in the Appendix.

ANALOG MODELS

Analog models for simulating aquifers in Illinois were patterned after analog models developed by H. E. Skibitzke,[3] and consist of regular arrays of resistors and capacitors. In the resistor-capacitor network, resistors are inversely proportional to the hydraulic conductivity of the aquifer and the capacitors store electrostatic energy, in a manner analogous to the storage of water within an aquifer. The electrical network bears a close resemblance in form to the aquifer and is a scaled-down version of the aquifer.

A resistor-capacitor network is characterized by "junctions" and discrete branches. Pegboard, masonite, 1/8-in. thick and perforated with holes on a 1-in. square pattern, is used in the construction of networks. The holes function as junctions or nodes; brass eyelets are inserted in the holes to provide terminals. For two-dimensional flow problems and in internal parts of the analog model, four resistors and a capacitor are connected to each terminal. The capacitor also is secured to the ground connection of the electrical system. The boundary of the network is adjusted in a step fashion to approximate the actual boundary of the aquifer. To do this, boundary terminals are connected to two or three resistors and a capacitor, depending on the geometry of the boundary. Exact duplication of the actual boundary of the aquifer, when the boundary does not coincide with the lines of the network grid, is possible through use of the "vector volume" technique.[4] In this method the magnitudes of resistors and capacitors in the proximity of the boundary are modified to take into consideration an irregularly shaped boundary, and resistors and capacitors are selected so that there is correspondence between network parameters and aquifer parameters.

Since aquifers are continuous phenomena, while a resistor-capacitor network consists of many discrete branches, the network is only an approximation of a true analog. However, it can be demonstrated mathematically that if the mesh size of the network is small in comparison to the size of the aquifer, the behavior of the network closely describes the response of an aquifer to pumping. To obtain solutions of uniform accuracy, it is often necessary to vary the net spacing from fine to coarse in areas where aquifer characteristics undergo sharp variations. Methods for effecting a variation in net spacing are available.[4]

Homogeneous and isotropic aquifers.—The partial differential equation[5] governing the nonsteady state, two-dimensional flow of groundwater in a homogeneous, isotropic, and infinite aquifer is

$$\frac{\partial^2 h}{\partial x^2} + \frac{\partial^2 h}{\partial y^2} = \frac{S}{T} \frac{\partial h}{\partial t} \qquad \qquad (1)$$

Equations of the same form occur in the theories of nonsteady state flow of heat and electricity.[5] Consider a finite difference approximation of Eq. 1.[3,6] The aquifer is subdivided into squares of equal area, $\partial x \, \partial y$. The sides of the squares, ∂x and ∂y, are equal and are of finite length, Δx and Δy, respectively. The square grid is shown in Fig. 1A.[3,6] The intersections of grid lines are called nodes. The infinitesimal $\Delta x \Delta y$ is approximated by a^2, in which a is the width of the grid interval. The area, a^2, is small compared with the area of the aquifer.

Examination of node 1 and Eq. 1 shows that the second differentials of head can be approximated by[6,7]

$$\frac{\partial^2 h}{\partial x^2} = \frac{\left(h_2 + h_4 - 2h_1\right)}{a^2} \qquad \qquad (2)$$

$$\frac{\partial^2 h}{\partial y^2} = \frac{\left(h_3 + h_5 - 2h_1\right)}{a^2} \qquad \qquad (3)$$

---

[4] "Analog Simulation," by W. J. Karplus, McGraw-Hill Book Co., Inc., New York, N. Y., 1958.

[5] "Flow of Ground Water," by C. E. Jacob, Engineering Hydraulics, Chapter 5, John Wiley and Sons, Inc., New York, N. Y., 1950.

[6] "Numerical Analysis of Regional Water Levels to Define Aquifer Hydrology," by R. W. Stallman, Transactions, Amer. Geophysical Union, Vol. 37, No. 4, 1956.

[7] "Relaxation Methods in Theoretical Physics," by R. V. Southwell, Oxford University Press, 1948.

Substitution of Eq. 2 and Eq. 3 in Eq. 1 results in

$$\frac{(h_2 + h_3 + h_4 + h_5 - 4h_1)}{a^2} = \frac{S}{T}\frac{\partial h}{\partial t} \quad\quad\quad\quad (4)$$

which is the finite-difference form[6] of the partial differential equation governing the nonsteady state, two-dimensional flow of groundwater in an infinite aquifer. Eq. 4 may be rewritten in abbreviated form as

$$T\left(\sum_{2}^{5} h_i - 4h_1\right) = a^2 S \frac{\partial h}{\partial t} \quad\quad\quad\quad (5)$$

in which h_1 is the head at node 1; h_i (i = 2, 3, 4, and 5) represents heads at nodes 2-5; a denotes the width of grid interval; T stands for the coefficient of transmissibility; and S equals the coefficient of storage.

A resistor-capacitor network with a square pattern as shown in Fig. 1(a) and network junctions at nodes as defined in Fig. 1(b) is set up. The junctions consist of four resistors and one capacitor connected to a common terminal; the capacitor is also connected to ground. The relation of electrical potentials in the vicinity of the junction, according to Kirchhoff;s current law, can be expressed as[3,8]

$$(V_3 - V_1)\frac{1}{R_A} + (V_2 - V_1)\frac{1}{R_B} + (V_5 - V_1)\frac{1}{R_C}$$
$$+ (V_4 - V_1)\frac{1}{R_D} = C\frac{\partial V}{\partial t} \quad\quad\quad\quad (6)$$

in which V_{1-5} is the electrical potential at ends of resistors; R_{A-D} denotes the resistance; and C represents the capacitance. If the value of the resistors are equal, Eq. 6 may be written as

$$(V_2 + V_3 + V_4 + V_5 - 4V_1)\frac{1}{R} = C\frac{\partial V}{\partial t} \quad\quad\quad\quad (7)$$

Eq. 7 may be rewritten in abbreviated form[3] as

$$\frac{1}{R}\left(\sum_{2}^{5} V_i - 4V_1\right) = C\frac{\partial V}{\partial t} \quad\quad\quad\quad (8)$$

in which V_i (i = 2, 3, 4, and 5) is the electrical potential at ends of resistors A-D.

---

[8] "Electronics," by Jacob Millman and Samuel Seely, McGraw-Hill Book Co., Inc., New York, N. Y., 1941.

Comparison of Eq. 5 and Eq. 8 shows that the finite-difference equation governing the nonsteady state, two-dimensional flow of ground water in an infinite aquifer is of the same form as the equation governing the flow of electrical current in a resistor-capacitor network. For every term in Eq. 5, there is a corresponding term of the same order of differentiation in Eq. 8.

The analogy between electrical systems and aquifer systems is apparent. The hydraulic heads, h, are analogous to electrical potentials, V. The coefficient of transmissibility, T, is analogous to the reciprocal of the electrical

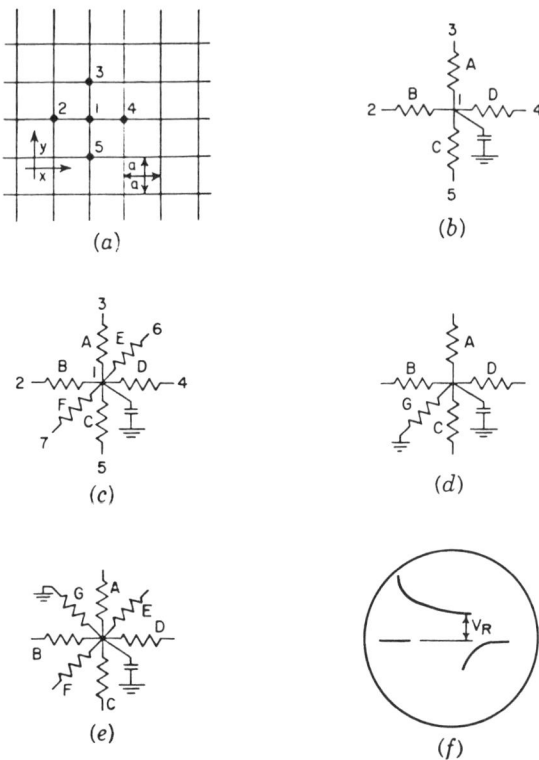

FIG. 1.—FINITE-DIFFERENCE GRID, RESISTOR-CAPACITOR NETS, AND PUMPING RATE OSCILLOSCOPE TRACE

resistance, $1/R$. The product of the coefficient of storage, S, and a^2 is analogous to the electrical capacitance, C.

Continuing the comparison, water moves in an aquifer just as charges move in an electrical circuit. The quantity of water is calculated in gallons while the charge is calculated in coulombs. The rate of flow of water past any point in the aquifer is expressed in gallons per day, while the flow of electricity is expressed in coulombs per second, or amperes. The hydraulic head loss between two points in an aquifer is expressed in feet, while the potential drop across a part of the electrical circuit is in volts.

Thus, there are four units that are analogous. There is necessarily a scale factor connecting each unit in one system to the analogous unit in the other system. Knowing the four scale factors, the hydrologist is able to relate electrical units associated with the analog model to hydraulic units associated with the aquifer. The four scale factors, K_1, K_2, K_3, and K_4, were defined[9] as follows:

$$q = K_1 Q \quad \ldots \ldots \ldots \ldots \ldots \ldots \ldots (9)$$

$$h = K_2 V \quad \ldots \ldots \ldots \ldots \ldots \ldots \ldots (10)$$

$$Q = K_3 I \quad \ldots \ldots \ldots \ldots \ldots \ldots \ldots (11)$$

$$t_d = K_4 t_s \quad \ldots \ldots \ldots \ldots \ldots \ldots \ldots (12)$$

in which q is gal; Q equals coulombs; Q denotes gal per day; I represents amp; h indicates ft; V corresponds to volts; t_d is equivalent to days; t_s describes sec; K_1 refers to gal per coulomb; K_2 stands for ft per volt; K_3 is gal per day per amp; and K_4 equals days per sec. The relation between scale factors K_1, K_3, and K_4 is expressed by

$$\frac{K_3 K_4}{K_1} = 1 \quad \ldots \ldots \ldots \ldots \ldots \ldots \ldots (13)$$

The values of resistors and capacitors required to model a given aquifer may be determined from Eqs. 9 to 13. To minimize computer costs, scale factors are chosen that allow the use of standard tolerance, low wattage, fixed carbon resistors with values from 10^2 ohms to 10^7 ohms, and capacitors of low voltage rating with values from 10^1 μuf to 10^5 μuf.[9]

The analogy between Ohm's law and Darcy's law is established by the fact that the coefficient of transmissibility is analogous to the reciprocal of the electrical resistance. Substitution of these laws in Eq. 11 yields

$$R = \frac{K_3}{K_2 T} \quad \ldots \ldots \ldots \ldots \ldots \ldots \ldots (14)$$

in which R is the resistance, in ohms; and T refers to the coefficient of transmissibility, in gpd per ft. Eq. 14 may be used to determine the value of the resistors of the interior parts of the analog model:[9]

---

[9] "An Electric Analog Method for Use in Quantitative Studies," by B. J. Bermes, U. S. Geol. Survey, 1960.

Eq. 15,[9] which may be used to determine the value of the capacitors of the interior parts of the analog model, is derived by taking into consideration the definitions of the storage and capacitance coefficients and the analogy between $(a^2 S)$ and C.

$$C = 7.48 \ a^2 \ S \ \frac{K_2}{K_1} \qquad \qquad (15)$$

in which C is the capacitance, in farads; a denotes the network spacing, in ft; and S represents the coefficient of storage, fraction.

The values of resistors and capacitors in the vicinity of boundaries can be computed, based on the "vector area" and "vector volume" techniques.[4]

Network spacings are selected to minimize the errors due to finite-difference approximation. For regional studies on a gross basis the network spacing can be 10,000 ft; for local studies involving details, a network spacing of 100 ft is adequate.[9]

Heterogeneous and anisotropic aquifers.—Thus far, only two-dimensional flow through homogeneous and isotropic aquifers have been considered. The hydraulic properties of the aquifer were assumed to be constant everywhere in space. In many problems, however, the variations of hydraulic properties with space are large and flow is three-dimensional. These variations and vertical components of flow must also be taken into account. A generalization of Eq. 5 and Eq. 8 permits accounting for differences in space of the coefficients of transmissibility and storage by varying resistors and capacitors.[3]

The partial differential equation governing the nonsteady state three-dimensional flow of ground water in a homogeneous, isotropic, and infinite aquifer[5] is

$$\frac{\partial^2 h}{\partial x^2} + \frac{\partial^2 h}{\partial y^2} + \frac{\partial^2 h}{\partial z^2} = \frac{S}{T} \frac{\partial h}{\partial t} \qquad \qquad (16)$$

The finite difference approximation of Eq. 16 based on a square grid[3] is

$$T \left(\sum_{2}^{7} h_i - 6h_1 \right) = a^2 S \frac{\partial h}{\partial t} \qquad \qquad (17)$$

in which h_1 is the head at node 1; h_i (i = 2, 3, 4, 5, 6, and 7) denotes the heads at nodes 2 to 7; and a equals the width of the grid interval.

A resistor-capacitor network with a square pattern and network junctions at nodes, as defined in Fig. 1C is constructed. The junctions consist of six resistors and one capacitor connected to a common terminal, with the capacitor connected to ground from the junction. If the values of the resistors are equal, the relation[3] of electrical potentials in the vicinity of the junction is expressed as

$$\frac{1}{R} \left(\sum_{2}^{7} V_i - 6V_1 \right) = C \frac{\partial V}{\partial t} \qquad \qquad (18)$$

in which V_i (i = 2, 3, 4, 5, 6, and 7) represents the electrical potential at ends of resistors A to F.

Comparison of Eq. 17 and Eq. 18 shows that the finite-difference equation governing the nonsteady state, three-dimensional flow of ground water in a homogeneous, isotropic, and infinite aquifer is of the same form as the equation governing the flow of electrical current in the resistor-capacitor network in Fig. 1(C). The resistor-capacitor junction for three-dimensional flow differs from the resistor-capacitor junction for two-dimensional flow (Fig. 1(b) and Fig. 1(c) only in that the resistor-capacitor junction for three-dimensional flow contains two more resistors (E and F).

A generalization of Eq. 17 and Eq. 18 permits accounting for differences in the coefficients of transmissibility, vertical permeability, and storage by varying resistors and capacitors. Substitution of Ohm's and Darcy's laws in Eq. 11 results in[8]

$$R_{EF} = \frac{K_3 m}{K_2 P_v a^2} \quad \ldots \ldots \ldots \ldots \ldots \ldots (19)$$

in which R_{EF} is the resistance in ohms; P_V equals the coefficient of vertical permeability, gpd per sq ft; a represents the width of grid interval, in ft; and m denotes the saturated thickness of aquifer, in ft. Eq. 14 and Eq. 15 may be used to determine the values of resistors A to D and capacitors of the interior parts of the analog model; and Eq. 19 may be used to determine the value of resistors E and F of the interior parts of the analog model.

Leaky Artesian Aquifers.—Leaky artesian conditions exist in many parts of Illinois[10] where aquifers are overlain by confining beds through which leakage occurs. Leakage through the confining bed into the aquifer is vertical and proportional to drawdown, and the hydraulic head in the deposits supplying leakage remains more or less uniform. Under these conditions, leakage can be simulated in an analog model by the addition of resistors connected to ground and to each node of the network. Network junctions for two-dimension and three-dimension flow in a leaky artesian aquifer are shown in Figs. 1(d) and 1(e), respectively.

Substitution of Ohm's and Darcy's laws in Eq. 11 results in[9]

$$R_G = \frac{K_3}{K_2 \left(\frac{P'}{m'}\right) a^2} \quad \ldots \ldots \ldots \ldots (20)$$

in which P'/m' is the coefficient of leakage, in gal per day per cu ft (gpd per cu ft);[11] P' is the coefficient of vertical permeability of confining bed, in gpd per sq ft; and m' denotes the saturated thickness of confining bed, in ft. Eq. 20 may be used to determine the value of the resistor G of the interior parts of the analog model.

---

[10] "Leaky Artesian Aquifer Conditions in Illinois," by W. C. Walton, Report of Investigation 39, Illinois State Water Survey, 1960.

[11] "Non-Steady Radial Flow in an Infinite Leaky Aquifer," by M. S. Hantush and C. E. Jacob, Transactions, American Geophysical Union, Vol. 36, No. 1, 1955.

Variations in the coefficient of vertical permeability of the confining bed may be corrected by adjusting the value of resistors connected to ground.

Recharge and Barrier Boundaries.—Some aquifers in Illinois are hydraulically bounded and connected to surface bodies of water. If the surface-water stage is constant, the effects of the recharge boundary can be simulated in an analog model by terminating the part of the network along the boundary in a short circuit. Barrier boundaries, (lines across which there is no flow) can be simulated by an open circuit. Network boundaries can be extended to infinity by the use of "termination strips."[4] Resistors connected to the nodes along the edge of the analog model and to the ground simulate horizontal leakage through boundaries of the aquifer. Multiple aquifer systems may be simulated with analog models, and the electric analogy can be extended to include cases in which hydraulic properties are time dependent.

Cost.—The cost of analog models largely depends on the price of resistors and capacitors embodied in the model and on the labor charge for the design and construction of the analog model. Fixed carbon resistors with tolerances of ± 10% can be purchased in bulk at a cost of approximately $0.04 each; ceramic capacitors with tolerances of ± 10% are available in bulk for approximately $0.05 each. Analog models of average size (4 ft by 8 ft or smaller) and complexity can usually be constructed in less than 80 man-hours. Computations involved in determining the values of resistors and capacitors of complex analog models usually can be performed in less than 40 man-hours.

EXCITATION-RESPONSE APPARATUS

Excitation-response apparatus,[12,13] force electrical energy in the proper time phase into the analog model and measure energy levels within the energy-dissipative, resistor-capacitor network. The excitation-response apparatus is comprised of three major parts, as shown in Fig. 2: A waveform generator, a pulse generator, and an oscilloscope. The waveform generator, which produces sawtooth pulses, is connected to the trigger circuits of the pulse generator and oscilloscope, thereby controlling the repetition rate of computation, and synchronizing the oscilloscope's horizontal sweep and the output of the pulse generator. The pulse generator, which produces rectangular-shaped pulses[14] of various duration and amplitude on command from the waveform generator, is coupled to that junction in the analog model representing the pumped well. The oscilloscope is connected to junctions of the analog model in which it is desired to determine the response of the analog model to excitation. An electron beam is swept across the cathode ray tube of the oscilloscope, providing a time-voltage graph that is analogous to the time-drawdown graph for an observation well.

Fig. 3 shows the simultaneous sequences of events that take place in the excitation-response apparatus during two computational cycles. The waveform generator sends a positive pulse to the oscilloscope (Fig. 3(a)) to start

---

12 "Electronic Analog Computers," by G. A. Korn and T. M. Korn, McGraw-Hill Book Co., Inc., New York, N. Y., 1952.

13 "Introduction to Electronic Analog Computers," by C. A. A. Wass, McGraw-Hill Book Co., Inc., New York, N. Y., 1955.

14 "Basic Pulses," by I. Gottlieb, John F. Rider Publisher, Inc., New York, N. Y., 1958.

its horizontal sweep; at the same time, it sends a negative sawtooth waveform to the pulse generator (Fig. 3(b). At a point along the sawtooth waveform (arbitrarily chosen as 25% of the waveform duration in Fig. 3(b) the pulse generator is triggered to produce a negative, rectangular-shaped pulse as shown in Fig. 3(c). The duration of this pulse is analogous to the pumping period, t_d, and the amplitude is analogous to the pumping rate, Q. This pulse is sensed by the oscilloscope as a function of the analog model components, boundary conditions, and node position of the junction connected to the oscilloscope. Thus, the oscilloscope trace (Fig. 3(d) and Fig. 3 (e) is analogous to the water-level fluctuation that would result after a step-function type pumpage change of known duration and amplitude. To provide data independent of the pulse repetition rate, the time interval between pulses is set at a multiple of the longest time constant in the analog model. The time constant

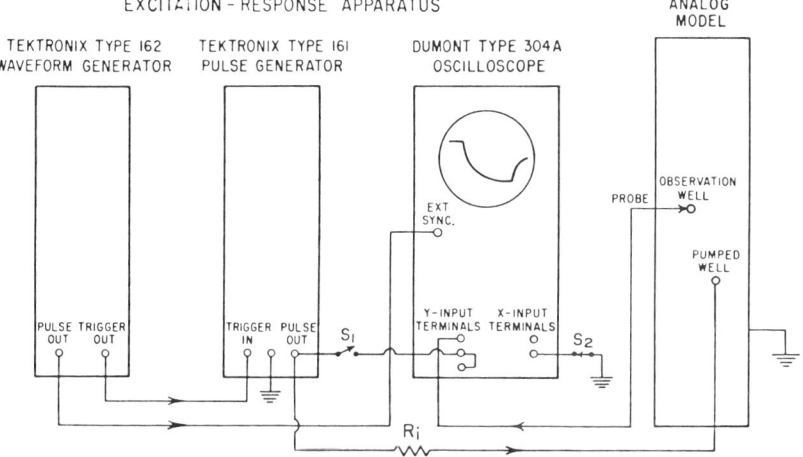

FIG. 2.—BLOCK DIAGRAM DESCRIBING ANALOG COMPUTER COMPONENTS

is the product of the capacitance at a point and the resistance in its discharge path.

A method of computing the pumping rate is incorporated in the circuit between the pulse generator and the analog model by the small resistor, R_i, in series, shown in Fig. 2. Substitution of Ohm's law in Eq. 11 results in

$$Q = \frac{V_R}{1.44 \times 10^3 R_i} K_3 \quad \ldots \ldots \ldots \ldots \ldots (21)$$

in which Q denotes the pumping rate, in gallons per minute, (gpm); V_R represents the voltage drop across the resistor R_i, in volts; and R_i describes the calibrated resistance, in ohms. Eq. 21 may be used to compute the pumping rate.

253

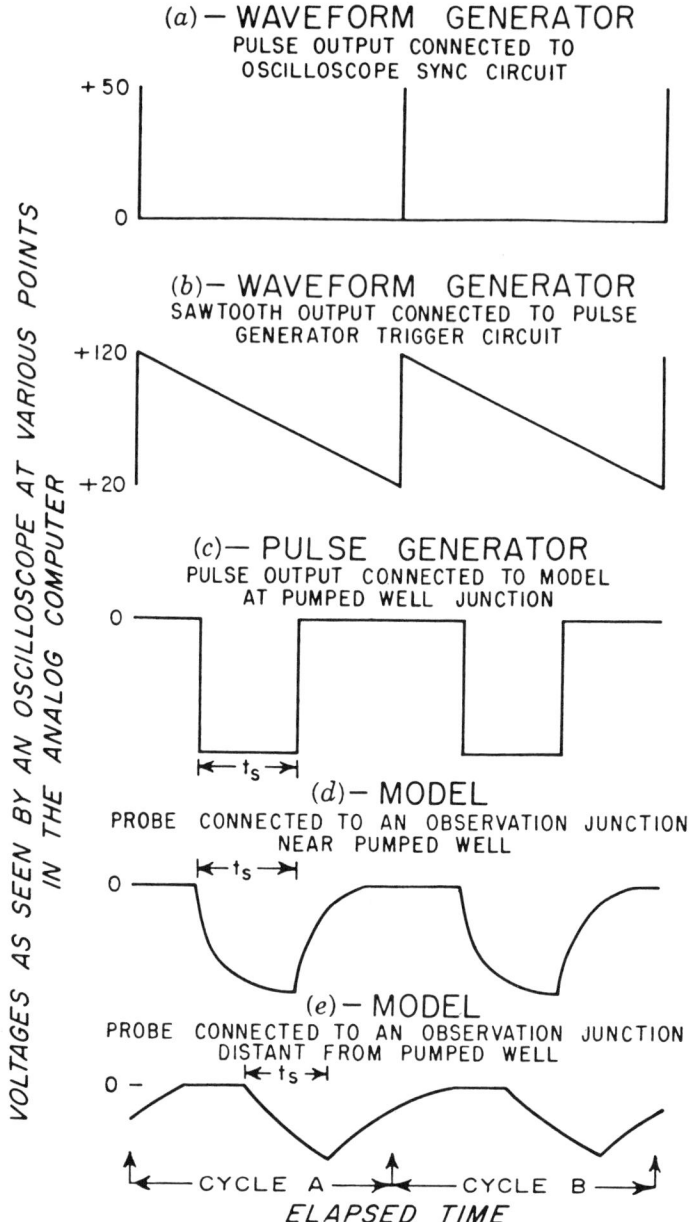

FIG. 3.—DIAGRAM SHOWING SEQUENCE OF EVENTS IN TWO COMPUTATIONAL CYCLES OF ANALOG COMPUTER

The voltage drop across the calibrated resistor is measured with the oscilloscope. Switches S_1 and S_2 in Fig. 2 are closed and opened, respectively, and the oscilloscope is connected to the pumped well junction. The waveform in Fig. 1(f) will appear on the cathode ray tube; the vertical distance as shown is the desired voltage drop, V_R.

The switches S_1 and S_2 are returned to their original positions. The oscilloscope is then connected to all junctions of the analog model representing observation wells. Time-voltage curves for junctions can be recorded by photographing oscilloscope traces. The screen of the oscilloscope is accurately calibrated so that voltage and time may be read on the vertical and horizontal axes, respectively. The time is in seconds; the value of each horizontal division on the screen is determined by noting the duration of the rectangular-shaped pulse and the number of divisions covered by the time-voltage trace for a junction adjacent to the pumped well. The time-voltage graphs obtained from the oscilloscope can be converted into time-drawdown graphs by means of Eq. 10 and Eq. 12 relating electrical units to hydraulic units. A catalog of time-drawdown graphs provides data for the construction of a series of water-level change contour maps. Thus, water-level changes are described everywhere in the aquifer for any desired pumping period. The pulse generator can be connected to many junctions, and a variety of pumping conditions can be studied.

The pulse generator can be coupled to junctions in the analog model representing a stream recharge boundary, and water-level fluctuations due to a step-function type change in stream stage can be determined with the oscilloscope. Factors such as well discharge and recharge, stream-stage changes, recharge from precipitation, and tidal fluctuations may be imposed on the analog model.[3]

The effects of complex pumpage changes on water levels may be determined by approximating the pumpage graph by a group of step functions and analyzing the effect of each step function separately. The total water-level change, based on the superposition theorem,[4] is obtained by summation of individual step-function water-level changes. Costly variable function generators are operational conveniences for imposing complex pumpage or boundary fluctuations on the analog model.

The pulse generator used by the Illinois State Water Survey has a maximum output of 50 and 20 ma; the pulse generator and oscilloscope have rise times of less than 1 μ sec and waveform durations hanging from less than 10 μ sec to 100 m sec. The performance specifications of the waveform generator, pulse generator, and oscilloscope are compatible with the following criteria desired for analog computers: Low power requirements, repetitive calculation at variable rates, and fast computing speeds. Excitation-response equipment can be purchased for less than $1500, depending chiefly on the quality of oscilloscope desired. Special apparatus also may be coupled to the excitation-response apparatus to reject some of the unwanted noise pick-up, thus permitting the measurement of smaller voltages.

ACCURACY AND RELIABILITY OF COMPUTER

The accuracy and reliability of the electric analog computer were assessed by a study of selected idealized aquifer situations that could be modeled and accurately analyzed both analytically and electrically. Analytical solutions involving mathematical models[15] were compared with results obtained with the analog computer.

In general, the accuracy of the analog solutions is a matter of the quality of resistors and capacitors, the effects of finite-difference approximation, and the signal-noise ratio. Inaccuracies due to the quality of resistors and capacitors are small, owing to the error-equalizing property of the resistor-capacitor network. Errors due to finite-difference approximation are minimum when the network spacing is small in comparison to the size of the aquifer; errors depend in part on the nature of the potential field being analyzed. Examples of analysis indicate that if the excitation-response apparatus in Fig. 2, inexpensive resistors and capacitors, and a reasonable network spacing are used, differences between analog and analytical solutions are not excessive. The over-all accuracy of the analog solutions can be increased 1% or 2% by using precision resistors and capacitors and refined excitation-response apparatus; however, the benefit-cost ratio of the increased accuracy is low, so that in most cases the precision is not warranted.

Examples of Analysis.—Analog models were constructed for three idealized aquifer situations that could occur in Illinois, in order to study the accuracy and reliability of the analog computer. The three aquifer situations are given in diagrammatical form in Fig. 4. The corresponding analog models are illustrated in Fig. 5. Details of analog model 1 will be described to illuminate model construction features and the use of scale factors.

A piece of 1/8-in. masonite pegboard (1 ft by 2 ft) was obtained, and aluminum angles (1 in. by 1 in.) were attached along the four edges with metal screws. The angles permit setting the model on a table without disturbing the capacitors, which are attached to ground wires (bus bars) secured with stand-offs on the bottom side of the pegboard. Then No. 3 brass laquered shoe eyelets were inserted in the holes of the pegboard to provide terminals for resistors and capacitors, and resistor wires were inserted in the eyelets and soldered in place. (The pegboard was turned over and resistor wires were trimmed to a length of about 3/8 in.). Capacitor wires were soldered to the group of resistor wires associated with each junction. Disc ceramic capacitors and 1/2-watt fixed carbon resistors, both having tolerances of ± 10% were used.

A model scale of 1 in. = 660 ft was chosen to minimize the finite-difference approximation. The electrical length of the model (2 ft) was selected so that there would be no water-level change beyond the analog model; therefore, the analog model would simulate an infinite strip aquifer. A total of 416 resistors and 225 capacitors were used.

Appropriate scale factors, the values of resistors and capacitors, and the pumping rate were determined by means of Eqs. 13, 14, 15, and 21. Sample computations follow.

Given T = 36,200 gpd per ft; S = 0.2; t_d = 180 days; a = 660 ft; and R_i = 9.77 x 10^{-4} ohms; let

---

[15] "Analyzing Ground-Water Problems with Mathematical Models and a Digital Computer," by W. C. Walton and J. C. Neill, Publication 52, Internatl. Assn. of Scientific Hydrology, 1961.

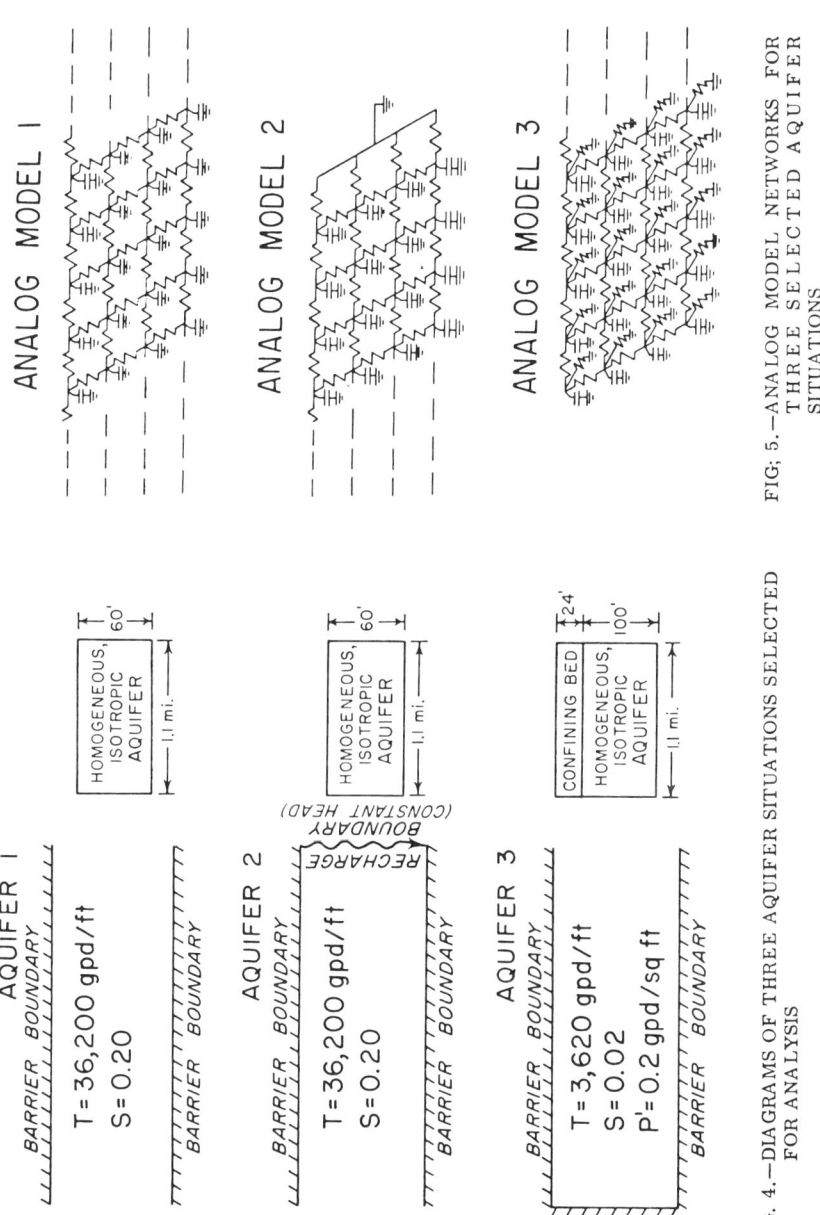

FIG. 4.—DIAGRAMS OF THREE AQUIFER SITUATIONS SELECTED FOR ANALYSIS

FIG. 5.—ANALOG MODEL NETWORKS FOR THREE SELECTED AQUIFER SITUATIONS

$$K_2 = 1 \text{ ft per volt}$$

$$K_4 = 1.8 \times 10^3 \text{ day per sec when } t_s = 0.1 \text{ sec}$$

$$K_3 = 3.62 \times 10^{10} \text{ gal per day per amp}$$

(These scale factors were chosen for convenience in reading voltages and so that resistors and capacitors would be simple values.)

From Eq. 13: $K_1 = K_3 K_4 = 1.8 \times 10^3 \left(3.62 \times 10^{10}\right)$

$$= 6.52 \times 10^{13} \text{ gal per coulomb}$$

From Eq. 14: $R = \dfrac{K_3}{K_2 T} = \dfrac{3.62 \times 10^{10}}{1 \left(3.62 \times 10^4\right)}$

$$= 10^6 \text{ ohms}$$

From Eq. 15: $C = 7.48 a^2 S \dfrac{K_2}{K_1} = \dfrac{7.48 \left(6.60 \times 10^2\right)^2 \; 2 \times 10^{-1} \; (1)}{6.52 \times 10^{13}}$

$$= 10^{-8} \text{ farads}$$

From Eq. 21: $Q = \dfrac{V_R}{1.44 \times 10^3 \; R_i} K_3 = V_R \dfrac{3.62 \times 10^{10}}{9.77 \times 10^4 \left(1.44 \times 10^3\right)}$

$$= 2.56 \times 10^2 \; V_R$$

The model was coupled to the excitation-response apparatus in Fig. 2, and the pulse generator was connected to the center junction of the analog model at one of the many selected locations of the pumped well. The oscilloscope was connected to each of the other junctions to obtain data for time-drawdown graphs. The drawdown contours for aquifer 1 and a selected pumping condition based on the analog solution and on exact analytical solution are shown in Fig. 6. The drawdown contours are for a pumping rate of 772 gpm and a pumping period, devoid of recharge, of 180 days. The effects of dewatering under water-table conditions are not considered.

Maps showing drawdown contours for aquifers 2 and 3 are given in Figs. 7 and 8. Close agreement between analog and analytical solutions is noted in each case. Differences in individual drawdowns at junctions based on analog

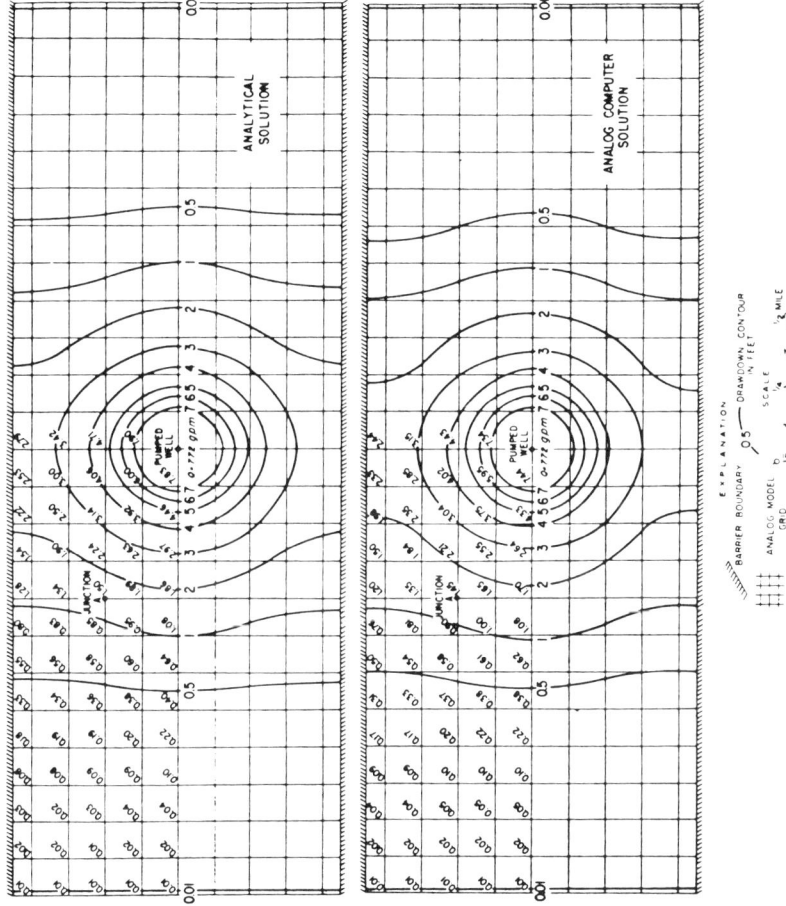

FIG. 6.—COMPARISON BETWEEN DRAWDOWN CONTOUR MAPS BASED ON ANALYTICAL AND ANALOG COMPUTER SOLUTIONS FOR AQUIFER 1

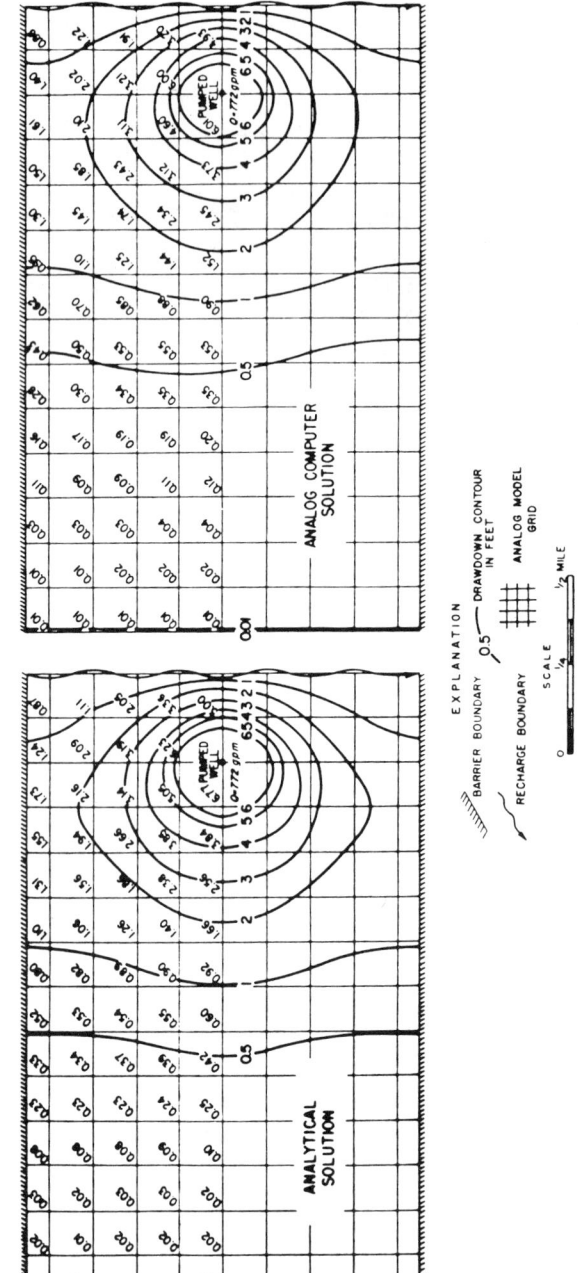

FIG. 7.—COMPARISON BETWEEN DRAWDOWN CONTOUR MAPS BASED ON ANALYTICAL AND ANALOG COMPUTER SOLUTIONS FOR AQUIFER 2

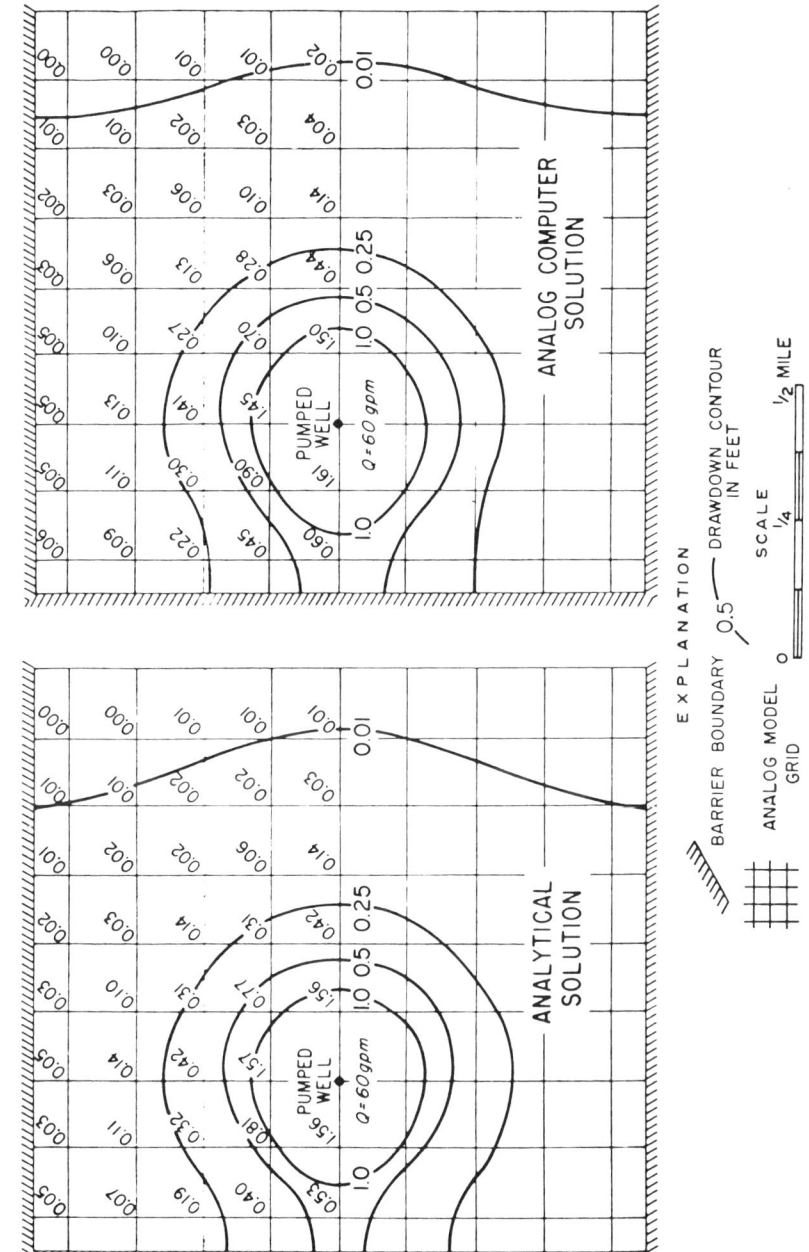

FIG. 8.—COMPARISON BETWEEN DRAWDOWN CONTOUR MAPS BASED ON ANALYTICAL AND ANALOG COMPUTER SOLUTIONS FOR AQUIFER 3

and analytical solutions are not significant when considered in relation to the accuracy and adequacy of geologic and hydrologic data.

A comparison of analog and analytical time-drawdown curves for a selected junction, A, in Fig. 6 is shown in Fig. 9. The time-drawdown graphs are for a pumping rate of 772 gpm. The reliability of transient water-level changes based on analog data is satisfactory for practical purposes. In general, differences between analog and analytical time-drawdown curves are greatest for junctions close to the pumped well and for short time intervals.[16]

ANALOG AND MATHEMATICAL MODELS FOR COMPLEX AQUIFERS

It has been suggested[17] that with sound professional judgment, hydrogeologic conditions often can be idealized with little sacrifice in the accuracy

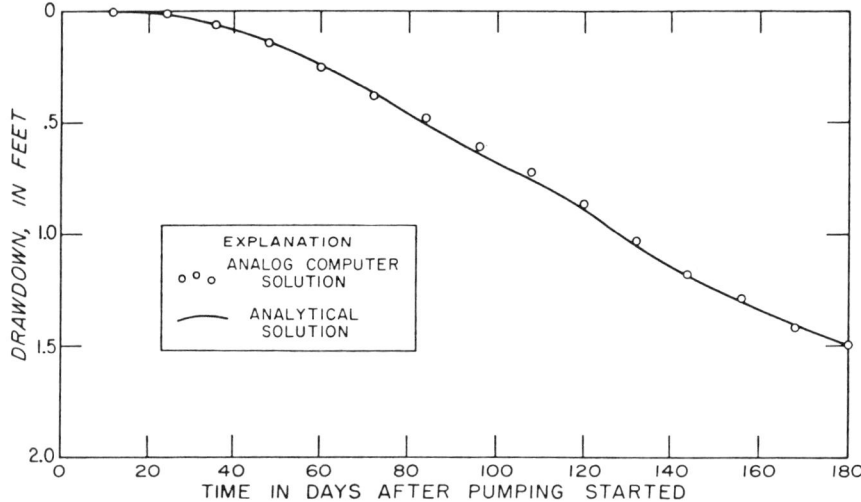

FIG. 9.—COMPARISONS BETWEEN TIME-DRAWDOWN GRAPHS BASED ON ANALYTICAL AND ANALOG COMPUTER SOLUTIONS FOR AQUIFER 1

of computations of the practical sustained yields of wells and aquifers. To test this conjecture the analog solution for a complex aquifer situation and the analytical solution for a corresponding idealized model aquifer were compared.

The complex aquifer situation and the idealized model aquifer and mathematical model selected for analysis are shown in Fig. 10. The idealized model aquifer is aquifer 1, discussed previously. Variations in the coef-

---

[16] "The Use of Analogs and Analog Computers in Heat Transfer," by M. Tribus, Publication 100, Oklahoma Engrg. Experiment Sta., 1958.

[17] "Evaluating Wells and Aquifers by Analytical Methods," by W. C. Walton and W. H. Walker, Journal of Geophysical Research, Vol. 66, No. 10, 1961.

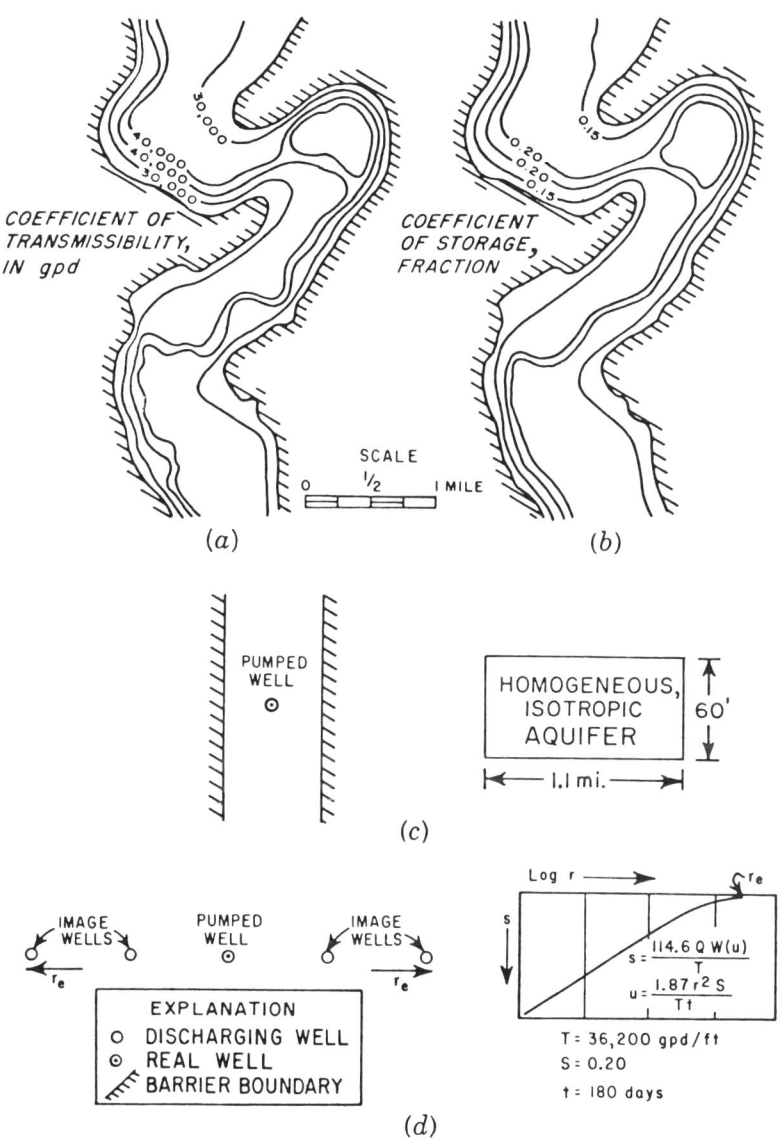

FIG. 10.—DIAGRAMS OF COMPLEX AQUIFER SITUATION AND CORRESPONDING IDEALIZED MODEL AQUIFER AND MATHEMATICAL MODEL

ficients of transmissibility and storage were taken into account in the analog model by varying resistors and capacitors.

Close agreement on a gross basis between the water-level change contour maps (see Fig. 6 and Fig. 11) based on the analog model and the idealized mathematical model is noted. However, on a detailed basis, changes in coefficients of transmissibility and storage do affect contours. Contour maps for two of many selected pumped well locations and the analog model are given in Fig. 11 to demonstrate that changing the position of the pumped well results in little change in water-level change contours on a gross basis. The

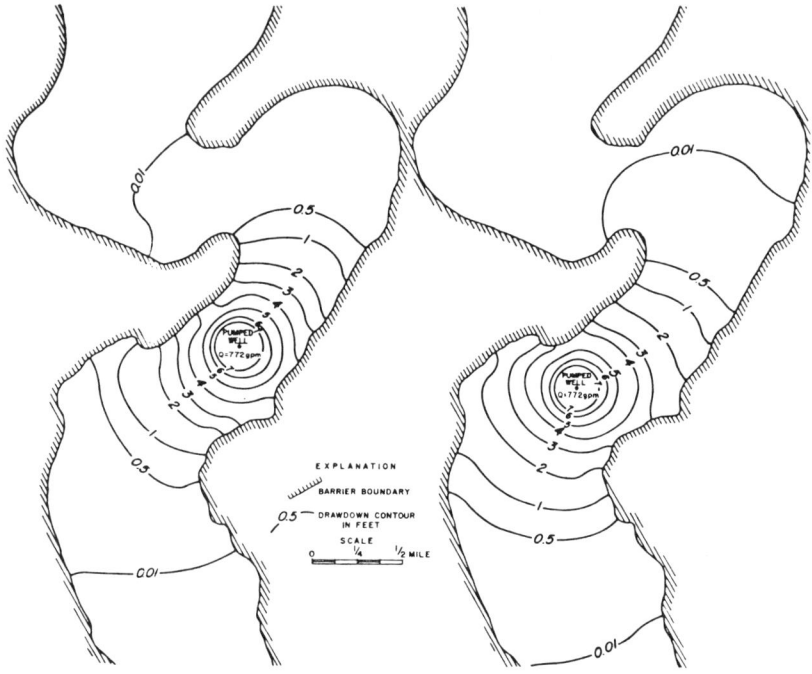

FIG. 11.—DRAWDOWN CONTOURS MAPS BASED ON ANALOG COMPUTER SOLUTIONS FOR SELECTED COMPLEX AQUIFER SITUATION

contour maps are for a pumping rate of 772 gpm. If there is close agreement on a gross basis between water-level change contours based on analog and mathematical models, there will be close agreement between the yields of wells and aquifers based on analog and mathematical models, because yield is proportional to drawdown. Thus, evaluations of the yields of wells and aquifers based on idealized mathematical models can be meaningful and useful. Simplified assumptions associated with analytical methods often are not so idealistic as to preclude reasonably accurate solutions. However, it is recognized that there are some complex cases in which highly detailed analog mapping of hydrogeologic conditions is necessary before meaningful quantitative estimates can be made.

As cited previously, the accuracy and adequacy of basic data are often insufficient to permit a precise description of aquifer conditions or the precise evaluation of the practical sustained yields of aquifers. However, the results of the studies described herein suggest that in many cases approximate solutions based on incomplete data and idealized mathematical models or simple analog models can be of great assistance in the proper development and management of ground-water resources.

Thus, the incompleteness of basic data need not necessarily dictate that quantitative evaluations of cause and effect relations be rarely made. Approximate solutions based on existing data, when properly qualified according to the quality and quantity of data, can be of great importance. Refinement and periodic re-evaluation of solutions after the collection of additional data will be necessary and should be foreseen. Actually, the study of cause and effect relations based on existing data illumines the deficiencies of information and the benefit-cost ratio for the collection of additional data.

CONCLUSIONS

Herein, it has been attempted to describe the characteristics of electric analog computers and to illustrate their application to ground-water resource evaluation problems by means of examples comparing analog computer and analytical results. The analog computer can deal with problems of much greater complexity than is practicable with analytical methods. The accuracy, flexibility, and speed of the analog computer give the hydrologist the ability to analyze rapidly almost any conceivable hydrogeologic system with a degree of accuracy compatible with the quality of existing data. Transient and complex conditions, forbiddingly complicated for analytical study, are easily handled by the analog computer.

The analog model is re-usable, easy to assemble, and bears a close correspondence to the hydrogeologic system, thus giving rise to comparatively simple problem formulation and computer set-up for relatively complicated situations. Unlike the digital computer, an analog computer does not require instruction in the proper mathematical method of solving a problem. In addition, analog computers are generally less expensive than digital computers and frequently permit much faster computation.

Methodology available for analyzing ground-water problems has been founded almost entirely on analytical solutions and has been inadequate in coping with complex aquifer conditions. Analog computers now provide a means for the analysis of nonhomogeneous systems having highly irregular shapes and a wide variety of head and discharge controls. With the advent of analog computers, quantitative studies of ground-water resources are not hampered by inadequate methods of analysis, but are limited by the magnitude and cost of the field effort in the collection of sufficient hydrogeologic data to properly describe aquifers.

Analog computers will not cause analytical methods or digital computers to become obsolete, but they will, when used in conjunction with these other tools available to the hydrologist, greatly improve the analysis of ground-water problems. With the aid of the analog computer, the Illinois State Water Survey is making quantitative appraisals of developed and undeveloped aquifers in Illinois.

ACKNOWLEDGMENTS

H.F. Skibitzke, A.E. Robinson, and their associates, of the staff of the U.S. Geological Survey in Phoenix, Ariz., spent several days in conference with the senior author, encouraging the use of the electric analog computer by the Illinois State Water Survey. They freely discussed and demonstrated the theory and practical application of the electric analog computer. A.E. Robinson assisted the authors in the construction of analog model 1 and assembled the excitation-response apparatus. E.A. Mueller, Electrical Engineer, Illinois State Water Survey, was frequently consulted on electrical matters and obtained the component parts of the analog computer.

This report was prepared under the general supervision of W.C. Ackermann, Chief, Illinois State Water Survey, and H.F. Smith, Head of Engineering Section. J.W. Brother prepared the illustrations for this report.

APPENDIX.—NOTATION

The following symbols have been adopted for use in this paper:

a = width of analog grid interval;

C = electrical capacitance;

h = hydraulic head;

I = electrical current;

K_1 = scale factor relating quantity of water and quantity of electric charge;

K_2 = scale factor relating hydraulic head loss and electrical potential drop;

K_3 = scale factor relating rate of water flow and electrical current;

K_4 = scale factor relating actual period of time and analog computer computation time;

m = aquifer thickness;

m' = saturated thickness of confining bed;

P_V = coefficient of vertical permeability of aquifer;

P' = coefficient of vertical permeability of confining bed;

Q = pumping rate;

Q = quantity of electrical charge;

q = quantity of water;

R = electrical resistance;

r_e = virtual radius of the cone of depression;

S = coefficient of storage of the aquifer;
T = coefficient of transmissibility of the aquifer;
t = time;
V = electrical potential drop; and
x, y, z = rectangular coordinates.

18

Copyright ©1955 by the American Society of Civil Engineers
Reprinted from *Am. Soc. Civil Engineers Proc., Jour. Hydraulics Div.*
81(separate):628-1-628-20 (1955)

GROUND-WATER FLOW IN RELATION TO A FLOODING STREAM

David K. Todd[1] J.M. ASCE

Abstract

Quantitative data on ground-water flow during storm runoff periods are scarce. This study presents some information on the time distribution and magnitude of such flows based on model measurements. Idealized hydrographs and aquifers were studied in a Hele-Shaw viscous fluid model. Results of the model analogy are interpreted in terms of bank storage volume and distribution, and magnitude and time distribution of ground-water flow. Seepage bordering the Sacramento River is discussed in relation to the model data.

INTRODUCTION

Little is known of the ground-water contribution of stream flow during flood periods. Data are difficult to obtain in the field because of surface runoff and rainfall contributions, lack of sufficient permeability data, and expensive recording well systems. Then, too, long and irregular intervals between isolated high flows discourage detailed studies.

Because engineering investigations require estimates of total stream flow, the usual procedure in practice is simply to assume some reasonable value for base flow during high flow periods. Various methods have been recommended[1,8,13][2] for this purpose. It has been stated, and rightly so, that as long as one method is used consistently in any study, errors will be kept to a minimum. Such procedures are justifiable since: (a) detailed, accurate information is lacking; (b) the base flow contribution during a flood period is usually only a small fraction of the total flow; and (c) the accuracy of most stream-flow records does not justify any greater precision.

Previous investigations provide some clues to the true ground-water contribution. Hursh and Brater[5] analyzed field data and used the sharp break of the recession-hydrograph to locate the maximum on the normal ground-water depletion curve. Suter[15] presented data on flood hydrographs and nearby well stages of the Illinois River, showing how well levels rose and fell with the river. In an analytical treatment of waves in unconfined aquifers Werner and Noren[18] showed that harmonic oscillations of a stream were transmitted to the adjacent ground-water table where they were rapidly damped. Jones,[7] in a detailed study of the Upper San Joaquin River Valley, showed by profiles of ground-water levels that ground-water levels up to about one mile from the river were influenced by river levels. Hursh and Barnes[4] discussed the need for distinguishing ground-water flow from surface runoff, and the difficulty of separating interflow from base flow. Also,

---

1. Asst. Prof. of Civ. Eng., Univ. of California, Berkeley, Calif.
2. Numbers in parentheses refer to references at the end of the paper.

the question of the amount of temporary bank storage and of its return as subsurface flow was brought out. Based upon watershed studies in Switzerland and laboratory experiments, Roessel[12] indicated that base flow may well be a major part of flood flows, and that different runoff rates may occur from the same quantity of ground-water storage, depending upon the distribution within the aquifer of the water. Both Riesbol[11] and Snyder[14] made important contributions to the subject by their detailed analyses of watershed runoff.

Model Analysis

Because of the absence of field data for analyzing ground-water flows near a stream during periods of storm runoff, a model study was undertaken for this purpose. A Hele-Shaw viscous fluid model was utilized. This model consists of a channel formed by two closely-spaced glass plates through which a mineral oil flows. It can be shown that the laminar flow conditions obtained are analogous to ground-water flow conditions. The model design and applications have been described elsewhere,[16,17] while a similar model has been employed for base course drainage studies.[3] Solitary sine waves of varying amplitudes and periods were generated in a regulating reservoir at one end of the channel and allowed to travel into the channel containing a uniform oil depth. The conditions may be thought of as a model of a flooding stream with a rising and falling stage and having a horizontal water table intersecting the stream channel. Figure 1 shows a schematic diagram of the field conditions reproduced by the Hele-Shaw model analogy.

The model arrangement requires the following assumptions for analogous field conditions: (a) the stage hydrograph is of a sinusoidal shape; (b) an unconfined aquifer with homogeneous isotropic permeability exists adjacent to the stream channel; (c) the stream bank is vertical (corresponding to the reservoir-channel boundary); (d) the flood crest elevation is less than the ground surface so that no overflow occurs; (e) an impermeable horizontal bedrock coincides with the elevation of the stream bed; (f) the stream flow is much larger than the ground-water flow so that the stage is independent of any ground-water flows; (g) the ground-water table is initially horizontal; and (h) compressibility effects are negligible. These assumptions present idealized field conditions; nevertheless they enable quantitative ground-water flow measurements to be made on a model analogy basis. In essence, the model presents by analogy a vertical cross-section perpendicular to a stream channel.

During the time when various sinusoidal oscillations within the regulating reservoir were in progress, lapse-time motion picture records were obtained of the movement of the free surface in the channel. It should be noted that an oscillation is here defined as a fluctuation beginning at the initial minimum oil elevation in the channel, rising to a maximum elevation, and returning to the original minimum level. This corresponds to a sine curve extending from -90° to +270°. The amplitude is the difference in elevation between the minimum level (-90°) and the maximum level (+90). Records of the oscillations showed that in the rising portion of the oscillation, oil flowed continuously into the channel. A reversal of flow in the second half of the oscillation became evident in that portion of the channel closest to the reservoir as the oil level in the reservoir decreased more rapidly than that in the channel. This process formed a wave crest which travelled from the reservoir into the channel and decreased in amplitude with time. Of interest here is the fact that the area, as viewed from the side of the channel, occupied by the oil above the original constant oil level is analogous to the volume of ground-water storage induced by a flooding stream. From a graph showing the variation of this area with time, the derivative, or slope, of this curve represents the rate of change of area with time, or the ground-water flow rate. To study this flow distribution, 14 independent solitary sinusoidal oscillations in the channel were recorded and analyzed. Oscillation periods ranged from 2 to 6 minutes and amplitudes from 1 to 6 inches (2.5 to 15.2 cm). Included in Table 1, which summarizes all of the basic data, are tabulations of amplitude H, in cm, and period T, in min, for the various runs.

Since the model measurements cover different amplitudes and period of the same sinusoidal oscillation, it was desirable to express the results in terms of a general curve which would be applicable to all of the data. Reasoning that the volume of ground-water (or bank) storage is some function of the oscillation amplitude and period, Figure 2 was prepared. Here the maximum recorded volume V_{max} (from Table I) of storage, expressed in two dimensions as an area with units of cm^2, is plotted against the product of amplitude and period. It can be seen that a linear relationship is reasonably well defined by the model data, indicating that the parameter, V_{max}/HT, may be assumed constant for the various runs. Now since the time from the beginning of the oscillation of V/HT must depend upon the period of the oscillation, the dimensionless time parameter t/T may be introduced. Here V is the volume measured at time t after the start of the oscillation. Curves of V/HT versus t/T were prepared for the 14 runs and superposed; close agreement resulted. A mean curve prepared from the group of curves is shown as the "observed" curve in Figure 3. This volume-time curve begins at zero (since the oil volume below the initial constant oil level is neglected) and rises to a maximum at t/T = 0.56. This instant follows shortly after the time of crest stage in the reservoir (or stream, when speaking of field conditions) at t/T = 0.50. The volume then recedes toward zero, displaying a typical hydrograph shape. The curve is cut off after t/T = 5.0 as model data were not carried beyond this time.

The slope, or derivative, of the volume-time curve represents the ground-water flow. Since the volume to any time t may be expressed as

$$V = \int_{t=0}^{t} Q \, dt \tag{1}$$

where Q is the ground-water flow occurring in any time increment dt, then the derivative of the curve in Figure 3 becomes

$$\frac{\frac{1}{HT} dV}{\frac{1}{T} dt} = \frac{1}{H} \frac{Q \, dt}{dt} = \frac{Q}{H} \tag{2}$$

Hence, a graph of the parameters Q/H versus t/T should enable the time distribution of the ground-water flow for all of the runs to be compared.

An inspection of the slopes of Figure 3 indicates how the ground-water flow curves will appear. The derivative begins at zero, reaches a positive maximum at the steepest point of rise, returns to zero at the instant of maximum, goes to a minimum at the steepest point on the recession portion of the curve, and finally returns toward zero. For purposes of the ground-water analogy, signs of all ground-water flow curves will be reversed. By so doing ground-water recharge, or stream-flow loss, is given a negative sign; ground-water

TABLE I – SUMMARY OF BASIC DATA FOR SOLITARY

SINUSOIDAL FLOWS

Run No.	Amplitude, H, cm.	Period T, min	Max. Vol., V_{max}, cm^2	Q_{min}, cm^2/min	Q_{max}, cm^2/min	t_{min} min.	t_o min.	t_{max} min.
1	7.6	3.0	379	-433	190	1.38	1.75	2.62
2	5.1	3.0	229	-389	129	1.25	1.50	2.25
3	7.6	3.0	419	-541	282	1.12	1.88	2.25
4	10.2	6.0	820	-606	184	2.25	3.20	5.25
5	10.2	2.0	445	-841	242	0.88	1.20	1.88
6	10.2	4.0	651	-587	352	1.50	1.70	3.50
7	5.1	3.0	256	-361	159	1.38	1.70	2.50
8	15.2	3.0	872	-1102	421	1.25	1.80	2.50
9	2.5	3.0	148	-230	30.5	1.12	1.50	2.75
10	10.2	3.0	535	-701	253	1.25	1.70	2.75
11	2.5	3.0	101	-120	35.0	1.38	1.50	2.75
12	12.7	3.0	714	-831	355	1.25	1.75	2.62
13	10.2	3.0	548	-741	268	1.25	1.70	2.50
14	10.2	5.0	742	-608	254	1.62	3.25	4.00

discharge, or stream-flow gain, becomes positive. This convention is in accord with the usual concept of treating ground-water flow into a stream as a positive contribution.

To verify the above indicated comparison, computed maximum and minimum discharges, Q_{max} and Q_{min}, respectively, were plotted against the oscillation amplitude, H. The minimum discharge may be regarded also as the maximum recharge. The results, using data from Table I, appear as Figure 4 and support the contention that the parameter Q/H may be assumed constant for the various runs. Similarly, if the times, t_{min}, t_o, and t_{max} (see Table I), correspond to Q_{min}, Q_o (zero flow between recharge and discharge), and Q_{max}, respectively, then these times should also bear linear relationships to the oscillation periods. The data plotted in Figure 5 substantiate this supposition—Q_{min} takes place at $t/T = 0.41$, Q_o at $t/T = 0.56$, and Q_{max} at $t/T = 0.86$. Based upon these comparisons among the different runs, the 14 discharge curves computed from the volume-time curves likewise showed substantial agreement when plotted in terms of Q/H versus t/T. The mean curve, and derivative of Figure 3, appears as Figure 6.

Although the exact magnitude of ground-water flow adjoining a stream always has been difficult to ascertain, the flow reversal shown by Figure 6 was recognized by earlier investigators. Brater[2] in 1940 stated that when a stream rises above the adjacent ground-water table, "---there must be a temporary reversal in the ground-water gradient and the ground-water contribution could conceivably be negative for a short interval." Other engineers (8, p. 399; 13, p. 266) also have correctly interpreted this phenomenon.

Regarding the scatter of data indicated in the previous figures, errors may be attributed largely to instrumental and analytic sources. The flow rate of the oil between the glass plates is a function of density and viscosity of the oil. Temperature modifies both of these properties; hence all data were corrected from the observed temperature to a standard of 70°F. However, if the oil temperature in the course of an oscillation changed, the single observed temperature would not correctly account for these effects. Errors also may have been introduced by small deviations from the true sinusoidal pattern, since the unsteady flows were manually controlled by inflow and outflow valves connected to the reservoir. Areas of oil were scaled from projections of the 35-mm film record by measuring ordinates at 6-inch intervals along the channel. A grid sheet attached to the back of the glass channel aided in scaling off the ordinates. Measurements were accurate to about 0.15 cm.

The far end of the channel, opposite to the reservoir, was left open except for an overflow lip fixed at the same elevation as the initially constant oil level. This end of the channel could not be sealed without violating the assumed field analogy (Figure 1). The fact that the channel was of a finite length and not of a semi-infinite extent as previously assumed, introduced erroneous volume and flow values. Fortunately, errors did not begin until the oil level had begun to rise and outflow ensued at the far end of the channel. This happened near the middle of the recession portion of the discharge curve. The nature of this limitation may be ascertained from a knowledge of the flow pattern. As the oscillation occurs within the reservoir a wave is propagated into the channel which damps out along the horizontal oil surface at a theoretically infinite distance. Practically, however, the damping rate is very rapid and changes of the oil level become negligible in relatively short distances. The effect of the outflow error was to reduce the oil area within the channel at a faster rate than would have resulted if a longer channel had been used, and to give too large a discharge in the recession portion of the discharge curve.

In an effort to evaluate the magnitude of the error, the recession portion of Figure 6 was plotted on logarithmic coordinates (Figure 7). This showed that the data closely fitted an equation

$$\frac{Q}{H} = 16.2(t/T)^{-2.76} \quad (3)$$

except for a sharp break near $t/T = 1.8$. Examination of the data showed that this time closely corresponded to the initial outflows at the far end of the channel. Hence it is reasonable to assume that if this flow had not occurred, the logarithmic recession curve, labelled "computed" on Figures 3, 6, and 7, would have continued. It is believed that this computed curve more nearly represents the true conditions pertaining to a semi-infinite aquifer. The flows indicated by the difference between the observed rate and projected logarithmic rate form a small hydrograph with rising and falling limbs. Outflow at the far end of the channel amounted to 9.8 percent of the total inflow volume up to $t/T = 5.0$.

Figure 8 shows typical dimensionless stage hydrographs of the oil (water table) at distances of 0, 0.5, 1.0, 2.0, and 5.0 feet from the reservoir. The height h occurs at the time t and is always less than H away from the stream channel. These curves illustrate the rapid damping of the oscillation as it travels from the source.

Model-Prototype Relations

To interpret properly the model results they must be expressed in terms of similar prototype conditions. This can be done by assuming length and time scale ratios, and solving for prototype velocities, discharge, and permeability.

The mean velocity of flow in the model may be expressed by

$$V_m = \frac{\rho_m b^2 g}{3 \mu_m} \frac{dh}{dx} \quad (4)$$

where ρ_m is the oil density, μ_m is the oil viscosity, b is the half-width of the channel spacing, g is gravity, and dh/dx is the surface slope.[17] Similarly, the velocity of flow in an aquifer may be expressed by the Darcy law as

$$V_p = \frac{\rho_p k g}{\mu_p} \frac{dh}{dx} \quad (5)$$

where ρ_p is the density of ground water, μ_p is its viscosity, and k is the aquifer permeability (expressed as an area). The ratio of these two velocities when governed by the same slope becomes

$$\frac{V_m}{V_p} = V_r = \frac{\rho_m b^2 \mu_p}{3 \rho_p k \mu_m} \quad (6)$$

where the subscript "r" denotes a model-prototype ratio. For the mineral oil used in the model, $\rho_m = 0.877$ gm/cc and $\mu_m = 1.016$ poise at 70°F. The plate spacing was 0.2096 cm, so that b = 0.1048 cm. At 70°F for water,

$\rho_p = 0.998$ gm/cc and $\mu_p = 0.00979$ poise. To obtain a value for k in equation (6), V_r must be known. This quantity depends upon the length L_r and time T_r, scale ratios. Values of $L_r = 0.01$ and $T_r = 0.0003$ may be assumed for illustration and to give information of the correct order of magnitude of prototype conditions. Thus, the model channel with a length of 10 feet represents a 1000-ft. distance on one side of and perpendicular to a stream channel, while a 3-inch oscillation is equivalent to a 25-ft river stage fluctuation. Similarly, a 3-min oscillation is comparable to a storm runoff duration of 6.9 days. Inserting these values to find V_r gives

$$V_r = \frac{L_r}{T_r} = \frac{0.01}{0.0003} = 33.33 \tag{7}$$

Using this value in equation (6), together with the other known data, gives a permeability of 9.29×10^{-7} cm^2, or 94.2 darcys. A 40-mesh sand, packed to a 40 percent porosity, would have roughly the same permeability (9, p. 113).

If field data happened to be available so as to provide an estimate of k, then V_r could be computed directly. Knowing V_r, L_r and T_r could be selected to study any desired flood amplitude and duration.

Bank Storage

Estimation of the volume of bank storage for assumed prototype conditions may be made from Figure 3. Direct conversion by means of the length scale cannot be made without a further assumption of specific yield. In the model channel the specific yield has been implicitly taken as 100 percent. This is true if the small adhesive oil film on the glass surfaces is neglected. A value of 10 percent specific yield may be assumed as a constant for field conditions. It has been pointed out by Jacob (13, p. 384) that the amount of water draining per unit volume of aquifer during the lowering of a ground-water table is a function of time. The maximum amount drained equals the specific yield. With the constant value of 10 percent specific yield, it is here assumed that the water will flow completely into and out of the aquifer with equal ease. The error incurred by this assumption will be small in this illustration because of the long time involved in the prototype ground-water flow. A large storm runoff of minimum duration would provide the least drainage time and hence give the largest error due to a variable storage coefficient.

As shown by Figure 2 the volume of ground-water recharge, or bank storage, is directly proportional to the amplitude-duration product of the storm runoff. For illustration the previously described 25-ft and 6.9-day runoff may be used. Converting the recharge and discharge from Figure 6 by means of the length and time scales, and multiplying by the specific yield, gives for the maximum recharge 0.00278 cfs per lineal foot of stream channel; 0.00109 cfs/ft for the maximum discharge. The total volume of recharge amounts to 421 ft^3 per lineal foot of stream channel.

Expressing the above results in more meaningful units and taking into account that the identical phenomenon is taking place within the opposite stream bank, the maximum ground-water recharge rate equals 29.3 cfs per mile of stream channel; the maximum discharge, 11.5 cfs/mi. It should be noted also that for this example, the peak stream flow occurs 3.45 days after the start, the maximum recharge occurs after 2.8 days, the flow reverses at 3.9 days, and the maximum discharge occurs after 6.0 days. Measurable return flow to the stream persists for several weeks. The total recharge volume amounts to 102 AF per mile of stream channel.

Bank storage occurs in that portion of the aquifer closest to the stream channel. Previous study[17] has indicated a very rapid decrease in the amplitude with distance inland. Rough calculations for the conditions described in the above illustration indicate that 50 percent of the total bank storage is contained within the first 175 feet of the channel, and 75 percent within 350 ft.

Sacramento River Seepage

An interesting field problem related to the topic of the model study is that of the Sacramento River seepage. During seasons of high runoff, Sacramento River stages below Shasta Dam exceed the adjacent land surface. An extensive levee system ordinarily confines the high flows to within the levees. Landowners of the strip of land adjoining the river and outside the levees, however, contend that "when the water surface is carried well above the outside land surface for any substantial number of days, the pressure forces the water to seep underground and saturate and waterlog the land to an extent that it cannot be worked or pastured and damages the orchard crops and kills the trees".[6] The Corps of Engineers has been studying the problem from a flood control standpoint, while engineers of the Bureau of Reclamation have made extensive measurements of the seepage because some parties have expressed the opinion that controlled releases of Shasta Dam have contributed to the problem. A committee of the California Legislature has recently looked into the problem,[6] and at the present time the California Division of Water Resources is analyzing all available data in preparation for a final report.

This particular problem provides a good illustration of the difficulties typically encountered in applying laboratory results to subsurface field investigations. Data collected by the U. S. Bureau of Reclamation provide an insight into the situation.[10] Shifting of the river channel in the past has produced irregular lenticular deposits of gravel, sand, silt, and clay. The surface material is predominantly fine-grained with scattered gravel and/or sand deposits to be found in the upper 100 feet. Because of the concern for seepage on agricultural land bordering the river, lines of wells were established by the Bureau at several points along the river. The lines extend inland from the levee and water table elevations were observed daily for long periods.

These observations showed that the ground-water table adjacent to the river was governed by rainfall, irrigation, river stage, and geology. The variability of these factors in either space or time indicate the complexity of the problem. The water table approximates the river stage. In summer and other periods with the river stage low, the water table is maintained by irrigation and is higher than the river. Drainage is then toward the river. During periods of high stages, the river level exceeds the water table elevation and recharge, or seepage, occurs. Water table elevations have been observed to change up to 10 feet within 5 days due to varying river levels. Amplitudes of fluctuations decrease inland, extending as far as 1/2 to 1-1/2 miles, depending upon locality. Depths to water table vary in extremes from zero to 25 feet in the river influence area.

Figure 9 shows a small portion of the data. River and nearby water table elevations at West Wilkins Slough for the spring of 1945 are shown. Well No. 1 is located only 100 feet from the levee, Well No. 4 is 600 feet, and Well No. 5 is 1100 feet. Precipitation and well logs are included for completeness. It can be seen that three major river rises occurred, affecting the nearest well most and the farthest well least. Until about April 1 a good correspondence between the river stage and the water table is apparent. Well fluctuations follow those of the river, delayed slightly with time and decreasing with amplitude inland. After April 1, Well Nos. 4 and 5 show a rise not indicated by either the river stage or Well No. 1. Presumably early irrigation is responsible for this deviation since rainfall is negligible.

The river stage continuously exceeded the ground water table elevation (Well No. 1) at this point in the 1944-45 season from November 3, 1944 to April 14, 1945. If the correspondence between the river and wells is taken as an indication of direct connection of permeable strata, a large recharge from the river may be expected. The available well logs, however, indicate rather tight formations so that a quantitative estimate of recharge is difficult to make. No direct application can be made of the model results in this instance because of the vertical recharge of the aquifer by rainfall and irrigation. Then, too, the inland water table may be regulated by Wilkins Slough, which parallels the well line and intersects the river about 900 feet southward, or by inland drainage systems.

Although the model results cannot be applied directly, an interesting comparison should be noted. Field studies by the Corps of Engineers have shown that the amount of seepage is roughly proportional to the number of days river stages are above natural bank levels and to the number of feet the river stage exceeds those levels (6, p. 25). This "foot-days" factor is analogous to the amplitude-period product which was shown by Figure 2 to be directly related to the total ground-water recharge.

Discussion

Although the model data lead to idealized results in terms of field conditions, they nevertheless describe for one of the simplest cases the distribution of ground-water flow. Variation of any of the controlling factors will modify the results herein presented. Storm rainfall is implicitly assumed to have fallen over some portion of the drainage basin upstream from the cross-section studied. Vertical percolation and interflow may be regarded as distinct stream-flow contributors and have been eliminated from the present study. Other hydrograph shapes will yield different ground-water storage and flow curves.

The model results showed that recharge and discharge were directly proportional to the amplitude-period product. In the field this may be interpreted by saying the ground-water recharge and discharge are directly proportional to a stage-duration factor. At a given point on a stream isolated storm events occurring over the drainage area above a given point on a stream would result in approximately constant flood durations, as indicated by the unitgraph theory. In such instances the ground-water recharge and discharge would depend only upon the flood stages, so that by means of a rating curve at that point, an approximate relationship between the stream flow and the ground-water flow could be established.

The model experiments required a horizontal ground-water table in equilibrium with the stream in the prototype situation. This would only rarely occur in the field. The more common situation where the ground-water table and stream are in contact would be an influent or an effluent stream. Here the ground-water table slopes upward or downward to the stream surface, respectively. As a first approximation for such cases the effects may be considered as additive, so that the ground-water flow curve would be shifted upward for effluent streams and downward for influent streams. Depending upon the relative magnitude and direction of the initial ground-water flow to the maximum recharge and discharge rates, flow reversals may or may not result from storm runoff.

The initial ground-water depth was taken as constant for all model runs. For completeness this depth also should be investigated because its variation affects flow rates. For these unsteady flows the ratio of amplitude to depth is a complex function of the flow, and no general statements can be made about the magnitude of this effect.

The ground-water contribution to a storm hydrograph at a given point on a stream is the combined effect of all ground-water flows upstream of that point. Thus the recharge-discharge pattern discussed herein is in phase with the crest stage, so that the flow depends upon the location of the flood crest. A complete analysis involves the integrating time distributions, aquifer and permeability variations, tributary effects, and changes in the storm hydrograph with travel. Although the generalized results presented in Figs. 3 and 6 are applicable only with respect to the assumptions stated, they nevertheless should aid toward a better understanding of ground-water flows bordering on a stream channel. Extending or modifying the analyses to suit particular problems may permit estimates of ground-water flow to be prepared.

ACKNOWLEDGMENTS

Robert G. Dean assisted in the computations and preparation of figures. Engineers of the U. S. Bureau of Reclamation, Sacramento, generously made available the Sacramento River seepage data, while engineers of the Sacramento offices of the California Division of Water Resources, U. S. Corps of Engineers, and U. S. Geological Survey were helpful through their discussions of the problem. Laboratory work was performed under the guidance of J. A. Putnam. J. W. Johnson provided valuable suggestions on the analysis and interpretation of the data.

REFERENCES

1. American Society of Civil Engineers, Hydrology Handbook, Manual of Engineering Practice No. 28, 184 pp., 1949.

2. Brater, E. F., The unit hydrograph principle applied to small watersheds, Trans. American Society of Civil Engineers, v. 105, pp. 1154-1192, 1940.

3. Casagrande, A. and Shannon, W. L., Base course drainage for airport pavements, Trans. American Society of Civil Engineers, v. 117, pp. 792-820, 1952.

4. Hursh, C. R. and Barnes, B. S., Appendix B—Report of sub-committee on subsurface-flow, Trans. American Geophysical Union, v. 25, pt. V, pp. 743-746, 1944.

5. Hursh, C. R. and Brater, E. F., Separating storm-hydrographs from small drainage-areas into surface and subsurface-flow, Trans. American Geophysical Union, v. 22, pt. 3, pp. 863-871, 1941.

6. Joint Committee on Water Problems of the California Legislature, Sacramento River seepage and erosion problems, Sixth Partial Report, pp. 22-29, Sacramento, 1953.

7. Jones, G. H., Hydrology of valley areas adjacent to the upper San Joaquin River, Trans. American Geophysical Union, v. 21, pt. I, pp. 58-78, 1940.

8. Linsley, R. K., Jr., Kohler, M. A., and Paulhus, J. L. H., Applied Hydrology, McGraw-Hill, New York, 689 pp., 1949.

9. Muskat, M., "The flow of homogeneous fluids through porous media," McGraw-Hill, New York, 763 pp., 1937.

10. Plummer, A. K., "Summary of basic data collected during Sacramento River seepage investigations," Sacramento Valley District, U. S. Bureau of Reclamation, Chico, California, 1949 (unpublished).

11. Riesbol, H. S., Some aspects of subsurface water in hydrologic research on agricultural watersheds, Proc. of the Hydrology Conference, Pennsylvania State College, State College, pp. 85-112, 1942.

12. Roessel, B. W. P., Hydrologic problems concerning the runoff in headwater regions, Trans. American Geophysical Union, v. 31, pp. 431-442, 1950.

13. Rouse, H. (ed.), Engineering Hydraulics, J. Wiley and Sons, New York, 1039 pp., 1950.

14. Snyder, F. F., A conception of runoff-phenomena, Trans. American Geophysical Union, v. 20, pt. IV, pp. 725-738, 1939.

15. Suter, M., Apparent changes in water storage during floods at Peoria, Illinois, Trans. American Geophysical Union, v. 28, pp. 425-437, 1947.

16. Todd, D. K., "Investigation of unsteady flow in porous media by means of a Hele-Shaw viscous fluid model," Ph. D. thesis, University of California, Berkeley, 85 pp., 1953.

17. Todd, D. K., Unsteady flow in porous media by means of a Hele-Shaw viscous fluid model, Trans. American Geophysical Union (in press).

18. Werner, P. H. and Noren, D., Progressive waves in non-artesian aquifers, Trans. American Geophysical Union, v. 32, pp. 238-244, 1951.

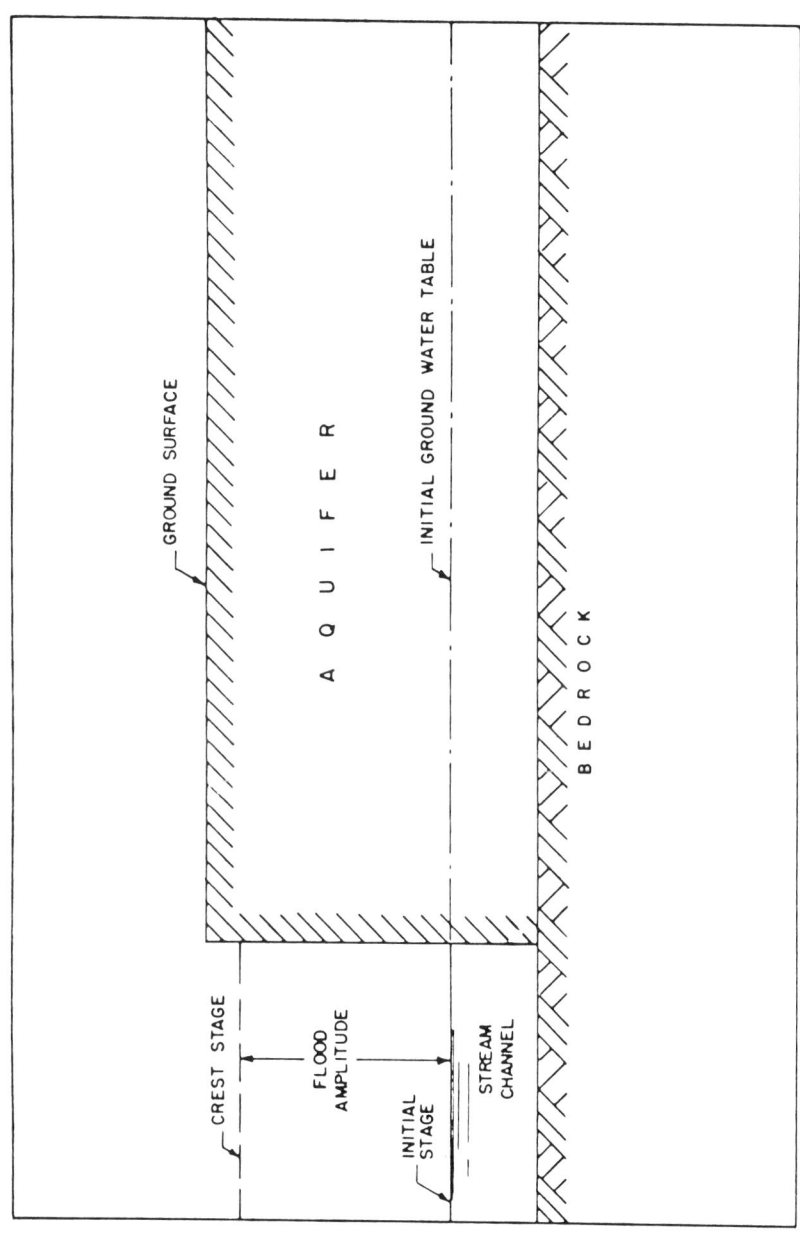

Fig. 1. Schematic diagram of vertical cross-section showing field conditions simulated by the Hele-Shaw model analogy.

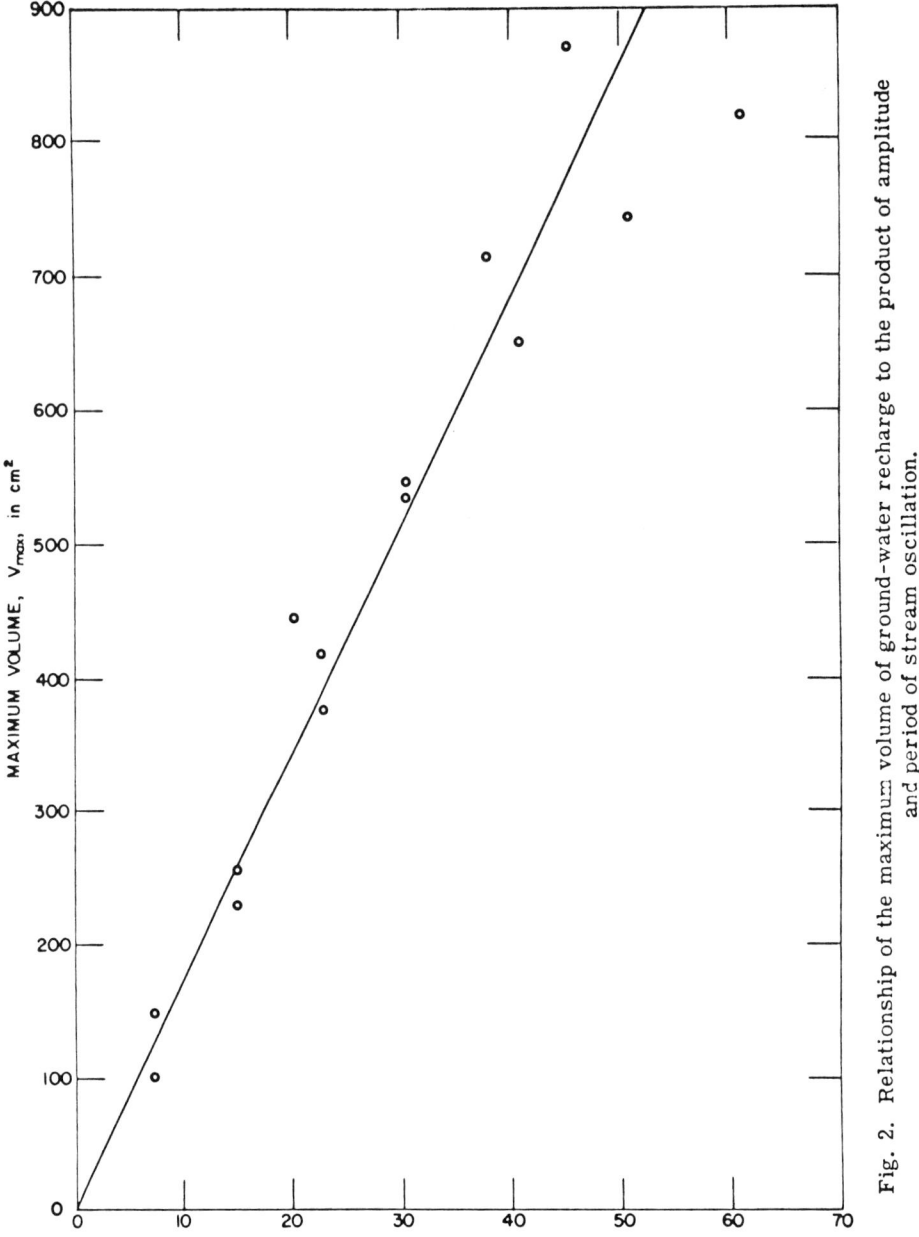

Fig. 2. Relationship of the maximum volume of ground-water recharge to the product of amplitude and period of stream oscillation.

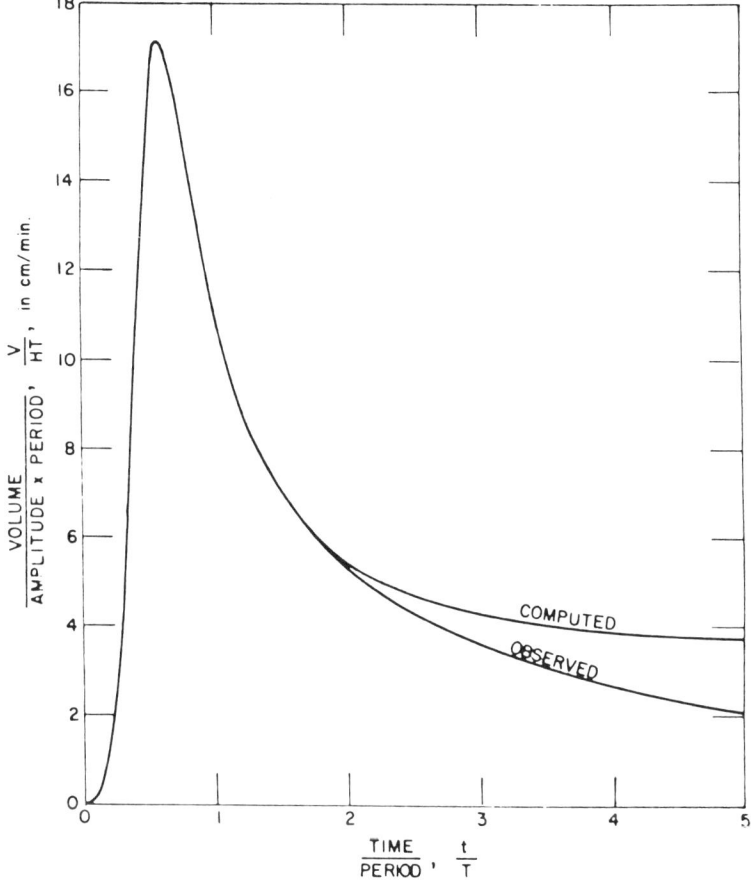

Fig. 3. Mean relationship between ground-water volume and time expressed as V/HT and t/T.

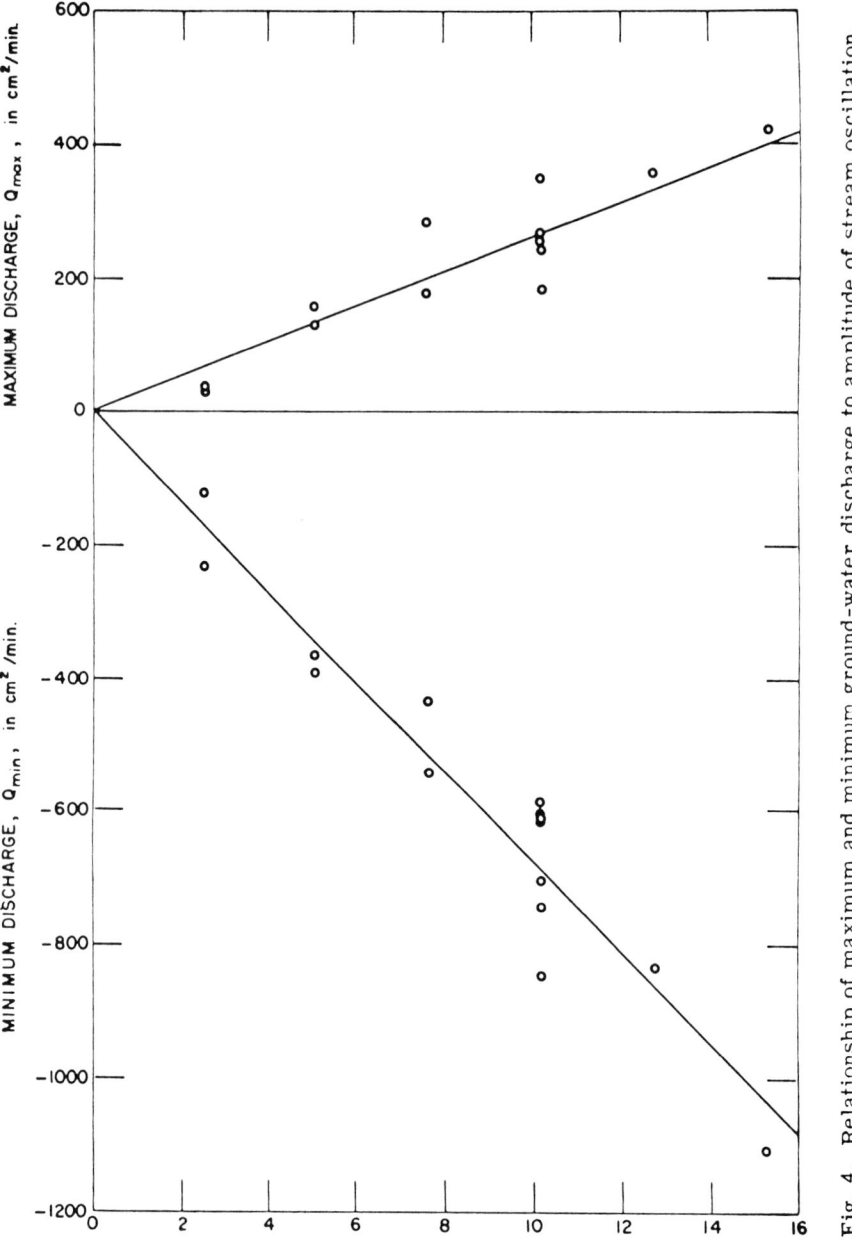

Fig. 4. Relationship of maximum and minimum ground-water discharge to amplitude of stream oscillation.

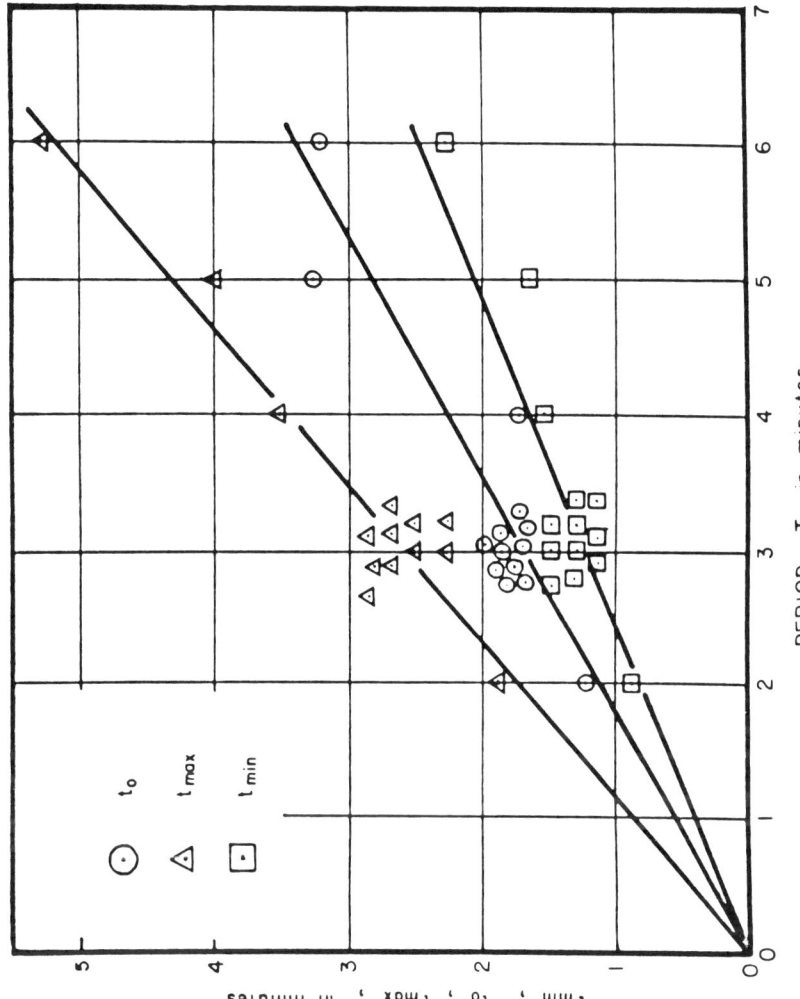

Fig. 5. Relationship of times t_{min}, t_o, and t_{max} to the period of stream oscillation.

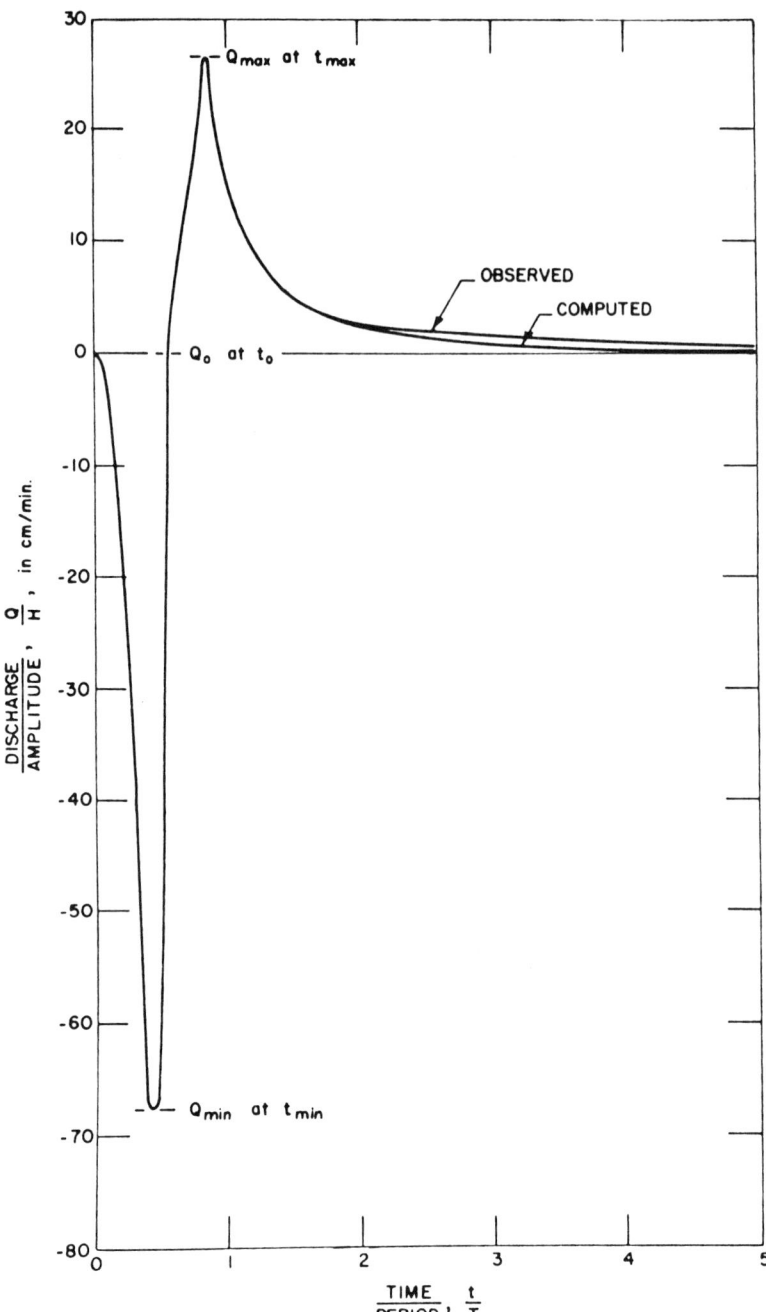

Fig. 6. Mean relationship between ground-water discharge and time, expressed as Q/H and t/T.

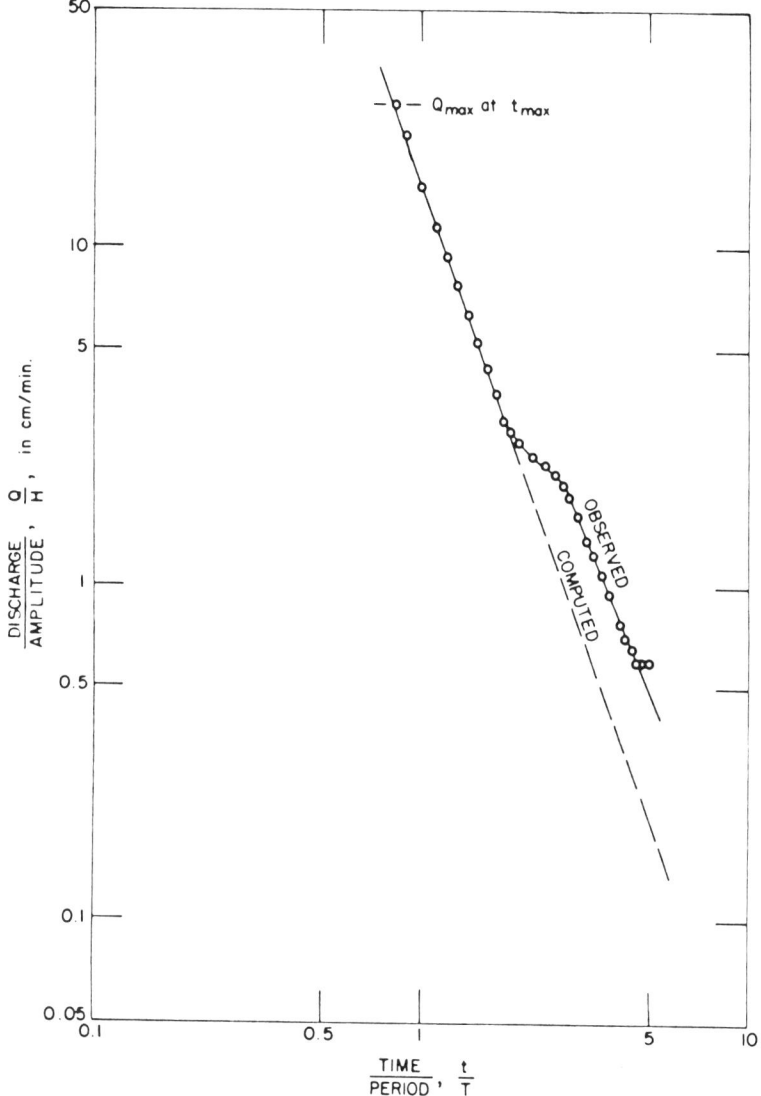

Fig. 7. Logarithmic plot of mean observed and computed ground-water recession curves, expressed as Q/H and t/T.

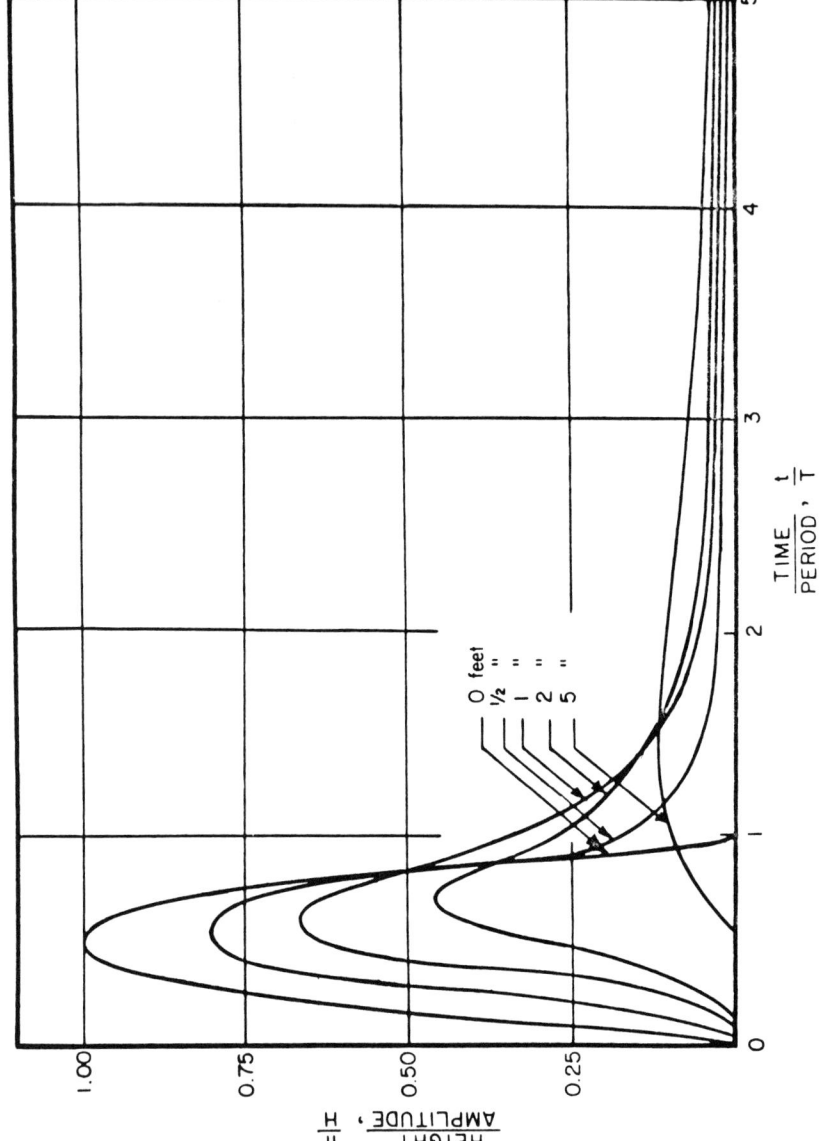

Fig. 8. Mean ground-water table hydrographs at various distances from the stream channel, expressed as h/H and t/T.

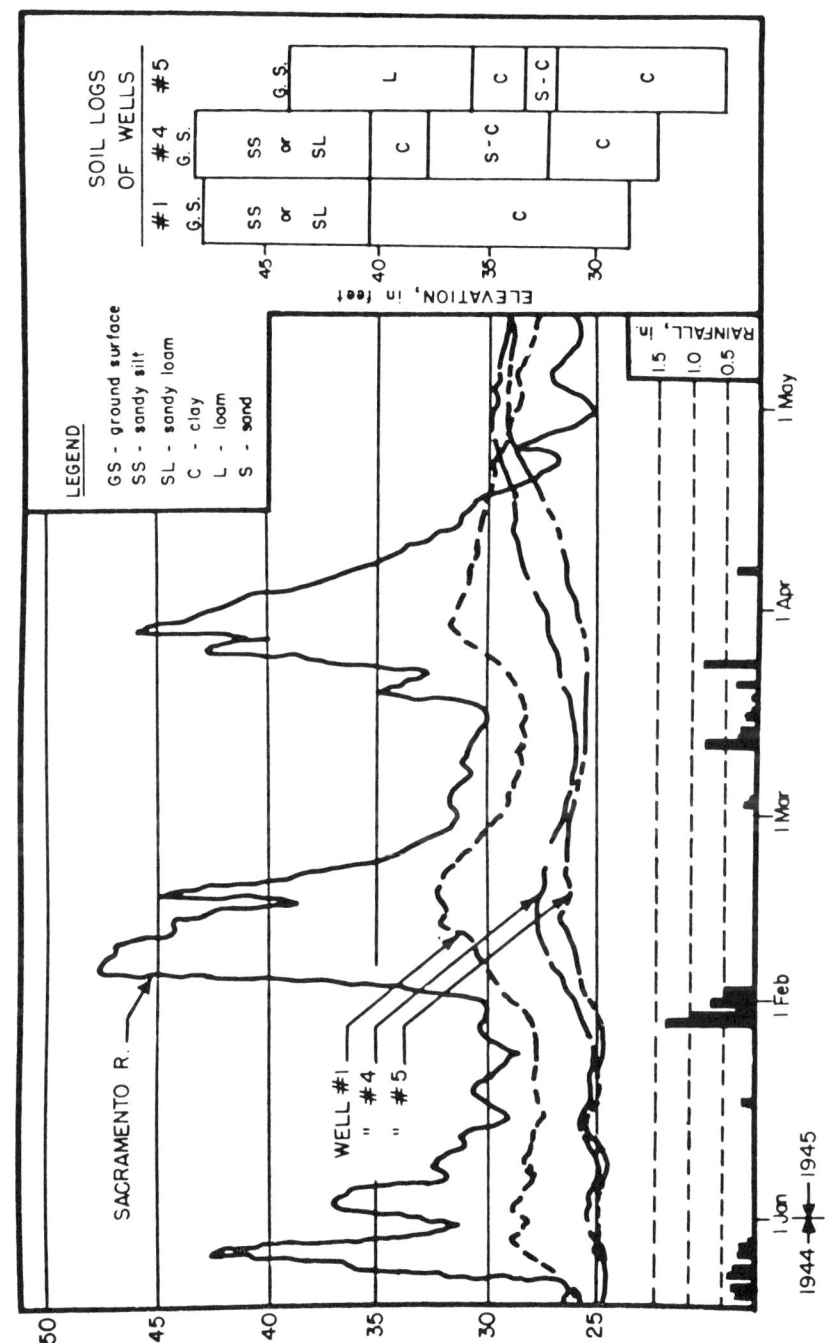

Fig. 9. Stage and ground-water table hydrographs, Sacramento River at West Wilkins Slough, California.

Part III
REGIONAL GROUNDWATER FLOW

Editors' Comments on Papers 19 Through 32

19 CHAMBERLIN
Excerpt from *Requisite and Qualifying Conditions of Artesian Wells*

20 DARTON
Excerpt from *Preliminary Report on Artesian Waters of a Portion of the Dakotas*

21 FIEDLER and NYE
Abstract from *Geology and Ground-Water Resources of the Roswell Artesian Basin, New Mexico*

22 BENNETT and MEYER
Abstract from *Geology and Ground-Water Resources of the Baltimore Area*

23 LOHMAN
Abstract from *Geology and Artesian Water Supply of the Grand Junction Area, Colorado*

24 NEWCOMB
Abstract from *Effect of Tectonic Structure on the Occurrence of Ground Water in the Basalt of the Columbia River Group of the Dalles Area, Oregon and Washington*

25 COOLEY et al.
Abstract from *Regional Hydrogeology of the Navajo and Hopi Indian Reservations, Arizona, New Mexico, and Utah*

26 TÓTH
A Theoretical Analysis of Groundwater Flow in Small Drainage Basins

27 FREEZE and WITHERSPOON
Theoretical Analysis of Regional Groundwater Flow. 2. Effect of Water-Table Configuration and Subsurface Permeability Variation

28 WALTON
Excerpts from *Leaky Artesian Aquifer Conditions in Illinois*

29 **DOMENICO and MIFFLIN**
   Water from Low-Permeability Sediments and Land Subsidence

30 **POLAND and DAVIS**
   Excerpts from *Land Subsidence due to Withdrawal of Fluids*

31 **LEGRAND and STRINGFIELD**
   Development of Permeability and Storage in the Tertiary Limestones of the Southeastern States, USA

32 **DAVIS**
   Excerpts from *Porosity and Permeability of Natural Materials*

The papers in Parts I and II emphasize the theory of groundwater flow and the methodology of well and aquifer hydraulics. These theoretical and methodological results were primarily intended for use in the assessment of groundwater resources on a regional scale. In Part III we have gathered together a suite of papers that illustrate the application of the principles and methods of groundwater flow analysis to regional hydrogeological environments.

Paper 19, an excerpt from Chamberlin's 1885 report, predates Slichter's theoretical treatise by three years and is generally recognized as the beginning of the science of hydrogeology in North America. In beautifully written prose, Chamberlin outlines the seven prerequisites for artesian flow and describes the hydrogeologic properties of water-bearing beds and confining beds. In passing he notes that there is no such thing as an impermeable formation and even relatively impermeable confining beds can be expected to provide leakage to water-bearing beds. In another early paper King (1898) provided an interpretation of groundwater movement through a geological formation under the influence of gradients in hydraulic head. King produced a flow net for an unconfined aquifer near Madison, Wisconsin, that showed recharge from precipitation and discharge to surface streams.

Between 1898 and 1912, a series of *U.S. Geological Survey Water-Supply Papers* appeared describing the results of hydrogeological field work in a variety of geologic terrains. These reports provided the first descriptions of the occurrence of groundwater in various rock types and stressed the importance of local and regional structure in controlling the continuity of aquifers. Among these early reports are those by Hill and Vaughan (1898) on the Edwards Plateau, Texas, and Mendenhall (1905), on the alluvial deposits of the San Bernardino Valley, California.

From a historical perspective, the most important early field studies were those carried out on the Dakota sandstone, the classic artesian aquifer of the north central plains. Paper 20 is an excerpt from the original 1897 work in which Darton pointed out that recharge to the aquifer takes place in the Black Hills of South Dakota and discharge from the system occurs some 300 to 500 km to the east. His report contains some classic examples of early hydrogeological mapping which, unfortunately, we have not been able to include. In a more complete treatise, Darton (1909) summarized the results of his studies on the Dakota sandstone, specifically mentioning for this case what Chamberlin had expressed in principle: confining beds can provide significant leakage to aquifers.

In the late 1920s a major debate arose with respect to the artesian conditions in the Dakota sandstone. Russell (1928) suggested an alternative explanation for the artesian pressures, invoking a mechanism based on the loading of lenticular sandstone bodies rather than one based on the relative differences between the topographic slope and the slope of the piezometric surface of the aquifer. Russell's paper appeared in the same year as Meinzer's treatise on the compressibility of artesian aquifers (Paper 6). Meinzer's concepts, which have had a more enduring appeal than Russell's, were based in large part on his own studies of the Dakota sandstone (Meinzer and Hard, 1925). The Dakota sandstone maintained its hydrogeological attraction as late as 1968 when Swenson suggested an interaction between the Dakota sandstone and the Madison limestone, which has significantly altered our understanding of the aquifer system and which may have an important impact on future development strategy.

In *Outline of Ground-Water Hydrology with Definitions,* Meinzer (1923) discussed the discharge end of the groundwater flow system in humid regions, emphasizing springs and the nature of influent and effluent streams. A later work by White (1932) clarified the relationship between evaporation and groundwater discharge in arid regions. By the 1940s, then, the role of aquifers and aquitards in regional flow systems was understood, as was the nature of their recharge and discharge. In addition, the methods of well and aquifer hydraulics outlined in Part II were in place to allow the analysis of aquifer yields. Since that time the Water Resources Branch and more recently the Water Resources Division of the U.S.G.S., along with state and provincial organizations, have carried out thousands of water resources studies in various parts of the country. Typical of these studies are those by Lonsdale (1935) on the Carizzo sandstone in Texas, Sayre and Bennett (1942) on the Edwards limestone in Texas, and Jacob (1945) on groundwater conditions on Long Island. The report by Davis et al. (1959) provides an excellent example of an alluvial valley-fill aquifer of a type common to the western United States.*

One of Meinzer's underlying principles in guiding groundwater investigations during the formative years was the need for both geologists and engineers to focus on the study of groundwater problems. Consequently, he often arranged to have an engineer and a geologist assigned to the same investigation. Paper 21 by Fiedler and Nye is an example of this early professional cooperation. Coastal aquifer studies also resulted in a great number of classic papers, among them the works of Bennett and Meyer (Paper 22) in the Baltimore area and Parker et al. (1955) in southeastern Florida. Lohman's work (Paper 23) on the Grand Junction area will long serve as a model for future investigators as it demonstrates the need for careful, detailed geologic mapping in order to understand the controls on the occurrence and movement of groundwater in regional settings. Extensive aquifers occurring in the basalt flows of the northwestern United States were the basis of outstanding investigations such as Paper 24 by Newcomb. Studies of the water resources of the Navajo reservation in northwestern Arizona culminated in a series of professional papers, among them Paper 25 by Cooley et al.

Winograd (1962), Fakin (1966), and Mifflin (1968) wrested an understanding of the Paleozoic carbonate aquifers at the Nevada test site from the complex geological environment encountered there. Winograd and Thordarson (1975) will long remain as the standard reference for the hydrogeological and hydrochemical conditions of the south-central Great Basin; it clearly demonstrates the type of investigations required to meet the needs of nuclear energy development in our society. Farvolden (1963) and Maxey (1964) provide detailed discussion of the geologic controls on regional groundwater flow.

Most of the pre-1960 work emphasized areally extensive horizontal aquifers; flow nets for these aquifers were usually presented in the

---

*The very nature of these investigations requires a comprehensive monograph to bring together the physics of groundwater flow and the results of geochemical studies, and place them within the geologic framework of aquifer occurrence. The strength of their contribution lies in this careful integration, which cannot be reflected using only excerpts, we have therefore decided to reprint a few abstracts (Papers 21 through 25) as an indication of the nature of these investigations, with the strong recommendation that for a clearer understanding of the true nature of hydrogeology, these papers must be examined in their entirety.

form of a piezometric surface. In the early 1960s, a Canadian group faced with aquifer-aquitard systems of sedimentary and glacial origin in which areally extensive aquifers were not well-developed began interpreting regional flow systems on vertical cross-sections. This approach is rooted in Hubbert's classic treatise on the theory of groundwater motion (Paper 4). Tóth (1962) produced a set of analytical solutions to the steady-state boundary value problem representing regional flow in a vertical profile. Later, Freeze and Witherspoon (1966) presented numerical solutions that allowed consideration of more complex water-table configurations and geologic environments. Papers 26 and 27 by Tóth, and Freeze and Witherspoon are the follow-up articles to their initial studies. The computer-generated flow nets in these two papers had considerable impact in clarifying the interpretation of hydraulic head measurements taken in the field. On the basis of field studies carried on during the same period, Meyboom (1966) described a widely recurring flow system, which he named the *prairie profile*, that has had considerable influence on the interpretation of many hydrogeological systems of the Great Plains.

Paper 28 is an excerpt from Walton's analysis of leaky aquifer conditions in Illinois, in which the early hydrogeological ideas of Chamberlin and Darton with respect to regional confining layers, and the pumping test methodology of Hantush and Jacob for leaky aquifers were first brought together. Walton applied his pumping test results on a regional scale in order to reach conclusions about areal recharge rates to the groundwater system.

One of the most important applications of transient groundwater flow theory is in the prediction of land subsidence due to subsurface withdrawals of groundwater. Fuller (1908) was apparently the first to suggest this mechanism of land subsidence, and Tolman and Poland (1940), working in the Santa Clara Valley, California, were the first to confirm the mechanism in the field. Domenico and Mifflin (Paper 29) provide one of the best analytical treatments of land subsidence, clarifying the interrelationship between Terzaghi's theory of consolidation and Jacob's theory of groundwater flow. Miller (1961) had earlier used Terzaghi's formulas to predict subsidence, and Lohman (1961) had developed a formula on the basis of Jacob's development; but Domenico and Mifflin were the first to integrate the two theories. Paper 30 by Poland and Davis is the classic paper in the field; it reviews the mechanisms of subsidence and summarizes the field evidence from various sites around the world. The excerpt reprinted here provides an updated analysis of conditions in the Santa Clara Valley.

As we move into the last decades of the twentieth century, hydrogeologists are more frequently asked to predict the likely response of groundwater systems to manmade stresses. These predictions are formulated with numerical simulations based on the theories, hydraulics, and regional concepts presented in the papers collected in this volume. Predictions, however, require more than theory; they also require data. Our final selection, (Paper 32), provides the best available summary of permeability and porosity data from a wide variety of geological materials. The five tables of values that are the heart of the paper and the reference list that provides the data sources are presented.

Paper 31 on the Floridan limestone aquifer is a shortened version of a more detailed paper by Stringfield and Legrand (1966). Both are the outgrowth of a long period of study that began with the mapping of the piezometric surface for the aquifer by Stringfield (1936). Paper 31 emphasizes the time dependency of permeability and the role of geologic processes on the development of aquifer properties and hydrogeologic environments.

REFERENCES

Darton, N. H., 1909, Geology and Underground Waters of South Dakota, *U.S. Geol. Survey Water-Supply Paper 227,* 156p.

Davis, G. H., J. H. Green, F. H. Olmstead, and D. W. Brown, 1959, Ground-Water Conditions and Storage Capacity in the San Joachin Valley, California, *U.S. Geol. Survey Water-Supply Paper 1469,* 287p.

Eakin, T. E., 1966, A Regional Interbasin Ground-Water System in the White River Area, Southeastern Nevada, *Water Resources Research* **2:**251-271.

Farvolden, R. N., 1963, Geologic Controls on Groundwater Storage and Base Flow, *Jour. Hydrology* **1:**219-249.

Freeze, R. A., and P. A. Witherspoon, 1966, Theoretical Analysis of Regional Groundwater Flow.1. Analytical and Numerical Solutions to the Mathematical Model, *Water Resources Research* **2:**641-656.

Fuller, M. L., 1908, Summary of the Controlling Factors of Artesian Flows, *U.S. Geol. Survey Bull. 319,* 44p.

Hill, R. T., and T. W. Vaughan, 1898, Geology of the Edwards Plateau and Rio Grande Plain Adjacent to Austin and San Antonio, Texas, with Special Reference to the Occurrence of Underground Waters, *U.S. Geol. Survey Annual Rept.* **18**(part 2):258-316.

Jacob, C. E., 1945, Correlation of Ground-Water Levels and Precipitation on Long Island, N.Y., *New York Dept. Conservation, Water Power and Control Comm. Bull. GW-14.*

King, F. H., 1898, Principles and Conditions of the Movements of Ground Water, *U.S. Geol. Survey Annual Rept.* **19**(part 2):207-245.

Lohman, S. W., 1961, Compression of Elastic Artesian Aquifers, *U.S. Geol. Survey Prof. Paper 424-B,* pp. 47-49.

Lonsdale, J. T., 1935, Geology and Ground-Water Resources of Atascosa and Frio Counties, Texas, *U.S. Geol. Survey Water-Supply Paper 676,* 90p.

Maxey, G. B., 1964, Hydrogeology, in *Handbook of Applied Hydrology,* V. T. Chow, ed., McGraw-Hill, New York, 38p.

Meinzer, O. E., 1923, Outline of Ground Water Hydrology with Definitions, *U.S. Geol. Survey Water-Supply Paper 494,* 71p.

Meinzer, O. E., and H. A. Hard, 1925, The Artesian Water Supply of the Dakota Sandstone in North Dakota with Special Reference to the Edgeley Quadrangle, *U.S. Geol. Survey Water-Supply Paper 520-E,* pp. 73-95.

Mendenhall, W. C., 1905, The Hydrology of San Bernardino Valley, California, *U.S. Geol. Survey Water-Supply Paper 142.*

Meyboom, P., 1966, Groundwater Studies in the Assiniboine River Drainage Basin: I. The Evaluation of a Flow System in South-Central Saskatchewan, *Canada Geol. Survey Bull. 139,* 65p.

Mifflin, M. D., 1968, Delineation of Groundwater Flow Systems in Nevada, *Desert Research Inst. Tech. Rept.,* ser. H-W, no. 4, 111p.

Miller, R. E., 1961, Compaction of an Aquifer System Computed from Consolidation Tests and Decline in Head, *U.S. Geol. Survey Prof. Paper 424-B,* pp. 54-58.

Parker, G. G., G. E. Ferguson, and S. K. Love, 1955, Water Resources of Southeastern Florida, with Special Reference to the Geology and Groundwater of the Miami area, *U.S. Geol. Survey Water-Supply Paper 1255.*

Russell, W. L., 1928, The Origin of Artesian Pressure, *Econ. Geology* **23:**132-157.

Sayre, A. N., and R. R. Bennett, 1942, Recharge, Movement, and Discharge in the Edwards Limestone Reservoir, Texas, *Am. Geophys. Union Trans.* **23:**19-27.

Stringfield, V. T., 1936, Artesian Water in the Florida Peninsula, *U.S. Geol. Survey Water-Supply Paper 773-C,* pp.115-195.

Stringfield, V. T., and H. E. Legrand, 1966, Hydrology of Limestone Terraces in the Coastal Plain of the Southeastern United States, *Geol. Soc. America Spec. Paper 93,* 46p.

Swenson, F. A., 1968, New Theory of Recharge to the Artesian Basin of the Dakotas, *Geol. Soc. America Bull.* **79:**163-182.

Tolman, C. F., and J. F. Poland, 1940, Ground-Water, Salt-Water Infiltration, and Ground-Surface Recession in Santa Clara Valley, Santa Clara County, California, *Am. Geophys. Union Trans.* **21:**23-24.

Tóth, J., 1962, A Theory of Groundwater Motion in Small Drainage Basins in Central Alberta, Canada, *Jour. Geophys. Research* **67:**4375-4387.

White, W. N., 1932, A Method of Estimating Ground-Water Supplies Based on Discharge by Plants and Evaporation from Soils, Results of Investigations in Escalante Valley, Utah, *U.S. Geol. Survey Water-Supply Paper 659-A,* 105p.

Winograd, I., 1962, Interbasin Movement of Groundwater at the Nevada Test Site, Nevada, *U.S. Geol. Survey Prof. Paper 450-C,* pp. 108-111.

Winograd, I., and W. Thordarson, 1975, Hydrogeologic and Hydrochemical Framework, South-Central Great Basin, Nevada-California; with Special Reference to the Nevada Test Site, *U.S. Geol. Survey Prof. Paper 712-C,* 126p.

19

REQUISITE AND QUALIFYING CONDITIONS OF ARTESIAN WELLS.[1]

By T. C. Chamberlin.

INTRODUCTION.

The basal principles of artesian wells are simple. The school-boy reckons himself their master. But the real problems they present are complex. It is a combination of varying conditions, rather than the application of simple principles, that determines success or failure. A clear statement of these conditions is as rare as a simple, but incomplete, exposition is familiar. This is perhaps not so much due to any special intricacy of the problem, or to any grave obstacle to a clear statement, as to the simple fact that it has been neglected.

It has not been the leading subject of any profession. Few drillers make the causes and conditions of artesian flow a special study, or find it within their province to master the geological elements of the question. Few geologists, among the multitude of more obtrusive resources pressed upon their attention, find themselves able to pursue the subject into its practical details. Few citizens have occasion more than once or twice in their lives to give the matter special consideration. This, however, is growing steadily less and less true. Drillers are developing the sinking of artesian wells into a specialty, and, through the aid of geological reports, are mastering the stratigraphical elements of the problem in their several regions. Geologists are solicited with increased frequency for advice and prognostic opinions. Citizens are becoming more widely interested in both the practical and the theoretical aspects of the subject.

Its importance does not need argument, though it may need emphasis. The problem of a pure and adequate water supply is among the gravest questions that now lay under tribute the thoughts of sanitarians. Artesian wells form one, but only one, source of such a supply; sometimes as

---

[1] The term artesian wells in this discussion is applied only to those that flow at the surface. Unfortunately the term is frequently used to denote deep wells that do not flow, a use that is to be condemned.

inadequate or impracticable as, at others, generous and beneficent. Were they an unfailing source, a universal possibility, so great a resource would command its own full measure of attention. Were they always inadequate, or unsatisfactory, they might be easily dismissed. This inconstant feature is an element that makes need for a discriminative discussion of the conditions that determine success or failure, since they are a valuable resource in certain regions and under certain limitations, while, in other regions and beyond assignable limits, they are only a lure to useless expenditure. On the one hand, large sums are needlessly spent in endeavors to obtain these fountains, where the essential conditions are altogether wanting, and, on the other, large possibilities of good have lain neglected, to the great sanitary and industrial loss of citizens.

It is the aim of this article to gather into a simple and convenient form such information relative to the necessary and qualifying conditions of artesian wells as may be capable of brief, general statement, and may seem to be serviceable alike to citizen, driller, and geologist. There will be no attempt, however, to make it an exhaustive exposition of the subject from the individual stand-point of either. The citizen will desire specific information concerning cost, quality of water, etc., that cannot be answered in a general discussion. The driller will desire constructive details and local particulars relating to the succession of beds, the texture and structure of the rock, etc., that only exhaustive geological descriptions, specially interpreted for his purpose, can supply. The geologist will desire a fuller discussion of the elements that enter into the formation of a professional opinion. It is obvious that, so far as these are dependent on local and varying conditions, satisfactory general answers are impracticable, and, in so far as they are specifically technical, they are here inappropriate. It is the ground of common interest, lying in the conditions of success, that I aim to cover. In addition to all that can be said here, special problems will need special study.

If there lingers in the mind any sense of marvel at the flow of artesian wells, it is best to cast it away at the outset. Artesian flow is but an expression of the common law of flowage, made a little unusual, it is true, by its special conditions. Any seeming strangeness springs from our partial observation. We see but a part of the stream. The rest lies hidden in the earth's depths, a realm which the imagination is prone to people with mysteries. Moreover, the part we do most see is a *rising* stream, that comes gushing up in the face of the dogma that water "never runs up hill." But water flowing up hill is one of the commonest facts of nature, an every day, an everywhere occurrence, illustrated in every brook, rill, and river, not to say spring. Stir up a little trash in the bottom of any deep pool in a brook, and see how readily it is borne up the slope of the pool-bed, and out over the shallows below. Disturb the bottom mud above a dam, and watch it ascend the steep slope, and pass over the weir.

Certain portions of the water of every stream are always "running up hill," though its average mass is moving down.

FIG. 7.—Longitudinal section of a stream, illustrating, in part, its upward currents, subordinate to the general downward flowage.

The bottom layer flows up and down according to the inequalities of the bed, while the top layer declines more uniformly with the surface slope.[2] In proportion as the stream is rapid and crooked, these layers exchange places, and there is a tortuous upward and downward flow, in addition to that directly enforced by the bottom inequalities, though all are largely due to these. The flowing up is merely a way of flowing down, i. e., the rising of a part permits the descent of a greater portion, or the greater descent of an equal or less part. No portion would rise, if it were not forced up by a superior portion pressing down. In the artesian stream, we only see the rising column issuing from the earth,

FIG. 8.—Section of the Chicago artesian stream, drawn to a nearly true scale. The dotted line represents the course of flow, exhibiting the long declining stream from the fountain-head in south-central Wisconsin, the short and *relatively* trivial rise in the wells at Chicago (C of the diagram), and the surface flow from these. The full line, G, representing the surface, shows the general decline of the stream. For the purpose of giving clear and striking illustrations, most diagrams of artesian wells are given enormously exaggerated vertical scales, and most of those of this paper are quite disproportionate in that respect, though the exaggeration has been greatly reduced. This diagram is introduced at the outset to forestall the erroneous impressions conveyed by such illustrations.

FIG. 9.—Ideal section illustrating the chief requisite conditions of artesian wells. A, a porous stratum; B and C, impervious beds below and above A, acting as confining strata; F, the height of the water-level in the porous bed A, or, in other words, the height of the reservoir or fountain-head; D and E flowing wells springing from the porous water-filled bed A.

and the brooklet that flows away. The more potent descending volume that forces the flow is concealed. This it is the mind's task to picture clearly.

---

[2] It is not a little surprising to note how frequently this simple fact, common to all streams, is neglected in the study of glacial flows. The traces of extinct glacial currents are those formed by the bottom of the stream, whose ascending and descending courses are no stranger than those of the brook. We study one from the top and neglect the bottom, the other from the bottom and forget the top.

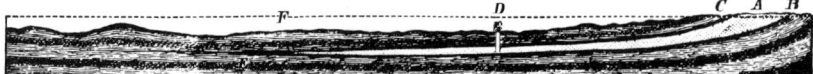

Fig. 10.—Section illustrating the thinning out of a porous water-bearing bed, A, inclosed between impervious beds, B and C, thus furnishing the necessary conditions for an artesian fountain, D.

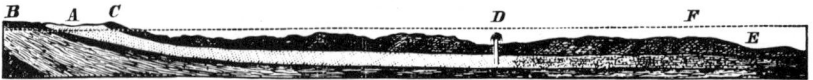

Fig. 11.—Section illustrating the transition of a porous water-bearing bed, A, into a close-textured impervious one. Being inclosed between the impervious beds B and C, it furnishes the conditions for an artesian fountain. D.

ESSENTIAL FEATURES OF ARTESIAN WELLS.

The artesian stream has its source, its underground water-way, its ascent through the well, and its final descent in the rill that runs away. It is peculiar mainly in its underground conditions. Upon these, chiefly, the ascending flow depends.

Typical examples.—To fashion a simple idea of the common class of flowing wells, picture to the mind a pervious stratum through which water can readily pass. Below this let there be a water-tight bed, and let a similar one lie upon it, so that it is securely embraced between impervious layers. Suppose the edges of these layers to come to the surface in some elevated region (save that they may be covered with soil and loose surface material), while in the opposite direction they pitch down to considerable depths, and either come up again to the surface at some distance, thus forming a basin (Fig. 9), or else terminate in such a way (Fig. 10) or take on such a nature (Fig. 11) that water cannot escape in that direction. Now, let rain-fall and surface-waters penetrate the elevated edge of the porous bed, and fill it to the brim. That such beds are so filled is shown by ordinary wells, which commonly find a constant supply in them at no great depth. Now it is manifest that if such a water-fat bed be tapped by a boring at some point lower than its outcrop, the water will rise and flow at the surface because of the higher head in the upper edge of the bed. If the surface-water continually supplies the upper edge as fast as the water is drawn off below, the flow will be constant.

Prerequisites.—The leading conditions upon which artesian flows depend are involved in this simple conception. Drawn out they are as follows:

I. A pervious stratum to permit the entrance and the passage of the water.

II. A water-tight bed below to prevent the escape of the water downward.

III. A like impervious bed above to prevent escape upward, for the water, being under pressure from the fountain-head, would otherwise find relief in that direction.

IV. An inclination of these beds, so that the edge at which the waters enter will be higher than the surface at the well.

V. A suitable exposure of the edge of the porous stratum, so that it may take in a sufficient supply of water.

VI. An adequate rain-fall to furnish this supply.

VII. An absence of any escape for the water at a lower level than the surface at the well.

These may be considered in detail, and then attention directed to some special practical questions.

THE WATER-BEARING BEDS.

There are two general methods by which water finds its way through the strata; in the one—the rock being close-textured—the water passes through fissures formed by fracture, or tubular channels formed by solution; in the other—the rock being open-textured—the water seeps through the pores, permeating the whole bed.[3]

1. *Fissured and channeled beds.*—Beds that offer only crevices and channels as water-ways are a very uncertain source of fountains, for open, continuous avenues of this class do not seem to be abundant in strata deeply buried, and the position of such as exist cannot be determined beforehand, so that there is no certainty of striking them.

The close-textured rocks that fall under attention here are chiefly the crystalline class (the granites, greenstones, etc.) and the limestones. The clay rocks (shales, etc.) are too compact to be in any available degree water-bearing; indeed, they form the chief confining strata.

The crystalline rocks are much fissured at the surface, and shrinkage has opened gaping crevices in them; but these avenues largely close up and disappear at moderate depths, and there are no sufficient grounds for supposing that they often afford facilities for a continuous, generous flow, where they have always lain under the protecting mantle of impervious strata, and have suffered the pressure of their weight. Experience confirms this. None of the igneous or metamorphic rocks are to be accounted available water-bearing strata unless some known local condition gives them exceptional possibilities.

The limestones are likewise much traversed by crevices near the surface, and are, besides, subject to the solvent action of the waters passing through them. These often form extensive underground channels, mammoth examples of which are found in the great elongated caves of

---

[3] Of course we are here considering only such generous supplies as are available for artesian purposes, not that slower penetration of rocks by water that is almost or quite universal.

Indiana and Kentucky. But, like the above, these prevail mainly in the superficial portion of the beds, and are chiefly confined to regions where the limestone stratum is not overlain by other rocks, and hence not available as a source of fountains. The reason of this is manifest, when it is considered that the solvent action is mainly accomplished by surface waters. These exhaust their solvent power before penetrating far. When the limestone is overlain by impervious beds, these surface waters are cut off, and hence solvent action is limited to such waters as entered at the distant outcropping edge. The cracks and cavities of deep-seated limestones are often found to be healed up with calcite, an index that the waters there are depositing, rather than solvent.

The grounds, therefore, for anticipating success in penetrating limestone for fountains, are not very flattering, though less adverse than in the crystalline rocks. However, limestones that have once been exposed to surface action, and thereby fissured and channeled, and subsequently buried beneath a thick mantle of drift clay, are not altogether unpromising. Not a few important flowing wells have been derived from them. But when the beds have always lain deeply buried beneath impervious strata, they have rarely been found productive, so far as my knowledge extends.

A little computation will show that even if such compact rocks were notably fissured and channeled, they would be a very questionable resource in deep, expensive wells. Suppose that vertical fissures or tubular channels traversed a given stratum at intervals of only 10 feet. It would be possible to sink twenty average bores between each two of them. If the fissures averaged as much as 6 inches, the chances of success would be about one in twenty, or only 5 per cent. of the whole, or, with a similar system of cross fissures, 10 per cent. If the case were really of this sort, however, connection with the channels might be made by firing torpedoes within the bore. But in fact deeply-buried limestones are to be esteemed a very uncertain dependence, while metamorphic and igneous rocks had better not be reckoned a dependence at all. Limestones that have been once at the surface, but afterward buried, may be locally serviceable.

2. *Porous beds.*—Quite in contrast with the close-textured beds, that are water-carriers only by virtue of fissures and channels, are the open-textured strata that constitute continuous water-filled sheets underspreading wide areas, and which can therefore almost certainly be tapped at the proper depth. Speaking in general terms, these are the only reliable sources of artesian wells.

To this class belong beds of sand, gravel, sandstone, conglomerate, and other less common rocks of loose granular texture. Some of the more porous chalks and granular limestones may be classed here. The common feature of the class lies in the construction of the rock from separate particles, loosely put together, leaving small open spaces between them. The porosity is of an interstitial, not vesicular, kind. A bed of sand is a typical illustration, and it will not be wide of the truth to speak of the whole class as sandstones. All sandstones, however, are not sufficiently porous to furnish a ready passage for water. In some the interspaces are filled with clayey or other impervious matter, that insinuated itself among the particles as they were being deposited originally; in others the pores were subsequently choked by deposits from solution and by compacting under pressure.

The degree of porosity is a very important consideration, when the volume of the flow is to be considered. If the entire rock is made up of sand-like particles the larger these are, in general, the greater the water-conveying capacity. Some sandstones are so fine-grained as to be almost impervious. The water merely oozes through them, and they are quite incapable of furnishing a generous flow. Others are so open-spaced that floods flow freely through them. In nature there is every gradation, from open gravels to close sandstones.

Besides, there is almost every possible intermixture of coarse and fine material. The constituents were not perfectly assorted. Fine sand and silt are interspersed among coarser grains, pebbles, and bowlders. This may reach almost any degree of influence upon the water-carrying capacity of the rock from a slight reduction to its almost complete extinction.

There is also a very wide range in the degree of consolidation after deposition. There are drift sands and gravels that are entirely loose and uncompacted, while, at the other extreme, are perfectly consolidated quartzites and analogous rocks. It is a general but quite unreliable rule that consolidation varies with the age of the formation, the Quaternary sands being looser than the Tertiary, the Tertiary looser than the Secondary, and so on. The rule is founded on reason; but there are many notable exceptions, as the Potsdam sandstone of the Upper Mississippi Valley, among the most ancient of the old life series, and yet among the most loose-textured, as well as the most generous of water-bearers. It is manifest, therefore, that we cannot rest with simple rules or general descriptions. The capabilities of each formation must be ascertained by direct observation of its constitution. This indicates one of the good offices subserved by critical descriptions of the texture of rocks, even of those which are coarsest and whose origin is most obvious. It is a suggestion that examinations be made more critical.

THE CONFINING BEDS.

No stratum is entirely impervious. It is scarcely too strong to assert that no rock is absolutely impenetrable to water. Minute pores are well nigh all pervading. To these are added microscopic seams, and to these again larger cracks and crevices. Consolidated strata are almost universally fissured. Even clay beds are not entirely free from partings.

But in the study of artesian wells we are not dealing with absolutes but with availables. A stratum that successfully restrains the most of the water, and thus aids in yielding a flow, is serviceably impervious. It may be penetrated by considerable quantities of water, so that the leakage is quite appreciable and yet be an available confining stratum. The nearest approach to an entirely impervious bed is furnished by a thick layer of fine, unhardened clay. In this case solidifying permits the formation of fissures and the clay rocks are less impervious than the original clay beds. The clayey shales rank next as confining strata, after which follow in uncertain order shaly limestones, shaly sandstones, the various crystalline rocks, and even compact sandstones. Paradoxical as it may seem water itself may form a confining agent. I shall presently attempt to show that it is even an agency of considerable importance. In a different and more forced sense an illustration may be found in the pool of the brook before cited. The water flows from the bottom of the pool up the slope and out, because it is pressed by the water behind and *confined* by that above. Each upper layer is a confining stratum to each lower one. This, however, is not the sense in which it is a recognizable agency in artesian flows.

1. *The confining stratum below.*—It is our habit to be less solicitous about the tightness of the cover of a water-bearing vessel than of its bottom. The reverse is the case in artesian wells. The confining stratum

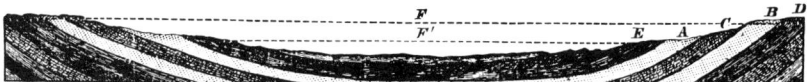

FIG. 12.—Section illustrating the usual order in which the strata of a basin come to the surface. A and B, porous beds; D and E, impervious beds; C, a half-impervious bed; F' and F the water levels of A and B, respectively.

beneath the porous bed demands less critical attention than that above, for if the layer next beneath the water-bed is defectively impervious, some lower layer will stop the water. It is only when the lower layers are so situated that they can carry the water out again to the surface, at a lower level than the water-bearing bed above, that they can discharge it, however leaky. This is not usually the case. As a rule, when beds are bent into a basin, or are inclined, the lower layers outcrop at

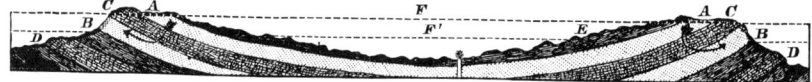

FIG. 13.—Section illustrating the possible effect of erosion upon strata originally like those in Fig. 6. A and B, porous beds; D and E, impervious beds; C, a half pervious bed; F and F' the water levels of A and B, respectively. If the stratum C is not practically a confining layer, the water from A will pass through it and escape at the edge of B, so that a flow cannot be obtained at a higher level than it, but may be had below the line F'.

higher elevations than those overlying them, as illustrated in Fig. 12. But this is not always so, and, even when so at first, unequal erosion of the surface may reverse the order of height of outcrop, as illustrated in Fig. 13. The consequence of a possible defect in the layers underlying a water-bed are illustrated in Figs. 13 and 14, and the descriptions appended to them. The confining strata below cannot, therefore, be

FIG. 14.—Section illustrating the failure of an artesian well, because of defects in a confining bed below. A and B, porous beds; D and I, impervious beds; C, a defective confining bed; E, the water level of the stratum B; G and H, wells that do not flow. The bed A might give a flow at G and H but for the defect in C, which permits the water to descend into B and escape through its outcrop, which lies below the surface of G and H.

neglected, though usually less imperative of critical attention than those above.

2. *The confining stratum above.*—The character of the strata that overlie the water-bearing bed is critically important, for the water, being under pressure, tends to rise through them, and, if they are in any degree penetrable, it will, to that extent, escape, and relieve the pressure, and thus reduce or prevent the flow. When the capacity of the water-bearing bed is great, and the fountain-head high, moderate defects in the cover-bed merely cause a reduction in the volume and height of flow; but when the conditions are closely balanced, either because of low head, feeble supply, or obstructed passage, he who assumes to foretell results has need to make careful estimate of the amount of leakage through the covering strata.

a. It may be noted that the leakage will be reduced in proportion as the pressure is lessened, so that, in nearly-balanced cases, the loss is less—other things being equal—than in cases of high head and free passage, *but it is more critical in determining success or failure.*

b. The most essential consideration, the nature of the rock, has already been discussed. In a more summary and general way it may be here restated that the effective imperviousness is somewhat nearly measured by the total amount of clayey constituents. But this is rather a convenient generalization than a safe rule.

c. Efficiency increases with thickness. If the cover-beds are of the highest impervious character, there is little need that they should be very thick, unless the fountain-head is so high that increased weight is needed to counterbalance the hydrostatic pressure—an improbable case. But when the degree of imperviousness is inferior, the element of thickness is not without consequence, in itself, and, taken in connection with the following point, may be decisive.

d. The element to be recognized here is, I believe, essentially new to discussions of the subject,[4] viz, the height of the surface of the common ground-water in the region between the proposed well and the

---

[4] I discussed this subject extemporaneously before the Wisconsin Academy of Science in December, 1880, but did not write it out for publication. The point is treated in Vol I, Geol. of Wis., p. 692.

fountain-head. It is a familiar fact that the common underground water stands at varying heights. Our common wells testify to this. The subterranean water-surface is almost invariably higher than the adjacent streams, and slowly works its way into them by springs, seeps, and invisible percolation. Speaking generally, the underground water-surface rises and falls with the rise and fall of the land-surface, only less in amount. Now, if the subterranean water in the region between the proposed well and its source—which we may call the cover-area—stands as high as the fountain-head (except at the well, where, of course, it must be lower), *there will be no leakage*, not even if the strata be somewhat permeable, for the water in the confining beds presses down as much as the fountain-head causes that of the porous bed to press up, since both have the same height. Capillarity does not disturb the truth of this. Under these conditions a flow may sometimes be secured when it would be impossible if the intervening water-surface were lower.

If the water between the well and fountain head is actually higher than the latter, it will tend to penetrate the water-bearing stratum, so far as the overlying beds permit, and will, to that extent, increase the supply of water, seeking passage through the porous bed, and will, by reaction, tend to elevate the fountain-head, if the situation permit.

FIG. 15.—Section intended to illustrate the aid afforded by a high water-surface between the fountain-head and the well. A, a porous bed; B, a confining bed below; and C, a confining bed above. The dark line immediately below the surface represents the underground water-surface. Its pressure downward is represented by the arrow m. The pressure upward due to the elevation of the fountain-head is represented by the arrow n. The line F represents the level of the fountain-head. There can be no leakage upward through the bed C except near the well D. There may be some penetration from the bed C into A, which would aid the flow.

I conceive that one of the most favorable conditions for securing a fountain is found when thick semi-porous beds, constantly saturated with water to a greater height than the fountain-head, lie upon the porous stratum, and occupy the whole country between the well and its source, as illustrated by Figure 15. This is not only a good, but an advantageous, substitute for a strictly impervious confining bed. Under these hydrostatic conditions, limestone strata reposing on sandstone furnish an excellent combination.

If on the other hand the underground water-surface between the proposed well and the source of supply is much lower than the fountain-head there will be considerable leakage, unless the confining beds are very close-textured and free from fissures. For example, if it be 100 feet lower there will be a theoretical pressure of nearly three atmospheres, or about 45 pounds to the square inch upward, greater than that of the underground water downward, disregarding the influence of capillarity, and this will be competent to cause more or less penetration of the water upward through the pores and crevices of the rocks, and consequent loss of head and forcing power.

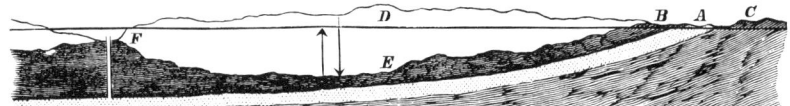

FIG. 16.—Double section illustrating the effects of high and low water-surface in the cover-area. (For explanation see text.)

Both of the above points may be illustrated by the accompanying double profile, in which A represents a porous stratum inclosed between the impervious beds B and C. The source of water-supply is at A, and the proposed well at F. Let E be supposed to represent the surface of the ground (and, for convenience, also, the surface of the common ground-water) in one of the two supposed cases, and D the surface in the other. The arrow springing from the surface E represents the upward tendency of the water in the porous bed, owing to pressure from the fountain-head, while the arrow depending from the line D represents the downward pressure of the ground-water whose surface is represented by D, and is, it will be observed, more than equivalent to the upward tendency due to pressure from the fountain-head. A flow at F could very safely be predicted if the surface were as represented by D, while it might be doubtful whether one could be secured if the surface were as represented by E.

My attention was first directed to this consideration by observing that where the intermediate country was elevated and had a high water-level, wells flowed at heights surprisingly near theoretical estimates, almost no deduction for obstruction and leakage being necessary, whereas in those cases where the opposite was true there was a very considerable falling short of theoretical estimates.

[*Editors' Note:* Material has been omitted at this point.]

20

Reprinted from pp. 609–617 of *U.S. Geol. Survey Annual Rept.* **17**(part 2):609–694 (1897)

PRELIMINARY REPORT ON ARTESIAN WATERS OF A PORTION OF THE DAKOTAS.

By N. H. DARTON.

INTRODUCTION.

The great artesian basin of the eastern portion of South Dakota and North Dakota has been so extensively developed by wells in the past decade that its general features are now well known. There is, however, great demand for detailed information as to the location of the water-bearing beds, the limits of the territory in which artesian flows may be expected, and the prospects for a continued water supply. There is also need for a statement of the extent and progress of well sinking and an account of the results of irrigation by the waters. In the autumn of 1895 I was directed to make a study of these questions, and this report gives the results of the preliminary investigations.

In Pl. LXIX are indicated the locations and depths of the deeper wells in the region studied, and also the area of artesian flow. On March 1, 1896, I estimated from all available data that the total number of the deeper artesian wells is about 400, of which over 350 are now flowing satisfactorily. The aggregate flow, as nearly as can be estimated, is 104,000 gallons per minute—232 cubic feet per second—or a little over 150,000,000 gallons per day. Eighty per cent of this flow is from 68 wells; of these, 31 have a flow of from 500 to 950 gallons per minute; 27 have a flow of from 1,000 to 1,950 gallons per minute, and 6 wells have a flow of from 2,000 to 2,900 gallons per minute. A well at Chamberlain is reported to have a flow of 4,350 gallons per minute; one at Springfield flows 3,292 gallons per minute; one at Yankton 3,000 gallons per minute; one on the Green farm in Brule County 3,000 gallons per minute; and one on the Yankton agency 3,000 gallons per minute. Much of this flow is shut off part of the time, but the figures indicate the present resources in artesian waters. These waters rise to the surface with pressures which are often over 100 pounds per square inch, and in a few cases over 150 pounds. At a number of localities this pressure is directly used for power to run large flouring mills, electric-light plants, sewage pumps, and other machinery. The principal use of the water is, however, for irrigation, and the greater number of the larger wells have been sunk for this purpose.

[*Editors' Note:* Plate LXIX and figure 59 have been omitted.]

THE NATURE OF THE INFORMATION.

In the preparation of this report I have endeavored to utilize all available data, including a considerable amount derived from other observers. The report by Col. E. S. Nettleton to the Department of Agriculture[1] has been the source of a larger part of the information concerning wells sunk prior to 1890 in the James River basin. In this report are given descriptive notes of all the deeper wells, a brief description of the conditions under which the waters occur, and vertical sections showing the relation of wells along a number of lines. The report also includes chapters by Prof. G. E. Culver and Maj. Fred. F. V. Coffin, which furnish important data.

Prof. J. E. Todd, State geologist of South Dakota, has presented some new "logs" and other information regarding wells in his Geology of South Dakota, which I have been glad to use. Information has also been derived from the report of Capt. C. S. Fassett, State engineer of irrigation, for 1893-94, and from the report of Mr. W. W. Barrett, State superintendent of irrigation and forestry in North Dakota.

From the 15th of September to the end of November, I had the assistance of Maj. Fred F. V. Coffin, who thoroughly canvassed large areas of the southeastern corner of South Dakota and obtained much valuable information, consisting of descriptions of wells, "logs," irrigation results, and data regarding the shallow waters. My own observations covered nearly all portions of South Dakota east of the Missouri River, the lower James River valleys of North Dakota, a trip into the Rosebud Indian Reservation, and a study of the geology from along the eastern slope of the Black Hills to the Cheyenne Valley. I had personal interviews with a number of well owners and well drillers, and a very large amount of information was obtained by means of blank schedules sent all over the artesian-well regions.

Considerable difficulty has been experienced in obtaining reliable information in regard to well borings and experience. For many of the wells no records could be obtained, and in some cases it was difficult or impossible to ascertain even depth and yield. The interpretation of well records supplied by various persons is always a difficult and uncertain task for the geologist, since many of the well drillers use haphazard terms of description for the materials which they penetrate in boring wells and trust much to memory when writing their "logs."

OUTLINE OF GEOLOGIC RELATIONS.

The geologic features of North Dakota and South Dakota are relatively simple, and they are mainly uniform over very wide areas. From the descriptions by Professor Todd and others we can obtain a

---

[1] Artesian and Underflow Investigation, Final Report of the Chief Engineer, Edward S. Nettleton, C. E., to the Secretary of Agriculture, with accompanying maps, profiles, diagrams, and additional papers, Part 2, 116 pages, 27 plates, Fifty-second Congress, first session, Senate Ex. Doc. 41, Part 2, Washington, 1892.

clear insight into the general relations, and as it is necessary to understand these relations in considering the conditions of underground water supply, a brief description of the geology is desirable. East of the vicinity of the Missouri River the surface is occupied to a greater or less thickness by gravel, clays, and sands, which were deposited by or in connection with the great continental glaciers of the Glacial period. These formations vary in thickness from 40 or 50 feet to over 100 feet, and they are so continuously spread over the surface that exposures of underlying formations are relatively few, particularly to the east and northeast. Lying beneath the drift and occupying the surface over the plains region westward in South Dakota are clays or shales of the Cretaceous period. Their thickness averages 1,000 feet over the greater part of the area, but they thin rapidly east of James River. In the region adjoining White River and extending thence southward through Nebraska this clay formation is overlain by a much younger series of Tertiary clays and sandy clays which occupy the surface in the Rosebud and Pine Ridge Indian reservations and in the Bad Lands. They have a thickness of between 300 and 400 feet in their greatest development. North of the Cheyenne River, and extending thence far to the north and eastward into North Dakota, the Cretaceous clays are overlain by sands and sandstones known as the Laramie formation.

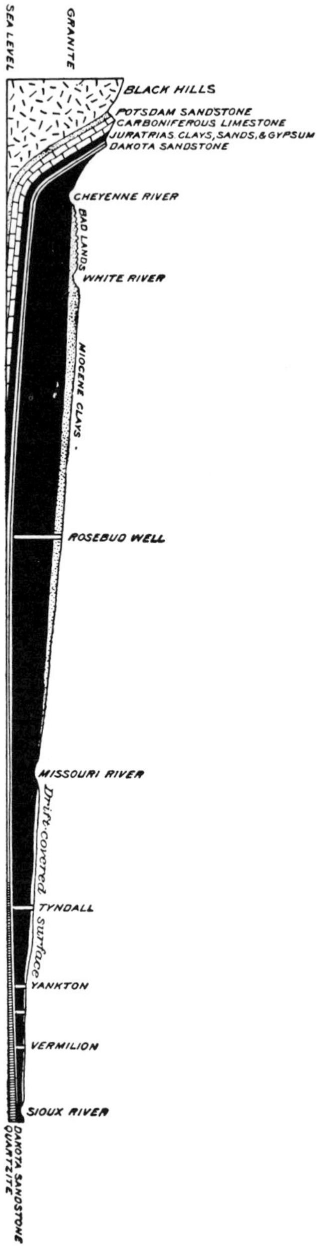

In the southern portion of South Dakota the Cretaceous clay series includes an extensive deposit of calcareous material about 300 feet above its base, which is known as the Niobrara chalk. This chalk formation outcrops extensively along the Missouri Valley below Chamberlain. The underlying beds are known as the Benton clays, and the overlying beds as the Pierre clays or shales.

Fig. 50.—Vertical section from eastern portion of the Black Hills across South Dakota, showing the attitude and relations of the water-bearing Dakota sandstones; looking north; 300 miles long; vertical scale considerably exaggerated.

Beneath the Cretaceous clay series lies a relatively thin but widely extended sheet of sands and sandstones, containing thin, irregular, intercalated beds of clays and iron pyrites. It is known as the Dakota formation, and it has proved to be a great water-bearer over wide areas in both States. To the southeastward it lies upon, or perhaps merges into, a pink quartzite, which attains great thickness in the vicinity of Sioux Falls. Adjacent to this area of quartzite the Dakota formation rises to the surface or lies under the drift at no great distance below, but to the north and west it is deeply buried below the great mass of Cretaceous clays, in addition to which, in the region south of the Cheyenne River, there is the overlying series of Tertiary clays, and to the north of that river the overlying Laramie sands and sandstones. In the Black Hills it is brought abruptly to the surface by an uplift in which the crust of the earth is sharply domed over an area of several thousand square miles. This doming or uplift also brings to the surface the formations underlying the Dakota formations, which are here seen to be a thick mass of sands, limestones, sandstones, granites, and schists. These limestones and sandstones are also known to underlie the Dakota formation in the eastern part of Nebraska and northward in Canada, but they appear not to extend far eastward across the Dakotas, where, over a wide area, the Dakota sandstone appears to lie directly on granite or quartzite. The granites and schists are the "bed rocks" of the entire region and extend to an indefinite thickness.

The geological structure of South Dakota from the Black Hills eastward is shown in the cross section, fig. 50. In this section are indicated nearly all the features above described excepting the Laramie formation, which lies north of the line of section and appears to thin out to the southward, so as to be absent from between the Tertiary clays of the White River region and the great Cretaceous clays series. Some of its relations in North Dakota are shown in fig. 59.

THE WATER HORIZONS.

The principal water-bearing bed in the Dakotas appears to be the eastward extension of the Dakota sandstones, but water is also found in smaller amounts in the Niobrara chalk beds above, in the basal beds of the drift formations, and in local beds of sand in the great series of clays lying above the Dakota formation. The Dakota formation consists of sands and sandstones which are particularly favorable in their character for the storage and transmission of water. The formation is not entirely sand and sandstone, but contains beds of clay irregularly disposed, and often streaks of iron pyrites and iron stone. The texture of its sands and sandstones being variable, the capacity for holding water is by no means constant. In every well of which I have heard that has penetrated the Dakota formation some water has been found in one bed, if not in others, and often several horizons of water-bearing sands have been penetrated. In many of the wells two or three, and in some as many as seven flows have been found, separated either by clays, beds of pyrites, or more compact sands or sand rock. In a few instances borings have penetrated sands whose physical characteristics appear to be favorable to a flow of water, but none has been found. In these cases it is probable that the nonwater-bearing bed was completely inclosed in clays or other impervious materials. In such instances water has been obtained by penetrating to lower beds to which the water has free access. It is by no means certain that the lowest beds in the artesian basin all belong to the Dakota formation, for they may be attenuated representatives of formations which in the Black Hills, Rocky Mountains, and Mississippi Valley lie between the Dakota beds and the crystalline rocks. None of the wells have, however, furnished definite evidence as to the age of the formations beneath the Dakota beds. Doubtless over wide areas the Dakota sandstones rest directly on the crystalline rock surface.

In an area of about 600 square miles lying mainly in eastern Sanborn and western Miner counties fairly abundant supplies of excellent water are obtained from wells in the chalk deposits. The flows have no great pressure, but the head is sufficient to carry the water over the more level farms, and the wells have proved to be very serviceable. Their depths average from 100 to about 350 feet. In some cases the water occurs near the top of the chalk, in others at various horizons below. A few wells derive their supply from sands lying entirely below the chalk.

Small artesian flows are obtained from the drift deposits in portions of Sanborn, Miner, and Hanson counties, in an area of which the outlines are indicated in Pl. LXIX. There is another small area along the valley of Turkey Ridge Creek, in Turner County, and there are a few scattered wells in lowlands at various points in other counties, as shown on Pl. LXIX. These wells have an average depth of about 100 feet, but some are not that deep and a few are much deeper. They derive their waters from the coarse sands and gravels which lie under the blue clay at the base of the drift deposits, but it is possible that this water is furnished originally by the chalk, which either immediately underlies these coarse beds or is separated from them by a short interval of more or less permeable beds.

There are many local occurrences of waters at various horizons in the clay formations that overlie the Dakota sands and sandstones. In some cases they are in sands intercalated in the clays lying between the chalk and the Dakota beds, and in others they lie at greater or less distances above the chalk. At Mandan, in North Dakota, the small flow of water in the 2,000-foot well is obtained from beds at a depth of about 350 feet, which are high up in the Pierre clays. Nearly all the wells have reported small flows in sands or sandy shales at horizons considerably above the Dakota formation, but I have been unable to ascertain whether they represent definite horizons. In many cases they furnish such a scanty supply that they are worthy of no consideration, but over quite a wide area north of Mitchell there are a number of wells with large flows and fairly high pressures which obtain their supply from sands in clays above the main Dakota horizon.

THE EXTENT OF THE ARTESIAN WATERS.

Almost the entire area of the States of North Dakota and South Dakota is underlain by the Dakota sandstone, and this formation appears throughout its extent to be more or less completely saturated with water. From its outcrops in the Black Hills it sinks deeply beneath the surface, and then gradually rises to the eastward, as shown in fig. 50. It finally abuts against the crystalline bed rock along the eastern border of the States at no great distance beneath the surface, where its waters are more or less free to escape and the eastern limit of the basin is determined. The waters have a head which is relatively great to the westward, but gradually diminishes to nothing at the eastern border of the basin. There are wide areas in which this head is sufficient to give surface flows in artesian wells, but, on the other hand, there are many thousand square miles which are too elevated for artesian flows. These elevated regions comprise the greater part of the highlands of the plains country west of the Missouri, the western Coteau north of latitude 44° 45′ north, and the eastern Coteau in South Dakota. On Pl. LXIX is shown the area in which artesian flows may be expected, both from the Dakota sandstone horizon and from certain higher horizons of local extent.

The restriction of the artesian area on the eastward is quite definitely proved by the experience of a number of wells which have been bored along the eastern Coteau. These wells are at Webster, Clark, De Smet, Brookings, Madison, Salem, and Bridgewater, and along the Big Sioux Valley. In some of these wells, as will be described more in detail in the following pages, waters have been found in fairly large supplies, and probably in the Dakota horizon, but they have not had sufficient head to rise to the surface as artesian flows. In the case of some of the wells, the head was sufficient to raise the water to altitudes considerably above the level of the wells in the lowlands westward, but they fall far short of the surface of the highlands into which they were bored.

The failure of the well at Millbank, in the Minnesota Valley, northeast of the eastern Coteau, to obtain a flow is due to the diminished head of the waters to the eastward, and it falls in line with the results at Brookings and in the Sioux Falls Valley. Various borings have been made at a number of localities within the known artesian area which have failed to find water, but it is now quite clearly apparent that they were not sunk sufficiently deep to reach the horizon in which waters are to be expected. For instance, at Fort Sully a 979-foot well was discontinued in the clays fully 200 feet above the water horizon, and the 600-foot boring at Bangor, at an altitude of about 2,000 feet, was abandoned still farther above the Dakota formation. The 2,000-foot boring at Mandan, N. Dak., probably did not reach the Dakota formation, and the boring at Bismarck was not sufficiently deep by about 1,000 feet.

The waters of the shallow artesian basin lying northeast of Mitchell afford a useful supply for farm and domestic purposes, but they have not the force and volume necessary for power and irrigation on a large scale. The extent of the basin is quite definitely known, for there are numerous well borings throughout its area and around its margin. Its area is shown on the map, Pl. LXIX. Another small area of similar waters occurs in the central part of Turner County, in the basin at Turkey Ridge Creek. It supplies water to over a score of farms and is similar in every respect to the larger area northeast of Mitchell. Several small detached areas occur in Hutchinson, Lincoln, and Moody counties. The waters in the chalk or Niobrara formation occur mainly within the same basin as the drift waters northeast and westward of Mitchell, and quite a number of wells in that basin receive large supplies of excellent water from the chalk beds, at depths averaging about 200 feet.

In the western Coteau area there is a wide tract of elevated country in which there appears to be no prospect for artesian waters. The area is undoubtedly underlain by the Dakota sands and sandstones, containing large volumes of water, but the head of these waters is not sufficient to carry them to the surface of this elevated region. As no wells have yet been bored within the area to a sufficient depth to ascertain the underground conditions, we have to depend on the evidence of wells in the surrounding lowlands for a basis of prediction as to the area in which a flow may be expected. The head which the water has at Jamestown is sufficient to raise the water to an altitude of 1,614 feet above sea level, but the land to the westward, from Windsor to beyond Sterling, rises several hundred feet higher than this. Of course if the pressure or head of the water increases westward, some parts of this elevated region may be within reach of artesian flows; but of this there is now no definite evidence, and the probabilities are strongly negative. At Ellendale the head is sufficient to raise the water to an altitude of 1,727 feet above sea level, or about halfway up the steep slopes of the highlands which extend westward from Lorraine to the edge of the Missouri Valley. The Ipswich well indicated sufficient head to elevate the water to an altitude of 1,775 feet, the Faulkton well to 1,664 feet, and the well near Orient to 1,865 feet, all of which would be insufficient to reach the surface in the wide, high area lying a short distance westward and extending nearly to the Missouri.

Along the line of wells extending from St. Lawrence to Pierre the head of the artesian waters presents some curious variations. About St. Lawrence and Miller it averages an amount sufficient to raise the waters to an altitude of 1,860 feet. At Highmore its altitude increases to 1,920 feet, at Harold it has decreased to 1,862 feet, and at Pierre is only 1,820 feet. The amount at Highmore is sufficient for the waters to reach the surface over a narrow area extending across the Coteau, but insufficient to carry them to the surface in the high region of Ree Heights, and probably also the ridge extending northward through Edwin, Sedgwick, Goudyville, and Cramer. The pressure at Pierre indicates that artesian waters may be expected over all the surrounding lands to an altitude of 1,820 feet. Whether the head increases westward under the higher lands or decreases as it does from Highmore and Harold to Pierre can not be stated. How far the artesian waters extend up the Missouri is not now known, but it is thought they may reasonably be expected far beyond Bismarck. The well which is soon to be sunk at the Cheyenne Agency, opposite Forest City, will give additional evidence as to the depth and pressure of the water in this direction.

In the series of wells extending along the Chicago, Milwaukee and St. Paul Railroad from Mitchell to Chamberlain there is a progressive increase of pressures to the westward as far as Kimball, and along this line, in the wells to the southward, the head is known to be sufficient to carry the water to the highest lands, excepting the summits of the Bijou Hills. The increase of pressure from east to west in this region may possibly continue far beyond the Missouri River, and if this is the case artesian flows may be had over wide areas of the now arid Plains region. Much of this region lies above an altitude over 2,000 feet, and to the southwestward the altitudes rapidly increase to about 2,700 feet in the vicinity of the Rosebud Agency. The deep well now in progress in the center of the Rosebud Reservation will indicate the conditions under which the artesian waters extend west of the Missouri River in that region, but we must wait for this evidence before any definite predictions can be made.

The eastern margin of the artesian area extends along the steep western slope of the eastern Coteau, crosses the western side of Miner County, extends up the Redstone Valley some distance, then, passing westward nearly to Mitchell, it extends along the eastern side of the James River Valley nearly to its mouth. In Clay County the pressure is only sufficient to give artesian flows in the lower portions of the valleys. At Vermilion the head is insufficient to carry the waters to the top of the table-land, and its amount declines gradually as the Missouri is descended, until at a point near Burbank the waters will not rise above the level of the river. At Elk Point extensive tests have been made which demonstrate beyond any question that the termination of the artesian flow is some distance westward. Along the Missouri bottom from Yankton to Burbank, up the Vermilion bottom lands to Centerville, and up the valley of Clay Creek nearly to the western boundary of Clay County, there are many wells which yield abundant water supplies.

The heads at Britton and Langford and Andover are insufficient to carry the waters to a level more than halfway up the slopes of the Coteau to the eastward, and at Dolan and Iroquois they fall short to a still greater degree. In the lowlands north of the end of the eastern Coteau artesian flows reach the surface at Rutland and Ransom with considerable head, and the area of available artesian flows appears to be extensive in this region. At Tower City the head may prove insufficient to carry water to the level of the higher points on the moderately high land of the ridge that extends throughout Alta, unless the increase of pressure westward, as indicated by the Jamestown well, begins near Tower City.

[*Editors' Note:* Material has been omitted at this point.]

Reprinted from page 1 of *U.S. Geol. Survey Water-Supply Paper 639,* 1933, 372p.

GEOLOGY AND GROUND-WATER RESOURCES OF THE ROSWELL ARTESIAN BASIN, NEW MEXICO

By ALBERT G. FIEDLER and S. SPENCER NYE

ABSTRACT

The Roswell artesian basin is in the Pecos Valley in southeastern New Mexico. The investigation, which covered a period of three years, 1925 to 1928, was made for the purpose of determining the available supply of artesian and other ground water within the area. The geologic formations of the region are of the Carboniferous (Permian series) and Quaternary systems. The Permian rocks consist of three units—an upper unit composed chiefly of clay, shale, and sand; a middle unit composed chiefly of limestone; and a lower unit composed chiefly of red beds, gypsum, and anhydrite. Most of the artesian water is obtained from the limestone beds of the middle unit, which has been designated the Picacho limestone.

Originally the area of artesian flow comprised 663 square miles; but largely on account of heavy draft upon the artesian reservoir, it decreased to 499 square miles in 1916 and to 425 square miles in 1925. The area irrigated by water derived directly or indirectly from the reservoir amounts to about 60,000 acres. The annual quantity of water derived from wells is about 200,000 acre-feet, and the total discharge at the surface from all sources is about 250,000 acre-feet. Recharge to the reservoir is derived from precipitation that falls on a catchment area of 4,000 square miles west of the artesian area.

In 1927 a law was passed by the State of New Mexico declaring underground waters to be public waters and subject to appropriation. This law was declared invalid because of a technicality, and in 1931 a new law was enacted, which furnishes a definite basis for the future regulation of ground waters in the area.

The investigation leads to the conclusion that no new land should be placed under irrigation with artesian water, but that the development of shallow ground water should be encouraged. The present decline of the artesian head is slight in comparison with that in earlier years, and there is ample evidence to show that the reservoir annually receives large quantities of recharge and that with proper conservation it will never be completely exhausted.

[Editors' Note: Material has been omitted at this point.]

GEOLOGY AND GROUND-WATER RESOURCES OF THE BALTIMORE AREA

BY

ROBERT R. BENNETT and REX R. MEYER

ABSTRACT

The Baltimore area comprises the city of Baltimore, and most of the area from the Susquehanna River south to Laurel, essentially between the Piedmont Plateau and the Chesapeake Bay. Most of the large ground-water developments are in the industrial districts in and near Baltimore; consequently that part of the area was investigated in greater detail. With the exception of the northern part of Baltimore, which is in the Piedmont Plateau, the area is chiefly in the Coastal Plain.

The Piedmont Plateau is underlain by pre-Cambrian crystalline rocks consisting mostly of gabbro, schist, granite, and gneiss. Owing to their greater resistance to erosion the land surface of the plateau is higher and more rugged than the Coastal Plain, which is underlain by soft unconsolidated sediments of Lower and Upper Cretaceous and Pleistocene ages. The land surface of the Coastal Plain slopes gently southeastward toward Chesapeake Bay. In some places estuaries, which are tributaries of Chesapeake Bay, extend northwestward across the Coastal Plain to the Piedmont Plateau.

The Coastal Plain sediments were deposited on the southeastward-sloping surface of the crystalline rocks. They form a wedge-shaped mass that thickens progressively from west to east. The strike of the formations of Cretaceous age is approximately parallel to the boundary between the Piedmont Plateau and the Coastal Plain (Fall Line). As these formations dip gently to the southeast they crop out as bands of irregular width trending northeast. The Pleistocene deposits are essentially flat lying and were deposited on the eroded surface of the pre-Cambrian and Cretaceous rocks.

In most of the area the sediments of Lower and Upper Cretaceous age, which consist essentially of irregular and lenticular beds of sand, gravel, and clay of continental origin, may be divided into three formations: the Patuxent formation of Lower Cretaceous age, and the Arundel clay and Patapsco formation of Upper Cretaceous age. Their combined thickness ranges from about 475 to 750 feet. The Pleistocene deposits, which consist chiefly of irregular beds of sand, gravel, and clay of continental and estuarine origin, are divided, in this report, into an upland and lowland unit. The combined thickness of these units ranges from nearly nothing to 175 feet.

Owing to the great difference in water-bearing properties of the crystalline rocks in the Piedmont Plateau and the unconsolidated sediments in the Coastal Plain, ground water occurs under two widely different sets of conditions. In the crystalline rocks the water is contained chiefly in joints and other fractures which are not uniform in size and gradually disappear with depth, consequently the water-bearing zones are very irregular and inhomogeneous. The sand and gravel in the Coastal Plain sediments are considerably more porous and permeable and form relatively uniform and widespread aquifers. In general water-table conditions occur in the outcrop areas of the crystalline rocks and the Coastal Plain sediments, but down dip from the outcrops the ground water occurs under artesian conditions. Because of their low permeability, where they underlie a substantial thickness of unconsolidated sediments the crystalline rocks are not considered to be an aquifer in most of the Coastal Plain area.

The reported yields from 106 industrial wells ending in the crystalline rocks in Baltimore show an average yield of 50 gallons a minute, and a median yield of 35; the mode, or most typical value, is 10 gallons a minute. The reported yields range from 0 to 350 gallons a minute.

Industrial wells in the Patuxent formation in and near its outcrop have yields of about 200 to 300 gallons a minute, whereas wells in this formation down dip in the southeastern part of the area have yields of about 500 to 900 gallons a minute. Pumping tests and flow net analysis show that the coefficient of transmissibility of the Patuxent formation, the principal water-bearing formation in the area, averages about 20,000 gallons a day per foot in the industrial districts near the outcrop, and about 50,000 in the industrial districts in the southeastern part of the area. The coefficient of storage of the Patuxent formation, under artesian conditions, is about 0.00026; under water-table conditions in the outcrop area, it is estimated to be 0.15 to 0.20.

The Patapsco formation is an important water-bearing formation in the southeastern part of the industrial area where it is separated by clay into a lower and upper aquifer. The lower aquifer yields as much as 500 to 750 gallons a minute to industrial wells. Its coefficient of transmissibility averages about 25,000 gallons a day per foot. The upper aquifer, which yields as much as 500 to 800 gallons a minute to wells, has a greater thickness than the lower aquifer, so that its coefficient of transmissibility probably is more than 25,000.

The upland unit of the Pleistocene deposits is thin and caps the hills and ridges and is not an important aquifer. In some places, the lowland unit is sufficiently thick and permeable to yield large quantities of water to wells.

The large ground-water supplies in the industrial area have been developed since about 1900. The pumpage increased progressively to a peak of about 47,000,000 gallons a day early in 1942. Late in 1942 the pumpage was decreased by about 13,000,000 gallons a day and in 1945 was 34,000,000 gallons a day. From 1942 to 1945 the pumpage outside the industrial area increased from about 3,000,000 to 5,000,000 gallons a day. The total pumpage in 1945 for the entire area was 39,000,000 gallons a day. The pumpage, in gallons a day, from each water-bearing formation, is approximately: pre-Cambrian crystalline rocks, 1,000,000; Patuxent formation, 30,000,000; Patapsco formation, 6,000,000; and Pleistocene deposits, 3,000,000.

Originally the artesian head in the aquifers generally was within a few feet of the land surface, but with the progressive increase in pumpage during 1940 to 1942 the artesian head in the Patuxent and Patapsco formations, in and near the centers of pumpage, declined respectively to as much as 160 and 190 feet below the land surface. The decrease in pumpage late in 1942 resulted in a rise in the artesian head; and at present, in most of the industrial area, the artesian head in the Patuxent and Patapsco formations ranges, respectively, from about 40 to 100 and 10 to 50 feet below the land surface. Detailed records of water-level fluctuations in observation wells show that during 1943-45 the general trend of water levels in most parts of the area was either slightly upward or essentially horizontal.

The ground water in the Baltimore area normally has a low mineral content, but the lowering of the water table or artesian head has caused salt water, chiefly from the Patapsco River estuary, to enter the aquifers and spread laterally throughout a large part of the industrial area.

Industrial wastes, chiefly sulfuric acid, also have contaminated the ground water in a small part of the industrial area.

The contamination of aquifers through leaking wells is a serious ground-water problem in the industrial area. As the artesian head in the Patuxent formation is lower than it is in the Patapsco formation or Pleistocene deposits, highly mineralized water may enter a well through a defective casing or move downward outside an improperly sealed casing and contaminate the water in the Patuxent formation. Although this problem is now serious, the repair of leaking active wells and effective plugging of abandoned wells would appreciably reduce this contamination in a relatively short time.

Practically all ground water pumped in the Baltimore area is derived from precipitation on the outcrops of the aquifers. The potential rate of recharge to the aquifers in the Coastal Plain exceeds the theoretical maximum quantity of water that can be transmitted through them. The rate of recharge, therefore, does not limit the quantity of water that can be pumped in the artesian part of the area.

The concentration of pumping has caused the water table or artesian head in the Patuxent formation to be so low in and near some of the centers of pumping that very little additional water can be developed from that formation in those parts of the industrial area. The water table or artesian head between the centers of pumping, however, is relatively high and if the wells were spaced at greater distances additional water could be pumped. The heavy pumping from the Patuxent formation, chiefly in and near its outcrop adjacent to the Patapsco River estuary, has caused local encroachment of salt water. However as pumping in and near that part of the outcrop area prevents most of the salt water now in the formation from moving to major well fields in the other industrial districts, it would not be advisable to decrease or discontinue this pumping.

Heavy pumping over a period of many years throughout the industrial area has caused local encroachment of salt water in the Patapsco formation, so that most of the pumping from the formation has been discontinued. The present pumpage is chiefly from the aquifer in the lower part of the formation in the southeastern part of the area where a large part of the water is derived from areas in which the Patapsco formation contains fresh water. Although this pumping is still causing encroachment of salt water, it would not be desirable to discontinue the pumping at least until all leaking wells drilled to the Patuxent formation are repaired or plugged. Even if the pumping were discontinued, many decades would pass before the salt water in the formation would be flushed out. It would not be advisable, however, to develop additional supplies of ground water from this formation in the industrial districts up dip as they are near the main source of contamination.

The economic value of ground water in the industrial area is different for various types of uses and there is little uniformity in the chemical quality of water that can be used. The application of the term safe yield in the sense that it represents a single rate of pumping for the entire industrial area would be unrealistic. It is apparent that owing to the contamination of the water the safe yield has been exceeded for some industries.

[*Editors' Note:* Material has been omitted at this point.]

Reprinted from pages 1-4 of *U.S. Geol. Survey Prof. Paper 451,* 1965, 149p.

GEOLOGY AND ARTESIAN WATER SUPPLY OF THE GRAND JUNCTION AREA, COLORADO

By S. W. Lohman

ABSTRACT

The Grand Junction area, as defined in this report, comprises about 332 square miles in the west-central part of Mesa County, Colo.; it includes the part of the northeastern flank of the Uncompahgre Plateau known as Piñon Mesa and the southwestern side of the Grand Valley including parts of Orchard Mesa and the Redlands. The area also includes the Colorado National Monument, noted for its colorful cliffs and deep canyons, and Grand Junction—the largest city in western Colorado.

The highest part of the area, on the flank of Piñon Mesa, has an altitude of about 8,200 feet. From the mesa the surface slopes gently northeastward to the Colorado River, which leaves the northwest corner of the area at an altitude of about 4,430 feet. The area as a whole has a relief of more than 3,700 feet. The northeastward-sloping surface is interrupted by a series of faults and monoclines and is cut by many deep canyons, some of which are 500 to 1,000 feet deep. Many of the canyon walls, particularly in the Colorado National Monument, are sheer cliffs of the Wingate Sandstone and, locally, even higher cliffs are formed by the Wingate and overlying formations.

The area is drained by the Colorado and Gunnison Rivers, at whose confluence Grand Junction is situated. Most of the tributaries are ephemeral because of the mild arid climate.

The varied flora and fauna include types adapted to climates ranging from arid to subhumid.

Soon after settlement of the area began in 1881 it was realized that crops could not be grown successfully without irrigation, and the first irrigation system was begun near Palisade in 1882. By 1960 nearly 100,000 acres of the Grand Valley was under ditches, mainly from the Colorado River but in part from the Gunnison River. Seventy to eighty percent of this area was irrigated most seasons. Much of the irrigated land is in fruit orchards, mainly peaches, but many other crops also are grown.

The geologic map accompanying this report (scale 1:31,680) is the first detailed geologic map of the Grand Junction area, and is the first geologic mapping in the area since the reconnaissance by members of the Hayden survey in 1875 and 1876.

The pre-Quaternary geologic formations exposed in the Grand Junction area range in age from Precambrian to Upper Cretaceous and include Precambrian schist, gneiss, granite, and pegmatite; Chinle Formation and Wingate Sandstone, Upper Triassic; Kayenta Formation, Upper Triassic(?); Entrada Sandstone (Slick Rock and Moab Members), Summerville Formation, and Morrison Formation (Salt Wash and Brushy Basin Members), Upper Jurassic; Burro Canyon Formation, Lower Cretaceous; and Dakota Sandstone and Mancos Shale, Upper Cretaceous.

A profound unconformity separates an almost smooth erosion surface on the Precambrian complex from the Chinle Formation and marks the absence from this area of part of the Precambrian, all the Paleozoic, and much of the Triassic, including most of the Chinle. This old erosion surface was formed on the San Luis-Uncompahgre highland, which was a mountainous area undergoing erosion from Pennsylvanian to Late Triassic time.

The 80 to 120 feet of the Chinle Formation present in the area, which has been correlated with the Church Rock Member, consists largely of soft red siltstone, but it also contains thin hard ledge-forming beds or lenses of red siltstone, limestone, and conglomerate, and thin layers of greenish siltstone. The Chinle yields no water to wells in this area.

The Wingate Sandstone, which conformably overlies the Chinle Formation, comprises about 215 to 370 feet of mainly buff to reddish-buff or red, very fine-grained sandstone and some fine-grained sandstone, silt, and clay. The Wingate is well cemented and generally forms cliffs or steep slopes, particularly in the Colorado National Monument. It is partly crossbedded and partly level bedded, and is in part of eolian origin and in part fresh-water laid. The Wingate is the thickest and lowermost of four artesian acquifers in the area; it yields small supplies of generally soft water to a few wells whose main supply comes from the overlying Entrada Sandstone.

In the western part of the area the Wingate Sandstone is overlain conformably by 16 to 80 feet of the Kayenta Formation, but in the southeastern part, where the Kayenta is absent, the Wingate is separated from the overlying Entrada Sandstone by an erosional unconformity. The Kayenta consists mainly of fluvial lenticular to irregularly bedded layers of fine- to medium-grained sandstone, irregular lenses of red, purple, or green siltstone, and a few lenses of conglomerate or conglomeratic sandstone. The Kayenta is not an aquifer in this area.

An erosional unconformity at the base of the Entrada Sandstone marks the absence from this area of much of the Entrada, all the Carmel Formation and Navajo Sandstone, most of the Kayenta Formation west of North East Creek, and all the Kayenta and possibly part of the Wingate Sandstone in and east of North East Creek Canyon.

In this area the Entrada Sandstone comprises the Slick Rock and Moab Members, which together form a distinctive and colorful series of cliffs in much of the lower part of the area, but which weather to more subdued forms at higher altitude. The beds of the Slick Rock are partly crossbedded and partly level bedded and are probably wholly continental eolian and water-laid deposits. The overlying, generally white or light-buff Moab Member is made up of thin, evenly bedded sandstones that generally weather to a series of steps or benches. The Moab may represent a beach deposit laid down on the margin of the Curtis sea, and is probably of Curtis age. The Entrada ranges in thickness from 100 to 200 feet in the western part of the area to about 60 feet in the eastern part, and consists mainly of fine to very fine grained sand, some medium-grained sand, and some silt and clay, all cemented by calcium carbonate.

The Slick Rock Member generally contains, particularly near the base, scattered grains or laminae of coarse-grained sand known as "Entrada berries." The Entrada is the principal artesian aquifer in the area, and yields small amounts of generally soft water.

The Summerville Formation conformably overlies the Moab Member of the Entrada Sandstone. The Summerville, which is only 40 to 60 feet thick in this area, consists mainly of thin beds of gray, blue-gray, greenish-gray, chocolate-brown, reddish-brown and red siltstone; thin beds of gray, yellow, greenish-gray, and reddish-gray fine- to medium-grained, hard, laterally persistent sandstone; thin beds of shale and mudstone; and, near the top, at least one thin bed or lens of limestone. The Summerville probably was formed as a marginal marine deposit in shallow water, possibly in or near a shallow arm of the Summerville sea. The Summerville yields no water to wells in this area, but it and the overlying Morrison Formation serve as a confining bed to artesian water in the underlying Entrada and Wingate Sandstones.

The Summerville Formation is overlain, probably conformably, by the Morrison Formation, which, in this area, includes only the Salt Wash and Brushy Basin Members. The Morrison comprises a varied and colorful assemblage of beds of siltstone, mudstone, sandstone, some conglomerate and limestone, and a little fresh and altered volcanic ash. The sandstones are highly lenticular and generally restricted to the Salt Wash Member, but locally the Salt Wash is nearly devoid of sandstone, and in other places a few sandstone lenses occur in the Brushy Basin Member. The Morrison, the lower one-third to one-half of which is formed by the Salt Wash Member, is 500 to 600 feet thick in this area; it has yielded fresh-water invertebrate fossils and many dinosaur remains, including the type specimen of *Brachiosaurus altithorax* Riggs. Sandstone lenses in the Salt Wash Member yield small supplies of soft water to a few flowing artesian wells.

The Burro Canyon Formation is virtually conformable on the Brushy Basin Member of the Morrison Formation, and locally the two units intertongue. In the western part of the area the Burro Canyon is 50 to 60 feet thick and consists mainly of green shale, but it includes a basal sandstone or conglomerate and one or more additional beds of sandstone; in the eastern part the Burro Canyon is as much as 120 feet thick and dominantly sandstone in most places. The Burro Canyon and Dakota yield small supplies of water to a few nonflowing artesian wells, but in most places the water is salty or brackish.

An erosional unconformity separates the Burro Canyon Formation from the overlying Dakota Sandstone. The Dakota, which probably is more than 200 feet thick, comprises a basal white sandstone or conglomerate, dark lignitic shale, lignite coal, and beds of buff sandstone. Some of the sandstone beds are fluvial but others are beach deposits formed in the gradually transgressing Mancos sea, and the lignitic beds were formed in coastal swamps.

The contact between the marine Mancos Shale and the Dakota sandstone is conformable and gradational, and locally the two formations intertongue. The Mancos, which is 3,800 feet thick in the general area, underlies most of the Grand Valley and forms most of the Book Cliffs, which border the valley on the northeast, but only the lowermost few hundred feet of the Mancos is present in the area mapped. It is a drab sequence of mainly soft olive-gray to gray-black fissile shale that contains a few sandy zones, thin beds of sandstone, and some light-buff to cream-colored chalky shale. The Mancos contains no usable shallow ground water, but it serves to confine artesian water in the Burro Canyon and Dakota Formations.

Although the Mancos Shale is the youngest pre-Quaternary formation in the Grand Junction area, deposits of late Mesozoic and Tertiary age remain in the Piceance Creek basin just to the northeast of the area, some of which probably formerly covered the Grand Junction area. There the marine Mancos is succeeded by the partly marine and partly continental Mesaverde Group and the wholly continental Paleocene and Eocene Wasatch Formation, the Green River Formation, and post-Green River basalt flows.

The Uncompahgre arch probably began to rise at about the close of the Cretaceous; it received renewed uplift and folding in post-Green River time, when the Green River Formation and older rocks were folded to form the Unita and Piceance Creek structural basins. Although late Tertiary vulcanism occurred in some nearby areas, events of Oligocene and Miocene times were not recorded in the Grand Junction area except for the outpouring of lava sometime after the Green River deposition.

The course of the Colorado River may have been established by superposition before or soon after extrusion of the post-Green River lavas, and, during epeirogenic uplifts in late Miocene to middle Pliocene time, the streams deepened their channels without regard to hardness of rocks or underlying structure. It seems likely that Unaweep Canyon was cut down to and probably into the Precambrian core of the Uncompahgre Plateau during this interval.

There is evidence that renewed differential uplift of the Uncompahgre arch occurred in Pliocene time, before abandonment of Unaweep Canyon, and again in latest Pliocene and earliest Pleistocene time, after abandonment of the canyon. Evidence is presented that abandonment of Unaweep Canyon was caused by successive captures of the superposed ancestral Colorado and Gunnison Rivers by a subsequent tributary of the ancestral Colorado that cut in the soft Mancos Shale around the northwestward-plunging Uncompahgre arch while downcutting by the ancestral Colorado was retarded by the hard rocks in the canyon. Capture of a tributary (East Creek), probably in the Pleistocene, completed the principal drainage changes.

The drainage divide in Unaweep Canyon stands about 2,500 feet above Gateway and Grand Junction. Studies of dissected pediments and other erosional features in and above Grand Junction suggest that this difference in altitude may include 600 to 800 feet of erosion and 1,700 to 1,900 feet of differential uplift of the Uncompahgre arch that occurred subsequent to abandonment of Unaweep Canyon in the Pliocene. The deep cliff-walled canyons in and near the Colorado National Monument were cut during this erosion interval. The nearly vertical, generally sun-facing cliffs probably were formed in part by daily alternate freezing and thawing in the winter, while the gentler northward-facing canyon walls remained frozen for long periods; the ephemeral streams in these canyons serve mainly as sewers to carry away the products of several types of erosion. During the latest period of erosion, pediments were cut in places and minor amounts of terrace deposits, pediment deposits, landslide deposits and alluvium were laid down.

The structure of the area is shown on the geologic map by structure contours drawn on top of the Entrada Sandstone, by one short cross section, by several stereoscopic pairs of aerial photographs, and by oblique aerial photographs. The strata on the northeastern flank of the Uncompahgre arch dip gently toward the Piceance Creek basin to the northeast, except where interrupted by a series of named major monoclines and faults generally parallel or nearly parallel to the axis of the uplift and by some minor structural features that trend in various directions. There probably were several successive periods of deformation, but many of the details are obscure. The monoclines, whose upper bends generally are sharper than the lower bends, probably are the result of lateral compression from the southwest or northeast. The faults all seem to be dip slip and are mainly normal. One fault (Redlands fault) is normal throughout most of its 6-mile length, but in two places it is a reverse fault that dips about 45° to the southwest, presumably because of rotation of a vertical fault by later compressive forces. Because the principal structural features have an important bearing on the recharge areas of the Entrada and Wingate Sandstones, they were examined in detail.

The total structural displacement of the Uncompahgre arch within or near the area is about 5,000 feet, 1,600 to 1,900 feet of which is presumed to have occurred in late Pliocene or early Pleistocene time, and about 3,100 to 3,400 feet of which occurred earlier, probably mainly in post-Green River time.

Unconfined ground water is relatively unimportant in the Grand Junction area, and its occurrence is discussed only briefly. Most of the Grand Valley is almost devoid of shallow ground water, and such meager supplies as are obtainable locally from soil, weathered rock, arroyo fill, or terrace deposits generally are too highly mineralized for most uses. Where thick, the alluvium along the principal streams should yield considerable water, but the water probably would be too hard for domestic use. Small supplies of unconfined water of reported good quality are obtained from the Entrada or the Wingate Sandstone in parts of Glade Park.

Confined, or artesian, ground water is obtained from four artesian aquifers in the Grand Junction area, which are, in order of importance and productivity: (1) the Entrada Sandstone, (2) the Wingate Sandstone, (3) lenticular sandstones in the Salt Wash Member of the Morrison Formation, and (4) the Dakota Sandstone and sandstones in the Burro Canyon Formation. These aquifers contain water under artesian pressure only in areas northeast of the principal faults and monoclines, where they are overlain by younger, relatively impermeable strata that serve as confining beds. In these areas, determination of depth to the two principal aquifers, the Entrada and the Wingate, is facilitated by use of the structure contours. The top of the Entrada has been reached at depths ranging from 188 to 1,555 feet, but in most wells it is reached at 600 to 800 feet.

The finding of water in the Morrison Formation is generally not predictable owing to the lenticularity of the sandstone beds. Water in the Dakota and the Burro Canyon Formations generally is of poor quality for most uses.

The coefficients of transmissibility (T) and storage (S) were determined in the field for 11 of the 48 artesian wells for which records are given, by flow tests using equipment and methods designed for this investigation; the T values were checked by the Theis recovery method and in part by laboratory determinations of outcrop samples. The average values of T and S for the Entrada Sandstone are 150 gpd per ft (gallons per day per foot) and 5×10^{-5}, respectively, and scanty data for the combined Entrada and Wingate Sandstones suggest values of about 300 gpd per ft and 10^{-4}. Field tests of wells tapping a sandstone lens in the Morrison Formation indicate T values of only 35–50 gpd per ft. No tests were made of wells in the Dakota and Burro Canyon Formations. Field and laboratory tests of the two principal aquifers suggest coefficients of permeability of about 1 gpd per sq ft (gallons per day per square foot) and 0.5 gpd per sq ft, respectively. The laboratory tests indicate that the permeability of these sandstones parallel to the bedding planes is much greater than at right angles to this direction.

The artesian aquifers are recharged mainly where streams cross the outcrops, but a small amount of recharge may result from precipitation on the outcrops. Except for the Gunnison River and North East Creek, the streams that produce recharge are all ephemeral. Because of the low permeability of the aquifers, the rate of recharge probably is very small and not readily determinable.

From known and assumed nondischarging conditions, an average velocity is computed for down-dip movement of water in the Entrada Sandstone to be only about 0.013 foot per day, or about 5 feet per year.

Natural discharge from the aquifers probably occurs throughout the Piceance Creek basin by slow leakage upward through relatively impermeable confining beds and possibly along faults. It is postulated that the amount of such natural discharge at any one place is too small to measure by conventional methods and that such water as may reach the surface probably is in a gaseous state.

The first deep wells in the Grand Junction area seemingly were drilled in the hope of finding oil or gas, but artesian water was found instead. The first well may have been drilled in 1903 or 1904, and by 1946 only 14 wells were in use, 13 of which were flowing wells. Twenty-seven additional wells were drilled during the 10-year period, 1947–56, but before the end of this period interference between wells had caused considerable decline in artesian heads and flows, some well owners had installed pumps, and enthusiasm for drilling additional wells had diminished. Thus, from 1956 to 1960, only seven additional wells were drilled.

Forty-three of the 48 wells for which records were obtained were drilled by the cable-tool method, and five wells were drilled all or in part by the hydraulic-rotary method. Most of the wells contain at least two casings, but some contain one to four. Many different types of commercial or homemade well seals, generally augmented by gravity or pressure cementing, were used, but in some wells water is leaking to the surface. Most of the wells are cased only to the well seal above the aquifer and are open holes through or into the aquifer, but a few have perforated pipe, and two wells have well screens surrounded by gravel. By 1960 most of the wells had been equipped with jet, turbine, or submersible pumps.

Because of the low permeability of the artesian aquifers, the wells have small yields by either natural flow or pumping, and average specific capacities of less than 0.1 gpm per foot of drawdown for the Entrada Sandstone, slightly more than 0.1 gpm per ft for the Entrada and Wingate Sandstones, and as low as about 0.01 gpm per ft for some wells in the Morrison Formation. Most of the wells are operated at rates of 5 to about 40 gpm; larger rates generally are not practicable. Such wells would be considered dry holes in many parts of the country but are valued in the arid Grand Junction area where water of good quality is scarce.

Curves are given in the report to show the amount of drawdown that might be expected at different distances from wells discharging at given rates for different periods of time from the two principal artesian aquifers. These curves show that there is considerable drawdown interference between wells in the same aquifer or aquifers, particularly in the most intensely developed areas.

Initial artesian heads ranged from near land surface to more than 150 feet above land surface, but overdevelopment and interference between wells has caused known declines in head of more than 50 feet in some wells and probably more than 100 feet in a few others.

Analyses of 26 samples of water from 23 wells are given to indicate the quality of water from the principal artesian aquifers.

The samples from the Entrada Sandstone or the Entrada and Wingate Sandstones were soft sodium bicarbonate water, most of which had a hardness of less than 50 ppm (parts per million) and some of which had a hardness of 10 ppm or less; three samples had a hardness ranging from 100 to 124 ppm. Samples from the Morrison were soft sodium bicarbonate-sodium sulfate water. These waters are of good quality for domestic use, but contain high percentages of sodium and may be harmful to certain plants or crops. No samples were obtained from the Burro canyon Formation or Dakota Sandstone, but reports from well drillers and owners indicate that the water generally is brackish or salty.

In all the water analyzed, the relative softness is attributed to natural softening by base exchange, whereby calcium and magnesium ions in the water are exchanged for sodium ions in certain minerals in the aquifers and thus remove part or most of the hardness-producing calcium and magnesium from the water. Petrographic and X-ray examinations of samples of the Entrada and Wingate Sandstones indicate that clay minerals cause the softening. There is an almost linear decrease in hardness of water in the Entrada with increased distance from the recharge area.

Most of the artesian water in the Grand Junction area is used for domestic purposes, either by the owner alone, by the owner and nearby homes connected by pipeline, or by hauling to homes equipped with storage tanks or cisterns. From 1 to as many as 30 tank-loads (1,100-gallon tanks) per day are hauled from 13 of the wells. Some of the water is used for watering livestock, filling a swimming pool, supplying a meat-packing plant, or watering small plots of lawn or shrubs.

Declines in artesian heads and flows indicate that the principal aquifer—the Entrada Sandstone and to a lesser extent the Wingate Sandstone—have been overdeveloped in parts of the Grand Junction area, but two relatively large areas are undeveloped or only slightly developed and would yield additional water to wells, preferably spaced more than a mile apart. One area comprises the southwest side of the Grand Valley and the Redlands, in and northwest from the northwestern part of the area. The other area comprises the southwestern side of the Grand Valley, parts of Orchard Mesa, and the lower part of the Gunnison River Valley in and southeast from the eastern part of the area.

Grand Junction and Fruita have municipal water supplies piped from distant surface-water sources. The Colbran Project of the U.S. Bureau of Reclamation will supply water for irrigation and power in Plateau Creek valley north of Grand Mesa and to the Ute Conservancy District for piping to several cities and towns and most rural residents in the Grand Valley, including those of the Redlands and Orchard Mesa. Completion of this water system should greatly reduce the draft on the artesian wells in in the Grand Junction area. This reduction should arrest the decline of the artesian head or should allow the head to recover gradually. Because of the small rate of recharge, however, the recovery in head will take considerable time.

[Editors' Note: Material has been omitted at this point.]

Reprinted from page 1 of *U.S. Geol. Survey Prof. Paper 383-C*, 1969, 33p.

HYDROLOGY OF VOLCANIC-ROCK TERRANES

EFFECT OF TECTONIC STRUCTURE ON THE OCCURRENCE OF GROUND WATER IN THE BASALT OF THE COLUMBIA RIVER GROUP OF THE DALLES AREA, OREGON AND WASHINGTON

By R. C. Newcomb

ABSTRACT

The 620-square-mile area studied lies across the boundary of the Cascade Range and Columbia Plateaus and affords geologic and hydrologic conditions typical of the basalt region.

The basalt of the Columbia River Group is the oldest rock exposed. More than 2,000 feet of accordantly layered lava underlies the area shown on the White Salmon, The Dalles, and Wishram quadrangle maps, and is at the surface in about half of the area. The overlying Pliocene Dalles Formation remains mostly in the two largest synclines. Its thickness reaches a maximum of about 2,000 feet, but beneath wide areas it is about 500 feet. Post-Dalles lavas and Pleistocene fluvial deposits in places cover the older rocks.

The rubbly tops of some of the lava flows and the brecciated flows within the sequence form aquifers that yield large to small amounts of water to wells and springs. The yield of the aquifers depends on the hydrologic conditions and structural situation.

The main tectonic structural features branch from the Cascade Range. They are the Bingen anticline and Columbia Hills anticlinorium and the accompanying Mosier and Dalles synclines. The folds and faults partly control the levels at which ground water occurs.

The natural channel of the Columbia River was graded between low falls and rapids. The Deschutes River is largely graded and the Klickitat River partly graded. The other streams are small and descend to river level through canyoned reaches cut in the basalt; some of the smaller tributaries are cut into or through the Dalles Formation in the broader upland parts of their valleys. Most of the stream valleys follow synclines and are consequent upon the tectonic structures; principal exceptions are at the Bingen and Rowena Gaps where the Columbia River crosses the major anticlines.

Ground water in the basalt aquifers occurs mainly under three situations: (1) beneath the regional, or main, water table near river level, (2) perched at intermediate altitude near base-leveled secondary streams, and (3) perched at high levels near the top of the basalt in the uplands.

Tectonic structure affects ground-water conditions in the basalt by the inclination of the aquifers, by formation of barriers to lateral percolation, by creation of avenues for some vertical movement, and by production of inlets for recharge and outlets for discharge.

The inclination of aquifers increases the opportunity for infiltration and vertical transfer, and also for lateral and vertical percolation along the permeable interflow zones.

The structural barriers to lateral percolation occur where the underlying impermeable strata rise in an anticline, where faults destroy or offset the permeable layers, or where combinations of faults and folds interrupt the aquifers. Together with stratigraphic discontinuities, the structural barriers have caused the impoundment of ground water in the basalt above the main water table. The levels of the outlets around barriers determine the uppermost level to which the obstructed ground water rises.

Ground-water is impounded behind a sheared anticline in Mosier Creek valley, upslope from a fault in Mill Creek valley, and behind a fault across Swale Creek valley. Two other such impoundments of ground water lie behind an anticlinal bulge across three creek valleys near Jap Hollow and isolated by a complex combination of fault and fold barriers that bound "The Dalles Ground Water Reservoir." These impoundments result in ground water occurring at levels many hundreds of feet above the ground-water level in nearby areas of structurally undisturbed basalt.

The discharge of ground water at springs is sufficiently controlled by structural elements that the interpretative process can be reversed; hydrologic features can indicate the geologic structure. For example, in places on long dip slopes, transverse synclines can be indicated by concentrations of springs and transverse courses of streams.

The close relation between the tectonic structure and the occurrence of ground water in the basalt can be used to make better predictions of drilling results. The structural criteria for ground-water storage can be projected elsewhere in the basalt region.

An accurate map of the geology, and particularly of the geologic structures, has been found imperative to the interpretation of the ground-water occurrences.

[*Editors' Note:* Material has been omitted at this point.]

25

Reprinted from pages 1-2 of *U.S. Geol. Survey Prof. Paper 521-A,* 1969, 61p.

REGIONAL HYDROGEOLOGY OF THE NAVAJO AND HOPI INDIAN RESERVATIONS, ARIZONA, NEW MEXICO, AND UTAH

By M. E. Cooley, J. W. Harshbarger, J. P. Akers, and W. F. Hardt

ABSTRACT

The Navajo and Hopi Indian Reservations have an area of about 25,000 square miles and are in the south-central part of the Colorado Plateaus physiographic province. The reservations are underlain by sedimentary rocks that range in age from Cambrian to Tertiary, but Permian and younger rocks are exposed in about 95 percent of the area. Igneous and metamorphic basement rocks of Precambrian age underlie the sedimentary rocks at depths ranging from 1,000 to 10,000 feet. Much of the area is mantled by thin alluvial, eolian, and terrace deposits, which mainly are 10 to 50 feet thick.

The Navajo country was a part of the eastern shelf area of the Cordilleran geosyncline during Paleozoic and Early Triassic time and part of the southwestern shelf area of the Rocky Mountain geosyncline in Cretaceous time. The shelf areas were inundated frequently by seas that extended from the central parts of the geosynclines. As a result, complex intertonguing and rapid facies changes are prevalent in the sedimentary rocks and form some of the principal controls on the ground-water hydrology. Regional uplift beginning in Late Cretaceous time destroyed the Rocky Mountain geosyncline and formed the structural basins that influenced sedimentation and erosion throughout Cenozoic time.

The rocks are characterized by the absence of severe deformation. The area has been relatively stable since late Precambrian time and was affected only moderately by the orogeny of Late Cretaceous and early Tertiary time, which produced a variety of folds. Later, in Tertiary and Quaternary time, the area was upwarp and locally faulted.

The reservations are divided into several hydrogeologic subdivisions on the basis of differences in the exposed sedimentary rocks, structure, and physiography. The occurrence of ground water in each subdivision is controlled principally by the geology.

The climate of the Navajo country varies widely, ranging from semiarid below 4,500 feet to relatively humid above 7,500 feet. Precipitation has a strong and fairly uniform relation to altitude and the orographic effects of the physiography. Mean annual temperature is affected by altitude and by the local physiographic position of the weather station. Dunes and other eolian deposits are common and were laid down intermittently throughout Quaternary and part of Tertiary time. The distribution and orientation of the dunes and the direction of crossbeds indicate that the prevailing wind during much of prehistoric time was from the southwest and therefore was similar to the present prevailing wind pattern.

Vegetation is divided into broad zones consisting of grass-shrub at altitudes below 5,500 feet, pinyon-juniper between 5,500 and 7,500 feet, and pine forest above 7,500 feet. In the grass-shrub and pinyon-juniper zones, some of the plant assemblages are controlled by the types of sedimentary rocks exposed.

The Colorado River developed in late Cenozoic time as a superimposed stream on the folded rocks of the Colorado Plateaus. Continuous downcutting by the river and its tributaries resulted in entrenchment of the entire system. Entrenchment was at a maximum during late Pliocene and Pleistocene time and was about 1,800 feet in Glen and San Juan Canyons and as much as 1,000 feet along the Little Colorado River. All runoff from the reservations is to the Colorado River. The Colorado and San Juan Rivers are perennial. All the other streams are ephemeral or intermittent except for short reaches downstream from large springs and where the streambed intersects the water table. Nearly one-sixth of the area drains internally.

The aquifers are composed of beds of sandstone between nearly impermeable layers of siltstone and mudstone. The main aquifers are in the Coconino Sandstone, Navajo Sandstone, and the alluvium; but all other units locally yield some water to wells and springs. The siltstone and mudstone layers act as aquicludes, thus confining the water in the underlying sandstone aquifers under artesian pressure in much of the area. For the most part, the aquifers in the consolidated sedimentary rocks are fine grained and do not transmit water rapidly. Coefficients of permeability are generally less than 10 gpd per sq ft (gallons per day per square foot), and many are less than 3 gpd per sq ft. Yields from wells in these aquifers usually are less than 25 gpm (gallons per minute). Most specific capacities computed from tests of wells range from 0.3 to 5.0 gpm per ft of drawdown.

The basins and uplifts control the movement of ground water in the sedimentary rocks; the other structural features affect the occurrence of ground water only locally. The larger folds divide the reservations into five general areas, which are considered as separate hydrologic basins—Black Mesa, San Juan, Blanding, Henry, and Kaiparowits hydrologic basins.

The main areas of recharge to the ground-water reservoirs are on the highlands—Defiance Plateau, Zuni Mountains, Mogollon Slope, San Francisco Plateau, and Navajo Uplands—along the structural divides between the hydrologic basins. Movement of the ground water in each hydrologic basin is downdip from the highlands and toward the Colorado, Little Colorado, or San Juan Rivers and their larger tributaries rather than toward the centers of the basins. Black Mesa basin is unique in that most water discharges into the Little Colorado River from Blue Spring and nearby springs, which flow at about 220 cubic feet per second. Natural discharge is from 1,000 springs and numerous seeps. Artificial discharge is from 1,300 drilled wells and 550 dug wells, which are used chiefly for domestic and stock purposes.

The ground water has a wide range in the type and amount of dissolved chemical constituents. Most water having less than 700 ppm (parts per million) of dissolved solids is either calcium or sodium bicarbonate, and water containing more than 700 ppm is sodium sulfate, calcium sulfate, or sodium chloride. The amount of fluoride in the ground water in parts of the Hopi Indian Reservation, the valley of the Little Colorado River, and the San Juan basin is excessive and may be more than 3.0 ppm. The dissolved-solids content of 1,300 water samples analyzed ranges from 90 to more than 25,000 ppm. Water from the flushed aquifers in the Navajo, Wingate, Coconino, and De Chelly Sandstones on the Mogollon Slope, San Francisco Plateau, Navajo Uplands, and Defiance Plateau usually contains less than 1,000 ppm of dissolved solids. Water having the greatest amount of dissolved solids is in deep aquifers in the Black Mesa and San Juan basins.

[*Editors' Note:* Material has been omitted at this point.]

A Theoretical Analysis of Groundwater Flow in Small Drainage Basins[1]

J. TÓTH

Abstract. Theoretically, three types of flow systems may occur in a small basin: local, intermediate, and regional. The local systems are separated by subvertical boundaries, and the systems of different order are separated by subhorizontal boundaries. The higher the topographic relief, the greater is the importance of the local systems. The flow lines of large unconfined flow systems do not cross major topographic features. Stagnant bodies of groundwater occur at points where flow systems meet or branch. Recharge and discharge areas alternate; thus only part of the basin will contribute to the baseflow of its main stream. Motion of groundwater is sluggish or nil under extended flat areas, with little chance of the water being freshened. Water level fluctuations decrease with depth, and only a small percentage of the total volume of the groundwater in the basin participates in the hydrologic cycle.

Introduction. Whereas it is important to have a *general understanding* of the motion of groundwater in dealing with groundwater problems, the careless and frequent use of the expression may subvert its basic meaning. Until certain characteristics of the flow systems involved are well defined, groundwater motion in a given area cannot be conceived to be generally known. Among the numerous features, a knowledge of which is indispensable to the understanding of groundwater motion in an area, the following are thought to be the most important: the locations and extent of recharge and discharge areas, the direction and velocity of flow at any given point in the region, and the depths of penetration of the flow systems. It is easy to appreciate the value of this information if one considers only the difficulties which may arise in connection with problems such as outlining areas of potentially equal yield, tracing contaminants, estimating baseflow of rivers, and establishing groundwater budgets.

The purpose of this paper is to present a theory by means of which groundwater flow in small drainage basins can be analyzed. Some of the properties of flow derived from this analysis are obvious and may be observed in the field, but others are hidden and may not be revealed even by expensive test programs. The neglect of these latter properties could lead to entirely wrong conclusions regarding groundwater flow in small basins either in general or in any particular case. Even if the theory is not used to obtain quantitative results, the qualitative application may still contribute to the general understanding of groundwater flow in small basins.

Before starting with the development of the theory a brief account will be given of the reasons why a theoretical analysis is believed to be best suited for an initial general study of groundwater flow in a given area.

General. The methods of studying groundwater motion can be either practical or theoretical. The group of practical methods includes field investigations based on the principles of geology, geophysics, geochemistry, and hydrology, and it is thus based on observations of phenomena controlling or related to the flow of groundwater in nature. The theoretical methods, on the other hand, make use of electrical analogs, scale models, and mathematical models to investigate phenomena resulting from idealized situations. In the final analysis the conclusions drawn from the data of both groups should be considered, and they must be in agreement. Nevertheless, the writer believes the results obtained by the application of the theoretical methods to be the more useful in the initial

---

[1] Contribution 185 from the Research Council of Alberta, Edmonton, Alberta. Presented at the Third Canadian Hydrology Symposium, Calgary, Alberta, November 8–9, 1962.

stages of an investigation. This view is based on the presumption that an observed phenomenon is usually related to only one feature of a given flow system, whereas it might be brought about by different causes in different situations; it may be the identical particular solution of several problems.

A few examples may prove helpful in clarifying this statement. For instance, a decrease in hydraulic head with depth, commonly observed in water wells, may be produced either by head losses due to the vertical downward-component of the water motion or by the water being perched in the permeable layers of a geological formation consisting of a series of more and less pervious beds [*Meinzer*, 1923, p. 41]. Another good example is a perennial body of impounded surface water. It may owe its existence either to poor underground drainage due to geologic conditions or to continuous groundwater discharge caused by the general pattern of the flow systems. The cause of a relatively low baseflow yield for a river may be even more uncertain because it could be explained by a basin-wide low permeability, by good surface drainage, or by the stream not being the only place of groundwater discharge in the basin. It is realized, of course, that the larger the variety of independent investigations, the more precisely the characteristics of the flow can be outlined. There still may remain the uncertainty, however, of whether some of the decisive features have been overlooked and whether there may be some characteristics that cannot be measured at all. Whereas it is practically impossible to observe separately all phenomena connected with a regime of groundwater flow, a correct theory discloses every feature and draws attention to the most important properties of the flow.

It is believed that small drainage basins are the most important units in the groundwater regime. A good understanding of groundwater movement in adjacent small basins makes possible an accurate representation of the motion of groundwater within the large basin that they form. In working from larger basins to smaller basins the weight of the uncertainties increases, and a vague and possibly unreliable analysis is obtained. Apart from this, a small basin is commonly much less complicated than a large basin with respect to geology and topography; therefore, it lends itself much better to both practical and theoretical studies.

The definition of a small drainage basin as it will be understood throughout this paper is: *an area bounded by topographic highs, its lowest parts being occupied by an impounded body of surface water or by the outlet of a relatively low order stream and having similar physiographic conditions over the whole of its surface.*

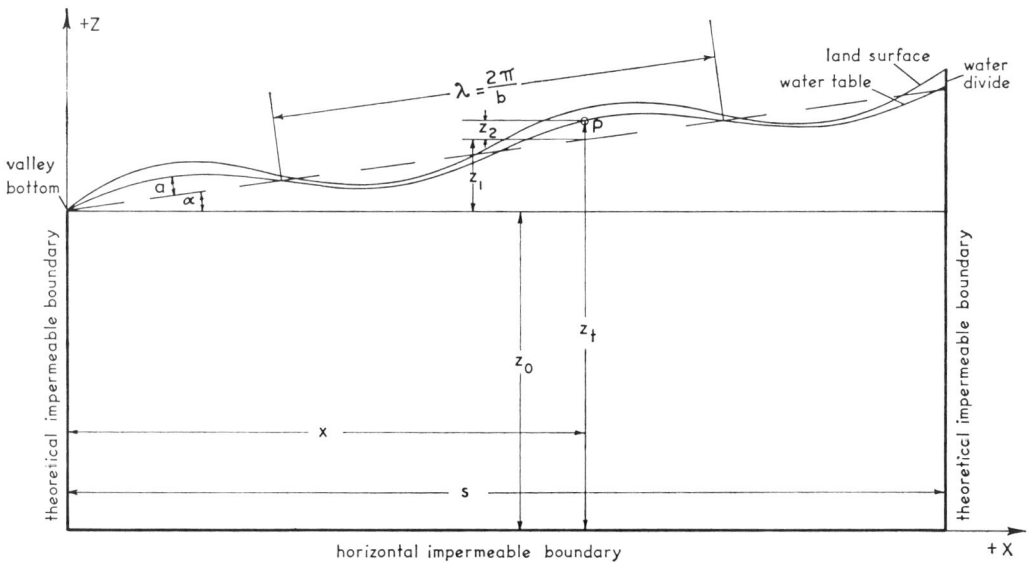

Fig. 1. Idealized cross section of a valley flank in a small drainage basin.

The upper limit of area for such basins is usually several hundred square miles.

Mathematical development. In the mathematical analysis the region of groundwater flow at one side of the valley is represented by a rectangular area (Figure 1). This area is limited by a horizontal impermeable boundary at its base, by two vertical impermeable boundaries extending downward from the stream and the water divide, and by a horizontal line at the elevation of the stream along which line the fluid-potential distribution is supposed to be the same as that for the real water table. The assumption of a horizontal impermeable boundary as the lower limit of the basin is justified because, in the interval above this in which no such boundary is known, all groundwater belongs to the flow region of the basin. If, however, a relatively impermeable boundary is present underlying the whole basin, the water systems under it will not significantly interfere with the systems within the basin. The assumption of the two vertical boundaries is, strictly speaking, correct only if the surface drainage pattern is symmetrical—that is, if the basin is bounded by two parallel and equally removed surface-water divides of equal topographic elevation. In this case the potential distribution at both sides of the stream is symmetrical and the impermeable boundaries may be drawn vertically at the stream and at the divides. It will be shown later that a small amount of asymmetry in the topography does not cause a significant deviation from the vertical of these boundaries.

The potential distribution along the theoretical surface, although identical with that of the water table, is along a horizontal surface, and this restricts the validity of the numerical results to small slopes of about 3° or less. For the topography within the basin, a sinusoidal shape has been chosen, the highs and lows of which are thought to be representative of the hills and depressions of the natural land surface.

The analysis is also based on the assumption that the geologic conditions in the basin are isotropic and homogeneous. Whether or not this assumption is justified depends on the extent to which a real case deviates from the ideal conditions.

The assumption that the problem can be treated as a two-dimensional one is supported by the recognition that in most small basins the slopes of the valley flanks greatly exceed the longitudinal slopes of the valley floors. This difference in slope causes the longitudinal component of the flow to become negligible compared with the lateral component.

The distribution of the fluid potential in a basin with boundaries as outlined above is derived from the general expression for the fluid potential [*Hubbert*, 1940, p. 802]:

$$\phi = gz + \int_{p_0}^{p} \frac{dp}{\rho} \quad (1)$$

where ϕ = fluid potential, g = acceleration due to the earth's gravity field, z = elevation above the horizontal impermeable boundary as standard datum, p_0 = pressure of the atmosphere, p = pressure in the flow region at any point, ρ = density of water.

If the water table is defined as *a specific piezometric surface in the groundwater region at which the gravity potential is a maximum and the pressure potential equals that of the atmosphere,* (1) reduces to

$$\phi_t = gz_t \quad (2)$$

for the water table, where z_t = the topographic elevation of the water table at any point in the basin. It has been observed in Alberta [*Meneley*, 1963, pp. 4–12] as well as elsewhere [*King*, 1892, pp. 15–18; *Meinzer*, 1923, p. 34; *Wisler and Brater*, 1959, Figure 85; *Wieckowska*, 1960, p. 64] that the water table is generally similar in form to the land surface. Thus z_t is found to consist of three components: z_0, z_1, and z_2 (Figure 1). z_0 is a constant, denoting the depth to the horizontal impermeable boundary from the stream bottom. $z_1 = x \tan \alpha$, where x is the horizontal distance of any point in the flow region from the valley bottom and α is the average slope of the valley flank. As long as α is small, z_2 may be approximated by

$$z_2 = a \frac{\sin (bx/\cos \alpha)}{\cos \alpha}$$

where a is the amplitude of the sine curve, $b = 2\pi/\lambda$ is the frequency, and λ is the period of the sine wave. With the three components known, the equation of the water table is obtained:

$$z_t = z_0 + x \tan \alpha + a \frac{\sin (bx/\cos \alpha)}{\cos \alpha}$$

Upon introducing the abbreviations $\tan \alpha = c'$, $a/\cos \alpha = a'$, and $b/\cos \alpha = b'$, the final form of z_t is written as

$$z_t = z_0 + c'x + a' \sin b'x \quad (3)$$

From (2) and (3) the potential at the water table is found:

$$\phi_t = g(z_0 + c'x + a' \sin b'x) \quad (4)$$

Owing to the natural equilibrium of the groundwater budget in a basin, the average level of the water table is assumed to be constant. The problem is thus a steady-state potential problem which may be solved by applying the Laplace equation:

$$\partial^2 \phi / \partial x^2 + \partial^2 \phi / \partial z^2 = 0$$

The four boundary conditions will be as follows:

$\partial \phi / \partial x = 0$ at $x = 0$ and s

for $0 \leq z \leq z_0$ (5a, 5b)

$\partial \phi / \partial z = 0$ at $z = 0$ for $0 \leq x \leq s$ (5c)

$\phi_t = g(z_0 + c'x + a' \sin b'x)$ at $z = z_0$

for $0 \leq x \leq s$ (5d)

where s is the horizontal distance between the valley bottom and the water divide.

The general solution of the Laplace equation can be written in the following form:

$$\phi = e^{-kz}(A \cos kx + B \sin kx) + e^{kz}(M \cos kx + N \sin kx)$$

The arbitrary constants A, B, M, and N can be found from the boundary conditions.

Upon performing the derivation we get the following final equation for the fluid potential:

$$\phi = g \left\{ z_0 + \frac{c's}{2} + \frac{a'}{sb'}(1 - \cos b's) \right.$$
$$+ 2 \sum_{m=1}^{\infty} \left[\frac{a'b'(1 - \cos b's \cos m\pi)}{b'^2 - m^2\pi^2/s^2} \right.$$
$$+ \left. \frac{c's^2}{m^2\pi^2}(\cos m\pi - 1) \right]$$
$$\left. \cdot \frac{\cos(m\pi x/s) \cosh(m\pi z/s)}{s \cdot \cosh(m\pi z_0/s)} \right\} \quad (6)$$

Equation 6 satisfies both the boundary conditions and the Laplace equation. By means of Darcy's law, (6) can be used to obtain the specific mass discharge in the direction of r [Hubbert, 1940, p. 842].

$$j_r = -\rho \sigma \, \partial \phi / \partial r \quad (7a)$$

or the total flow vector:

$$\mathbf{j} = -\rho \sigma \, \text{grad} \, \phi \quad (7b)$$

where $\sigma = k\rho/\eta$ (k = coefficient of permeability, η = viscosity of fluid).

Numerical computations. To analyze the effect of the geometry of the basin on the groundwater flow, (6) has been solved for various parameters. To facilitate visualization of the flow, the numerical values of the potential are expressed in 'head of water above standard datum.' The potential distributions and flow patterns for the various cases are shown in Figures 2a to 2i.

The horizontal distance between the water divide and the valley bottom is 20,000 feet in all computed cases. This distance seems to be fairly representative for the half-width of a small basin.

Three values have been assumed for the depth to the impermeable boundary at the valley bottom: 1000, 5000, and 10,000 feet. The 1000-foot case is likely to be encountered in nature, whereas a relatively homogeneous body of sediment 10,000 feet thick is a rather hypothetical case. Flow patterns have, however, been evaluated for this situation, for several reasons: first, it represents an extreme case, and therefore the general validity of the conclusions arrived at by employing (6) can be checked; second, the general features of the flow patterns are more conspicuous in the deeper boundaries than in the shallow ones; and third, the measure that is the most characteristic in determining the potential distribution is the *ratio* $n = (z_0/s)$ of the depth of the impermeable boundary to the horizontal distance between divide and stream. By employing the three values of z_0, a wide variety of potential distributions for values of n up to 0.5 can be inferred at least qualitatively.

Flow systems in small basins based on interpretation of the mathematical results. Upon inspection of Figures 2a to 2i we recognize a certain grouping of the flow lines. (In the figures the solid lines are called lines of force. Under isotropic conditions the lines of force coin-

GROUNDWATER FLOW IN SMALL BASINS

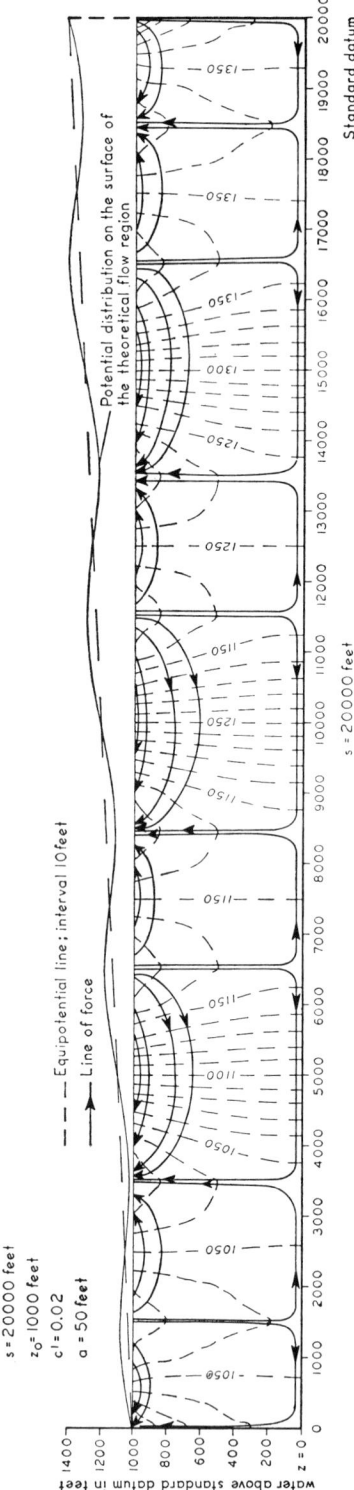

Fig. 2a. Potential distribution and flow pattern as obtained by equation 6.

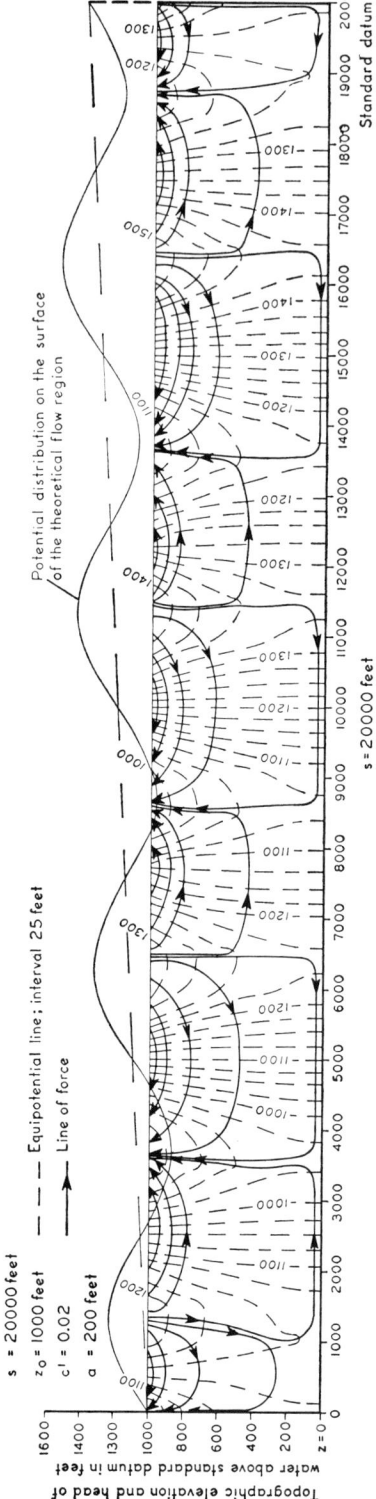

Fig. 2b. Potential distribution and flow pattern as obtained by equation 6.

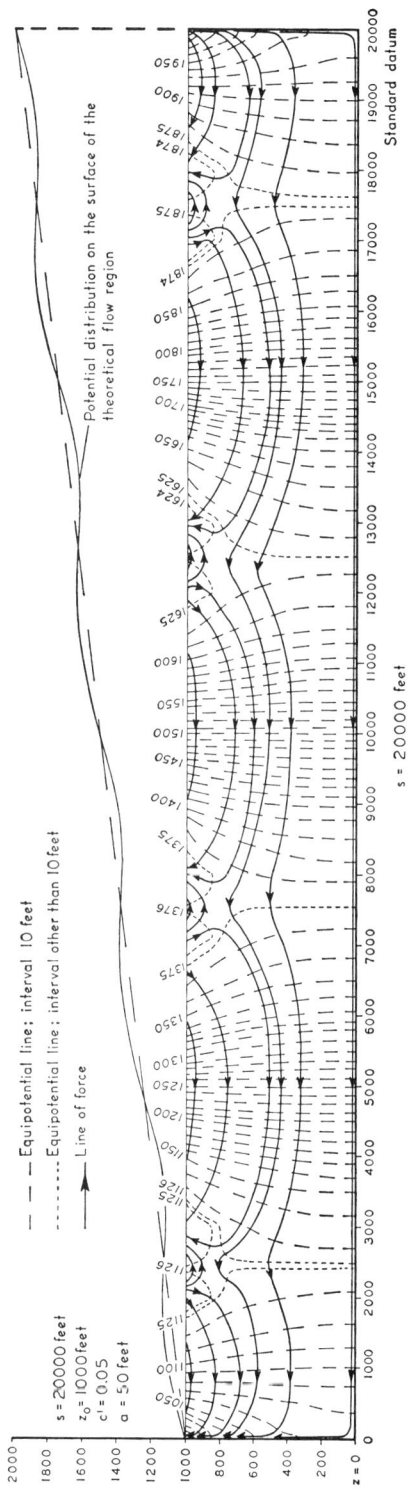

Fig. 2c. Potential distribution and flow pattern as obtained by equation 6.

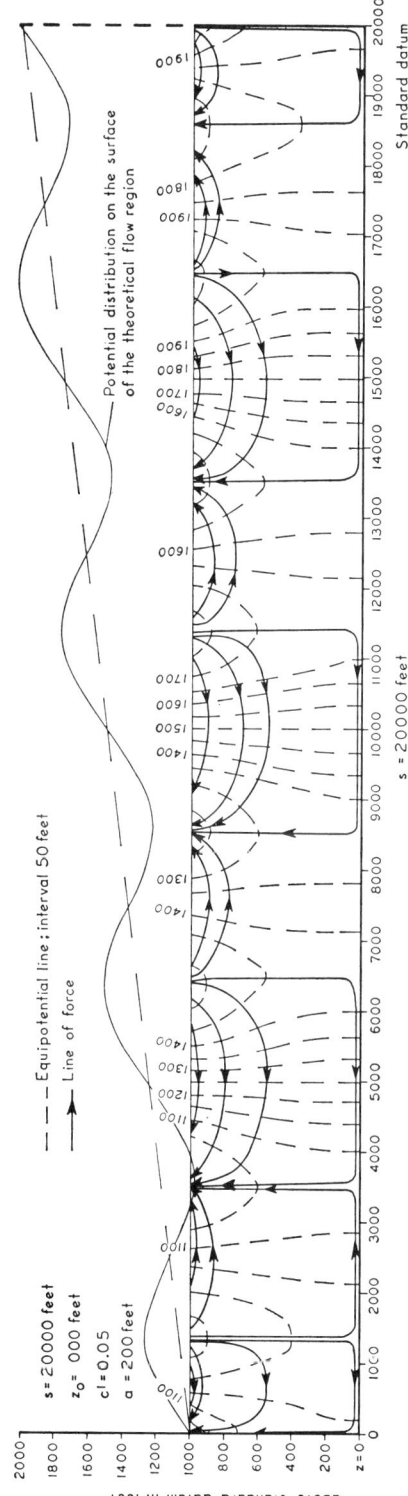

Fig. 2d. Potential distribution and flow pattern as obtained by equation 6.

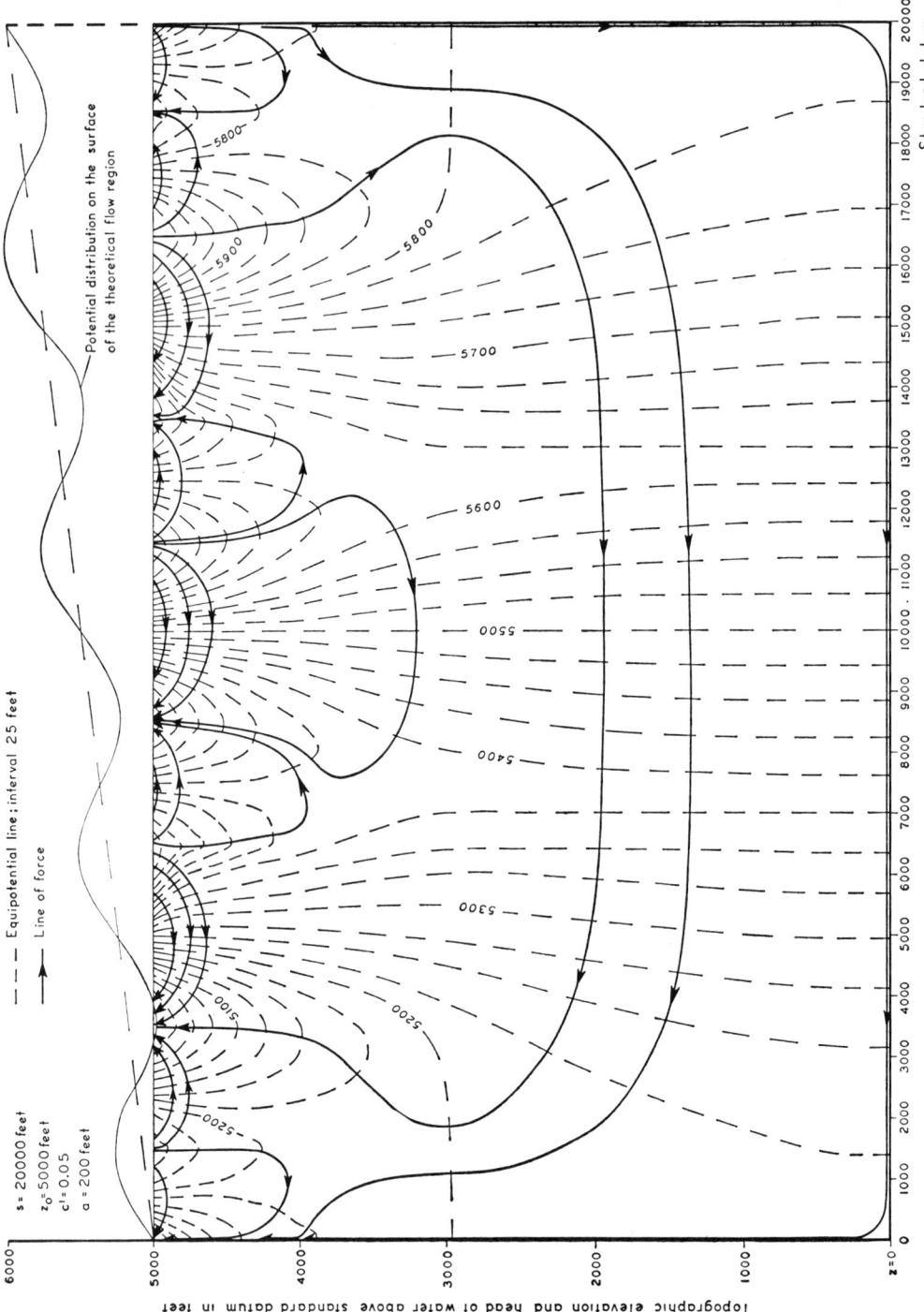

Fig. 2e. Potential distribution and flow pattern as obtained by equation 6.

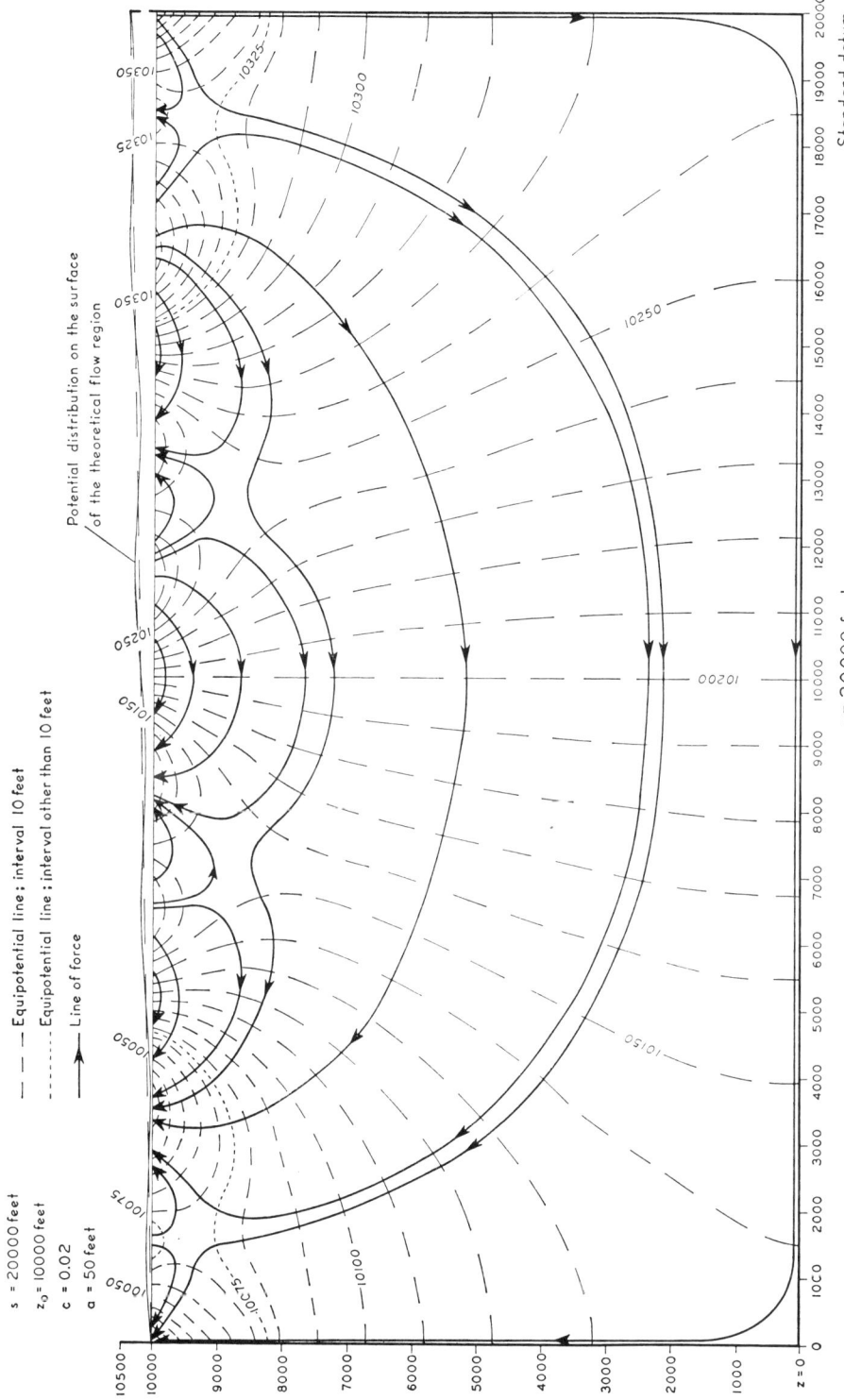

Fig. 2f. Potential distribution and flow pattern as obtained by equation 6.

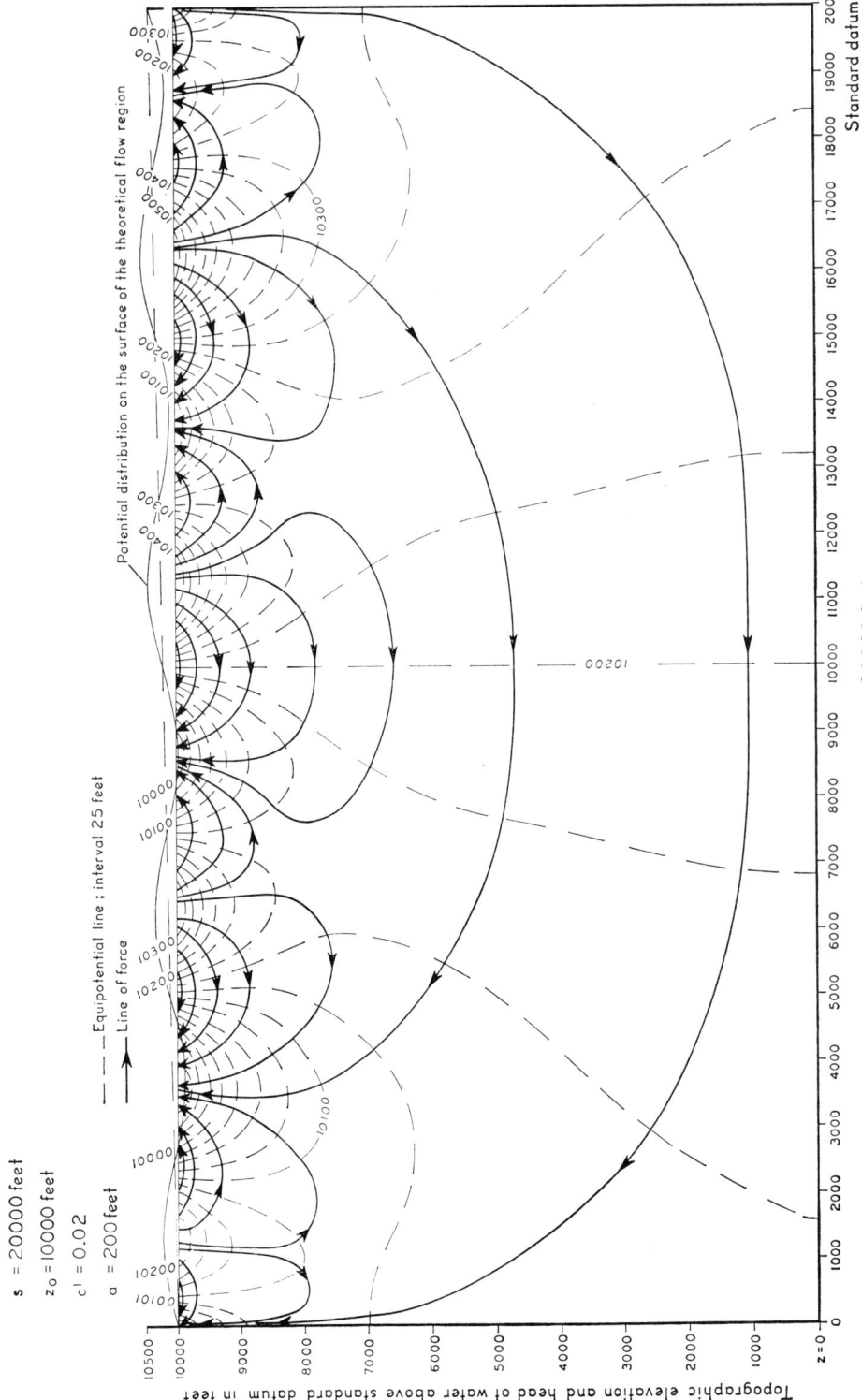

Fig. 2g. Potential distribution and flow pattern as obtained by equation 6.

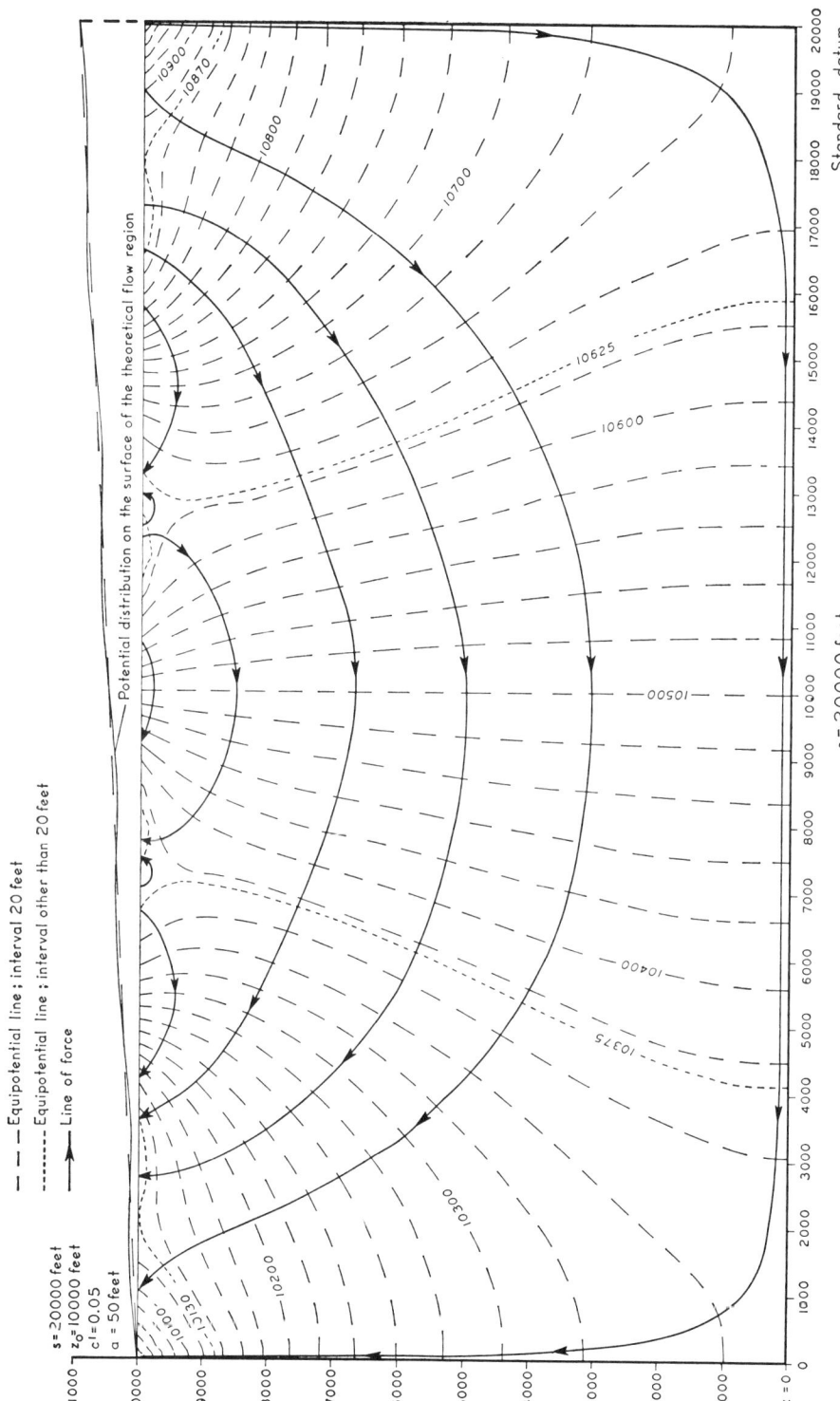

Fig. 2h. Potential distribution and flow pattern as obtained by equation 6.

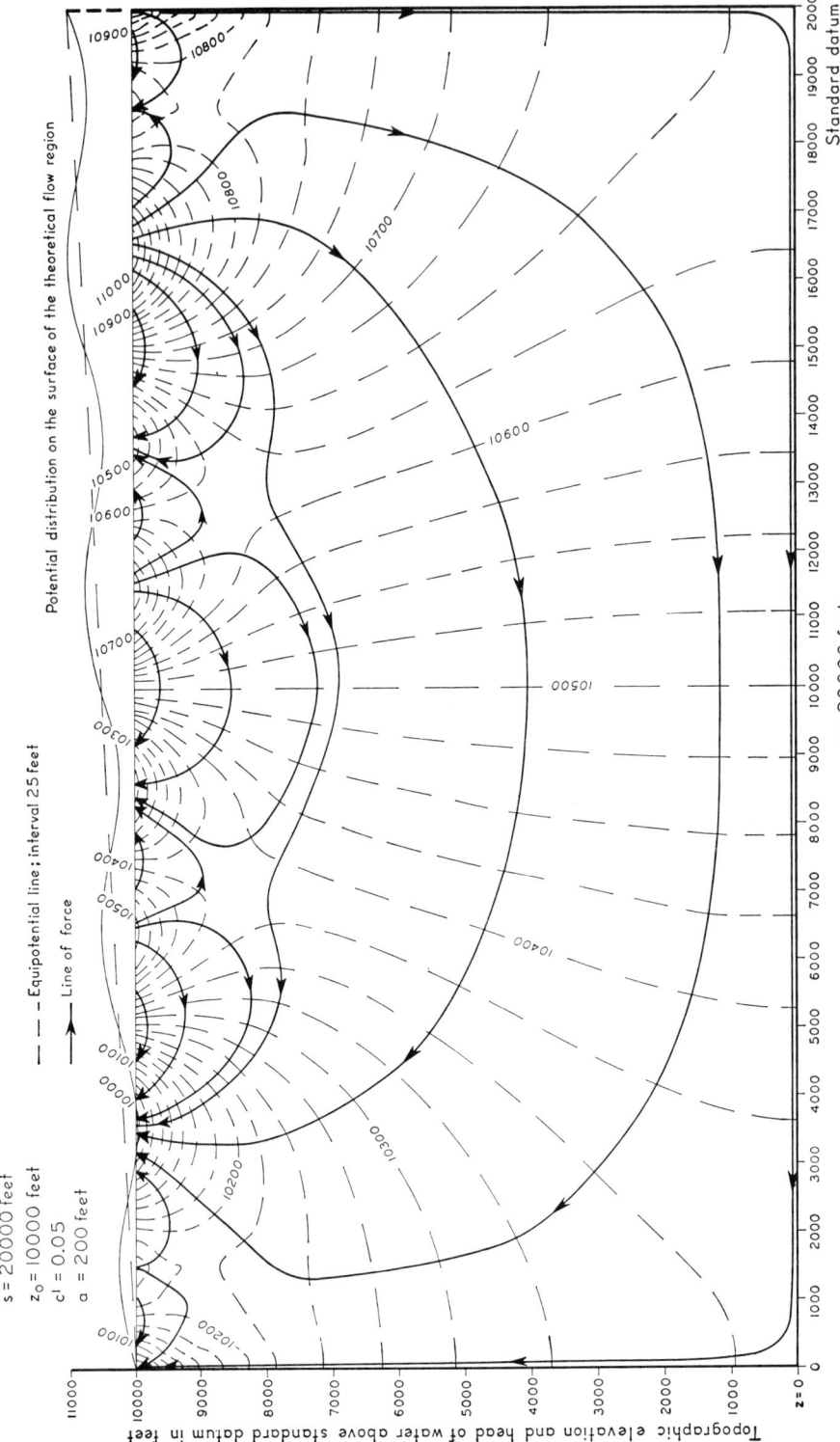

Fig. 2i. Potential distribution and flow pattern as obtained by equation 6.

cide with the flow lines.) Such a group of flow lines is said to form a flow system if it satisfies the following definition: *a flow system is a set of flow lines in which any two flow lines adjacent at one point of the flow region remain adjacent through the whole region; they can be intersected anywhere by an uninterrupted surface across which flow takes place in one direction only.*

Further investigation of the figures shows that three distinctly different types of flow systems can occupy a basin, namely, local, intermediate, and regional systems (Figure 3). A *local system* of groundwater flow has its recharge area at a topographic high and its discharge area at a topographic low that are located adjacent to each other. Local systems can be readily observed on each diagram of Figure 2. The major characteristic of an *intermediate system* of groundwater flow is that, although its recharge and discharge areas do not occupy the highest and lowest elevated places, respectively, in the basin, one or more topographic highs and lows may be located between them. Very-well-defined intermediate systems can be seen in Figures 2e, f, g, h, and i. The apparent lack of intermediate systems in those cases for which z_0 is 1000 feet does not mean that no such systems may exist in basins of relatively shallow depth. As soon as the real land surface departs from the regularity of the sine curve, the symmetrical flow pattern of Figure 2c, for instance, will be somewhat modified, and flow will occur between intermediate highs and lows also. A system of groundwater flow is considered to be *regional* if its recharge area occupies the water divide and its discharge area lies at the bottom of the basin. Regional systems can be observed in all the deep cases and in Figure 2c, where z_0 is 1000 feet.

Whereas theoretically the boundaries between different flow systems are very well defined, they do not signify an abrupt change of any of the physical properties of the flow. Relatively rapid changes of the chemical composition of the water across the boundaries, however, could be expected because of the different locations of the recharge and the different lengths of the flow paths of the different systems. In a small basin of moderate relief the amount of recharge water is directly proportional to the area of recharge. With this in mind it is obvious from Figure 2 that the greatest flow-line densities are found at shallow depths of the local systems. Except at places where local stagnant bodies of water occur, the density of the flow lines decreases rapidly with depth and with the transition from the local to the intermediate region and reaches its minimum in the regional system, provided the latter exists.

This interpretation of the theoretical results is very much in agreement with views expressed by *Norvatov and Popov* [1961, p. 21]. They recognize 'three well pronounced vertical zones of groundwater flow':

1. 'upper zone of active flow, whose geographical zonality coincides with climatic belts. The lower boundary of this zone coincides with the local base levels of rivers;
2. 'medium zone of delayed flow, subject to lesser climatic effect but also geographically zonal. The lower boundary of this zone is the base level of large rivers;
3. 'lower zone (of relatively stagnant water), geographically azonal and lying below the base level of large stream systems.'

Taking into account the extent of the recharge areas of the regional systems (which are small relative to those of the local systems), we see that flow in the regional system is influenced by climatic effects to a much lesser degree than flow in the upper zone. Climatic or geographical zonalities are, therefore, a straightforward consequence of the present theory.

In the next few paragraphs an analysis will be given of the effects of geomorphological factors on the flow of groundwater. These factors, or parameters in (6), are (*a*) the ratio n of the depth z_0 to the impermeable boundary to the half-width s of the basin (for convenience, in discussing the effect of n, only the depth to the impermeable boundary will be referred to, s being the same in all cases); (*b*) the average slope of the valley flanks; and (*c*) the local relief.

In analyzing the effect of z_0 on the flow of groundwater, a comparison of the diagrams of Figure 2 is helpful. Let those diagrams be considered for which all parameters but z_0 are equal, for instance, Figures 2d, e, and i. It appears that the spacing of the equipotential lines is *closer* in the shallow case than in the deeper ones. The flow lines are more arcuate as depth

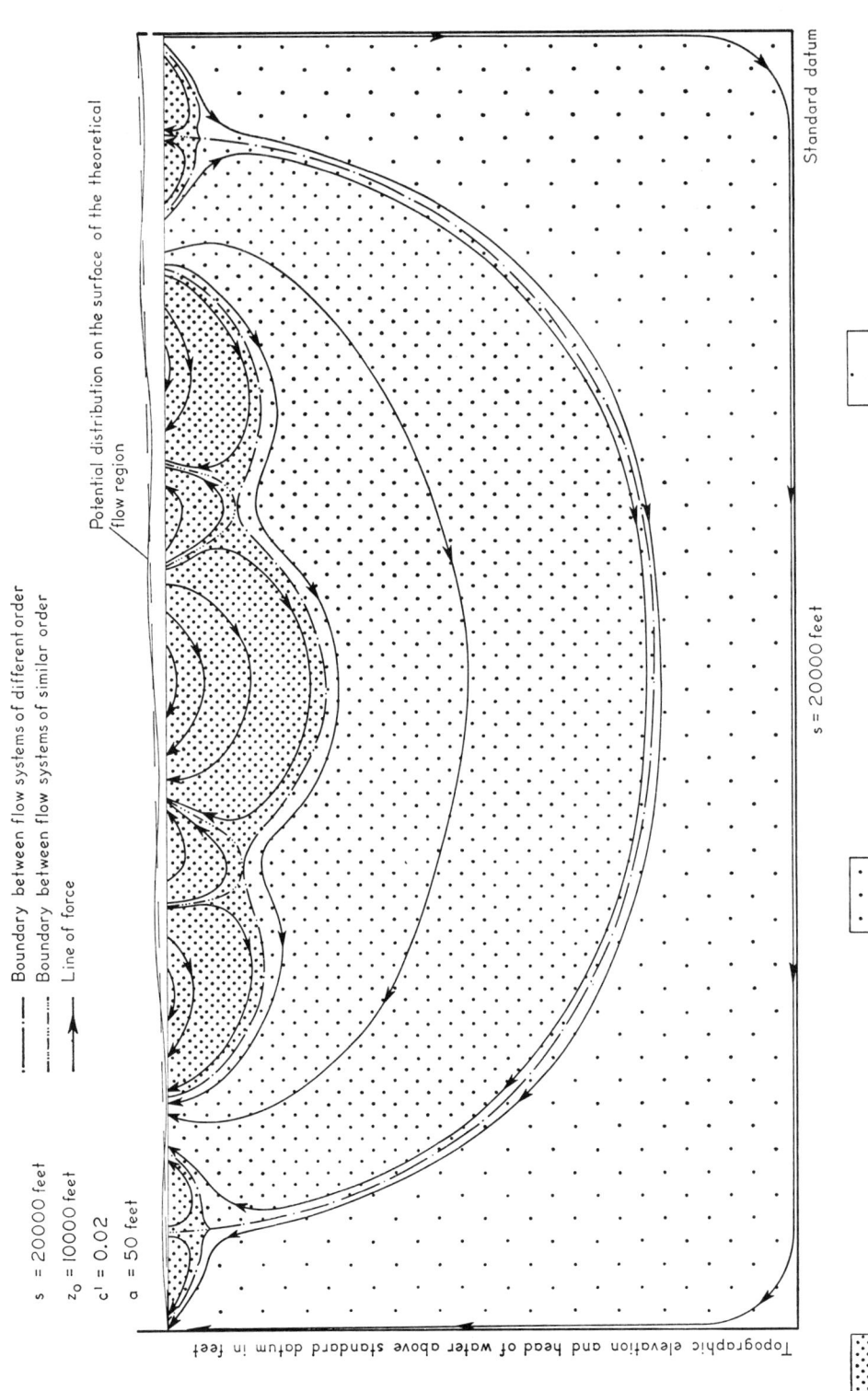

Fig. 3. Theoretical flow pattern and boundaries between different flow systems.

increases and there is room for intermediate and regional systems to form in those cases for which z_0 is 5000 or 10,000 feet. These features indicate a more evenly distributed flow, therefore a less intense motion if z_0 increases. A comparison of Figures 2c and 2h shows that a regional system is also possible in a relatively shallow case but that, with other parameters remaining the same, a much larger amount of water is transmitted through the regional system if z_0 is large. To summarize the effect of z_0 on the flow system it can be stated that, as the depth of the flow region increases, the water movement will slow down. A slow motion will probably result in higher mineralization of the groundwater.

A general increase in the slope of the valley flank will result in an increased lateral flow toward the bottom of the valley. The surficial areas of those systems whose flow lines are directed toward the center of the basin extend, and those of the other systems decrease. Local flow systems may degenerate to stagnant areas and may even vanish, thus allowing intermediate or regional systems to form. Figures 2a, 2c, 2f, and 2h illustrate this change well. Owing to the generally increased velocity of motion, a larger range of fluctuation of the piezometric surface is to be expected for the steeper slope of the valley flank.

Increasing topographic relief will tend to increase the depths and the intensities of the local flow systems. Whereas in Figure 2c a well-developed regional flow system is observed, in Figure 2d the entire basin is occupied by local systems. The depth ranges of the local system in Figure 2f are approximately 1000 feet shallower than those in Figure 2g. In the extreme case where the relief is negligible no local systems will form. However, since well-defined local flow systems are found even with the relatively low relief of approximately 100 feet per 2500 feet (a gradient of 0.04), neglecting their existence in theoretical or practical problems can hardly be justified.

Consequences of the theory. The major features of groundwater flow and flow systems derived from the theory presented here will be discussed in the following paragraphs.

1. Recalling the effects of the general slope and local relief on the flow, we see that under extended flat areas groundwater movement is retarded; neither regional nor local systems can develop. Groundwater may be discharged only by evapotranspiration; discharge of this type will possibly result in water-logged areas. If a relation between mineralization of the water and velocity of flow can be assumed, water in those areas will have high concentrations of soluble salts.

2. If the local relief is negligible and if there is a general slope only, a regional system will develop. Theoretically, if the line located on the surface midway between and parallel to the valley bottom and the water divide is called the 'midline,' the recharge and discharge areas of this regional system are located between the midline and divide, and between the midline and valley bottom, respectively [*Tóth*, 1962]. Because of the decreasing velocities, a gradual increase of dissolved mineral constituents with depth is to be expected. A good example of potential distribution and flow pattern in this situation has been produced by measurements in Long Island, New York, in connection with attempts to solve contamination problems. Figure 4 shows the results of the measurements [*Geraghty*, 1960, p. 38].

3. If the topography has a well-defined relief, local flow systems originate. The higher the relief, the deeper are the local systems. At the boundary between two adjacent systems the flow is subvertical and downward or upward in direction. Such a boundary is located under the highest and lowest elevated parts of local hills and depressions, respectively. Thus, whereas in the basin the underground drainage is not strictly symmetrical, imaginary impermeable boundaries may be thought to be located at local lows and highs, at least to the depths of the local systems. Figure 5 is a good example of an inferred local system [*Back*, 1960, p. 94].

4. The very pronounced effect of the relief on the formation of local systems suggests that no extended, unconfined regional systems of flow can extend across valleys of large rivers or highly elevated watersheds.

5. As a result of the local systems, alternating recharge and discharge areas are found across a valley. This means that the origins of waters obtained from closely located places may not even be related. Rapid change in chemical quality may thus be expected.

6. At points where three flow systems meet

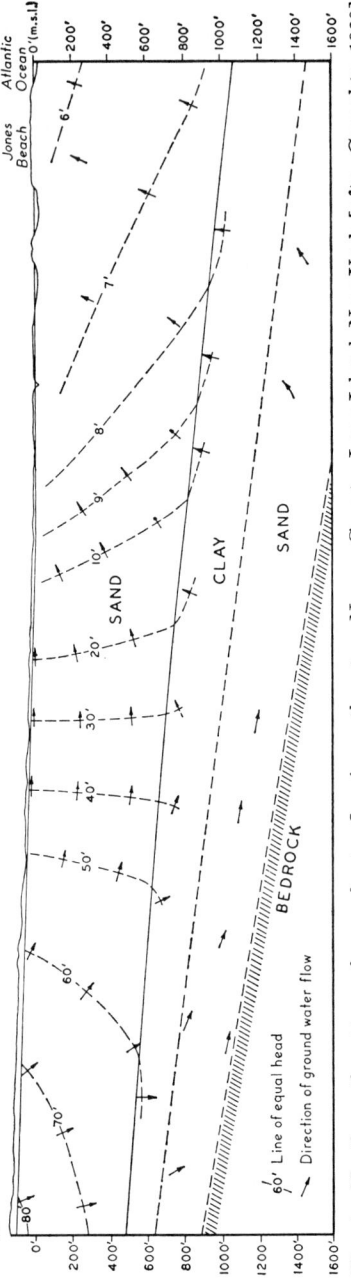

Fig. 4. North-south pattern of groundwater flow in southeastern Nassau County, Long Island, New York [after *Geraghty*, 1960].

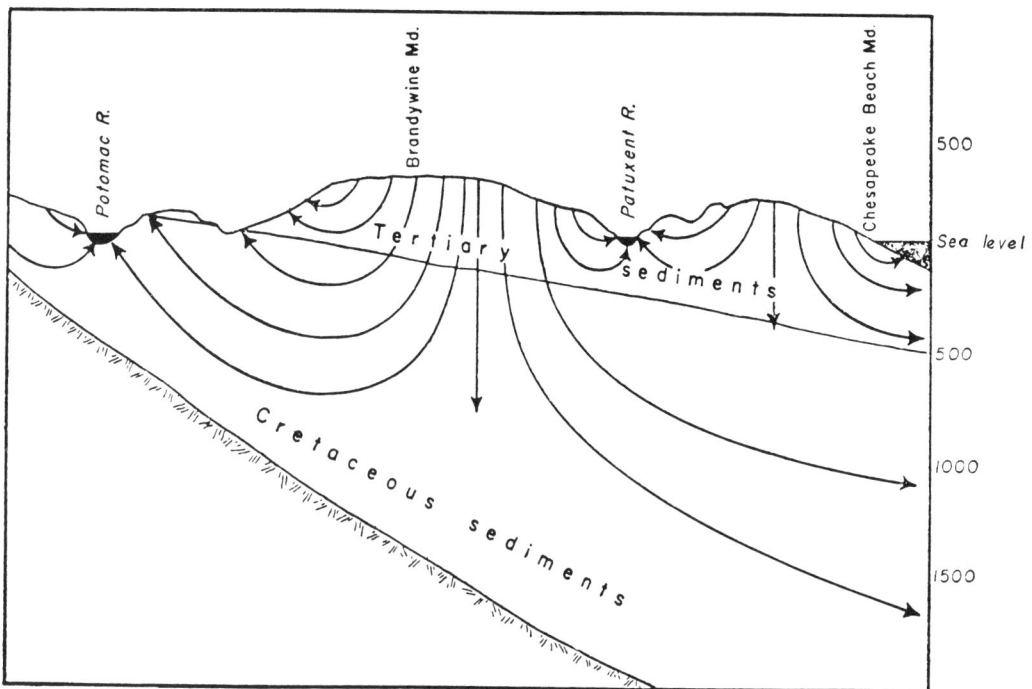

Fig. 5. Diagrammatic cross section through southern Maryland showing the lines of groundwater flow [after *Back*, 1960].

(Figure 2c), an area of stagnant water is formed. A high accumulation of mineral constituents is probable at these places. Below such a stagnant body of groundwater, flow occurs again and may result in a better quality of water than that from shallower depths.

7. Decreasing potential with depth in recharge areas and increasing potential in discharge areas are direct consequences of the theory and can be observed in all diagrams of Figure 2. It should be noted, however, that such a configuration of the equipotential lines is pronounced only in the immediate vicinities of the highs and lows. The midlines and their vicinities are locations of relatively straight, vertical equipotential lines.

8. From Figure 2 it can safely be stated that the major stream of the basin receives groundwater contributions only from the adjacent topographic highs and from possible regional flow. The latter is probably unimportant in most cases because of the low rate of flow. It is conceivable, then, that the methods in which baseflow data are used for computation of basinwide characteristics (average recharge, permeability, etc.) are misleading or erroneous. Even if the two flanks of a basin are of low relief, so that there are no local systems, the bulk of the basin discharge will take place between the midline and the valley bottom and only a small portion will appear as baseflow.

9. A further consequence of the theory is that the water levels at shallow depths are the most affected by seasonal recharge and discharge. The small intake and outlet areas of the intermediate and regional zones prevent the water levels from fluctuating widely. *Plotnikov and Bogomolov* [1958, p. 90] make a distinction between two zones on the basis of fluctuation of the water levels. They call the first 'zone of oscillations of underground water levels.' According to them the volume of water that occupies the zone of oscillation undergoes seasonal variations. This volume would control groundwater discharge and therefore they call it 'control reserves of underground waters.' Their second zone includes all the water that is below the zone of oscillation, both that in the deeper, still homogeneous parts of the basin and that in artesian aquifers; these are the 'secular resources.' It seems that the 'zone of oscillation' and the portion of the 'secular resources' that

is above the first impermeable boundary coincide very well with the 'local systems' and with the 'intermediate and regional systems' of the present paper.

10. Another result of the analysis is that only a small portion of the total amount of water occupying the basin participates in the hydrologic cycle. The deeper the basin, the smaller is this portion. This is easily conceived when one considers that the greatest part of the surface of the basin is occupied by the recharge and discharge areas of local systems which are usually shallow. But even when the local systems reach the horizontal impermeable boundary (Figures 2a, b, d), approximately 90 per cent of the total recharge water never penetrates deeper than 250 to 300 feet. A similar view is expressed by *Ubell* [1962, p. 96] who believes that about 80 to 90 per cent of the 'static supply does not participate in the natural hydrological cycle.' His experiments, on the basis of which this conclusion was drawn, indicated that 'below a certain depth in loose sedimentary rocks . . . water does not move in the voids until their state of stress is disturbed by boring.'

Summary. It is the writer's belief that in drainage basins, down to depths at which basin-wide extended layers of contrasting low permeability are found, groundwater motion may be treated as an unconfined flow through a homogeneous medium. On the basis of this principle a mathematical model of a small drainage basin (as defined in the paper) has been constructed. Potential distributions have been computed for basins of different geometrical parameters. These computations have led to a number of conclusions regarding features of the groundwater flow.

In the most general case, groundwater flow in a basin can be thought to be apportioned among three types of flow systems, the regional, intermediate, and local systems. The three systems, being the results of combinations of three particular solutions of the Laplace equation, can be superimposed on one another. If the local variations in topography are negligible the flow consists of the combination of only two particular solutions, and no local systems occur. This case has been found in nature by *Geraghty* [1960] and has been theoretically treated in detail elsewhere [*Tóth*, 1962].

The emphasis in the present paper has been on the general situation for which local topography plays a part in controlling groundwater motion. The distribution of the flow systems will, in turn, have its effect on the chemical quality of local occurrences of groundwater. The areally unrelated origin of local systems, associated with local topographic highs and lows, may result in abrupt changes in the chemical composition of relatively shallow groundwater. Vertical changes in quality may be the result of local stagnant bodies and of the vertical arrangement of different flow systems.

It is thought that (6) may be used for obtaining quantitative information about groundwater flow in an area, the surface of which can be approximated by a harmonic function. It is hoped also that the results of the above analysis will be useful in test programs planning well fields, solving pollution and tracer problems, making baseflow studies, and setting up water budgets.

Acknowledgments. I am greatly indebted to Messrs. R. Newton and P. Redberger, of the Petroleum Division, Research Council of Alberta, who adapted the flow equation described in the paper for solution by digital computer. Only the availability of the computer solutions made possible the derivation of the many flow diagrams.

REFERENCES

Back, W., Origin of hydrochemical facies of ground water in the Atlantic coastal plain, *Intern. Geol. Congr., 21st, Copenhagen, 1960, Rept. Session, Norden,* part 1, pp. 87–95, Copenhagen, 1960.

Geraghty, J. J., *Movement of Contaminants through Geologic Formations,* 90 pp., Technical Division Activities, National Water Well Association, Urbana, Ill., 1960.

Hubbert, M. K., The theory of groundwater motion, *J. Geol., 48*(8), part 1, 785–944, 1940.

King, F. H., Observations and experiments on the fluctuations in the level and rate of movement of ground-water on the Wisconsin Agricultural Experiment Station Farm, *U. S. Dept. Agr. Weather Bureau, Bull. no. 5,* 75 pp., 1892.

Meinzer, O. E., Outline of ground-water hydrology with definitions, *U. S. Geol. Surv. Water Supply Paper 494,* 71 pp., 1923.

Meneley, W. A., The occurrence and movement of groundwater in Alberta, in *Early Contributions to the Groundwater Hydrology of Alberta, Res. Council Alberta, Can., Bull. 12,* 123 pp., Edmonton, 1963.

Norvatov, A. M., and O. V. Popov, Laws of the formation of minimum stream flow, *Bull. Intern. Assoc. Sci. Hydrol.,* vol. 6, no. 1, pp. 20–27, Louvain, 1961.

Plotnikov, N. A., and G. B. Bogomolov, Classification of underground waters resources and their reflection on maps, *Intern. Assoc. Sci. Hydrol., General Assembly of Toronto*, vol. 2, no. 44, 525 pp., Gentbrugge, 1958.

Tóth, J., A theory of groundwater motion in small drainage basins in Central Alberta, Canada, *J. Geophys. Res.*, *67*(11), 4375–4387, 1962.

Ubell, K., A felszin alatti vízkészlet, *Hidrol. Kozl.*, *42*(2), 94–104, Budapest (Hungary), April 1962 (English summary).

Wieckowska, H., Zones geographiques des eaux phreatiques, *Intern. Assoc. Sci. Hydrol., General Assembly of Helsinki*, no. 52, 609 pp., Gentbrugge, 1960.

Wisler, Ch. O., and E. F. Brater, *Hydrology*, 2nd ed., 408 pp., John Wiley & Sons, New York, 1959.

(Manuscript received February 2, 1963; revised May 29, 1963.)

Theoretical Analysis of Regional Groundwater Flow.

2. Effect of Water-Table Configuration and Subsurface Permeability Variation

R. ALLAN FREEZE

P. A. WITHERSPOON

Abstract. Details of steady-state flow in regional groundwater basins can be investigated using digital computer solutions of appropriately designed mathematical models. The factors that must be considered are: (1) ratio of depth to lateral extent of the basin; (2) water-table configuration; and (3) stratigraphy and resulting subsurface variations in permeability. The results of this study provide a theoretical basis for the following properties of regional flow systems: (1) groundwater discharge will tend to be concentrated in major valleys; (2) recharge areas are invariably larger than discharge areas; (3) in hummocky terrain, numerous sub-basins are superposed on the regional system; (4) buried aquifers tend to concentrate flow toward the principal discharge area, have a limiting effect on sub-basins, and need not outcrop to produce artesian flow conditions; (5) stratigraphic discontinuities can lead to distributions of recharge and discharge areas that are difficult to anticipate and that are largely independent of the water-table configuration. (Key words: Groundwater; computers, digital; drainage basin characteristics)

INTRODUCTION

Regional groundwater flow in a nonhomogeneous, anisotropic system can be investigated by means of mathematical models. In Part 1 of this study [*Freeze and Witherspoon*, 1966], steady-state solutions have been obtained by analytical and numerical methods. The numerical methods, which involve computer solutions to finite-difference equations, proved to be superior to the analytical methods and were recommended for both two-dimensional and three-dimensional models of groundwater basins.

In this study, potential nets obtained from two-dimensional hypothetical models are examined qualitatively to show the effect of variations in the controlling parameters on the regional groundwater flow system. We have chosen only a limited number of examples from a much more comprehensive study [*Freeze*, 1966] to illustrate some important points. Much more of this kind of work can be done, but we believe that these examples will demonstrate the usefulness of the digital computer approach to investigations of regional groundwater flow.

In Part 3 of this study, the quantitative significance of the mathematical model approach will be discussed with field examples.

It is first necessary to define some of the terms that appear throughout this study. Natural groundwater *recharge* refers to water that percolates down through the unsaturated zone to the water-table and actually enters the dynamic groundwater flow system. This definition excludes that portion of the moisture surplus that enters the ground and increases the soil moisture content but does not enter the flow pattern itself. The term is not to be confused with the actual areal precipitation which, in some cases, may lead to groundwater recharge and in other cases may not. Natural groundwater *discharge* is water that is discharged from the dynamic groundwater flow system by means of stream baseflow, springs, seepage areas, and evapotranspiration. We have not considered the effects of artificial discharge (wells and well fields) in this study.

A *discharge area* is an area where the direction of groundwater flow is toward the water-table. A *recharge area* is an area where the direction of groundwater flow is away from the water-table. It is convenient to introduce the concept of a *hinge-line*, which is a line on the surface of the water-table that separates a discharge area from a recharge area. Its projection on a two-dimensional vertical section will be a point on the water-table.

A *groundwater basin* is a three-dimensional closed system that contains the entire flow paths followed by all the water recharging the basin. The flow pattern within a given basin may be simple, involving only one recharge area and one discharge area, or complex, involving many. A two-dimensional section through a groundwater basin is representative of the three-dimensional basin if it is taken parallel to the direction of dip of the water-table slope.

In this study the *water-table* is considered to be an imaginary surface beneath ground level at which the absolute pressure is atmospheric. It is assumed to be the upper boundary of the saturated flow system [*Freeze and Witherspoon*, 1966].

The potential function used throughout this study is the hydraulic head φ, which is defined by

$$\varphi = \phi/g$$

where

ϕ = hydraulic potential [*Hubbert*, 1940]
g = acceleration due to gravity

At any point in the potential field, φ equals the elevation above a standard datum of the liquid level in a piezometer inserted at that point.

The manner in which any flow system is defined by a potential field is governed by the interrelations between three governing factors: (1) shape of the region in which the potential field is defined; (2) existing boundary conditions; and (3) nature of the inhomogeneities in the flow properties of the region.

In our case, the potential field is a region represented by a two-dimensional, vertical cross section through a groundwater basin. The region is roughly rectangular with vertical impermeable sides, a horizontal impermeable base, and an irregular upper boundary (the water-table). There are two ways by which we can control the shape of this region: (1) by changes in the water-table configuration, which result in small but important changes in shape, or (2) by changes in the ratio of depth to lateral extent of the basin. By varying this ratio, we can examine all cases from that of a deep basin of limited lateral extent to the more usual case of a shallow groundwater basin of broad extent.

At the external boundaries, the controlling conditions are explicitly defined by the mathematical model; only the water-table configuration can be altered. Within the region, the property of the medium that affects the nature of the potential field is, of course, the permeability. We must, therefore, examine the effect of inhomogeneity and anisotropy of permeability on groundwater flow patterns.

Tóth [1962, 1963] has used an analytical solution to arrive at results for the homogeneous case that show the effect of varying the 'depth : lateral extent' ratio. For this study, we have therefore chosen a fixed basin size and attempted to isolate the effects of water-table configuration and subsurface permeability variations.

Each of the diagrams in Figures 1 through 6 depicts the potential pattern in a vertical cross section through a hypothetical groundwater basin. In some cases, random flowlines are included to clarify the direction of flow. The diagrams are true scale and dimensionless, i.e., all dimensions are given in terms of s, the total length of the basin. The diagrams could thus represent flow patterns for systems covering only a few acres or for those extending over many hundreds of square miles.

For this discussion, we have chosen a basin with a 'depth : lateral extent' ratio of 1:12. Depth was arbitrarily taken as measured at the shallowest point, i.e., at the left-hand edge of each diagram. The total relief on the water-table surface was arbitrarily set at ⅛ the basin depth for all water-table configurations. The machine-plotted equipotential lines were all constructed using a constant $\Delta\varphi$, which was 1/50 of the total available head.

Permeability contrasts are noted on most

diagrams. Where a single value is shown within a given formation, the formation is isotropic. In anisotropic cases, both horizontal and vertical permeabilities are indicated. Where no permeability is given, the formation is homogeneous and isotropic.

These permeabilities may also be considered dimensionless, because it is the permeability ratio that controls the nature of the potential field. For example, in Figure 2A the same potential net would result from permeabilities of 10 and 100 as exists for 1 and 10. The quantity of flow through the basin would, of course, be different.

In some diagrams the closeness of the equipotential lines was such that only every other or every third line has been reproduced to avoid a maze of lines. In these instances, the position of the missing lines is indicated by tick marks, and the resulting gaps in the potential fields must not be interpreted as sudden changes in gradient.

EFFECT OF WATER-TABLE CONFIGURATION

The investigation of the effect of the water-table configuration on regional groundwater flow patterns was begun by Tóth [1962, 1963]. He considered two cases: a constant gentle regional slope such as one would expect to find in the flat prairie, and a water-table with the configuration of a sine curve, as one might expect in hummocky terrain. With the increased versatility of the methods introduced in Part 1 of this study, we can now investigate water-table configurations of a more irregular nature, and ones more representative of actual field conditions.

In Figure 1, the effect of three different water-table configurations on the flow through a homogeneous isotropic medium is shown. The following general comments can be made:

1. In the recharge areas, the equipotential lines meet the water-table obliquely with the acute angle on the upslope side. In the discharge areas, the acute angle is on the downslope side. At the hinge-line, the equipotential meets the water-table at right angles.

2. The existence of a high in the water-table configuration, whether it be a major regional high or a minor reversal in slope, results in recharge at that point and for some distance

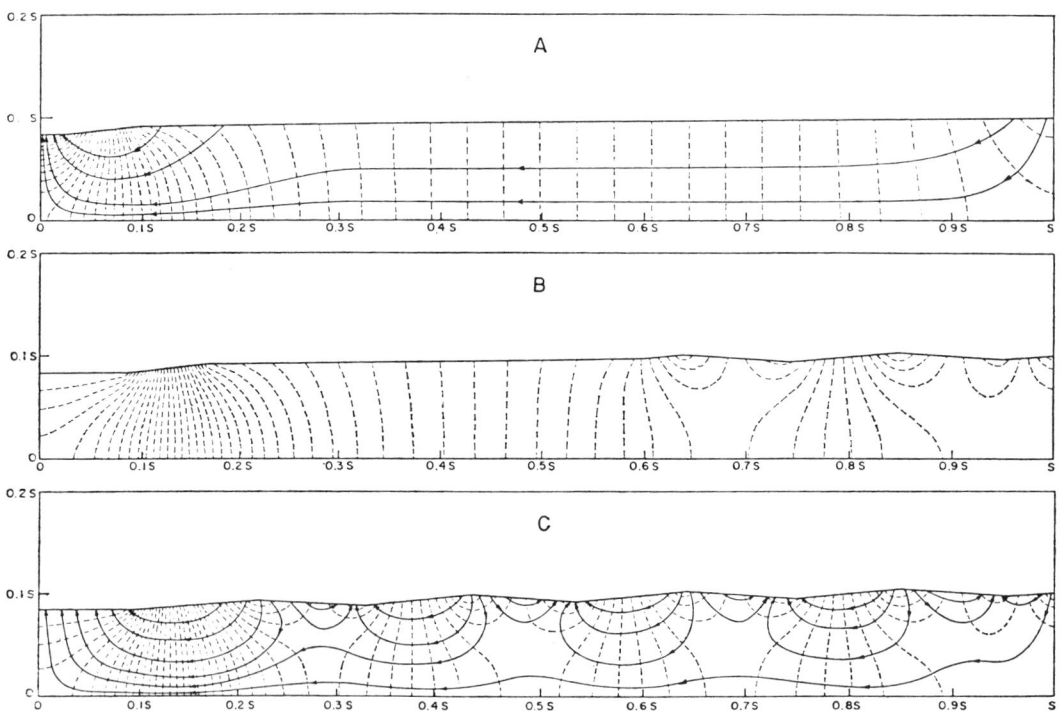

Fig. 1. Effect of water-table configuration on regional groundwater flow through homogeneous isotropic mediums.

on either side. The existence of a low results in discharge at that point and for some distance on either side.

3. If the spacing between equipotential contours is x feet, then an equipotential line must meet the water-table at every point along its length that represents an increase in elevation of x feet. Steep water-table slopes therefore result in many equipotential lines and high gradients near the water-table (and indeed to some depth). Shallow slopes are conducive to low gradients. Where the water-table is flat, it represents an equipotential surface, and vertical gradients result.

4. *Tóth* [1962] showed that a gentle constant regional water-table slope over a homogeneous medium results in flow which is essentially horizontal. Recharge is concentrated at the upstream end of the recharge area; discharge, at the downstream end of the discharge area. The hinge-line is at the midpoint. By contrast, when there is a major valley present (Figure 1A), the hinge-line occurs midway up the steep valley flank. Discharge is concentrated in the valley. Recharge occurs over the entire upland area but is concentrated in two locations: at the upstream end of the recharge area and at the break in slope above the steep valley flank.

5. The existence of a hummocky water-table configuration (Figure 1C) results in numerous sub-basins within the major groundwater basin. Water that enters the flow system in a given recharge area may be discharged in the nearest topographic low or may be transmitted to the regional discharge area in the bottom of the major valley. In essence, Figure 1C shows the effect of the addition of a major valley to *Tóth*'s [1963] sine curve configuration (although the hummocks in Figure 1C are not sine curves but straight line segments). Figure 1B shows a composite situation involving a major valley, an area of constant slope, and a hummocky water-table in the upstream end of the region.

It should be noted that the second comment above holds only for two-dimensional sections taken parallel to the direction of dip of the water-table slope. Section EE', Figure 9, in Part 1 of this study [*Freeze and Witherspoon*, 1966] provides a counter example of a random two-dimensional section through a three-dimensional basin in which a topographic low (near the intersection of CC') acts as a recharge area. Section EE' crosses this valley in its upstream portion, where longitudinal components of flow parallel to the direction of the valley are more important than lateral flow components parallel to the section.

EFFECT OF SUBSURFACE PERMEABILITY VARIATION

Figures 2 through 6 consist of eighteen potential nets designed to display the diversity of regional flow patterns that can arise from the consideration of a variable stratigraphy and its resulting subsurface permeability variations. The principles are best shown using the simplest realistic water-table configuration, that of Figure 1A. Thirteen of the diagrams utilize this configuration, whereas the other five show the effect of stratigraphy on flow in hummocky basins.

Layered Cases

Comparison of Figure 2A with Figure 1A shows the effect of the introduction of a layer with a permeability 10 times that of the overlying beds. The lower formation is, in effect, an aquifer with essentially horizontal flow that is being recharged from above. As a consequence, the vertical component of flow in the upper layer is much more pronounced than it was in the homogeneous case (Figure 1A). One should also note the downstream increase in gradient within the aquifer of Figure 2A, caused by an increasing number of flowlines that enter the aquifer from the upper layer. Discharge is concentrated in the valley bottom; the entire constant regional slope is a recharge area.

These effects can be further altered by considering a higher permeability contrast. Figures 2B and 2C show the effect of increasing the permeability of the basal aquifer. As the permeability ratio increases, the following changes can be noted:

1. The vertical upward or downward flow through the overlying low-permeability layer becomes more pronounced. For example, in the upstream recharge areas at the right-hand side of the diagrams, the flow becomes more vertical, the vertical flow exists over a larger area, and the vertical gradient increases.

2. The horizontal gradient in the aquifer de-

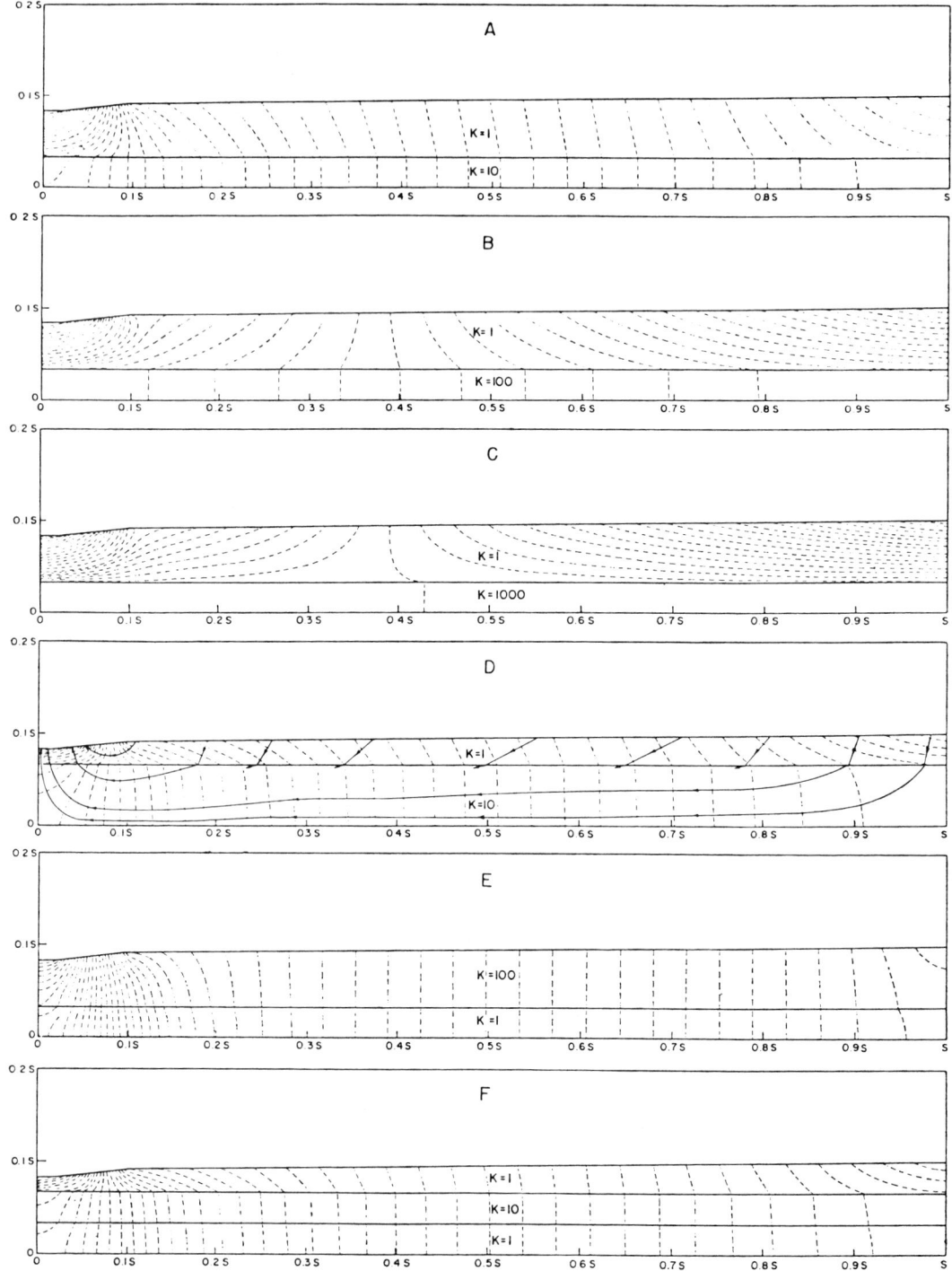

Fig. 2. Regional groundwater flow through layered mediums with a simple water-table configuration.

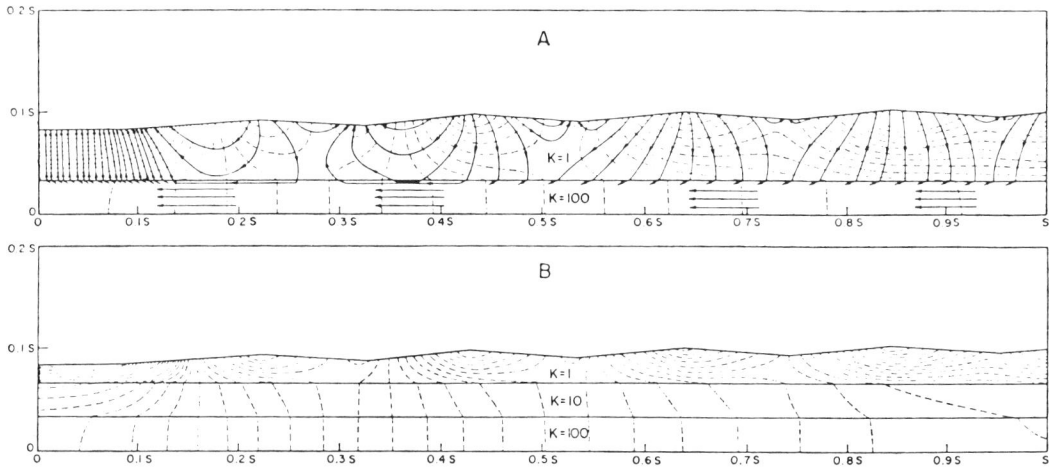

Fig. 3. Regional groundwater flow through layered mediums with a hummocky water-table configuration.

creases, but the quantity of flow (which can be calculated using Darcy's Law) increases.

3. The hinge-line moves upslope, creating larger discharge areas. This is a result of the increased quantity of water flowing through the basal aquifer, which must escape as the influence of the left-hand vertical impermeable boundary is felt. The magnitude of this effect may not be entirely realistic, as it is possible that, for permeability ratios of 1000:1 (Figure 2C), the 'major valley' at the left of the diagram would not create an imaginary impermeable boundary. Horizontal flow through the aquifer might proceed to the left until a more pronounced topographic influence was encountered.

The thickness of the basal aquifer has little effect on the nature of the flow pattern, as shown by a comparison of Figures 2D and 2A. The quantity of water flowing through the system represented by Figure 2A would, of course, be about half that flowing through the system of Figure 2D.

Comparison of Figures 2E and 1A shows that the flow pattern resulting from a two-layer case, when the upper layer has the larger permeability, is almost identical to that of the homogeneous case. The quantity of flow is, of course, considerably different in each case. The fact that a geologic configuration exists that results in a potential pattern identical to that of the homogeneous case emphasizes the fact that good permeability data are necessary before results of piezometer installations can be interpreted successfully in terms of quantitative estimates of regional flow.

Figure 2F shows a three-layer case with the aquifer in the middle of the section. Again we see that a low permeability layer beneath the aquifer has little or no effect on the potential pattern (compare Figures 2E and 2F).

Figure 3 shows the effect of stratigraphic variations on regional groundwater flow when the water-table has a hummocky surface. As shown in Figure 3A, the effect of the basal aquifer is to intensify the downward flow through the upper layer. The aquifer provides a highway for flow that passes under the upper layer and restricts the sub-basins in the hummocky region considerably. Figure 3A should be compared with Figure 1C. In effect, the relative importance of the discharge area in the major valley has been increased many fold owing to the presence of the buried aquifer.

It should be noted that the conditions of artesian flow inferred by the term 'confined aquifer' are met by several of the potential nets presented in Figures 2 and 3. For example, if a piezometer were sunk into the basal aquifer in Figure 2C, the static water level would lie considerably above the top of the aquifer. In the vicinity of the major valley, flowing artesian wells would occur.

Partial Layers and Lenses

The effect of lenticular bodies of high permeability and the particular importance of their position in the basin are shown in Figure 4.

The presence of a partial basal aquifer in the upstream half of the basin (Figure 4A) results in a discharge area that occurs in the middle of the constant regional slope. The occurrence of such a discharge area under strictly topographic control would, of course, be impossible. The majority of the flow that has entered the system in the upper half of the basin is discharged at this point. What was originally a single basin in the homogeneous case has become two basins under the influence of the partial aquifer.

When the partial basal aquifer occurs in the downstream half of the basin (Figure 4B), the central discharge area does not exist and, indeed, recharge in the region over the aquifer is concentrated. The zone of most intensive recharge is thus shifted from the upstream portion of the basin in the homogeneous case to the downstream portion because of the presence of the partial aquifer.

Figure 4C shows the aquifer as a stratigraphic lens in the regional basin. In this case, there is recharge over the upstream end of the lens and discharge over the downstream end. There is horizontal flow through the lens as well as in the low permeability layer beneath. Figure 4D shows the effect of a lens on the potential field in a hummocky basin.

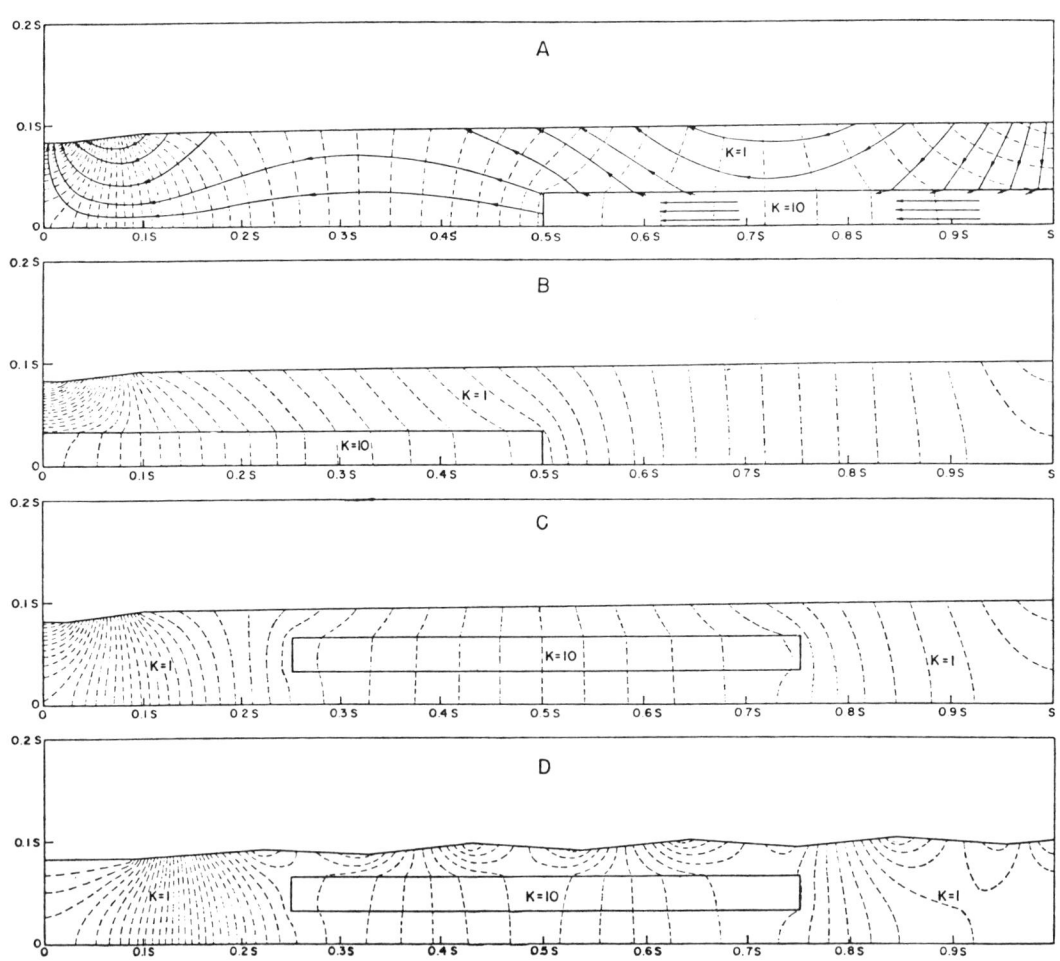

Fig. 4. Regional groundwater flow through partial layers and lenses.

Sloping Stratigraphy

Figure 5 presents three examples of the effect of sloping stratigraphy. In Figure 5A there is an increased recharge where the $K = 10$ layer outcrops, which, of course, is to be expected. The point where the upper boundary of this layer meets the water-table is a major hinge-line with a large discharge area downstream. Above the break in slope, a second recharge area is evident, and as usual, the major discharge is concentrated in the valley.

An interesting pair of flowlines to consider is shown in Figure 5B. Here, the difference of a few feet in the point of recharge will make the difference between the water entering a minor sub-basin or the major regional system of groundwater flow. In effect, what was thought to be a single basin when we set up this model is, in reality, two basins. If the sloping bed is reversed as in Figure 5C, a considerably different potential field is developed.

Anisotropic Formations

When anistropy exists, the problem of analyzing regional groundwater flow becomes more complex. The digital computer approach is well suited to the analysis of such problems, and Figure 6 shows three simple cases to illustrate the method.

Figures 6A and 6B show the effect of anisotropy on the regional groundwater flow pattern in a homogeneous medium bounded by the same simple water-table slope of Figure 1A. In Figure 6A the horizontal permeability is 10 times the vertical. For illustrative purposes, this situation is reversed in Figure 6B. The effect of these permeability configurations is best realized by comparing the flowlines of these figures with those of Figure 1A for the isotropic case. Figure 6C is a two-layer case that combines the anisotropy conditions of Figures 6A and 6B in one system. This system could be representative of a vertically fractured formation overlying a horizontally stratified layer.

Considerable care must be exercised in the construction of flowlines in anisotropic mediums, since the flowlines will not in general intersect the equipotentials at right angles. Two methods are available. The first [*Maasland*, 1957] utilizes the transformed section, whereby

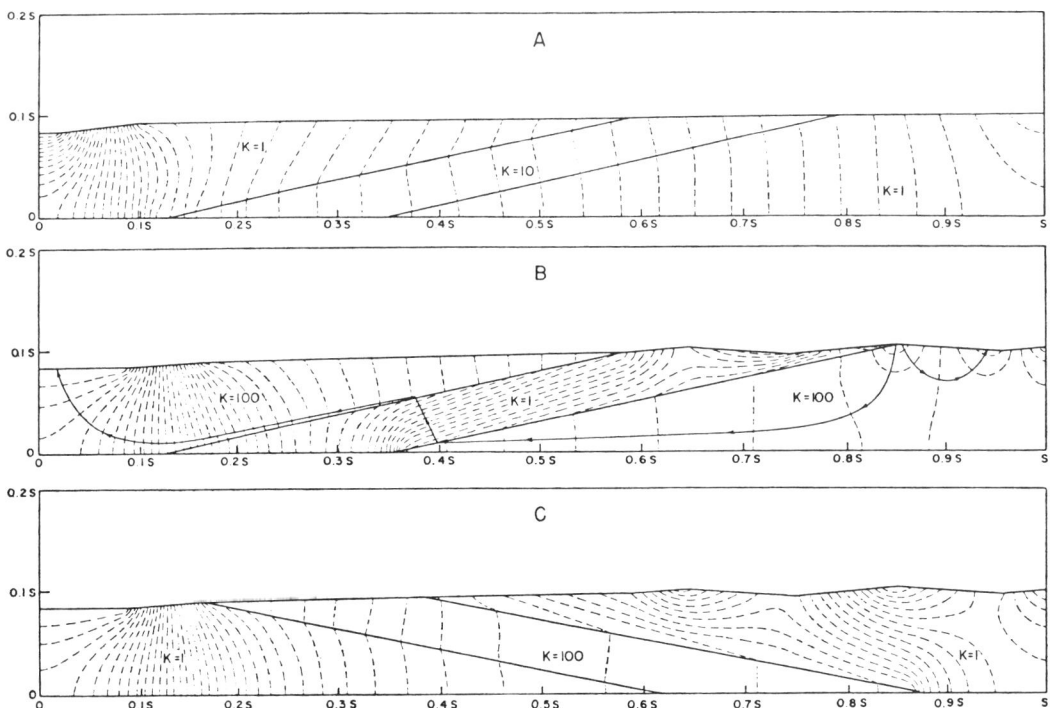

Fig. 5. Regional groundwater flow in regions of sloping stratigraphy.

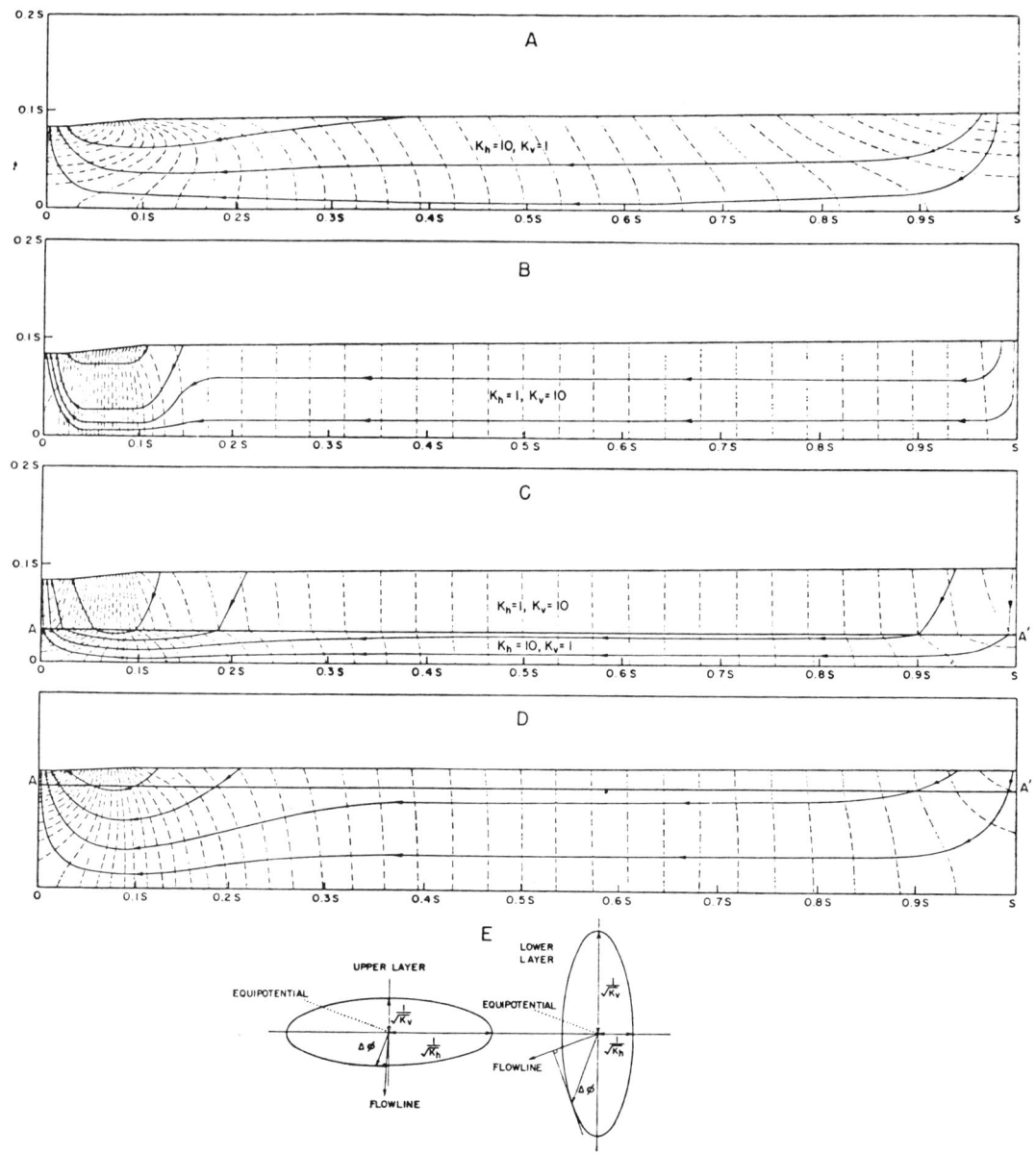

Fig. 6. Effect of anisotropy on regional groundwater flow.

an equivalent homogeneous isotropic system is obtained by suitably expanding or shrinking the coordinates of each point in the anisotropic medium. The transformation is

$$x' = (K_0/K_h)^{1/2} x$$
$$y' = (K_0/K_v)^{1/2} y$$

where x and y are the original coordinates; x' and y' the transformed coordinates; K_h and K_v the horizontal and vertical permeabilities; and K_0 is an arbitrary constant having the dimensions of K_h and K_v.

Figure 6D shows the transformed section for the two-layer case of Figure 6C. In the upper layer, we choose $K_0 = 1$; then $x' = x$, $y' = y/\sqrt{10}$, and the vertical dimension is reduced by a factor of $\sqrt{10}$. In the lower layer, we choose $K_0 = 10$; then $x' = x$, $y' = \sqrt{10}\, y$, and the vertical dimension is expanded by a factor of $\sqrt{10}$. The position of the interlayer boundary AA' is shown on both diagrams.

Once the equipotentials have been transferred from the real (Figure 6C) to the transformed (Figure 6D) section, a homogeneous isotropic flownet can be drawn in the transformed section and the flowlines transferred back to the true case. Several random flowlines are shown to illustrate the direction of flow.

The method of the transformed section indicates that the effect of varying the anisotropic ratio in a homogeneous medium is identical to that of varying the 'depth : lateral extent' ratio. The diagrams presented by *Tóth* [1963], which were designed to show the effect of the 'depth : lateral extent' ratio in a homogeneous basin bounded by a hummocky water-table, could therefore also be interpreted in terms of the effects of anisotropy.

A second method has recently been described [*Liakopoulos*, 1965] whereby the direction of flow at any point in an anisotropic medium can be determined with the aid of the permeability ellipse and without the necessity of a transformed section. Figure 6E shows the permeability ellipses for both upper and lower layers of Figure 6C. The direction of flow at any point can be obtained graphically as follows:

1. Draw a vector in the direction of the hydraulic gradient (i.e., perpendicular to the equipotential at the point in question).

2. Draw a tangent to the ellipse at the point where the vector cuts the ellipse.

3. The direction of flow is perpendicular to the tangent line.

In the constructions shown on the two ellipses in Figure 6E, the direction of the hydraulic gradient is the same in each case. The resulting direction of flow, however, is radically different and is dependent on the prevailing direction of anisotropy.

It should be noted that all the cases treated in Figure 6 have axes of anisotropy that coincide with the coordinate directions. The more general case of a skewed anisotropy requires a more complicated mathematical approach utilizing the concept of permeability in tensor form. Numerical solutions employing the finite element method [*Zienkiewicz et al.*, 1966] appear to be well suited to this problem.

DISCUSSION OF RESULTS

The water-table configuration and the variations in subsurface permeability have been identified throughout this study as the broad governing factors that control regional groundwater flow within a basin of given size. Both these properties of the basin can exist in an infinite variety, but certain generalizations are evident on the basis of the potential patterns that have been developed in this study.

Distribution of Discharge and Recharge Areas

Areas of groundwater discharge occur under the influence of at least six distinguishable cases of water-table configuration and geologic setting.

1. The existence of a valley of sufficient magnitude to create an imaginary vertical impermeable boundary that extends to the full depth of the basin will cause concentrated groundwater discharge into the valley.

2. Minor topographic lows will also cause discharge areas. The sub-basins may be sufficient to capture the flow from the entire depth of the basin [*Tóth*, 1963] or may be restricted to the upper layers in a stratified system (Figure 3A).

3. A break in slope even though both slopes are positive may be sufficient to cause small quantities of groundwater discharge just below the steep components of slope. This phenomenon is illustrated in Figure 1B at the base of the first hummock near the middle of the diagram.

4. Discharge areas that are entirely the result of geologic control can occur at the surface above the pinchout of a buried high-permeability aquifer. The extent of the discharge area and the intensity of discharge depend on: (a) the position of the partial aquifer within the basin; and (b) the permeability contrast between the aquifer and the surrounding medium.

5. A discharge area can be created below the outcrop of a downstream sloping aquifer (Figure 5A).

6. A discharge area can occur at the outcrop of an upstream sloping aquifer (Figure 5C).

As can be seen in the various illustrations, areas of recharge are widespread. However, concentrations of recharge can be expected in the following situations.

1. Near the topographic divide at the upstream end of a homogeneous isotropic basin with a simple water-table configuration (Figure 1A).
2. In the upstream half of basins with continuous layered stratigraphy (Figures 2B and 2C).
3. On and just above steep valley slopes (Figure 1).
4. On water-table highs in regions with a hummocky water-table configuration (Figures 1C, 3A, and 3B).
5. In the area overlying the upstream portion of a partial aquifer (Figure 4).
6. In the outcrop area of a downstream sloping aquifer (Figure 5A).

It is clear that the distribution of recharge and discharge areas is affected by the presence of anisotropy. From the limited number of examples that we have examined, few generalizations can be made. However, two observations are worthy of note: (1) it appears that where the horizontal permeability is significantly greater than the vertical, the quantities of groundwater entering the system are more evenly distributed over the recharge area; and (2) Figure 6B brings out the paradoxical fact that high vertical permeabilities do not necessarily lead to large rates of vertical recharge.

In general, discharge areas are smaller than recharge areas. In the two-dimensional hypothetical models of this study, the discharge area occupies between 7% and 40% of the total length of the basin. Results of studies with three-dimensional models show that the percentage in actual groundwater basins is near the lower end of this range.

Depth and Lateral Extent of Groundwater Basins

It has been shown that hummocky water-table configurations are conducive to the establishment of small sub-basins within the major basin. Under these circumstances, the concept of a total basin yield is negated, and each component basin must be considered separately. It is logical, therefore, to examine the factors that control the depth and lateral extent of these sub-basins.

In the homogeneous case (Figure 1C) the majority of flow takes place near the surface in small sub-basins, but a certain amount of flow by-passes these near surface systems to enter the major flow system. At least one flow path traverses the entire basin. *Tóth* [1963] has shown that the influence of the hummocks increases as: (1) the amplitude of the hummocks increases, and (2) the 'depth : lateral extent' ratio decreases. The groundwater basin may even be broken up into a series of small sub-basins with no flow traversing the entire basin.

The effect of introducing a high-permeability aquifer into the system is to create a highway for groundwater flow such that the percentage flow traversing the entire basin increases. The percentage of total flow that enters the basin-wide flow system depends on three parameters:

1. The permeability ratio between the aquifer and other formations;
2. The 'depth : lateral extent' ratio of the basin;
3. The percentage of the total depth taken up by the high-permeability layer.

There is very little generalization possible regarding the effect of irregular geologic configurations. It is safe to say, however, that the introduction of discontinuities (partial layers, lenses, and sloping beds) will result in the formation of small sub-basins that did not exist in the homogeneous case.

Effective Depth to a Basal Impermeable Boundary

One of the basic assumptions of this study [*Freeze and Witherspoon*, 1966] requires the presence of a horizontal impermeable boundary at some depth. Another assumption suggests that there is no such thing as a completely impermeable formation. The resolution of this seeming paradox lies in the fact that there are certain geologic configurations that have the same effect on the potential pattern as an impermeable boundary.

Figure 2A shows a two-layer case with a simple water-table configuration. The equipotential lines cross the $K = 10$ layer vertically and meet the assumed basal impermeable boundary perpendicularly, as they must. In Figure 2F, another low permeability layer has been added beneath the aquifer. The effect on the flow pattern is negligible. The equipotential lines still cross the aquifer vertically and meet its lower boundary perpendicularly. They are not refracted (except at the extremities of the flow net) and are vertical over most of the $K = 1$ layer down to the base of the model. Thus, the lower boundary of the $K = 10$ layer has the same effect as an impermeable boundary.

It is concluded that when one is designing models of groundwater basins, there is some depth in regions of reasonably horizontal sedimentation below which the equipotential lines will remain vertical. On the basis of our hypothetical two-dimensional models, this depth appears to be the lower boundary of the deepest aquifer whose permeability significantly exceeds that of underlying beds. To be on the safe side, preliminary studies should probably begin with a greater basin depth than would seem necessary. If one finds vertical equipotentials as suggested by the above, the basin depth can be limited accordingly.

The twenty-one potential nets presented in this paper have been chosen from a much larger number that appear in the original study [*Freeze*, 1966]. Such idealized hypothetical models provide an excellent insight into the nature of groundwater flow, but it must be recognized that it would take a much larger number of models to cover even a fraction of all of the possible combinations of water-table configurations and subsurface permeability variations. In the ultimate practical use of our method, a mathematical model must be designed for each individual groundwater basin under investigation.

CONCLUSIONS

1. The factors that affect steady-state regional groundwater flow patterns within a nonhomogeneous, anisotropic basin are: (a) 'depth : lateral extent' ratio; (b) water-table configuration; and (c) the stratigraphy and resulting subsurface variations in permeability.

2. It is possible to study the influence of these factors by utilizing digital computer solutions to appropriately designed mathematical models.

3. The presence of a major valley will tend to concentrate discharge in the valley. Where the regional water-table slope is uniform, the entire upland area is a recharge area. In hummocky terrain, numerous sub-basins will be superposed on the regional system.

4. The presence of a buried aquifer of significant permeability will have a profound effect on regional groundwater flow. It acts as a highway that transmits water to the principal discharge area and affects the magnitude and position of the recharge areas. It can also have a significant effect on the extent and flow capacity of sub-systems within the regional basin. Buried aquifers need not outcrop to produce artesian flow conditions.

5. Certain nonhomogeneous basins may have potential fields that are indistinguishable from those of the homogeneous case.

6. Stratigraphic pinchouts at depth can create recharge or discharge areas where they would not be anticipated on the basis of the water-table configuration.

7. There is some depth in regions of reasonably horizontal sedimentation below which equipotential lines remain vertical. This condition allows the effective depth to the basal impermeable boundary to be selected when designing a mathematical model.

REFERENCES

Freeze, R. A., *Theoretical analysis of regional groundwater flow*, Ph.D. Thesis, University of California, Berkeley, 1966.

Freeze, R. A., and P. A. Witherspoon, Theoretical analysis of regional groundwater flow, 1. Analytical and numerical solutions to the mathematical model, *Water Resources Res.*, 641–656, 1966.

Hubbert, M. K., Theory of groundwater motion, *J. Geol.*, *48*, 785–944, 1940.

Liakopoulos, A. C., Variation of the permeability tensor ellipsoid in homogeneous anisotropic soils, *Water Resources Res.*, *1*, 135–141, 1965.

Maasland, M., Soil anisotropy and land drainage, in: *Drainage of Agricultural Lands*, J. N. Luthin, ed., American Society of Agronomy, Madison, Wisconsin, 216–285, 1957.

Tóth, J., A theory of groundwater motion in small drainage basins in Central Alberta, Canada, *J. Geophys. Res.*, *67*, 4375–4387, 1962.

Tóth, J., A theoretical analysis of groundwater flow in small drainage basins, *J. Geophys. Res.*, *68*, 4795–4812, 1963. (Also published in *Natl. Res. Council Canada, Proc. Hydrol. Symp. 3*, Groundwater, 75–96, 1963).

Zienkiewicz, O., P. Mayer, and Y. K. Cheung, Solution of anisotropic seepage by finite elements, *J. Engr. Mech. Div., Am. Soc. Civil Engrs., Proc.* Paper 4676, *92*, EM 1, 111–120, 1966.

(Manuscript received September 27, 1966.)

…

LEAKY ARTESIAN AQUIFER CONDITIONS IN ILLINOIS

By

WILLIAM C. WALTON

Abstract

Leaky artesian conditions exist in many parts of Illinois where aquifers are overlain by deposits or confining beds which impede or retard the vertical movement of ground water. Under leaky artesian conditions, the cone of depression developed by a pumping well is influenced by the vertical permeability of the confining bed in addition to the hydraulic properties and geohydrologic boundaries of the aquifer.

The vertical permeability of a confining bed often can be determined from the results of pumping tests by using the nonsteady-state leaky artesian aquifer equation derived by Hantush and Jacob (1955). A time-drawdown type curve method for analyzing pumping test data under nonsteady-state conditions is described in detail. A distance-drawdown type curve method for analyzing pumping test data under steady-state conditions devised by Jacob (1946) is also described. These two methods are applied to available pumping test data for Illinois. The results of a test made near the village of Dieterich in Effingham County are presented to illustrate the analysis of data. A summary of the leaky artesian test data collected to date indicates that the vertical permeability of glacial drift deposits in the southern half of Illinois ranges between 0.08 and 1.6 gallons per day (gpd) per square foot.

Effects of leakage closely resemble the effects of a recharge boundary if the effects of partial penetration are excluded. The data for the Dieterich pumping test are used to show that recognition of leaky artesian conditions is critically important in predicting the water supply potential of wells and aquifers.

A form of Darcy's law is applied to data on the piezometric surface of the Cambrian-Ordovician Aquifer to determine the order of magnitude of the vertical permeability of the Maquoketa Formation. The Maquoketa Formation has a maximum thickness of about 250 feet, consists largely of beds of dolomitic shale, and confines water in the Cambrian-Ordovician Aquifer under artesian pressure. The Cambrian-Ordovician Aquifer is encountered at an average depth of 500 feet below the surface at Chicago, has an average thickness of 1000 feet, consists mainly of beds of sandstone and dolomite, and is the most highly developed source of large ground-water supplies in northeastern Illinois. Computations indicate that the average vertical permeability of the Maquoketa Formation in northeastern Illinois is about 0.00005 gpd per square foot. Leakage in 1958 through the Maquoketa Formation in northeastern Illinois is estimated to be about 8,400,000 gpd or about 11 per cent of the water pumped from deep wells.

Introduction

In Illinois, ground water is obtained for municipal, institutional, commercial, and industrial supplies largely from (1) thick and extensive Ironton-Galesville and Mt. Simon Sandstones of Cambrian age and the Glenwood-St. Peter Sandstone of Ordovician age; (2) sand and gravel deposits of Pleistocene age which in comparison to the sandstones mentioned above are thin and limited in areal extent; and (3) thick and extensive dolomites of Silurian age and the Galena-Platteville Dolomites of Ordovician age. Minor amounts of ground water are derived from sandstones and limestones of Mississippian, Devonian, and Pennsylvanian age. In all of these aquifers, ground water occurs under leaky artesian conditions at many places.

Leaky artesian conditions exist where an artesian aquifer is overlain or/and underlain by deposits (confining bed) which impede or retard the vertical movement of ground water. If a well is drilled through the confining bed and into the aquifer, the water in the well will rise above the top of the aquifer. Water may or may not flow over the top of the well. The surface to which water will rise under leaky artesian conditions, as defined by water levels in a number of wells, is the piezometric surface. When the pressure head, and hence, the piezometric surface is lowered by the pumping or free flow of wells, the aquifer is not dewatered but is still completely full. The water discharged from the well is derived by the compaction of the aquifer and associated beds, by the expansion of the water itself, and by vertical leakage through the confining bed into the aquifer.

Artesian conditions differ from leaky artesian conditions in that under artesian conditions the confining bed overlying and/or underlying an aquifer is assumed to prevent the movement of water and the confining bed is referred to as an *aquiclude*. In the majority of cases, geologic deposits are capable only of impeding the movement of ground water rather than preventing it.

Chamberlin (1885, pp. 131-173) made the following pertinent statements concerning confining beds.

> "No stratum is entirely impervious. It is scarcely too strong to assert that no rock is absolutely impenetrable to water . . . . But in the study of artesian wells we are not dealing with absolutes but with avail-

ables. A stratum that successfully restrains the (sic) most of the water, and thus aids in yielding a flow, is serviceably impervious. It may be penerated by considerable quantities of water, so that the leakage is quite appreciable and yet be an available confining stratum. The nearest approach to an entirely impervious bed is furnished by a thick layer of fine, unhardened [sic] clay. In this case solidifying permits the formation of fissures and the clay rocks are less impervious than the original clay beds. The clayey shales rank next as confining strata, after which follow in uncertain order shaly (sic) limestones, shaly sandstones, the various crystalline rocks, and even compact sandstones."

Hall, Meinzer, and Fuller (1911, p. 52) described leaky artesian conditions in Minnesota. Pertinent remarks made by these writers are:

"In localities where the water from the deeper beds rises to a level below that of the surficial ground-water table, the two bodies of water are not in equilibrium, and if the material separating them is at any point not entirely impervious water will pass from the surficial layer into the deeper beds. This relation is the general one throughout southern Minnesota .... Confining layers of till are not sufficiently impenetrable to prevent the escape of waters upward from the confined beds when the pressure is outward. Neither can they prevent the passage of water downward into these beds in localities where the balance of pressure favors movement in this direction."

Most if not all of the so called artesian aquifers in Illinois are actually leaky artesian aquifers. If the permeability of the confining bed is very low, vertical leakage may be difficult to measure within the average period (8 to 24 hours) of pumping tests. However, since the cone of depression created by pumping a well tapping a leaky artesian aquifer continues to expand until discharge is balanced by the amount of induced leakage, it does not follow that vertical leakage is of small importance over extended periods of time. As the cone of depression grows in extent and depth, the area of leakage and the vertical hydraulic gradient become large. Accordingly then, with long periods of pumping, contribution by leakage through a confining bed may be appreciable even though the vertical permeability is very low. If a source is available to replenish continuously the confining bed, the cone of depression developed by a well pumping for long extended periods will be influenced by the vertical permeability of the confining bed in addition to the hydraulic properties and geohydrologic boundaries of the main aquifer. Any long-range forecast of well or aquifer yield must include the important effects of leakage through the confining bed. The vertical permeability of a confining bed often can be determined from the results of pumping tests as described in the following section.

[Editors' Note: Material has been omitted at this point.]

Leakage Through Confining Beds in Illinois

Glacial Drift

Confining beds of Pleistocene age are well known throughout most of central, eastern, and much of southern Illinois. Large areas in western, south central, and southern Illinois are covered by glacial drift of Illinoian age. The drift cover is relatively thin and seldom exceeds 75 feet in thickness. The bedrock beneath the drift is shale, sandstone, and limestone of Pennsylvanian age which yield only small amounts of water to wells. Large deposits of water-yielding sand and gravel are scarce in the glacial drift and they occur chiefly in existing or buried valleys and as lenticular and discontinuous layers. The sand and gravel aquifers are commonly overlain by deposits of till that contain a high percentage of silt and clay and have a low permeability. In many areas, recharge to the aquifers is derived from vertical leakage through the till.

In the area of the Wisconsinan glacial drift in the east central and northern parts of Illinois, drift is thicker and consequently may contain more aquifers. The glacial drift is several hundred feet thick in deeply buried preglacial valleys such as the Mahomet Valley in east central Illinois. The outwash sand and gravel deposits partly filling these ancient valleys exceed 100 feet in thickness at places. Permeable deposits are commonly interbedded and overlain by layers of till which greatly retard the vertical movement of water. Permeable glacial deposits also occur on bedrock uplands and are often covered with till.

A typical glacial drift aquifer and its confining bed near Mattoon in east central Illinois were described by Foster (1952, pp. 85–94.) The bedrock in the vicinity of the city of Mattoon is immediately overlain by Illinoian drift as shown in figure 7. Bedrock consists of shales, thin limestones, sandstones, and coals of Pennsylvanian age which yield small amounts of water to wells. The relief of the bedrock surface is not great except southwest of Mattoon where two bedrock valleys occur.

The Illinoian drift was described by Foster (1952, p. 89) as "a grey or grey-green calcareous clay till." The upper surface of the Illinoian drift is gently undulating; the thickness of the till is greatest in bedrock valleys where it is about 90 feet. At many places the top of the Illinoian till is marked by the Sangamon soil zone or peat deposits.

A widespread layer of permeable sand and gravel probably of ice-contact origin and Wisconsinan age overlies the Sangamon soil to points about five miles south of Mattoon. Here complex fan and outwash materials occur above the Sangamon zone south of the Shelbyville moraine that marks the limit of Wisconsinan glaciation. Post-Shelbyville deposits include the Cerro Gordo moraine. The Cerro Gordo moraine contains only scattered thin lenses of sand.

The aquifers in the Mattoon area are the ice-contact deposits of sand and gravel north of the limit of Wisconsinan glaciation and the fan and outwash deposits south of the Shelbyville moraine. The textural composition of the ice-contact sand and gravel varies from place to place. Fine sand with a maximum thickness of 10 feet overlies the Sangamon soil at places and sometimes constitutes the entire aquifer. Coarse, clean gravels interbedded with silty sand exceeding 15 feet in thickness occur over wide areas. The fan and outwash deposits have thicknesses up to 65 feet and are in general more permeable than the ice-contact deposits.

The till deposits of the Shelbyville and Cerro Gordo moraines constitute the confining bed which overlies the ice-contact sand and gravel aquifer. The confining bed has an average saturated thickness of about 30 feet in the Mattoon area. The general textural composition of the confining bed and the aquifer below is described by the correlated driller's log of a well in sec. 30, T. 13 N., R. 7 E. given below:

Formation	Depth (feet)
Cerro Gordo-Shelbyville deposits (confining bed)	
Top soil	0– 2
Yellow clay	2–12
Gravelly yellow clay	12–14
Gravelly blue clay	14–26
Shelbyville sand and gravel (ice contact)	
Medium sand	26–28
Clean coarse sand and gravel	28–39
Very dirty sand and gravel	39–45
Sangamon soil	
Peat	45

The nonpumping water level in the well was about seventeen feet below land surface in May, 1954.

The vertical permeability of the Cerro Gordo-Shelbyville deposits is estimated from pumping test data to be 0.63 gpd per square foot. The coefficients of transmissibility and permeability of the Shelbyville sand and gravel are estimated from pumping test data to be 25,600 gpd per foot and 1600 gpd per square foot respectively.

Prior to 1935 the water supply for the city of Mattoon was obtained from wells (Doran well field) penetrating the Shelbyville ice-contact sand and gravel aquifer in sec. 30, T. 13 N., R. 7 E. The sustained yield of the Doran well field was not great because of the following reasons: (1) the sand and gravel aquifer was relatively thin (average thickness in the Doran well field area is about 16 feet), (2) recharge to the aquifer was limited by the vertical permeability of the till, and (3) during extended dry periods recharge from precipitation was not sufficient to replenish the till bed continuously, and heavy pumping ultimately drained the till and part of the aquifer.

Summary of Test Data for Illinois

A summary of the leaky artesian test data collected to date in Illinois and the coefficients of transmissibility, storage, and vertical permeability computed therefrom are given in table 1. Values of P' given in the table range between 0.08 and 1.6 gpd per square foot. The least permeable glacial drift confining bed is the clayey material overlying the aquifer at Winchester. The most permeable confining bed is the sandy clay materials at Cowden. The coefficient of vertical permeability of the confining bed overlying the Doran well field at Mattoon is high compared to most of the values of P' given in table 1.

Most of the confining beds listed in table 1 are less than 15 feet thick. To facilitate the planning of future pumping tests in areas where confining beds exceed 15 feet in thickness, theoretical time-drawdown curves for several leaky artesian conditions thought to exist in Illinois are presented in figure 8. It is apparent from the curves that as the thickness of the confining bed increases, or the coefficient of transmissibility increases, or P' decreases, the time that it takes for leakage to affect drawdown data increases and the effects of leakage decrease. The influence of T, P', and m on the cone of depression is shown by the theoretical distance-drawdown graphs in figure 9. As T decreases, the cone of depression deepens. As T increases, the virtual radius of the cone of depression decreases. Both the depth and virtual radius of the cone of depression increase as P' decreases or as m' increases.

Maquoketa Formation

The Maquoketa Formation overlies the Cambrian-Ordovician Aquifer in large parts of northeastern Illinois, including the Chicago region, and to great extent confines the water in the aquifer under artesian pressure. As described in a detailed report on the ground-water resources of the Chicago region (Suter, et al, 1959), the Cambrian-Ordovician Aquifer is the most highly developed source of large ground-water supplies in northeastern Illinois and consists in downward order of the Galena-Platteville Dolomite, Glenwood-St. Peter Sandstone, and Prairie du Chien Series of Ordovician Age; the Trempealeau Dolomite, Franconia Formation, and Ironton-Galesville Sandstone of Cambrian Age. The sequence, structure, and general characteristics of these rocks are shown in figure 10a. The Cambrian-Ordovician Aquifer is underlain by shale beds of the Eau Claire Formation which have a very low permeability. Available data indicate that on a regional basis, the entire sequence of strata, from the top of the Galena-Platteville to the top of the shale beds of the Eau Claire Formation, essentially behave hydraulically as one aquifer.

As shown in figure 10b, the Maquoketa Formation has a maximum thickness of about 250 feet and thins to the north and west to less than 50 feet. The formation dips regionally to the east at a uniform rate of about 10 feet per mile. Bergstrom and Emrich (see Suter, et al, 1959, p. 33) divided the Maquoketa Formation into three units; lower, middle, and upper. As described by Bergstrom and Emrich,

"the lower unit, is normally a brittle, dark brown, occasionally gray or grayish brown, dolomite shale grading locally to dark brown, argillaceous dolomite. The middle unit is dominantly brown to gray, fine-to-coarse-grained, fossiliferous, argillaceous, speckled dolomite and limestone. It is commonly interbedded with a fossiliferous brownish gray to gray, dolomitic shale. The upper unit is a greenish gray, weak, silty, dolomitic

Table 1—Results of Leaky Artesian Pumping Tests in Illinois

Physical Data

| | | | | | Lithology | |
| | | | | | Aquifer | Confining bed |
Owner	Location	Date of test	Duration (hrs)	Pumping rate (gpm)	(Driller's Log)	
Village of Beecher City	Effingham Co.	2/5-6/51	23	15	sand, fine & some gravel	clay, sandy
Village of Dieterich	Effingham Co.	7/2-3/51	20	25	sand, fine & some gravel	clay, sandy, hardpan
Village of Cowden	Shelby Co.	10/22/54	4	141	sand & gravel	sandy clay
City of Assumption	Christian Co.	5/29/58	3	32	sand, some gravel	clay, sandy
City of Mattoon	Coles Co.	2/24/54	4	156	sand & gravel	clay, gravelly
City of Barry	Pike Co.	7/4-3/56	21	207	sand & gravel	clay, sandy
City of Winchester	Scott Co.	11/9/49	8	100	sand & gravel	clay

Results

| | | Aquifer | | | | Confining bed | | |
Owner	Location	T (gpd/ft)	m (feet)	P (gpd/sq. ft.)	S	P' (gpd/sq. ft.)	m' (feet)	(P'/m') (gpd/cu. ft.)
Village of Beecher City	Effingham Co.	1,220	7	175	0.0003	0.25	12	2.1×10^{-2}
Village of Dieterich	Effingham Co.	1,500	8	188	0.0002	0.10	14	7.1×10^{-3}
Village of Cowden	Shelby Co.	39,000	25	1,560	0.0080	1.60	7	2.3×10^{-1}
City of Assumption	Christian Co.	4,900	12	408		0.19	8	2.4×10^{-2}
City of Mattoon	Coles Co.	25,600	16	1,600	0.0015	0.63	12	5.2×10^{-2}
City of Barry	Pike Co.	119,000	36	3,300	0.0030	0.15	16	9.4×10^{-3}
City of Winchester	Scott Co.	10,000	26	384	0.0003	0.08	16	5.0×10^{-3}

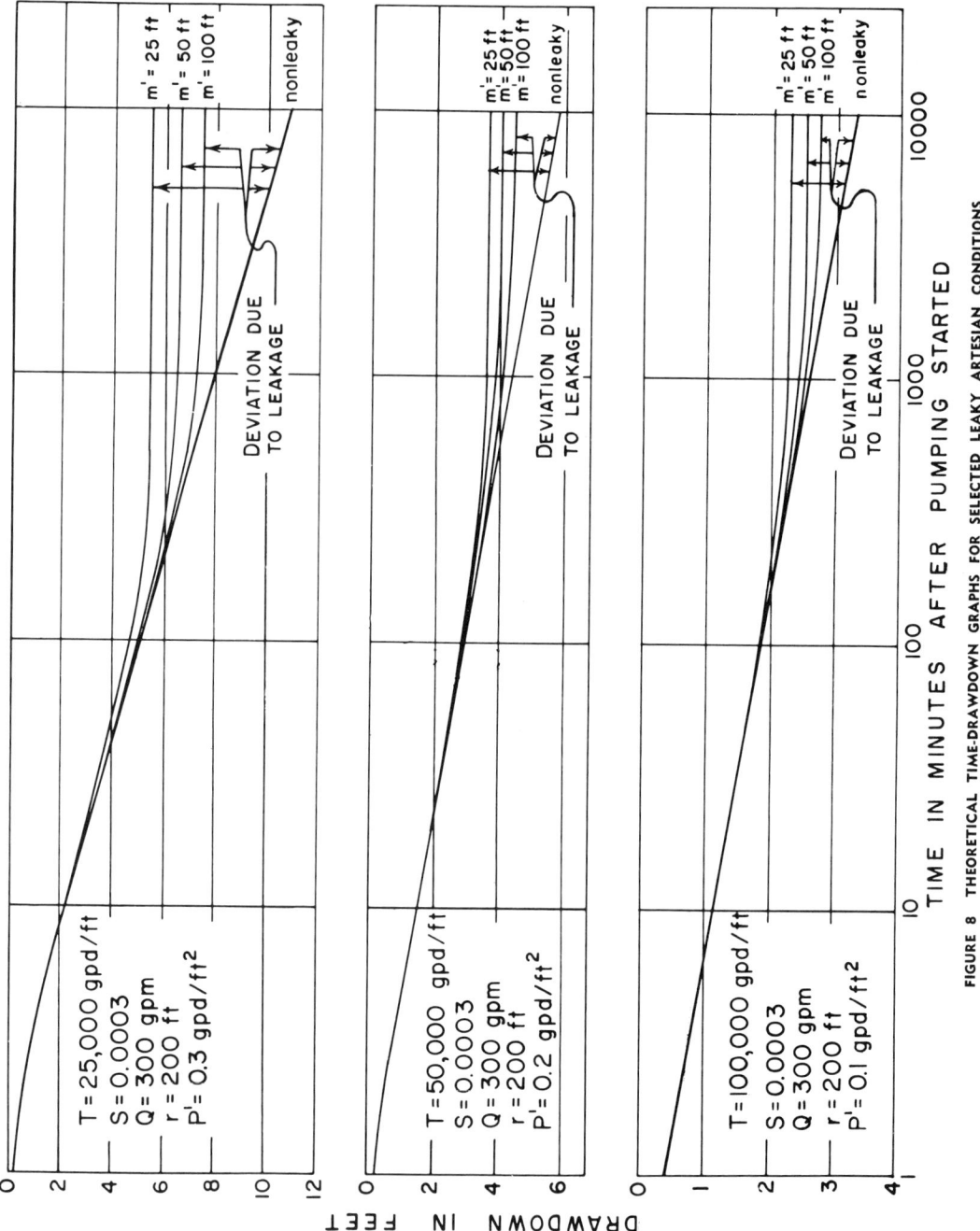

FIGURE 8 THEORETICAL TIME-DRAWDOWN GRAPHS FOR SELECTED LEAKY ARTESIAN CONDITIONS

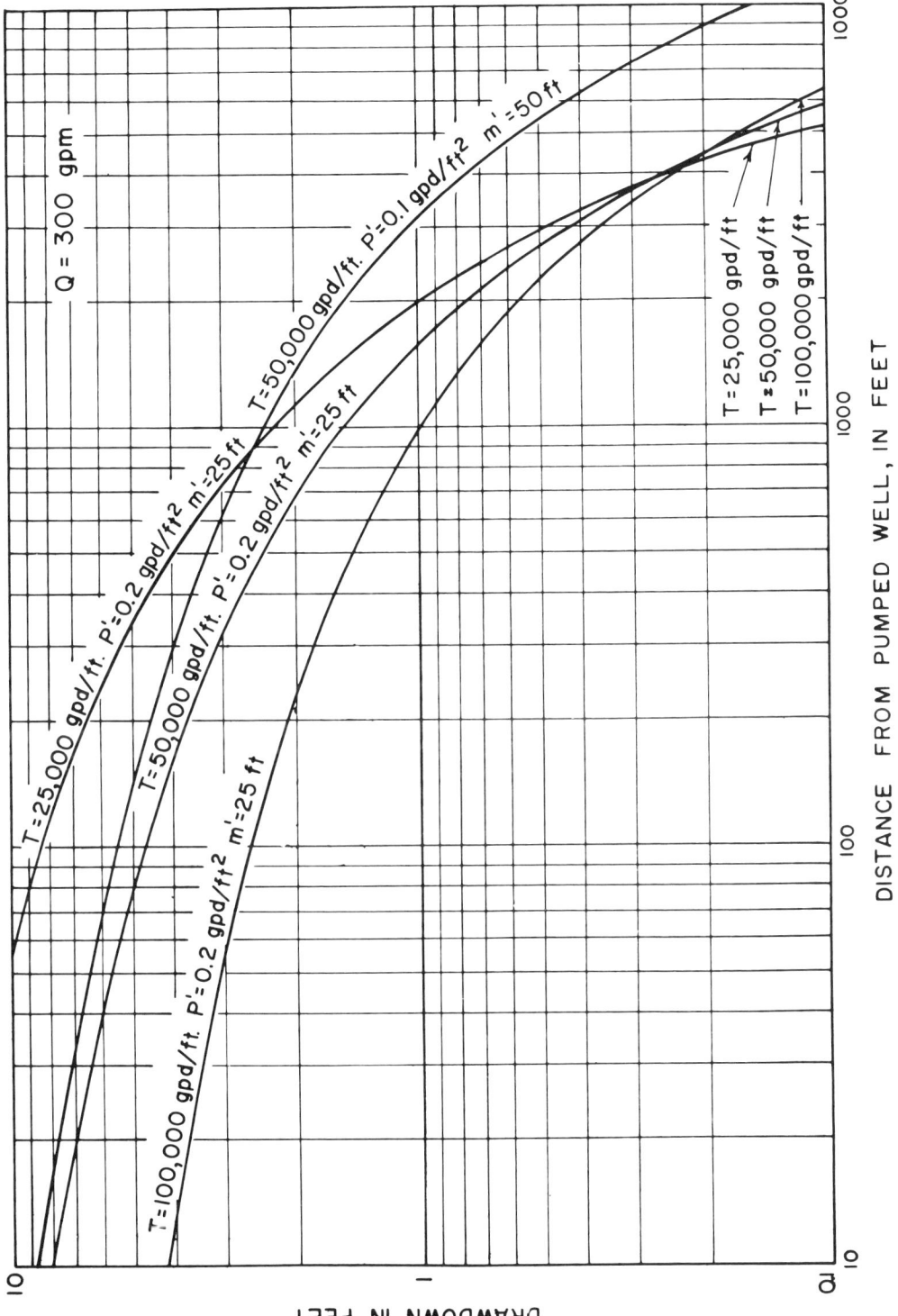

FIGURE 9 THEORETICAL DISTANCE-DRAWDOWN GRAPHS FOR SELECTED LEAKY ARTESIAN CONDITIONS

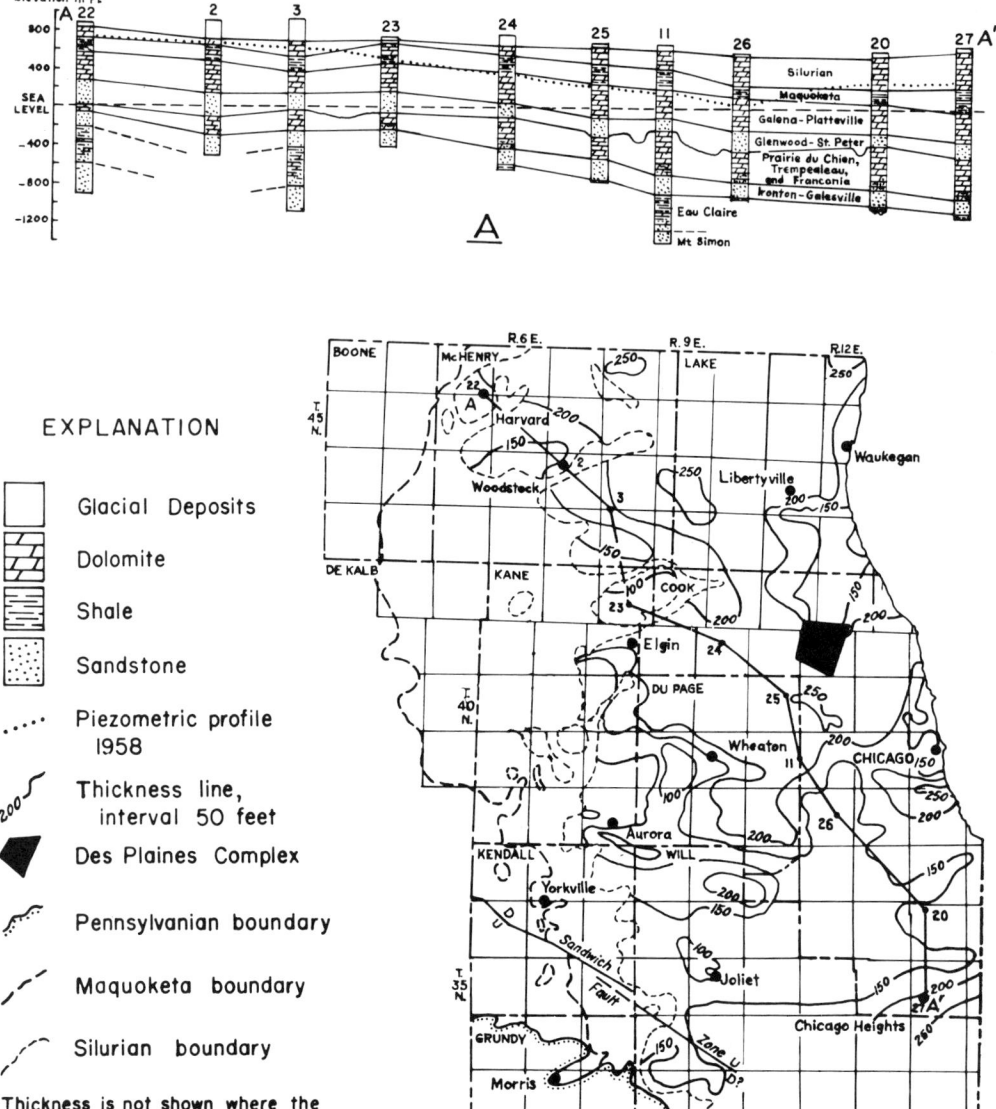

FIGURE 10 GEOHYDROLOGIC CROSS SECTION AND THICKNESS OF THE MAQUOKETA FORMATION IN NORTHEASTERN ILLINOIS

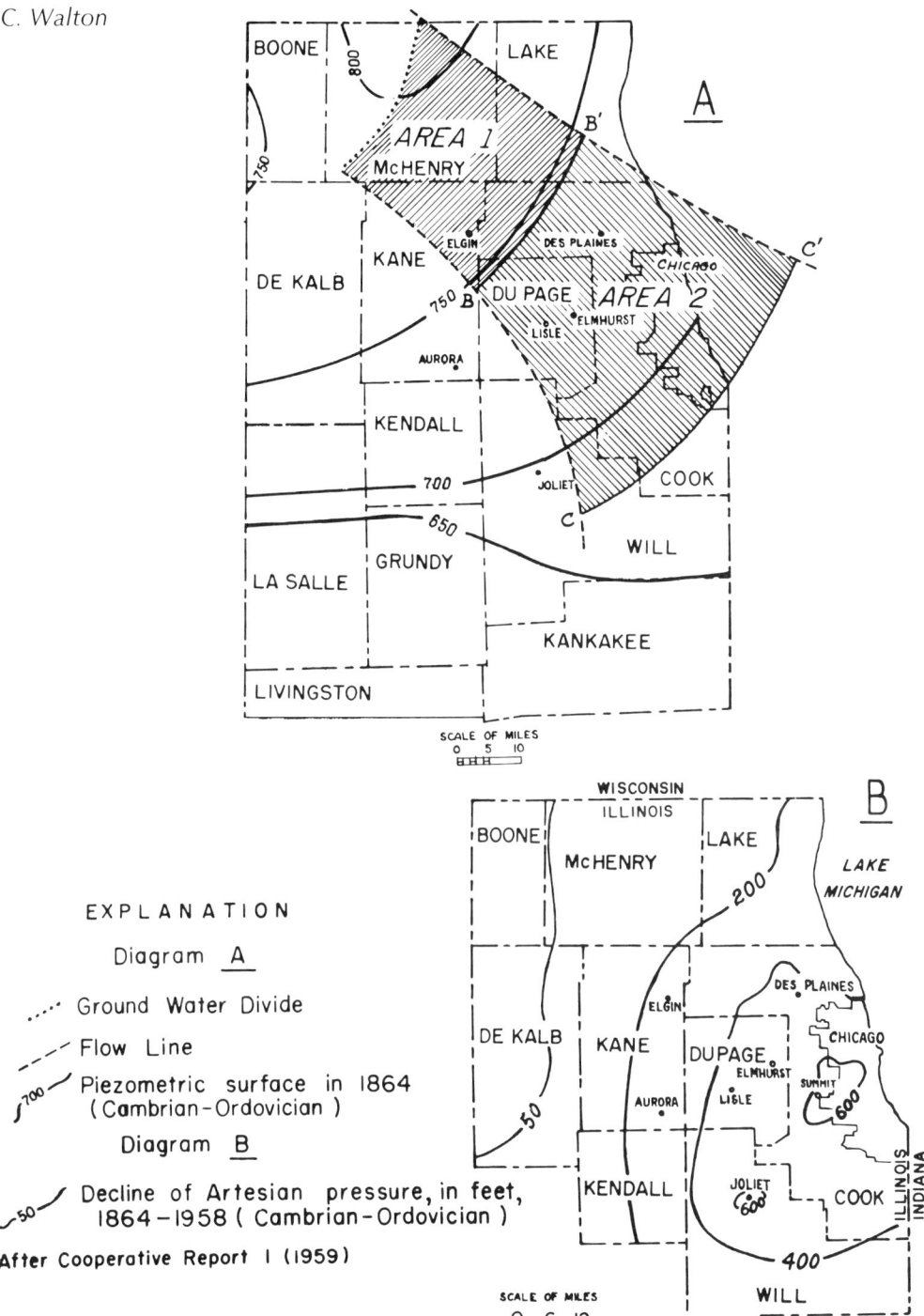

FIGURE 11 PIEZOMETRIC SURFACE OF CAMBRIAN-ORDOVICIAN AQUIFER IN NORTHEASTERN ILLINOIS, ABOUT 1864, AND DECLINE IN ARTESIAN PRESSURE IN CAMBRIAN-ORDOVICIAN AQUIFER, 1864-1958

shale that grades into very argillaceous, greenish gray to gray dolomite. The lower unit is thicker in Cook and Will Counties where it exceeds 100 feet. It thins to the north and west to less than 50 feet. The middle unit is thicker to the west where it is more than 100 feet locally and thins to the east. The upper unit ranges in thickness from less than 50 feet in the west to more than 100 feet in parts of Cook and Will Counties. The lower dense shale unit is the most impermeable unit. Dolomite beds in the middle unit yield small quantities of ground water."

The Cambrian-Ordovician Aquifer beneath the Maquoketa Formation receives water from overlying glacial deposits mostly in areas west of the border of the Maquoketa Formation shown in figure 10b where the Galena-Platteville Dolomite, the uppermost unit of the aquifer, is directly overlain by glacial deposits. Recharge of the glacial deposits in turn is derived from precipitation that falls locally. The piezometric-surface map for the Cambrian-Ordovician Aquifer in the year 1864 (see figure 11a) indicates that under natural conditions water entering or recharging the aquifer was discharged in areas to the east and south by vertical leakage upward through the Maquoketa Formation and by leakage into the Illinois River valley.

The changes in artesian pressure produced by pumping since the days of early settlement have been pronounced and widespread. Pumpage from deep wells has increased from 200,000 gpd in 1864 to about 78 mgd in 1958. Figure 11b shows the decline of artesian pressure in the Cambrian-Ordovician Aquifer from 1864 to 1958 as the result of heavy pumping. The greatest declines, more than 600 feet, have occurred in areas of heavy pumpage west of Chicago, at Summit and at Joliet. In 1958, the piezometric surface of the Cambrian-Ordovician Aquifer was several hundred feet below the water table in most of northeastern Illinois, and downward movement of water through the Maquoketa Formation was appreciable under the influence of large differentials in head between shallow deposits and the Cambrian-Ordovician Aquifer. The vertical permeability of the Maquoketa Formation and the quantity of leakage through the confining bed in 1958 are discussed in the following sections.

Vertical Permeability of Maquoketa Formation

The quantity of leakage through a confining bed into an aquifer can be computed from the following form of Darcy's law:

$$Q_c = \left(\frac{P'}{m'}\right) \Delta h \, A_c \qquad (9)$$

where:

Q_c = leakage through confining bed, in gallons per day
P' = vertical permeability of confining bed, in gallons per day per square foot
m' = thickness of confining bed through which leakage occurs, in feet
A_c = area of confining bed through which leakage occurs, in square feet
Δh = difference between the head in the aquifer and in the source bed above the confining bed, in feet

The quantity (P'/m') was termed the leakage coefficient by Hantush (1956, p. 702). Values of the leakage coefficient determined from pumping test data are given in table 1.

Equation 9 may be rewritten as:

$$P' = \frac{Q_c m'}{\Delta h \, A_c} \qquad (10)$$

Thus, the vertical permeability of a confining bed may be determined if Q_c, m', A_c, and Δh are known.

Equation 10 was used to determine the order of magnitude of the vertical permeability of the Maquoketa Formation. Figure 11a shows the piezometric surface of the Cambrian-Ordovician Aquifer before extensive development occurred in the Chicago region. Flow lines were drawn from the ground-water divide in McHenry County toward the northern and southern boundaries of Cook County at right angles to the estimated piezometric surface contours for 1864. The part of the aquifer (area 1) which is enclosed by the flow lines, the ground-water divide, and section B-B', was considered. In 1864 the piezometric surface was below the water table and downward leakage through the Maquoketa Formation into the aquifer was occurring in area 1. Leakage was equal to the quantity of water percolating through section B-B'. At section B-B' the hydraulic gradient of the piezometric surface was about two feet per mile, and the distance between limiting flow lines was about twenty-five miles. Based on data given by Suter, et al (1959, p. 50) the average coefficient of transmissibility of the Aquifer at section B-B' is about 19,000 gpd per foot.

The quantity of water percolating through a given cross section of an aquifer is proportional to the hydraulic gradient (slope of the piezometric surface) and the coefficient of transmissibility. It can be computed by using the following modified form of the Darcy equation (see Ferris, 1959, p. 148):

$$Q = TIL \qquad (11)$$

in which Q is discharge, in gallons per day, T is coefficient of transmissibility, in gallons per day per foot, I is hydraulic gradient, in feet per mile, and L is width of cross section through which discharge occurs, in miles.

Using equation 11, the quantity of water moving southeastward through the aquifer at section B-B' was computed to be about 1,000,000 gpd. Leakage downward through the Maquoketa Formation in area 1 was therefore about 1,000,000 gpd in 1864. As measured from figure 11a, area 1 is about 750 square miles. The average Δh over area 1 was determined to be about 85 feet by comparing estimated elevations of the water table and the piezometric surface contours given in figure 11a. The average thickness of the Maquoketa Formation over area 1 from

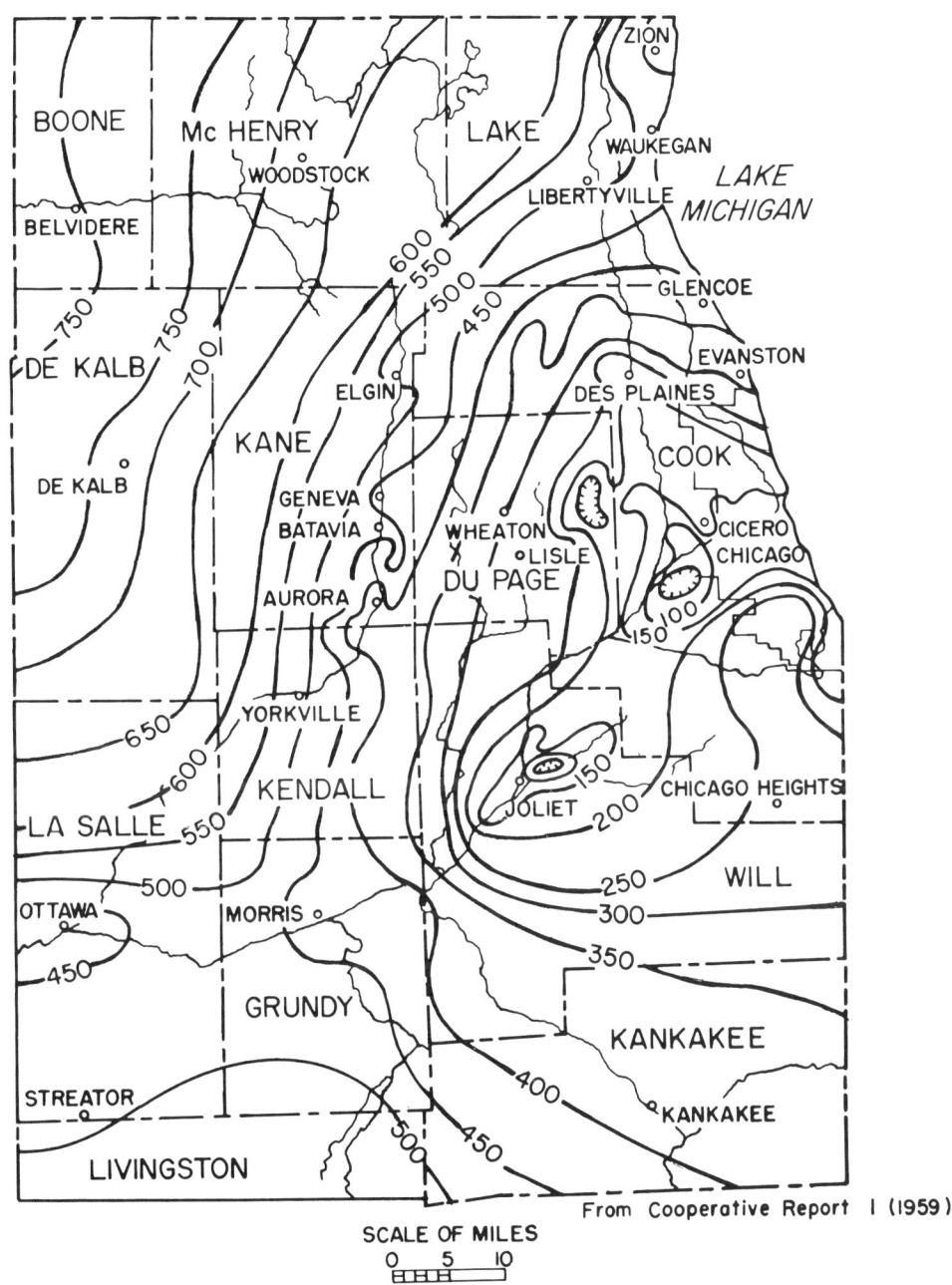

FIGURE 12 PIEZOMETRIC SURFACE OF CAMBRIAN-ORDOVICIAN AQUIFER IN NORTHEASTERN ILLINOIS IN 1958

figure 10b is about 175 feet. Substitution of these data in equation 10 indicates that the average vertical permeability of the Maquoketa Formation in area 1 is about 0.0001 gpd per square foot.

In 1864 the piezometric surface was above the water table southeast of section B–B', and the quantity of water entering the aquifer in area 1 was discharged by leakage up through the Maquoketa Formation in the areas between the limiting flow lines southeast of section B–B' in northeastern Illinois and northwestern Indiana.

Figure 11a indicates that appreciable flow was occurring through section C–C' near the Illinois-Indiana state line and that leakage between sections B–B' and C–C' was something less than 1,000,000 gpd. Therefore, the average vertical permeability of the Maquoketa Formation in area 2 is something less than that computed by substituting a value of 1,000,000 gpd for Q_c in equation 10. Area 2 was scaled from figure 11a and is about 1500 square miles. The average Δh over area 2 was determined to be about 70 feet by comparing estimated elevations of the water table and the piezometric surface contours given in figure 11a. The average thickness of the Maquoketa Formation over area 2 from figure 10b is about 200 feet. Substitution of these data in equation 10 indicates that the average vertical permeability of the Maquoketa Formation in area 2 is less than 0.00007 gpd per square foot.

Actually the leakage over area 2 is equal to the difference between the quantities of water moving through sections B–B' and C–C'. The quantity of water moving through section C–C' cannot be determined with any degree of accuracy because the location of the 650-foot piezometric surface contour is largely conjectural and the average coefficient of transmissibility of the aquifer at section C–C' is uncertain. On the basis of available data, the flow through section C–C' is estimated to be at least one third of the flow through section B–B' and the average vertical permeability of the Maquoketa Formation in area 2 is estimated to be not less than 0.00002 gpd per square foot. A value of 0.00003 gpd per square foot was selected as the best estimate of P for area 2.

Computations indicate that the average vertical permeability of the Maquoketa Formation increases to the north and west. Available geologic information supports this conclusion. The lower unit of the Maquoketa Formation, probably the least permeable of the three units (Bergstrom & Emrich, personal communication), thins to the west. In addition, the Maquoketa Formation is the uppermost bedrock formation below the glacial deposits in a large part of area 1 and locally may be completely removed by erosion.

A comparison of the average vertical permeability of the Maquoketa Formation with data in table 1 indicates that the glacial drift confining beds for which test data are available are at least 800 times as permeable as the Maquoketa Formation.

Leakage through Maquoketa Formation in 1958

Even though the vertical permeability is very low, leakage in 1958 through the Maquoketa Formation was appreciable. The area of the confining bed within the part of Illinois shown in figure 10b through which leakage occurred (4000 square miles) and the average head differential between the piezometric surface of the Cambrian-Ordovician Aquifer and the water table (300 feet) were great (figures 11a and 12). Computations made using the data given above, and assuming a m' of 200 feet and a P' of 0.00005 gpd per square foot, indicate that leakage through the Maquoketa Formation within the part of Illinois shown in figure 10b was about 8,400,000 gpd or about 11 per cent of the water pumped from deep wells in 1958.

Conclusions

The vertical permeability of a confining bed often can be readily determined from pumping test data with the nonsteady-state leaky artesian equation derived by Hantush and Jacob. The solution of nonsteady drawdown distribution caused by pumping a well in an infinite leaky artesian aquifer is simplified by using a family of leaky artesian type curves. The vertical permeability of a confining bed and the quantity of leakage into an aquifer also can be determined from forms of Darcy's law by using data on the piezometric surface of the aquifer.

Although the vertical permeabilities of only seven glacial drift confining beds in southern Illinois have been computed from the results of pumping tests, they represent a good start in cataloging the vertical permeabilities of confining beds in the state. The values probably can be applied to geologically similar areas, at least in making rough quantitative investigations in the southern half of Illinois. Test data under leaky artesian conditions are not available for the northern half of Illinois. Controlled pumping tests should be made in the future in areas of northern Illinois where leaky artesian conditions exist. The results of tests in many areas throughout Illinois will provide a means for detecting local and regional changes in the vertical permeability of glacial desposits and will aid greatly in the interpretation of the geology of Illinois as it relates to the quantitative appraisal of the State's water resources.

The vertical permeability of the Maquoketa Formation may vary greatly from place to place and locally other beds may serve to confine the Cambrian-Ordovician Aquifer, however, the computed average vertical permeability of the Maquoketa Formation indicates the order of magnitude of the parameter. No great accuracy is implied in the computations of the quantity of leakage through the Maquoketa Formation. The intent of quantitative studies is to better understand the relationship of the Maquoketa Formation to the Cambrian-Ordovician Aquifer and to estimate leakage with greater accuracy than mere expression of concept.

References

Chamberlin, T. C., 1885, Requisite and qualifying conditions of artesian wells: U.S. Geol. Survey 5th Ann. Rept., pp. 131–173.

De Glee, G. J., 1930, Over grondwaterstroomingen bij wateronttrekking door middel van putten, J. Waltman Jr., Delft, 175 pp.

Ferris, J. G., 1959, ed. Wisler, C. O., and Brater, E. F., Hydrology, Chap. VII, Ground Water: New York: John Wiley & Sons, Inc., p. 148.

Foster, J. W., 1952, Major aquifers in glacial drift near Mattoon, Illinois: Illinois State Geological Survey, Circular No. 79, pp. 85–94.

Hall, C. W., Meinzer, O. E., and Fuller, M. L., 1911, Geology and underground waters of southern Minnesota: U.S. Geol. Survey Water Supply Paper 256, p. 52.

Hantush, M. S., and Jacob, C. E., 1955, Non-steady radial flow in an infinite leaky aquifer: Am. Geophys. Union Trans., Vol. 36, No. 1, pp. 95–100.

Hantush, M. S., 1956, Analysis of data from pumping tests in leaky aquifers: Am. Geophys. Union Trans., Vol. 37, No. 6, pp. 702–714.

Jacob, C. E., 1940, On the flow of water in an elastic artesian aquifer: Am. Geophys. Union Trans., pt. 2, p. 582.

Jacob, C. E., 1946, Radial flow in a leaky artesian aquifer: Am. Geophys. Union Trans., Vol. 27, No. 2, pp. 198–205.

Rorabaugh, M. I., 1948, Groundwater resources of the northeastern part of the Louisville area Kentucky: City of Louisville, Louisville Water Supply Company, p. 63.

Suter, Max, Bergstrom, R. E., Smith, H. F., Emrich, G. H., Walton, W. C., Larson, T. E., 1959, Preliminary report on groundwater resources of the Chicago Region, Illinois: Coop. Report 1, State Water Survey and State Geological Survey.

Theis, C. V., 1935, The relation between the lowering of the piezometric surface and the rate and duration of discharge of a well using ground water storage: Am. Geophys. Union Trans., pt. 2, pp. 519–524.

Water from Low-Permeability Sediments and Land Subsidence

P. A. DOMENICO AND M. D. MIFFLIN

Abstract. Seepage pressures are part of the neutral or nondeformative stresses acting in a groundwater basin. The reduction of these pressures gives rise to a stress transfer from neutral to effective. The increase in effective stresses is exclusively responsible for measurable deformations of the land surface. The amount of land subsidence or groundwater recovery from compressible confining layers depends upon the specific storage of the strata and the average head change within them. Expressions for the specific storage are obtained from both consolidation theory and conservation principles of compressible flow. Average head changes are identified on a depth-pressure diagram in terms of head changes in adjacent aquifers and are referred to as 'effective-pressure areas.' The geometry of the effective-pressure area is shown to depend upon the thickness of the compressible strata, the magnitude of artesian pressure decline, the manner in which the basin is developed, and time. These factors are embodied in equations that quantitatively describe the release of stored water from compressible confining layers resulting from their vertical compression in areas of land subsidence.

INTRODUCTION

The conservation principles of classical physics upon which most engineering analysis is based typically lead to one or more differential equations describing the performance of the system. Analytical treatment of compressible flow through compressible mediums, for example, is aided by the continuity equation

$$\text{Inflow} - \text{outflow} = -\left[\frac{\partial(\rho v_x)}{\partial x} + \frac{\partial(\rho v_y)}{\partial y} + \frac{\partial(\rho v_z)}{\partial z}\right] \Delta x\, \Delta y\, \Delta z \quad (1)$$

which, for negligible changes in horizontal length, can be restated as the time rate of change in fluid mass

$$\frac{\partial(\Delta m)}{\partial t} = \left[\frac{\rho \theta \partial(\Delta z)}{\partial t} + \frac{\rho \Delta z \partial \theta}{\partial t} + \frac{\Delta z \theta \partial \rho}{\partial t}\right] \Delta x\, \Delta y \quad (2)$$

and, ultimately, the descriptive differential equation [*Jacob,* 1950]

$$\frac{\partial^2 h}{\partial x^2} + \frac{\partial^2 h}{\partial y^2} + \frac{\partial^2 h}{\partial z^2} = \frac{\beta \theta \gamma_\omega + \alpha \gamma_\omega}{k} \frac{\partial h}{\partial t} \quad (3)$$

In (1) inflow minus outflow is the net inward flux in a porous element, ρ is fluid density, v (x, y, z) are velocity components, Δx, Δy, and Δz are dimensions of the element; in (2) $\partial(\Delta m)/\partial t$ is the time rate of change in fluid mass and θ is porosity; and in (3) h is total head, k is the coefficient of permeability, and the quantity $\beta \theta \gamma_\omega + \alpha \gamma_\omega$ is the specific storage (s_s), where β is the compressibility of water, α is the vertical compressibility of aquifer material, and γ_ω is the unit weight of water.

Solutions to the descriptive differential equation stated above have two general applications in hydroscience. In groundwater problems, expressions for the two-dimensional form of (3) allow determination of the volume of water released from storage from a compressible aquifer owing to a decrease in hydraulic pressure caused by pumping [*Theis,* 1935; *Jacob,* 1950]. In soil mechanics problems, expressions are available for an equation similar to (3) but for one-dimensional flow that allow determination of the rate of consolidation due to an increase in hydraulic pressure and subsequent drainage from compressible clays [*Terzaghi,* 1925]. Terzaghi's work implies a depletion factor or change in water content in compressible units with time due to changes in their stress-strain relationships. In this paper, these principles are translated into more familiar hydrologic terminology for study of land subsidence.

SPECIFIC STORAGE OF COMPRESSIBLE CONFINING LAYERS

The derivation of (3) as it pertains to unsteady flow of water in a compressible aquifer is well understood by hydrologists. For the case of unsteady vertical flow in compressible confining layers, it is only necessary to consider Jacob's treatment of the transient change in fluid mass equation (2). According to *Jacob* [1950] the first term of (2) deals with change in vertical direction (Δz) and is therefore related to vertical compressibility (α). The second term deals with porosity (θ) of the element and is also related to vertical compression, since the dimensional change in the x and y directions is assumed to be negligible. The third term refers to change in fluid density and depends upon fluid compressibility (β). Hence, the first and second terms of the transient change in fluid mass equation are functions of effective or intergranular pressure, and both terms express the vertical compression of the element. Further, in a compressible confining layer, the volume of water obtained from expansion of water is negligible compared with that obtained through a change in porosity, so that the third term of (2) can be neglected. The descriptive differential equation is then expressed

$$\partial^2 h/\partial z^2 = (\alpha \gamma_\omega / k')(\partial h/\partial t) \quad (4)$$

where k' is vertical permeability.

The head (h) in (4) is an expression for total head, or sum of the static and transient pore-water pressures. In confined aquifer systems, only heads due to transient pore-water pressures cause flow that results in consolidation. By considering the static head invariant with respect to time, (4) can be rewritten

$$\partial^2 u/\partial z^2 = (\alpha \gamma_\omega / k')(\partial u/\partial t) \quad (5)$$

where u is the excess in pore-water pressure.

In this equation, $\partial^2 u/\partial z^2$ denotes change in excess pore-water pressure gradient, $\partial u/\partial t$ denotes rate of change of excess pore-water pressure, and $(k'/\alpha \gamma_\omega) \partial^2 u/\partial z^2$ denotes the volume of water expelled from the voids per unit surface area per unit time.

The specific storage is defined as the volume of water that a unit volume of confining layer releases from storage, owing to its compression when the average excess pressure within the unit volume undergoes a unit decline, and is expressed

$$s_s' = \alpha \gamma_\omega \quad (6)$$

From (3) and (5), the specific storage of a confining layer is similar to the specific storage of an adjacent aquifer, differing only in that compressibility of water has been neglected. Further, from the descriptive differential equations, the ratio s_s'/k' influences the response of excess pore water in a confining layer in the same manner as the ratio s_s/k influences the response of a groundwater system to a pumping stress. For a compressible aquifer, the smaller this ratio is the shorter is the duration of time required to develop a cone of depression. For a confining layer, the time required for development of a 'cone of depression' due to vertical movement of water out of the layer, more appropriately thought of as time required to achieve full consolidation, also depends upon the ratio s_s'/k'. The larger the compressibility or smaller the permeability, the greater is the time required to re-establish steady-flow conditions in the confining layer.

Stress-strain relationships. The concept of storage in clay beds implies acceptance of a storage factor or coefficient for such units, and solution requires analysis of changes in stress-strain relationships over long periods of pumping. Consequences of these changes are well documented in the literature: progressive land subsidence in the San Joaquin Valley, California [*Poland and Davis*, 1956; *Poland*, 1961]; the upper Gulf coastal region, Texas [*Winslow and Wood*, 1959]; the Savannah area, Georgia [*Davis et al.*, 1963]; and Las Vegas Valley, Nevada [*Domenico et al.*, 1964]. Other prominent localities include Mexico City [*Cuevas*, 1936] and London [*Wilson and Grace*, 1942]. Analytical description of this phenomenon where it is controlled by pore compressibility is best accomplished by examination of the stress-strain relationships in compressible clays.

With expulsion of water from the voids of an elemental clay volume, the volume decreases, resulting in a vertical shortening or decrease in thickness. The change in vertical dimension has been shown to be related to the compressibility of the element. By definition

$$\alpha = 1/E_c \quad (7)$$

where E_c is the bulk modulus of compression. The bulk modulus of compression can be identified as a stress-strain ratio

$$E_c = (\Delta \text{ stress})/(\Delta \text{ strain})$$
$$= [\Delta \bar{\sigma}/(\Delta H/H_0)] \quad (8)$$

where $\Delta \bar{\sigma}$ is change in effective stress and $\Delta H/H_0$ is change in elemental volume for a unit element, or simply the ratio of change in height (ΔH) to original height (H_0).

Confining layers, whether of unconsolidated or indurated rock, follow Hooke's law of deformation for very small loading increments. With reference to the analogy of Hooke's law, (8) can be restated

$$(\Delta H/H_0) \text{ (strain)} = (1/E_c) \Delta \bar{\sigma} \text{ (stress)} \quad (9)$$

where E_c is a constant of proportionality. As will be demonstrated, measurable strains of a compressible confining layer resulting from an instantaneously applied stress are time dependent.

Figure 1 is a diagrammatic representation of the void ratio of a unit element of confining layer. The dimensions of the element consist of the height of solids (H_s) and the height of voids (H_v). The original void ratio is defined as the quotient of the volume of voids divided by the volume of solids. Since the solids are assumed to be incompressible, changes in elemental height (ΔH_0) are proportional to changes in void ratio (Δe). The relative compression of the element or relative amount of water loss per unit height can be expressed

$$\Delta H/H_0 = \Delta e/(1 + e) \quad (10)$$

The change in void ratio is assumed to be directly proportional to change in effective pressure, or $\Delta e = a_v \Delta \bar{\sigma}$ where a_v is the coefficient of compressibility, or rate of change of void ratio with respect to rate of change of effective pressure causing consolidation. In soil mechanics usage, a_v is the slope of the line obtained by plotting void ratio versus pressure for test specimens [*Terzaghi and Peck*, 1948]. As a_v is inherently variable for any given diagram, its practical use is limited. However, it is a useful soil characteristic and can be considered constant for small changes in pressure.

Substituting $a_v \Delta \bar{\sigma}$ for Δe in (10) and placing the new expression in (9) results in

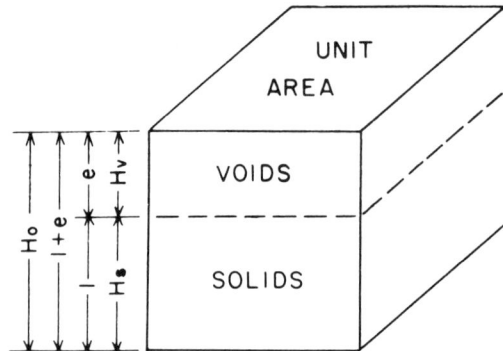

Fig. 1. Schematic representation of void ratio for an element of confining layer.

$$1/E_c = a_v/(1 + e) = \Delta e/[\Delta \bar{\sigma}(1 + e)] \quad (11)$$

For a one-dimensional system, the quantities expressed in (11) represent the height of a pore-water column expelled from a unit element when the effective pressure is increased by one pressure unit. The specific storage of the element can then be expressed by any of the following equalities

$$s_s' = \alpha \gamma_\omega = \frac{\gamma_\omega}{E_c} = \frac{a_v \gamma_\omega}{1 + e} = \frac{\Delta e \gamma_\omega}{\Delta \bar{\sigma}(1 + e)} \quad (12)$$

In soil mechanics practice, the expression $a_v \gamma_\omega/(1 + e)$ is the favored term but is seldom utilized. Instead, soils engineers recognize that both compressibility and permeability of a clay layer in the range between its initial and ultimate void ratio are only two of numerous factors that determine the time to attain a certain degree of consolidation. To reduce the number of variables, the coefficient of consolidation (c_v) has been introduced to include combined effects of compressibility and permeability, and its value is determined in the laboratory. The parameter c_v is given as $k'(1 + e)/a_v \gamma_w$. Therefore, in soil mechanics literature, (5) is generally expressed as

$$\partial^2 u/\partial z^2 = (1/c_v)(\partial u/\partial t) \quad (13)$$

However, it is clear that c_v is expressed equally well by the product of permeability and the reciprocal of any one of the expressions for specific storage.

Specific storage of natural sediments. Assuming that Hooke's law applies, the specific storage of a perfectly homogeneous confining bed is a constant. As demonstrated in the preceding section, notations available to express this parameter are numerous, depending mainly on the discipline involved. In soil mechanics literature, $a_v \gamma_w/(1 + e)$ is used almost exclusively. In groundwater hydrology, the notation γ_w/E_v is possibly preferable because of use of the bulk modulus of compression in groundwater literature [*Jacob*, 1940]. Variations in the bulk modulus of compression and specific storage based on compressibility alone are shown in Table 1.

Although a wide range for supposedly similar lithologies is indicated, it appears that considerable water is stored in sediments of low permeability that commonly constitute the bulk of aquifer-aquitard systems. Special attention should be given to comparison of values for dense sands, sandy gravels, and clays. The given values for coarse clastics compare favorably with the specific storage of compressible aquifers, a parameter easily calculated from field tests. On the other hand, values for the specific storage of clay beds can be obtained from consolidation tests on undisturbed samples or from a modification of the theory of leaky aquifers developed by *Hantush* [1960]. Because of the time required for pumping to affect a confining layer such that its full thickness contributes to the replenishment of a pumped aquifer, there is likely to be little resemblance between values given in Table 1 and values obtained by Hantush's model.

Recalling that the coefficient of consolidation equals the product of the permeability and the reciprocal of specific storage, the specific storage can be stated as

$$s_s' = k'/c_v \quad (14)$$

This expression is particularly useful because of the fund of laboratory determinations of the parameters involved. For example, data reported from results of an extensive drilling and testing program carried out by the U. S. Geological Survey in the Los Banos-Kettleman City area, California [*Johnson and Morris*, 1962], are suited to this analysis. Undisturbed samples from test holes drilled to depths of 1000 to 2200 feet were tested by the Survey for coefficients of consolidation and permeability and classified according to the Unified Soil Classification System. When these data are used in (14), calculations of specific storage are between 1 and 2 orders of magnitude smaller than those reported in Table 1. The discrepancy occurs because permeability values calculated from consolidation-test data are small owing to large porosity decline under loads required to consolidate the samples. As pointed out by Johnson and Morris, permeability determinations are most meaningful if they are computed at the void ratio existing under effective overburden pressure.

In the same study, undisturbed samples immediately adjacent to samples subjected to laboratory consolidation were tested by the Survey in a variable-head permeameter under no load. These data, when used in (14) with consolidation-test data, provide a range in specific storage of the order of magnitude reported in Table 1. For six samples classified as CL (inorganic clays of low to medium plasticity), the specific storage ranged from 3.3×10^{-4} to 1.9×10^{-3} and averaged 8.9×10^{-4}. For nine samples classified as CH (inorganic clays of high plasticity), the specific storage ranged from 1.4×10^{-4} to 2.7×10^{-3} and averaged 4.6×10^{-3}. Based on average values of specific storage, the bulk modulus of compression for CL and CH material is calculated to be 7.0×10^4 and 1.4×10^4, respectively.

TABLE 1. Range in Values for the Bulk Modulus of Compression (E) and Specific Storage (γ_ω/E) (Modified after *Jumikis*, 1962)

Material	E (lb/ft^2)	γ_ω/E (lb/ft)
Plastic clay	1×10^4 to 8×10^4	6.2×10^{-3} to 7.8×10^{-4}
Stiff clay	8×10^4 to 1.6×10^5	7.8×10^{-4} to 3.9×10^{-4}
Medium hard clay	1.6×10^5 to 3×10^5	3.9×10^{-4} to 2.8×10^{-4}
Loose sand	2×10^5 to 4×10^5	3.1×10^{-4} to 1.5×10^{-4}
Dense sand	1×10^6 to 1.6×10^6	6.2×10^{-5} to 3.9×10^{-5}
Dense sandy gravel	2×10^6 to 4×10^6	3.1×10^{-5} to 1.5×10^{-5}
Rock, fissured, jointed	3×10^6 to 6.25×10^7	2.1×10^{-5} to 1×10^{-6}
Rock, sound	Greater than 6.25×10^7	Less than 1×10^{-6}

In summary, decreases in porosity unquestionably affect permeability, but not to the extent permitted in the consolidation apparatus. Therefore, the values of specific storage calculated from conventional consolidation test data are too small. On the other hand, recognizing that the no-load permeameter operates at the other extreme, i.e., immaterial effect on voids, the calculated values of specific storage given, and therefore those reported in Table 1, may be too high for application in subsidence problems.

PORE-WATER-PRESSURE DECAY AND ULTIMATE SUBSIDENCE

From soil mechanics studies it is known that settlement-time rates depend upon numerous factors. Mathematically, the theory can be stated from (5)

$$\partial u/\partial t = (E_c k'/\gamma_\omega)(\partial^2 u/\partial z^2) \quad (15)$$

This equation states that the rate of change of excess pore-water pressure in a confining bed equals the amount of water expelled from the voids of the bed per unit area per unit time. When u is solved as a function of z and t, the equation describes the exact shape of a family of curves (isochrones) showing the proportion of effective and neutral stresses in a confining bed from time $t = 0$, when u begins declining, to time $t = \infty$, when steady-flow conditions are re-established.

Taylor [1948] derived a solution to (15) by assigning boundary values to the internal pressure within a consolidating mass and the external pressure at its upper and lower boundary. For boundary conditions stipulated by Taylor, an identical solution is available from heat-flow theory for conditions of linear flow of heat in the region of a solid $-L < x < L$ with zero initial temperature (v) and with the surfaces $x = \pm L$ kept at constant temperature (V) for $t > 0$ [*Carslaw and Jaeger*, 1959]. The solution is in the form of a family of curves in two dimensions and demonstrates values of v/V and x/L for various values of a dimensionless time factor (T) (Figure 2). The temperature ratio v/V represents the temperature buildup in the slab, where v is the temperature at any given time and V is the temperature applied to the surfaces. At time equal to zero, v/V equals zero; at time equal to ∞, v/V equals unity.

The ratio v/V is analogous to the pore-water-pressure decay, stated for this problem as $1 - u/u_i$, where u_i represents the initial value of the pore-water pressure and u its value at any later time. At time equal to zero, u equals u_i and $1 - u/u_i$ equals zero; at time equal to ∞, u equals 0 and $1 - u/u_i$ equals unity. In other words, as the excess pressure (u) varies from its initial value (u_i) to zero, the excess-pressure decay that takes place over the same interval of time varies from 0 to 100%. The dimensionless parameter x/L in Figure 2 is analogous to z/H, where z is the depth to any given horizontal plane through the layer and H is the total thickness of the layer.

The curves in Figure 2 represent conditions in the slab after periods of time have elapsed since the first application of a constant temperature at its surfaces. The numbers on the curves are values of $\kappa t/L^2$, where the constant κ is the diffusivity or thermometric conductivity of the substance. The diffusivity in heat flow is analogous to the product of the permeability and the reciprocal of specific storage in consolidation. The actual time for a given isochrone to be reached is a constant multiple of some dimensionless time factor T. For the case of consolidation of a clay stratum

$$T = k't E_c/H_L^2 \gamma_\omega \quad (16)$$

where H_L is distance to a drainage face. For the case of an interbedded clay member where drainage is possible from both an upper and lower surface, H_L equals $H/2$. In single drainage, i.e., drainage from an upper or lower surface only, H_L is taken as the thickness of the compressible member. This relationship and Figure 2 offer further proof that for any given thickness, the greater the specific storage or the smaller the permeability, the greater the time that is required to re-establish a steady-flow condition in the confining layer.

The principle of effective stress is intimately related to the process of pore-water decay described in the thermal model. This principle was best expressed by *Terzaghi* [1936] who originated the concept:

> The stresses in any point of a section through a mass of soil can be computed from the total principle stress σ which acts in this point. If the voids of the soil are filled with water under a stress u, the total principle stress

consists of two parts. One part, u, acts in the water and in the solid in every direction with equal intensity (neutral stress). ⋯ The balance $\bar{\sigma} = \sigma - u$ represents an excess over the neutral stress u and has its seat exclusively in the solid phase of the soil.

This fraction of the total principle stress will be called the effective stress. ⋯ A change in the neutral stress u produces practically no volume change and has practically no influence on the stress conditions for failure. Porous materials (such as sand, clay, and concrete) react to a change of u as if they were incompressible and as if their internal friction were equal to zero. All the measurable effects of a change of stress, such as compression, distortion, and a change of shearing resistance are exclusively due to changes in the effective stress $\bar{\sigma}$.

A depth-pressure diagram is commonly used to illustrate the phenomenon of stress transfer from pore water to effective. Figure 3 depicts a common set of conditions in an artesian basin where the flow potential varies with depth. The term 'artesian' is used in the sense that the head at any depth, when calculated from the top of the saturated zone, exceeds hydrostatic pressure. For the situation shown, the potential increases with increasing depth, and water is slowly seeping upward, demonstrated by the loss of fluid pressure across the low-permeability confining layers and the different heads in each aquifer. The loss of head occurring across each aquifer is considered negligible. A water table is assumed at the surface, and the hydrostatic pressure increases linearly with depth. The total pressure acting on any plane is the sum of the neutral and effective pressures. The neutral pressure consists of the hydrostatic and seepage pressure.

If the lower aquifer in Figure 3 is developed by closely spaced wells, the head will be covered, and the seepage pressure in aquifer 1 will be reduced by $\Delta h \gamma_\omega$ where Δh is the amount of lowering. Owing to this lowering, effective- and neutral-pressure fractions acting throughout the overlying confining layer must seek a new equilibrium compatible with the surrounding environment. The resulting process is one of

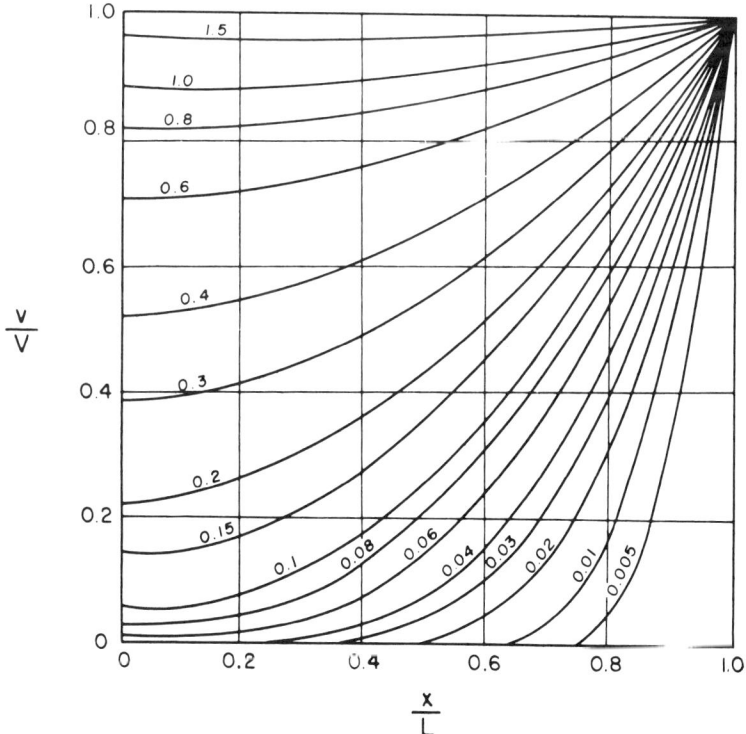

Fig. 2. Temperature distribution at various times in the slab $-L < X < L$ with zero initial temperature and surface temperature V. The numbers on the curve are the values of $\kappa t/L^2$ (after *Carslaw and Jaeger*, 1959).

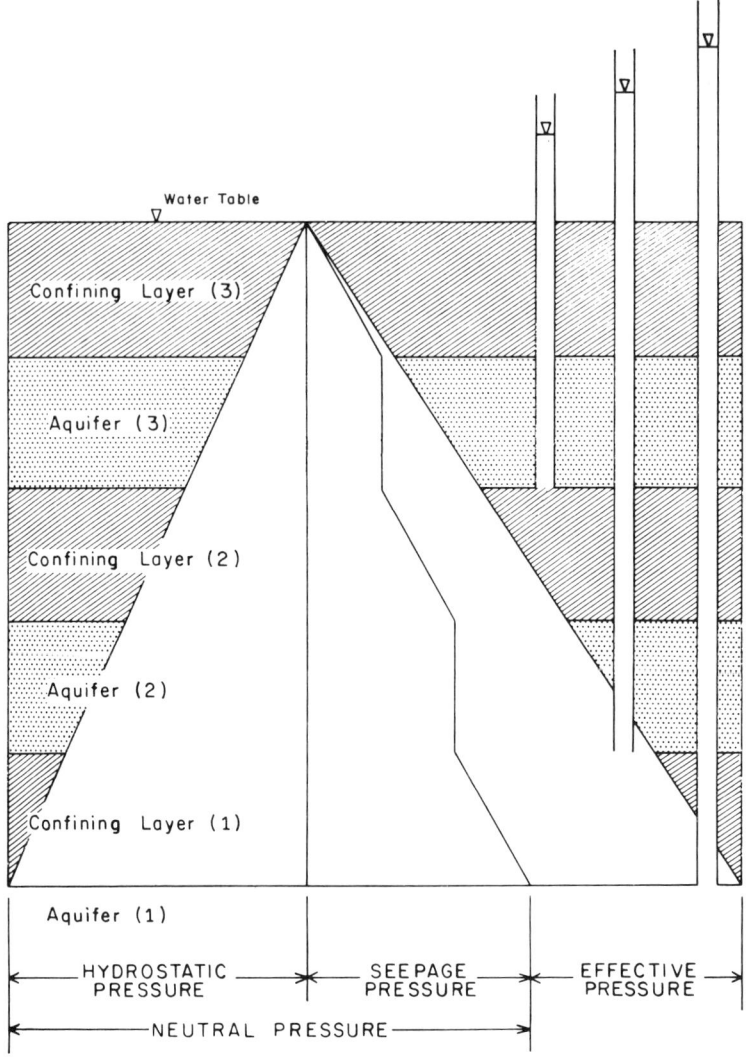

Fig. 3. Idealized depth-pressure diagram for part of an artesian basin.

stress transfer from neutral to effective as shown by the isochrone development from $t = 0$ to $t = \infty$ in Figure 4A. As previously mentioned, all measurable compressions are due to changes in effective stress. Hence, the migration of isochrones over the limits of time can be visualized as a progressive increase in the effective pressure or consolidation of the stratum as it drains. The water released is derived from storage in the confining layer.

Here, one important feature of the system must be considered. After a new equilibrium is achieved ($t = \infty$, Figure 4A), the total loss in head across the confining layer varies linearly from Δh_1 at the bottom to zero at the top. It follows that the change in head within the confining layer at any depth z below the top of the layer at $t = \infty$ can be expressed

$$\delta h = \Delta h(z/H) \qquad (17)$$

Hence, the change in pressure and therefore the volume of water derived from each element of confining layer are not constants but variables depending mainly on the depth z of the element below the top of the layer.

A more general expression is derived by considering changes in aquifer pressure both above and below a confining layer. This is demon-

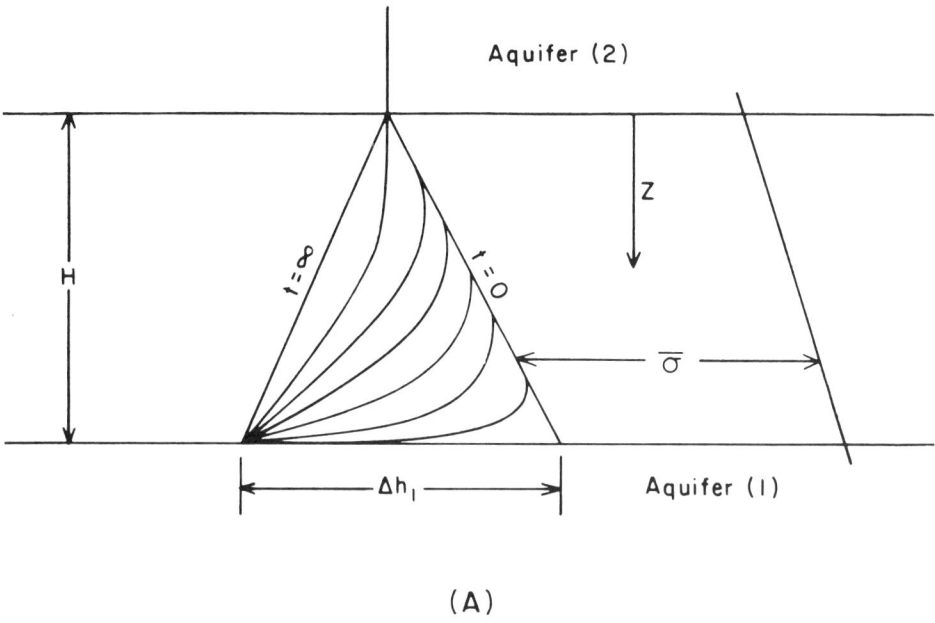

(A)

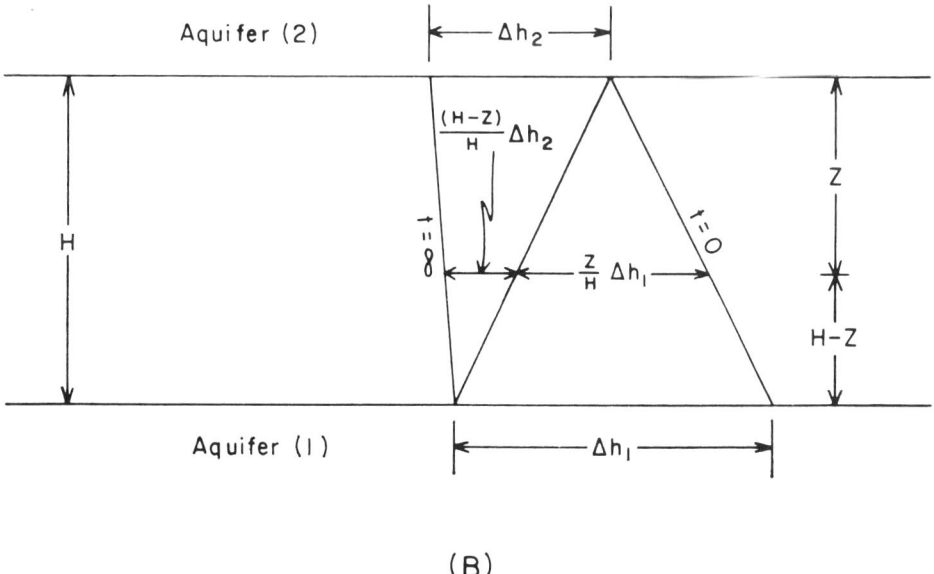

(B)

Fig. 4. Depth-pressure diagrams of a confining layer illustrating effective-pressure areas when (A) the head is lowered in one bounding aquifer, and (B) the heads are lowered in both bounding aquifers.

strated in Figure 4B where Δh_1 and Δh_2 are the pressure changes in the lower and upper aquifers, respectively. The total pressure change experienced by the confining layer at depth z at equilibrium is, from (17)

$$\delta h_1 = \Delta h_1 (z/H) \quad (18)$$

plus

$$\delta h_2 = \Delta h_2 [(H - z)/H] \quad (19)$$

or

$$\delta h = (\Delta h_1 - \Delta h_2)(z/H) + \Delta h_2 \quad (20)$$

The coefficient of storage of a compressible confining layer, if defined as the volume of water removed from a prism with a 1-ft² base per unit lowering of excess pore-water pressure throughout the prism, is expressed as the product of specific storage and thickness. As previously demonstrated, the pressure change within each element in Figure 4 is not constant but variable depending mainly on the depth z of the element.

The specific discharge (dq) of an element, being related to the variable pressure change, is introduced as the product of the specific storage and the average pressure change within that element after a new equilibrium is reached

$$dq = (\gamma_\omega/E_c)[(\Delta h_1 - \Delta h_2)(z/H) + \Delta h_2]\, dz \quad (21)$$

The total amount of water (q) removed from the full prism after a new state of equilibrium can be expressed by summing the elemental discharges

$$q = \frac{\gamma_\omega}{E_c} \int_0^H \left[(\Delta h_1 - \Delta h_2)\frac{z}{H} + \Delta h_2 \right] dz \quad (22)$$

or

$$q = \frac{\gamma_\omega}{E_c} \frac{(\Delta h_1 + \Delta h_2)}{2} H \quad (23)$$

It is noted that (23) expresses a volume per unit area and therefore has the dimensions of a length. This equation is a general expression for the vertical shortening of a prism of confining layer when steady-flow conditions are re-established.

For the case shown in Figure 4A, Δh_2 is zero, and (23) can be restated as

$$q = (\gamma_\omega/E_c)(\Delta h_1/2) H \quad (24)$$

Examination of (24) shows that, at equilibrium, the vertical shortening of the confining layer equals the product of the specific storage and the final area described by the increase in effective pressure on the two dimensional depth-pressure diagram (Figure 4A), in this case a triangle of height H and base Δh_1. This triangle can be appropriately termed 'effective-pressure area.' Hence, the total volume of water removed from the prism of confining layer is equal to the product of specific storage and final effective-pressure area as determined by a depth-pressure diagram.

If the heads in aquifers 1 and 2 are lowered an identical amount, the final effective-pressure area on the depth-pressure diagram is a parallelogram (Figure 5A), and the general equation 23 reduces to

$$q = (\gamma_\omega/E_c)\, \Delta h H \quad (25)$$

When heads in aquifers adjacent to a compressible confining layer are lowered an unequal amount (Figure 5B), a trapezoidal effective-pressure area results (½ the sum of the bases times the height), and the general equation 23 expresses the vertical shortening at equilibrium.

For all situations discussed, the coefficient of storage of the confining layer is the volume of water removed from the prism per unit change in average head across the prism, or from (23)

$$q/[(\Delta h_1 + \Delta h_2)/2] = (\gamma_\omega/E_c) H \quad (26)$$

which, like a confined aquifer, is the specific storage multiplied by the thickness.

From the preceding discussion, two relationships are apparent. First, for confining layers of equal specific storage and thickness in Figure 5, equal changes in seepage pressures will produce at equilibrium the same amount of vertical shortening and recovery of stored water. If the head remains unchanged in one of the aquifers, the vertical shortening and recovery of stored water at equilibrium will be half as great. This is demonstrated by the triangular effective-pressure area shown in Figure 4A. Second, confining layers shown in Figure 5 will drain from both an upper and lower surface but not equally, as assumed in the double-drainage expression for (16). This unequal drainage is due to the variable initial pressure in the cases cited as opposed to the constant initial temperature in the thermal model from which (16) was

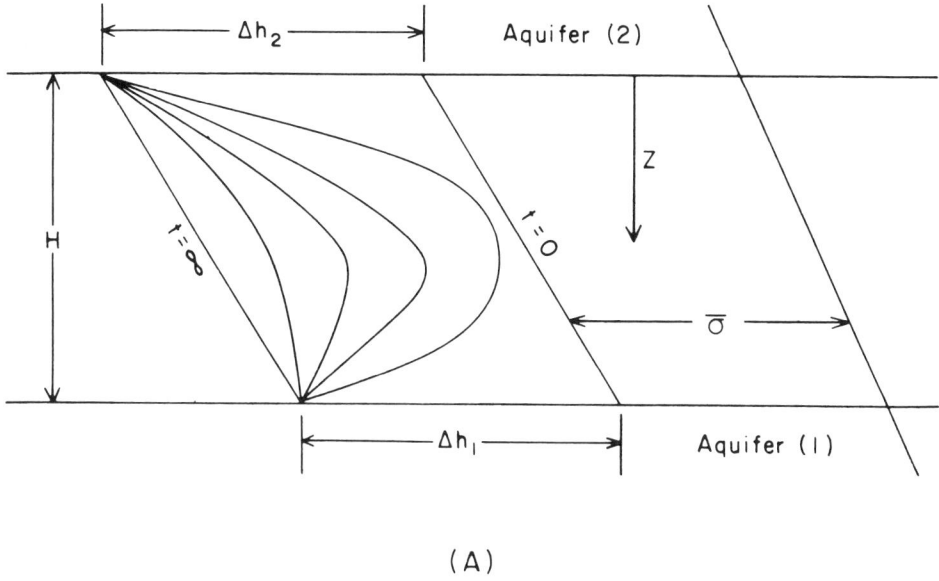

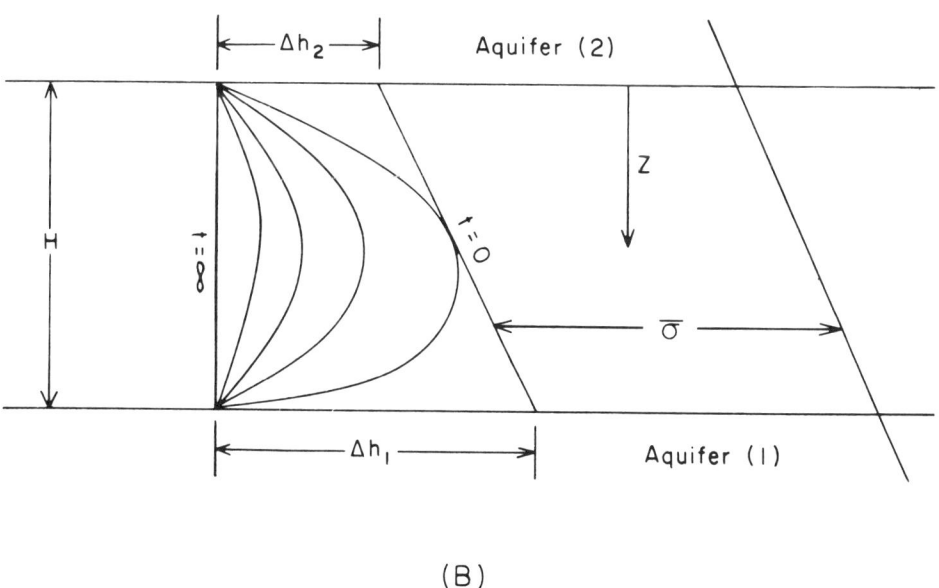

Fig. 5. Depth-pressure diagrams of a confining layer illustrating (A) parallelogrammic and (B) trapezoidal effective-pressure areas.

derived. For this same reason, the so-called single-drainage case represented by Figure 4A indicates upward as well as downward flow during most of the transient period of stress transfer. Hence, the absolute distances H and $H/2$ are valid only in situations where initially there was no flow across the confining layer, and then only if the final effective-pressure areas are either triangular or rectangular. *Carslaw and Jaeger* [1959] discuss solutions to other than initial conditions of constant temperature, including linear and linear-plus-constant temperature distributions for a region $0 < x < L$ with no flow at $x = 0$ and zero temperature at $x = L$. However, an imposed no-flow boundary at the top of a confining layer ($x = 0$) is not compatible with the cases discussed and requires a different geometry. Further, *Taylor* [1948] compared the effect of constant and variable initial pressure on time of consolidation and concluded that the differences are not great. In view of the approximate nature of the consolidation theory as a whole, he proposed acceptance of the symmetrical double- or absolute single-drainage conditions. Although certain refinements may be in order for specific situations, this idealization is utilized in the following discussion.

Discharge–time relationship. In areas of land subsidence, the volume of water removed from compressible confining layers equals the volume of measured land subsidence (ignoring compression of the aquifers). This equation was first stated for a practical problem by *Poland* [1961]. As demonstrated in this paper, the volume of water removed from compressible confining layers equals the product of their specific storage and the effective-pressure area developed across them. The utility of this concept hinges on conditions in a given basin. Equation 26 is valid only if the change in effective pressure and accompanying vertical shortening of the confining beds is complete under a given reduction in artesian head. In other words, if the amount of land subsidence is used in the numerator of (26), it is inherently assumed that the stress transfer from excess pore water to effective is complete under the average artesian-head decline expressed in the denominator. Anything less than ultimate shortening merely reflects a value of the transient effective-pressure area, demonstrated in Figures 4A and 5 by the migration of isochrones over the limits of time.

The process of stress transfer from pore water to effective for an ideal symmetrical pattern of pore-water decay can be related to the increase in total discharge from the beds and to time with Figure 6. This figure, taken from *Carslaw and Jaeger* [1959], shows average values of the temperature ratio depicted on Figure 2 and is used by soils engineers in predicting settlement-time rates. For application in areas of land subsidence, the temperature ratio (ordinate) is replaced by a ratio expressing percentage of total discharge from a clay prism, where S'' is total discharge per unit head decline across the prism at any given effective-pressure area for times less than ∞, and $\gamma_w H/E_c$ is total discharge per unit head decline across the prism at the final effective-pressure area at time equal to ∞, designated as S'. The abscissa of Figure 6 represents values of the time factor as defined in (16). Hence, Figure 6 shows the average percentage of completed stress transfer from excess pore water to effective for various periods of time where t is expressed as a constant multiple of a dimensionless time factor. The storage coefficient or average storativity of the beds is thus related to time through the principle of effective stress.

A corollary to the relationship shown in Figure 6 may be derived by recognizing that the section of the curve from $S''/S' \leq 0.6$ very closely approximates a parabola with the equation

$$(\pi/4)\,(S''/S')^2 = k'E_c t/H_L^2 \gamma_\omega \qquad (27)$$

Substituting $\gamma_\omega H/E_c$ for S' and replacing $H/2$ for H_L for the ideal symmetrical double-drainage case results in

$$(\pi/4)\,(S'')^2 = 4k't\gamma_\omega/E_c \qquad (28)$$

For the case of absolute single drainage, this relationship is expressed

$$(\pi/4)\,(S'')^2 = k't\gamma_\omega/E_c \qquad (29)$$

From (28) and (29), representing double and single drainage respectively, for $S''/S' \leq 0.6$, equal values of S'' entail different values of time, other factors being equal. The length of time required to develop a certain effective-pressure area in the single-drainage case is 4 times greater than in the case of double drainage. On the other hand, for equal values of time, S'' corresponding to double drainage is twice as great as S'' for single drainage.

APPLICATIONS

The principles of groundwater flow in compressible confining layers may be useful to analytical study of the phenomenon of land subsidence caused by fluid withdrawals. As has been demonstrated, the amount and rate of land subsidence or groundwater recovery from

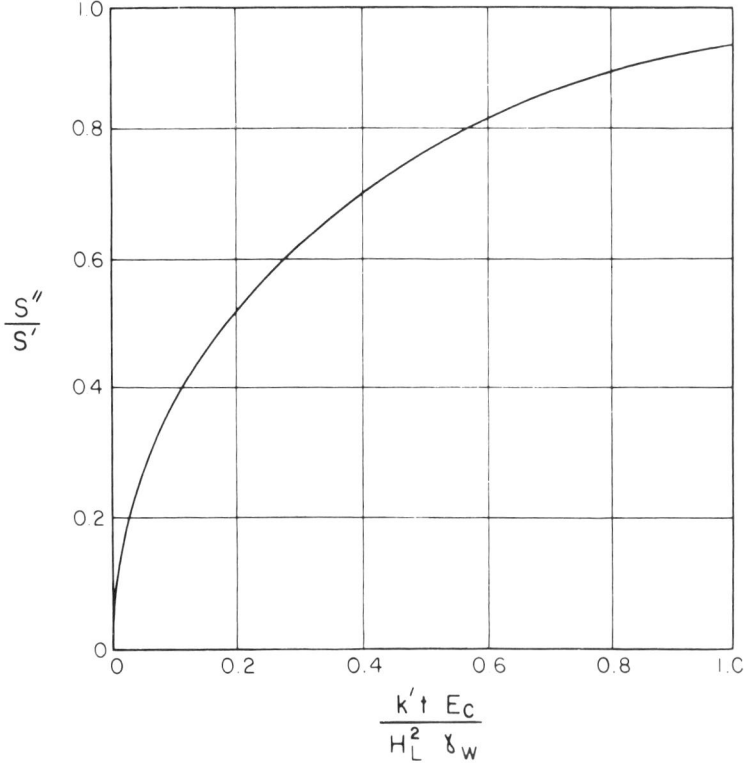

Fig. 6. Average storativity ratio for a confining layer (modified after *Carslaw and Jaeger,* 1959).

compressible confining layers is a function of the specific storage of strata and the development of an effective-pressure area across them. The time lag inherent in release of stored water from compressible layers and their subsequent compaction is demonstrated in Figure 6. Strict interpretation of this figure must concur with stipulations of the heat-flow model from which it was developed. The figure can be used to find the time when a certain percentage of water has been released (or subsidence completed) only if the stress was originally applied instantaneously. That is, if computed times are to have meaning, this method can only be applied to situations where pressure levels in bounding aquifers were lowered rapidly and then maintained at a relatively constant level. More often, pressure levels in groundwater basins undergoing subsidence remain in a state of progressive decline for years or even decades. Hence, the effective-pressure area is continually expanding. The amount of subsidence at any given benchmark for a given period of time equals the product of specific storage and the effective-pressure area for that time. If the specific storage and thickness of the beds are known, pressure change in units that have contributed to subsidence can be computed and compared with pressure change in the aquifers. This will demonstrate, at least theoretically, the percentage of completed stress transfer.

In areas where the subsurface is relatively well known and land subsidence has been temporarily halted at some time over the period of available record, measurements of potential at various times and depths in the aquifers will define the distribution of the effective-pressure area. Equations presented in this paper can be used to determine a field value of the bulk modulus of compression. This value will enable evaluation of the regional properties of clayey sediments and may be used to predict the response of the system to a new cycle of applied stress. The elastic compression of aquifers may be determined by the method of *Lohman* [1961]. When dealing with clay beds subjected to recurrent loading, however, the bulk modulus of compression is not the same for each cycle of load. From (8), the change in height per unit original height decreases with recurrent cycles of equal load, resulting in an apparent increase in the bulk modulus. From the formulas developed, this increase results in less subsidence per unit pressure change. This situation occurs in nature when water levels in a subsiding basin are lowered, allowed to recover due to cessation in pumping, and lowered once more. The second drawdown phase results in considerably less subsidence than the first, owing to the preloading effect produced by the first period of water-level decline. Physically, what occurs is an increase in the bulk modulus of compression, which according to (23) has an inverse effect on amount of subsidence. This phenomenon accounts for large differences in the ratio of land subsidence to water-level decline reported by *Green* [1964] for two drawdown phases of approximate equal magnitude in the Santa Clara Valley, California. A laboratory testing program conducted to simulate as closely as possible actual field conditions should provide information on the apparent change in the bulk modulus of compression due to recurrent cycles of of load.

An artesian basin often can be subdivided into two or more aquifer zones separated by one or more relatively impervious confining layers. The aquifer zones in turn consist of interbedded clays and sands in which the clay units contribute to most of the subsidence. In cases such as these, changes in fluid potential across various aquifer zones not only delineate the approximate depth of the seat of subsidence but define one dimension of the ultimate effective-pressure area to be reached. The geometry of the effective-pressure area may influence the time and rate of subsidence in at least two localities where land subsidence has been reported. In the Savannah area [*Davis et al.*, 1963] the reported ratio of land subsidence to water-level decline is about 0.0033, or 1 foot of subsidence for 304 feet of head decline. In the London area [*Wilson and Grace*, 1942] a similarly low ratio is implied from discontinuous records. For these areas, 'underdrainage' of compressible units overlying aquifers is thought to be one of the major factors causing the subsidence. Although several other factors are responsible for the small reported ratio, it is likely that underdrainage of a clay layer and subsequent development of a triangular effective-pressure area are significant controls on amount and rate of subsidence. Control of the geometry of the effective-pressure area across selected compressible units may provide a means for regulating rate and amount of subsidence within tolerable limits. The prospects for success of any such plan would be aided greatly by electrical simulation, or the process by which a system is modeled, tested, and improved, in whole or in part, with the aid of an analogous system [*Skibitzke*, 1960; *Domenico and Clark*, 1964].

Finally, translation of available soil-mechanics constants into terms correlative with hydrologic systems may be useful in the over-all analysis of aquifer-aquitard systems.

Acknowledgments. The writers are indebted to R. N. Farvolden, Assistant Professor of Geology, University of Illinois, for his encouragement and suggestions, most of which have been incorporated in this paper. They also are indebted to J. D. Bredehoeft and I. S. Papadopulos, U. S. Geological Survey, for their critical review of the manuscript and helpful suggestions.

This study was undertaken as part of contract P-16143 under the sponsorship of the National Science Foundation.

REFERENCES

Carslaw, H. S., and J. C. Jaeger, *Conduction of Heat in Solids*, pp. 92–102, Oxford at the Clarendon Press, 1959.

Cuevas, J. A., Foundation conditions in Mexico City, *Proc. Intern. Conf. Soil Mech., 3*, Cambridge, Mass., 1936.

Davis, G. H., J. B. Small, and H. B. Counts, Land subsidence related to decline of artesian pressure in the Ocala limestone at Savannah, Georgia, *Eng. Geol. Case Histories, 4, Geol. Soc. Am.*, 1–8, 1963.

Domenico, P. A., and G. Clark, Electric analogs in time-settlement problems, *Am. Soc. Civil Engr., 90*(SM3), 33–51, 1964.

Domenico, P. A., D. A. Stephenson, and G. B. Maxey, Ground water in Las Vegas Valley, *Desert Research Institute Tech. Rept. 7*, pp. 34–35, University of Nevada, 1964.

Green, J. H., The effect of artesian-pressure decline on confined aquifer systems and its relation to land subsidence, *U. S. Geol. Surv. Water-Supply Paper 1779-T*, 1964.

Hantush, M. S., Modification of the theory of leaky aquifers, *J. Geophys. Res., 65*(11), 3713–3725, 1960.

Jacob, C. E., On the flow of water in an elastic

artesian aquifer, *Trans. Am. Geophys. Union,* *21,* 574–586, 1940.

Jacob, C. E., Flow of ground water, in *Engineering Hydraulics,* edited by Hunter Rouse, pp. 321–386, John Wiley & Sons, New York, 1950.

Johnson, A. I., and D. A. Morris, Physical and hydrologic properties of water-bearing deposits from core holes in the Los Banos-Kettleman City area, California, *U. S. Geol. Surv. open-file rept.,* 1962.

Jumikis, A. R., *Soil Mechanics,* p. 384, D. Van Nostrand Book Company, New York, 1962.

Lohman, S. W., Compression of an elastic artesian aquifer, *U. S. Geol. Surv. Prof. Paper 424-B,* B47–B49, 1961.

Poland, J. F., and G. H. Davis, Subsidence of the land surface in the Tulare-Wasco (Delano) and Los Banos-Kettleman City areas, San Joaquin Valley, California, *Trans. Am. Geophys. Union,* *37,* 287–296, 1956.

Poland, J. F., The coefficient of storage in a region of major subsidence caused by compaction of an aquifer system, *U. S. Geol. Surv. Prof. Paper 424-B,* B52–B54, 1961.

Skibitzke, H. E., Electronic computers as an aid to the analysis of hydrologic problems, *Intern. Assoc. Sci. Hydrol.,* *2* (52), 347–358, 1960.

Taylor, D. W., *Fundamentals of Soil Mechanics,* pp. 225–238, John Wiley & Sons, New York, 1948.

Terzaghi, K., *Erdbaumechanik auf Bodenphysikalischer Grundlage,* Franz Deuticke, Vienna, 1925.

Terzaghi, K., The shearing resistance of saturated soils, *Proc. First Intern. Conf. Soil Mech.,* *1,* 54–56, 1936.

Terzaghi, K., and R. B. Peck, *Soil Mechanics in Engineering Practice,* pp. 64–65, John Wiley & Sons, New York, 1948.

Theis, C. V., The relation between the lowering of the piezometric surface and the rate and duration of discharge of a well using groundwater storage, *Trans. Am. Geophys. Union,* *2,* 519–524, 1935.

Wilson, G., and H. Grace, The settlement of London due to underdrainage of the London clay, *J. Inst. Civil Eng.,* *19* (2), 100–127, 1942.

Winslow, A. G., and L. A. Wood, Relation of land subsidence to groundwater withdrawals in the upper Gulf Coastal region of Texas, AIME Mining Section, *Mining Eng.,* 1030–1034, 1959.

(Manuscript received April 28, 1965.)

30

Reprinted from pages 187, 238-239, and 252-262 of *Geol. Soc. America Rev. Engineering Geol.* **2**:187-269 (1969).

LAND SUBSIDENCE DUE TO WITHDRAWAL OF FLUIDS*

J. F. POLAND AND G. H. DAVIS

U. S. Geological Survey, Sacramento, California, and Washington, D. C.

ABSTRACT

Land-surface subsidence due to the withdrawal of fluids by man has become relatively common in the United States since 1940 and has been described at several other places throughout the world. This paper reviews the known examples of appreciable land subsidence caused by fluid withdrawal. Those related to exploitation of oil and gas fields include Goose Creek, Texas; Wilmington, California; Lake Maracaibo, Venezuela; Niigata, Japan; and the Po Delta in Italy. The areas of major subsidence related to ground-water withdrawal include areas in Japan; Mexico City, Mexico; and Texas, Arizona, Nevada, and California. The areas of greatest extent and maximum subsidence are in California.

The principles involved in the compaction of sediments and of aquifer systems, basically the increase in effective stress, are examined briefly, together with their application to subsidence problems involving head decline both under water table and confined conditions. The amount of compaction that a confined aquifer system will experience is a function of compressibility. Other factors that influence compaction (and, in part, compressibility) include particle size and shape, clay mineralogy, geochemistry of pore water in the clayey beds and of the water in contiguous aquifers, and secondary compression.

Land subsidence has caused great damage in some areas. At several of these places, subsidence problems are being alleviated in one or more of several ways; these include (1) cessation of withdrawal and (2) increase or restoration of reservoir pressure by reduction in production rate, artificial recharge, or repressuring by injection of water. The greatest subsidence control measures are being taken at Wilmington, California, where subsidence that had reached 27 feet at the center now is nearly stopped; in addition, significant rebound has occurred.

[*Editors' Note:* Material has been omitted at this point.]

AREAS IN CALIFORNIA

Introductory Comment

Development of ground water has been intensive in California. Most of the water is withdrawn for irrigation although ground water is also pumped extensively for domestic and industrial supply. In 1955, about 13 million acre-feet of ground water was pumped for irrigation from ground-water basins in California, slightly more than one-third of the total pumpage for irrigation in the United States.

The ground-water withdrawals in California are chiefly from intermontane basins in which the valley fill of late Tertiary and Quaternary age is chiefly alluvial deposits but also, in places, is of lacustrine and shallow marine origin. In many of the basins, the water bodies tapped are semiconfined or confined below depths ranging from 100 to 600 feet, and, thus, much of the withdrawal is from confined aquifer systems.

The intensive and long-continued pumping has drawn down water levels more than 100 feet in many of the basins—maximum drawdown of 400–500 feet has been on the west side and in the southern end of the San Joaquin Valley. Thus, in California, substantial lowering of water level has occurred in young, unconsolidated, compressible deposits that contain extensive semiconfined to confined aquifer systems. It is not surprising, therefore, that land subsidence due to ground-water withdrawal has developed in many areas. Subsidence in all areas shown on Figure 31, except for those in the Wilmington oil field and in the Sacramento-San Joaquin Delta, is due to ground-water withdrawal and decline of artesian head. Subsidence in most of these areas has been described briefly by Poland (1958). About 11,000 square miles (7 million acres) of land in California is irrigated, and nearly two-thirds of this is irrigated wholly or in part by ground water. Of this 11,000 square miles, at least 30 per cent has subsided 1 foot or more due to artesian-head decline. Doubtless, many other areas in addition to those shown on Figure 31 have subsided at least a few tenths of a foot, but leveling control is not available to define the magnitude and extent of subsidence in most of these.

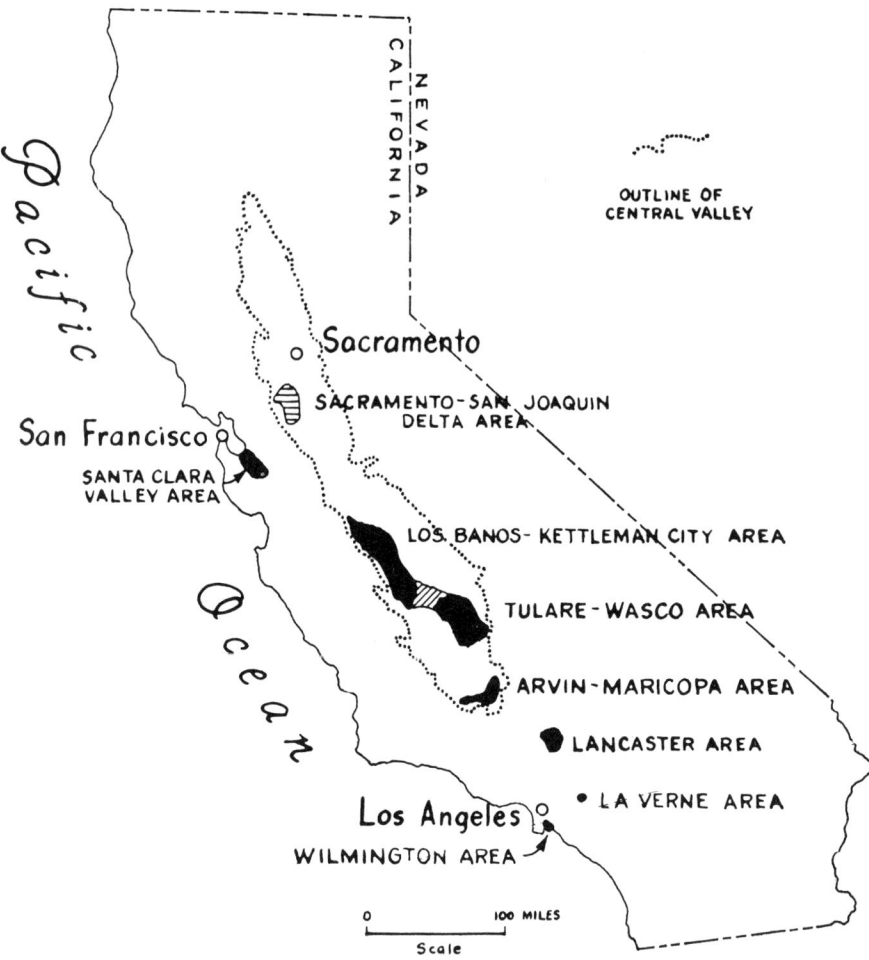

Figure 31. Areas of land subsidence in California. Major subsidence due to fluid withdrawal shown in black; subsidence in Delta caused by oxidation of peat.

The following description is limited to the major subsiding areas—in the San Joaquin Valley and the Santa Clara Valley south of San Francisco. The peat lands in the Sacramento–San Joaquin Delta have subsided as much as 15 feet in the last century as a result of drainage (lowering of the water table) for cultivation (Poland, 1958, p. 1774–1775). The primary cause of this subsidence, however, is oxidation of the peat (Weir, 1950), rather than compaction of an aquifer system due to increase in effective overburden load. The subsidence of peat lands is discussed by Alice Allen elsewhere in this Review Volume.

[Editors' Note: Material has been omitted at this point.]

Santa Clara Valley

Land subsidence in the central part of the Santa Clara Valley, beneath the southern part of San Francisco Bay and south past San Jose, has been recognized for 30 years (Rappleye, 1933; Tibbetts, 1933). It was first noted when releveling in 1932–1933 of a line of first-order levels established by the U. S.

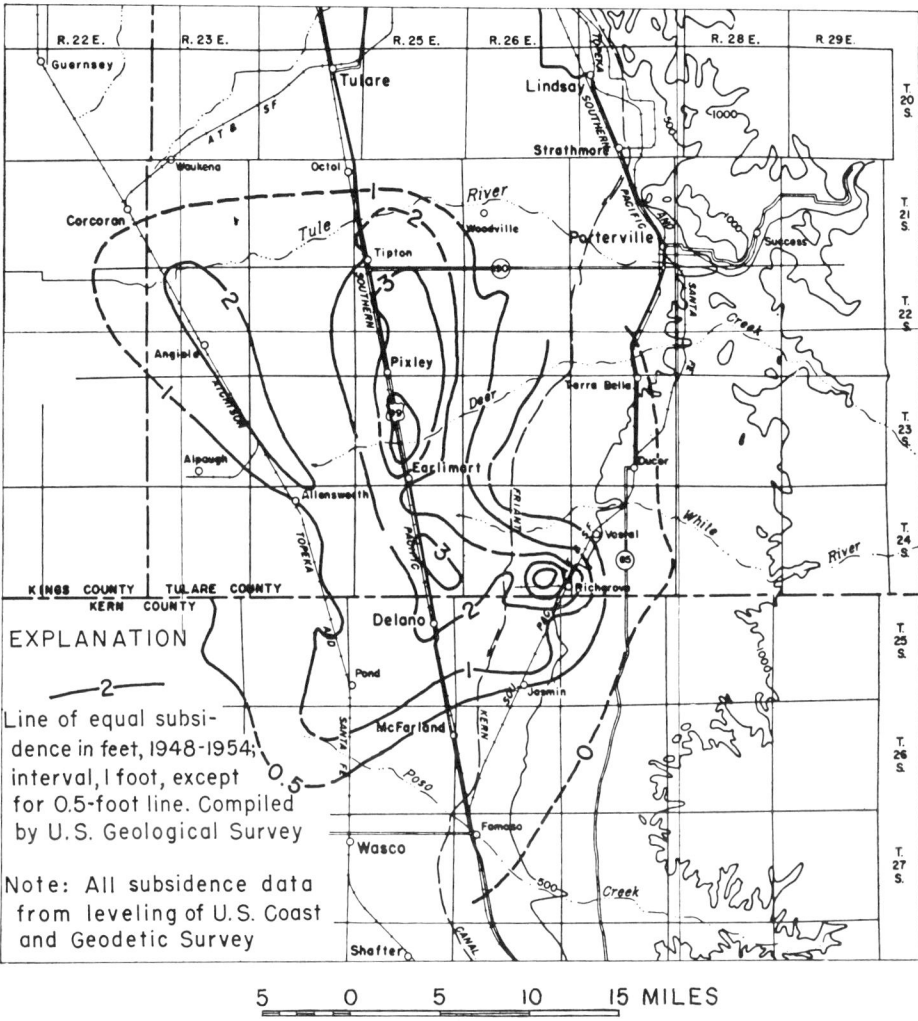

Figure 41. Land subsidence, 1948–1954, in Tulare-Wasco area, California.

Coast and Geodetic Survey in 1912 showed about 4 feet of subsidence at San Jose (Tolman and Poland, 1940, p. 29). Tolman and Poland described the subsidence to 1939 and concluded that decline in artesian head due to intensive and continued ground-water withdrawal was the principal cause. Poland and Green (1962) reported on the subsidence through 1954 and confirmed the earlier conclusions concerning the principal cause of subsidence.

The subsiding area extends southward about 25 miles from Redwood City and Niles to and beyond San Jose, has a maximum width of about 13 miles, and includes about 250 square miles (Fig. 45). The following summary is chiefly from the cited reports, from subsequent investigation of the subsidence by J. H. Green of the Geological Survey from 1958 through 1962, and from later leveling of the U. S. Coast and Geodetic Survey.

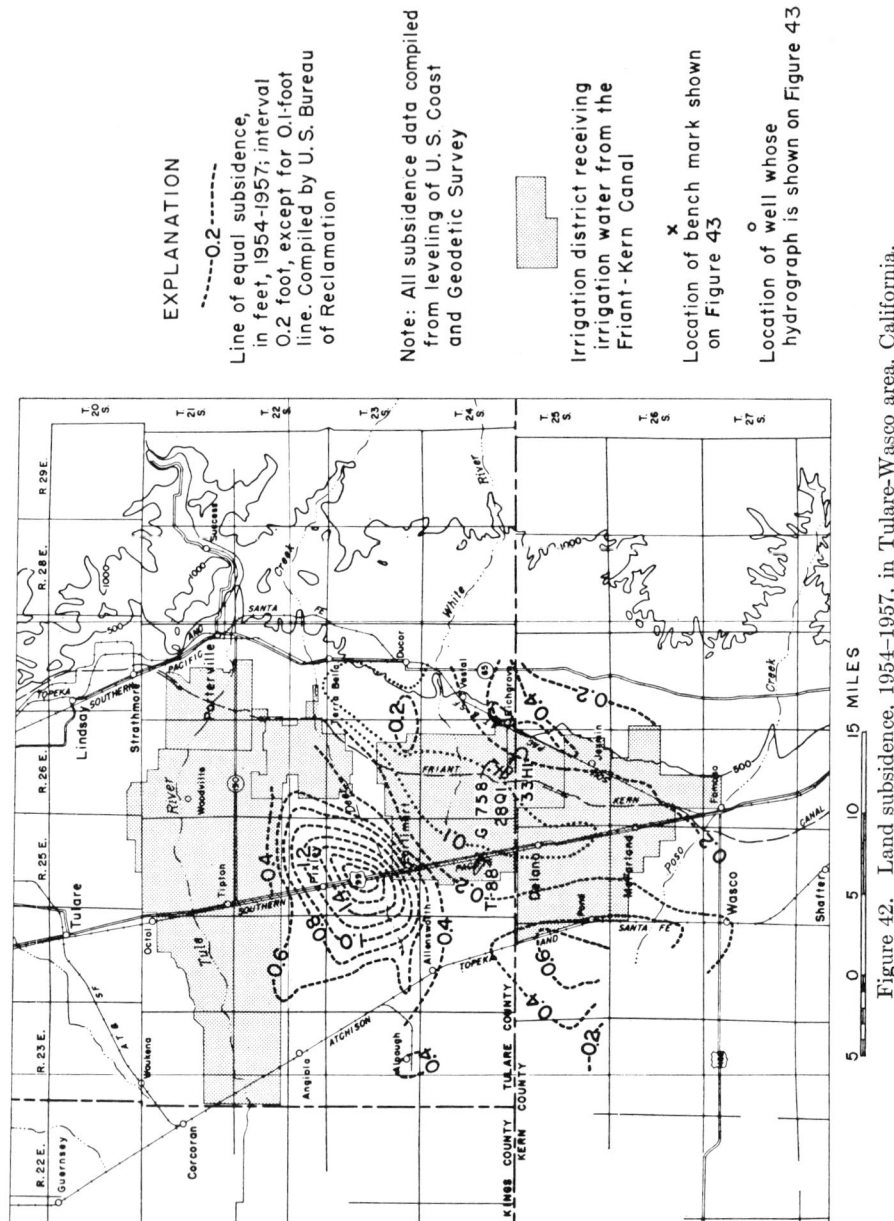

Figure 42. Land subsidence, 1954–1957, in Tulare-Wasco area, California.

Geology. The Santa Clara Valley is a structural trough, bounded on the southwest by the Santa Cruz Mountains and on the northeast by the Diablo Range. Consolidated bedrock forms the floor and sides of the trough. Overlying this bedrock are the fresh-water-bearing deposits, which are the semiconsolidated Santa Clara Formation of Pliocene and Pleistocene age and the unconsolidated alluvial and bay deposits of Pleistocene and Recent age. The contact between

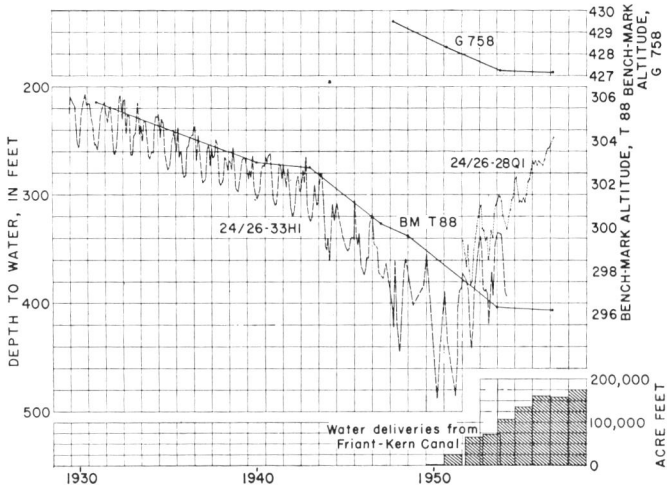

Figure 43. Subsidence and change in water level near Delano, California. Location shown on Figure 42.

these deposits cannot be recognized in well logs. Fresh water is obtained from wells that tap these alluvial and bay deposits and underlying deposits of the Santa Clara Formation; these deposits have a combined thickness of more than 1000 feet and possibly as much as 1500–2000 feet. Water wells range in depth from 200 to 1000 feet. Beneath the valley floor, the deeper wells probably tap the upper part of the Santa Clara Formation.

Where it is exposed around the valley margin, the Santa Clara Formation consists of semiconsolidated conglomerate, sandstone, siltstone, and claystone. These deposits, where they are tapped by wells in and near the outcrop area, have a low transmissibility and, at most places, do not yield enough water for irrigation wells.

The unconsolidated alluvial and bay deposits of clay, silt, sand, and gravel contain the most productive aquifers of the ground-water reservoir. Layers and tongues of silt and clay in the alluvial deposits are irregular but extensive, and they confine the ground water in the coarser lenses and tongues of sand and gravel. Thus, artesian conditions prevail beneath most of the valley, and initially, wells flowed near San Francisco Bay and south to San Jose (Clark, 1924).

Studies of samples from two 1000-foot core holes at the centers of subsidence in Sunnyvale and San Jose indicate that montmorillonite composes about 70 per cent of the clay-mineral assemblage in these deposits (R. H. Meade, June 1963, written commun.).

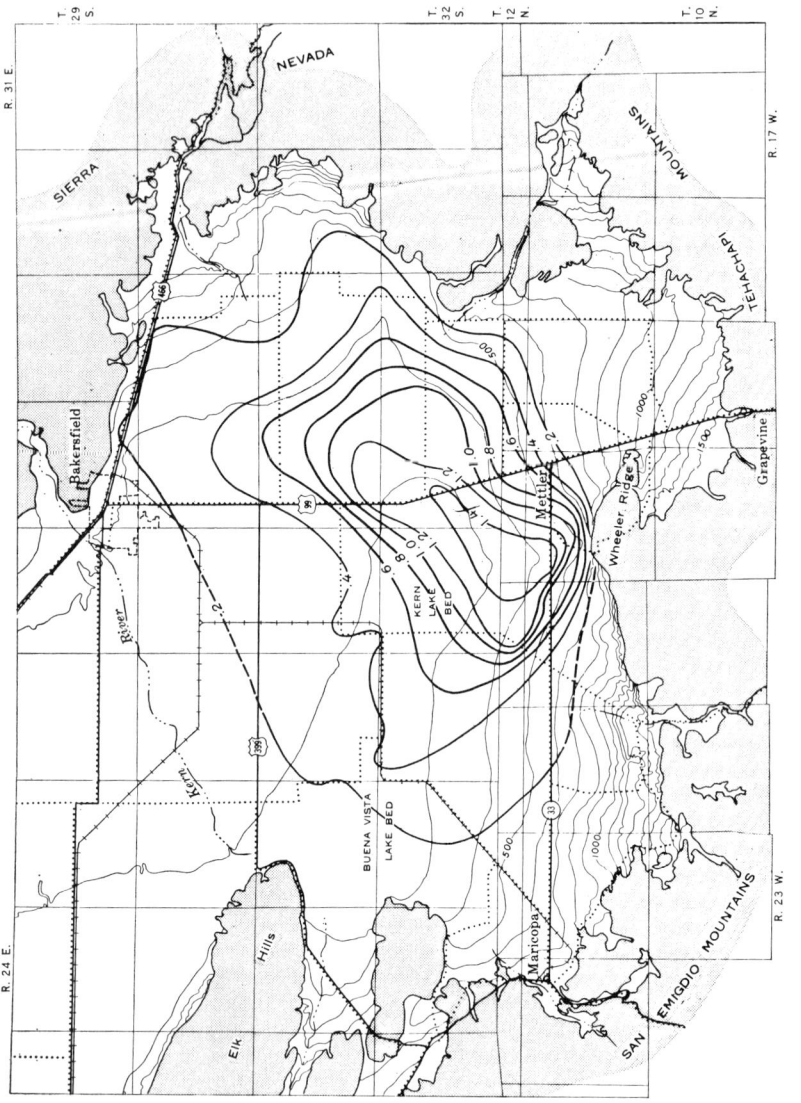

Figure 44. Land subsidence, 1959–1962, in Arvin-Maricopa area of San Joaquin Valley, California. Heavy contours are line of equal subsidence, in feet; shaded areas are deformed rocks bordering the valley. (*After* Lofgren, 1963, Fig. 47.7.)

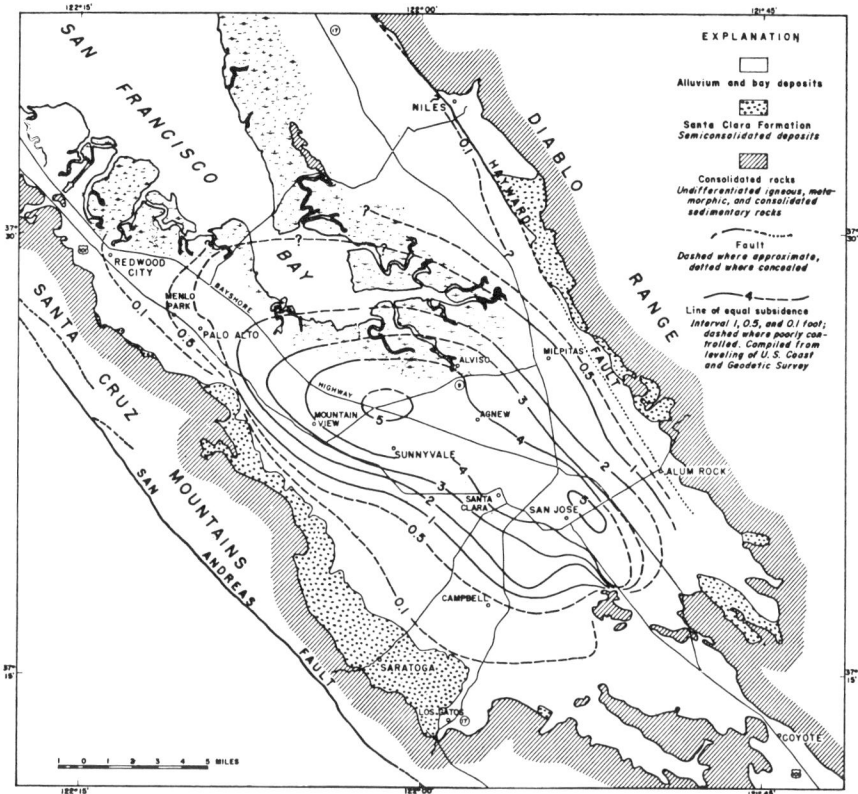

Figure 45. Land subsidence, 1934–1960, in Santa Clara Valley, California. (*Compiled by* J. H. Green.)

Hydrology. Development of ground water for irrigation began about 1900, and by 1934, about all irrigable land was under irrigation (Hunt, 1940, p. 14). Prior to 1942, pumpage for domestic and industrial use was small compared to irrigation draft, but with the rapid urbanization of the area during and since World War II, it has increased rapidly. Average pumpage in the fifties for all purposes has been about 200,000 acre-feet a year.

The intensive pumping of ground water has drawn down water levels throughout the valley. The hydrograph for well 7S/1E–7R1 in San Jose (Fig. 46) illustrates the change in water levels in the artesian area. The artesian head in this well, which was above the land surface in 1915, declined about 100 feet by 1934, recovered about 75 feet by 1943 during a period of above-normal rainfall, and then declined about 140 feet by 1962.

Land subsidence. The subsidence in the Santa Clara Valley has been well defined in time and magnitude by leveling of the U. S. Coast and Geodetic Survey. The level net established by that agency in 1933 (Fig. 47), which is tied to bedrock at several places, was completely leveled for the first time in 1934, was releveled seven times from November 1934 to April 1940, and again in 1948, 1954, and 1960; a partial releveling of the net from Palo Alto to bedrock south of San

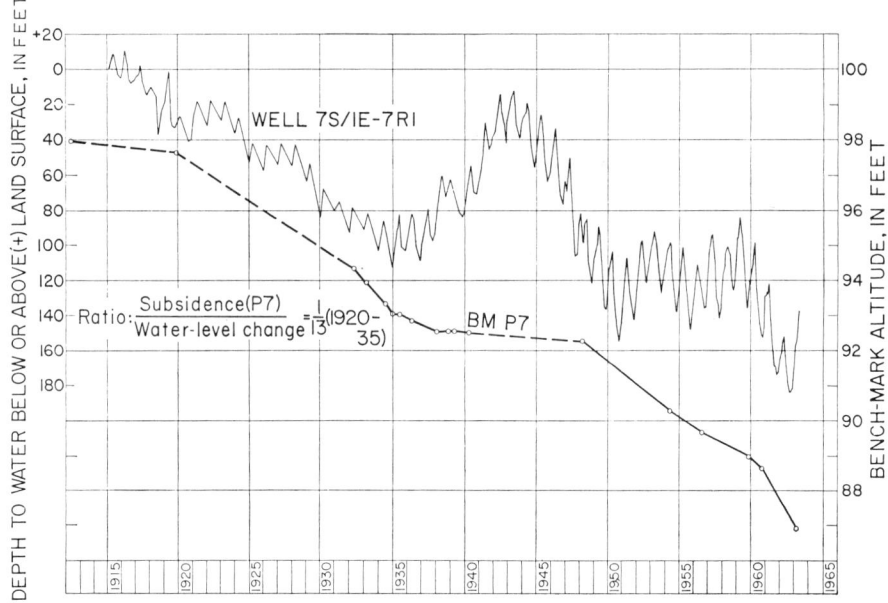

Figure 46. Change in altitude at bench mark P7 in San Jose, California, and change in artesian head in nearby well.

Jose was done in February 1963. The U. S. Coast and Geodetic Survey (1956) has published a compilation of adjusted elevations for all bench marks in the network through the 1954 leveling. Subsidence from 1934 to 1960 (Fig. 45) was a maximum of 5 feet at centers in San Jose and near Sunnyvale.

The longest record of subsidence is at bench mark P7 at the Hall of Records in San Jose (Fig. 46); from 1912 to February 1963, this bench mark subsided 11 feet. The rate of subsidence was greatest during periods of rapid decline in artesian head (*see* hydrograph for nearby well 7S/1E–7R1); the maximum rate was 0.73 foot per year from October 1960 to February 1963.

Profiles of land subsidence extending southeast near Redwood City to Coyote (Fig. 45) for 10 levelings between 1912 and 1963 (Fig. 48) indicate that maximum subsidence along the profiles occurred at bench mark P7 in San Jose although from 1934 to 1963, subsidence at bench mark J111 Reset in Sunnyvale (7 feet) was about equal to the maximum in San Jose (at bench mark B149).

Cause of subsidence. Within the artesian area (area of confined water), there is general correlation in time and rate of the subsidence and of the artesian-head decline as Figure 46 illustrates. Around the margins of the valley, however, where the deposits are mostly sand and gravel and are chiefly unconfined, substantial decline of the water table has not caused appreciable subsidence (Poland and Green, 1962, Fig. 12). These facts are positive evidence that the principal cause of the subsidence is the decline in artesian head and the corresponding increase in the grain-to-grain load on the aquifer system.

Quantitative comparison of land subsidence and compaction of the water-

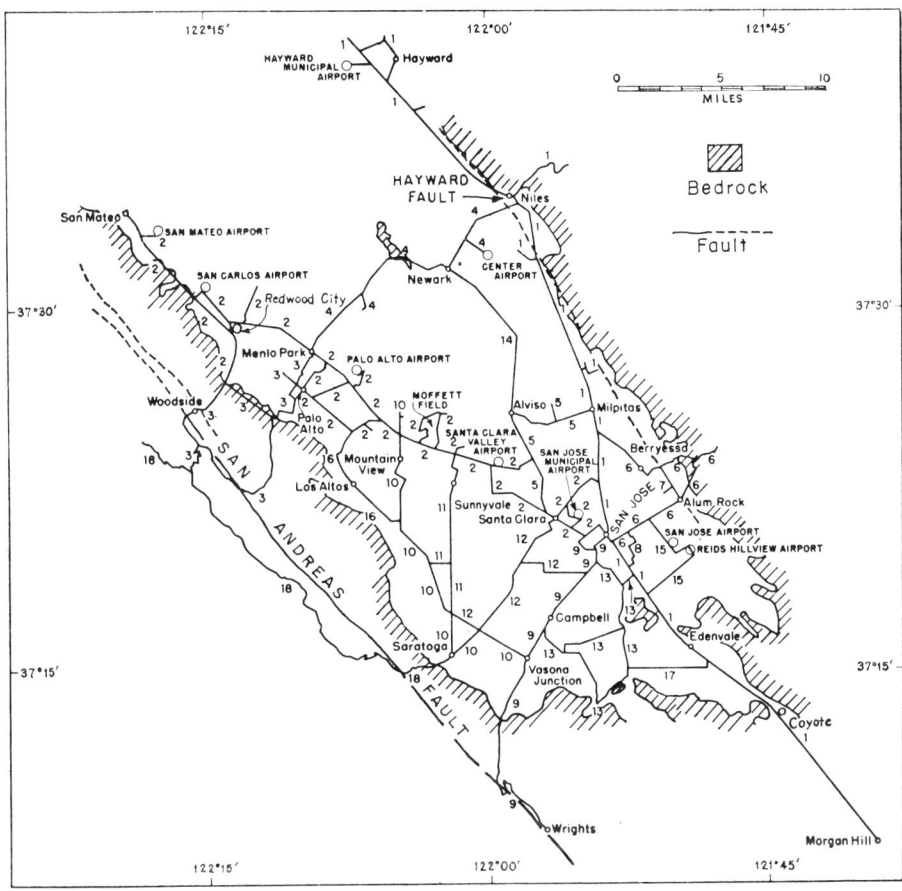

Figure 47. Map showing network of level lines in San Jose subsidence area, Santa Clara Valley, California. (*Modified after* U. S. Coast and Geodetic Survey; numbers identify level lines according to 1956 tabulation.)

bearing deposits tapped by wells has been obtained. In 1960, the Geological Survey installed compaction recorders in two core holes drilled at the centers of subsidence in Sunnyvale and San Jose. The purpose of the equipment was to measure the rate and magnitude of compaction occurring between the land surface and the bottom of the test well. The equipment is similar to that installed in the San Joaquin Valley (Fig. 38). Figure 49 shows the compaction measured in the Sunnyvale core hole and in two shallower companion test wells. The core hole (24C7) was drilled to a depth of 1000 feet, which is the maximum depth of water wells in the area, and measured compaction in this well from October 1960 to February 1963 was 0.83 foot. Subsidence of the land surface at nearby bench mark J111 Reset during the same period, based on relevelings of the U. S. Coast and Geodetic Survey, was 0.82 foot. Thus, the land subsidence at this site is demonstrated to be equal to the compaction of the water-bearing deposits within the depth tapped by water wells, and the decline in artesian head is proved to

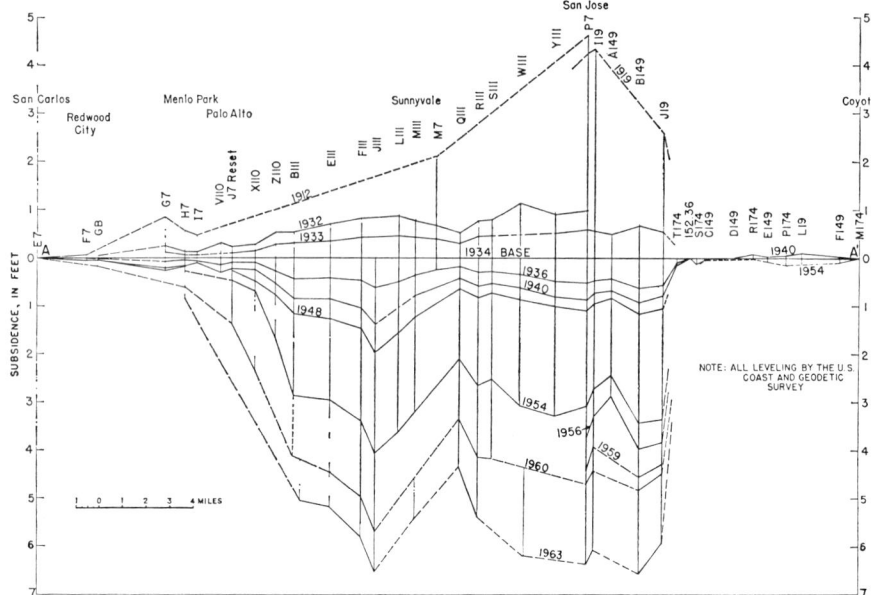

Figure 48. Profiles of land subsidence, 1912–1963, Redwood City to Coyote, California.

be the sole cause of the subsidence. Comparison of measured compaction in the other 1000-foot core hole in San Jose with subsidence of a nearby bench mark in the same period confirms this conclusion.

Green (1962) utilized results of consolidation and other laboratory tests on cores from the Sunnyvale and San Jose core holes to compute the compaction of the water-bearing deposits that theoretically should have occurred under the known change in artesian head from 1915 to 1960. The computed compaction agreed reasonably closely with the observed subsidence in the same period.

Protrusion of well casings has been common in the sinking area (Tolman, 1937, p. 344–345 and Figure 122). Many of the casings rose 2–3 feet above the land surface but usually were cut off before rising higher. This protrusion registers compaction between the land surface and the bottom of the casing but often is accompanied by compression and rupture of the casing and thus supplies only a minimum value for subsidence, not an absolute measure.

Relationship of subsidence to head decline. As a part of a study of the subsidence by the Geological Survey, J. H. Green (July 1962, written commun.) utilized subsidence and water-level change maps for the period 1948–1960 to map the ratio of subsidence to head decline in that period. The ratio increased from zero in the water-table areas around the valley margin to 0.10 in a central reach of the artesian area, extending from eastern San Jose to Sunnyvale. Thus, in this central reach, 1 foot of subsidence occurred for each 10 feet of water-level decline in the 12-year period. The ratio of subsidence at bench mark P7 to head decline at nearby well 7S/1E–7R1 (Fig. 46) was 0.08. From 1920 to 1935, when both subsidence and head decline were occurring at almost uniform rates, the ratio at P7 was also about 1:13 or 0.08.

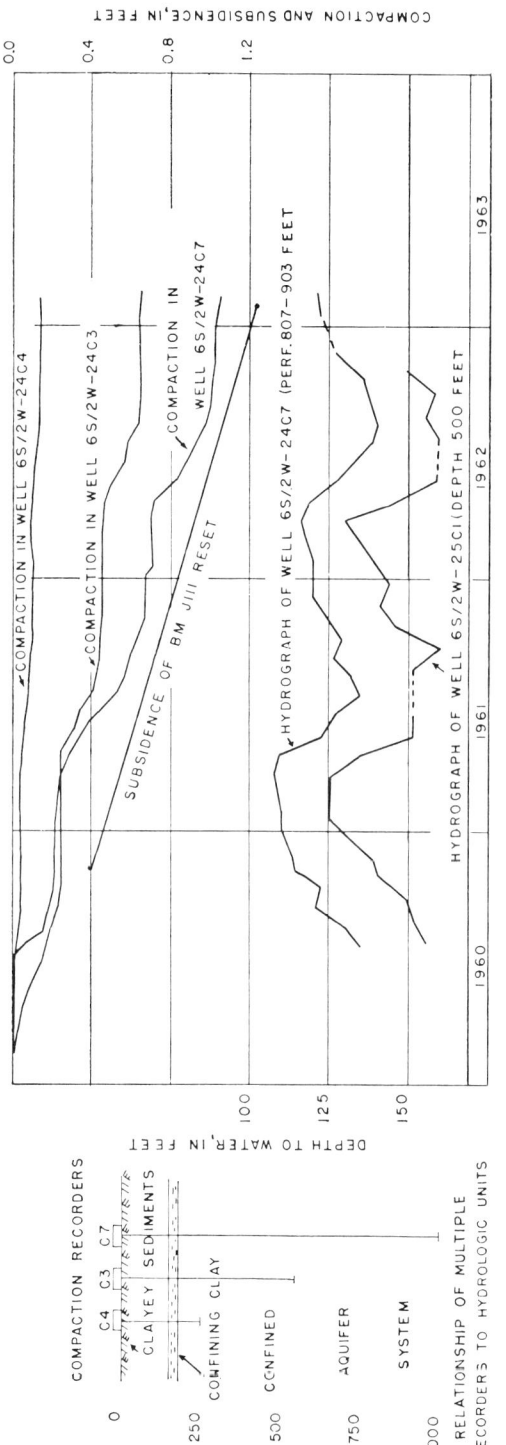

Figure 49. Measured compaction in wells 6S/2W-24C4, 24C3, and 24C7; water-level change; and subsidence of bench mark J111 Reset, in Sunnyvale, California.

Time lag in compaction, which affects the ratio of subsidence to head decline, is demonstrated clearly in Figure 46. From 1952 through 1958, the average artesian head was about uniform; however, bench mark P7 continued to subside throughout that 7-year period at only a slightly decreasing rate. The hydrograph registers the head change in the coarse permeable segments of the aquifer system; increase in effective stress in these segments occurs simultaneously with the head decline and is equal in magnitude. In the clay interbeds and confining beds (aquitards), however, the low permeability retards the escape of water, the reduction of pore pressure, and the compaction. Evidently, substantial excess pore pressures still existed in the aquitards in 1958 because the rate of compaction from 1956 through 1959 was two-thirds as great as from 1948 to 1954. This suggests that decay of excess pore pressure in the aquitards to approximate equilibrium with pore pressure in the aquifers would require many tens of years. Obviously, this time lag affects the ratio of subsidence to head decline; for example, if the average artesian head of 1952–1958 were maintained uniformly to 1970, the ratio of subsidence to head decline at this site for the 1948–1970 period would be substantially larger than the ratio obtained for the 1948–1960 period.

The volume of subsidence represented by the lines of equal subsidence from 1934 to 1960 (Fig. 45) is about 330,000 acre-feet (J. H. Green, January 1963, written commun.). This represents reduction in the pore volume of the ground-water reservoir, but the reduction has been principally in the fine-grained compressible clayey layers; therefore, it should not affect the storage capacity of the coarser grained sand and gravel aquifers appreciably. On the other hand, the quantity of water squeezed out of the aquifer system by permanent compaction is water available only once and thus is a nonreplenishable resource.

The extent to which the land surface would recover in response to substantial recovery of artesian head is not known. The artesian head at well 7S/1E–7R1 recovered about 75 feet from 1934 to 1943 (Fig. 46). Relevelings between 1934 and 1940 showed cessation of subsidence at bench mark P7 but no recovery. Leveling control was not available thereafter until 1948; thus, minor recovery of land surface may have occurred by 1943. Based on study of laboratory consolidation and rebound tests made on fine-grained clayey samples from the core holes, however, it is concluded that if initial artesian head were restored and the recharge water were chemically compatible with the water in the aquifers, recovery of the land surface would not be more than a few per cent of the total subsidence that has occurred.

[*Editors' Note:* Material has been omitted at this point.]

REFERENCES CITED

[*Editors' Note:* Only the references cited in the preceding excerpts are reproduced here.]

Clark, W.O., 1924, Ground water in Santa Clara Valley, California: U.S. Geol. Survey Water-Supply Paper 519, 209 p.

Green, J.H., 1962, Compaction of the aquifer system and land subsidence in the Santa Clara Valley, California: U.S. Geol. Survey Prof. Paper 450-D, art. 172, p.D175–D178

Hunt, G.W., 1940, Description and results of operation of the Santa Clara Valley Water Conservation District's project: Am. Geophys. Union Trans., pt. 1, p. 13–22

Poland, J.F., 1958, Land subsidence due to ground-water development: Am. Soc. Civil Engineers Proc., Jour. Irrigation and Drainage Div., Paper 1774, v. 84, no. IR3, 11 p.

Poland, J. F., and Green, J. H., 1956, Subsidence in the Santa Clara Valley, California—A progress report: U.S. Geol. Survey Water-Supply Paper 1461, p. 142–146

Rappleye, H. S., 1933, Recent areal subsidence found in releveling: Eng. News-Rec., v. 110, p. 848

Tibbetts, F. H., 1933, Areal subsidence: Eng. News-Rec., v. 3, p. 204

Tolman, C. F., 1937, Ground Water: New York, McGraw-Hill Book Co., p. 341–345

Tolman, C. F., and Poland, J. F., 1940, Ground-water, salt-water infiltration and ground-surface recession in Santa Clara Valley, Santa Clara County, California: Am. Geophys. Union Trans., pt. 1, p. 23–24

U.S. Coast and Geodetic Survey, 1956, San Jose area, California, 1912–1954 leveling (190A Calif.): Washington, D.C., U.S. Coast & Geod. Survey, 42 p.

Weir, W. W., 1950, Subsidence of peat lands of the Sacramento-San Joaquin Delta, California: Hilgardia, California Univ. Agr. Expt. Sta., v. 20, no. 3, p. 37–56

DEVELOPMENT OF PERMEABILITY AND STORAGE IN THE TERTIARY LIMESTONES OF THE SOUTHEASTERN STATES, USA*

H.E. LEGRAND and V.T. STRINGFIELD
Geologists US Geological Survey, Washington, DC

ABSTRACT

Permeability and storage characteristics in the Tertiary limestone system of southern United States have developed progressively but non-uniformly as circulation of water and solution in the limestone have changed during the geologic and hydrologic history.
The limestone formations, predominantly of Eocene age and subordinately of Oligocene and Miocene age, are widespread at and beneath the surface. They commonly dip gently seaward and are covered in coastal areas by Miocene to Recent clays and sands. Sinkholes and other karst features are common, but topographic relief is generally not great.
Circulation of water under water-table conditions when the limestone was exposed to meteoric weathering, before middle Miocene time, resulted in development of secondary permeability as solution channels in near-surface parts of the limestone. Marine deposition of middle and late Miocene clays and later emergence converted part of the water-table circulation system to the present great artesian system. Later, Pleistocene changes in sea level caused changed in places where water discharged, which in turn caused changes in rates of circulation and changes in rates and positions of solution of limestone. Both present and past circulation of water have contributed to changes in permeability and storage of this limestone system.

INTRODUCTION

It is generally agreed that limestone terranes (1) range greatly in permeability and storage characteristics and (2) may have inherent permeability increased by the interdependent processes of (*a*) circulation of water and (*b*) solution of limestone. Two questions logically follow these two accepted facts. What conditions encourage the development of permeability? What is the role of geologic and hydrologic history in the development of permeability? The purpose of this paper is to answer these questions as they may be asked of the Tertiary limestone system of the southeastern United States.

Limestone formations of Tertiary age underlie all of Florida and parts of adjacent areas in Alabama and Georgia; they extend northward through Georgia and South Carolina and into North Carolina. They are at or near land surface through much of the Coastal Plain province in these States. Collectively they represent one of the most productive aquifer systems in the United States, although they show significant patterns of variations in hydrologic conditions.

The Coastal Plain, beneath which the limestone formations have a widespread occurrence, ranges in altitude from sea level at the coast to as much as 600 feet along the inner margin. Relatively flat interstream areas are characteristic of coastal regions, and the remnants of Pleistocene coastal terraces extending to an elevation of 270 feet have been only slightly dissected in most places. Karst topography and other features associated with solution of the limestone are present in large areas, chiefly where the limestone is at or near the surface (fig. 1). Sinkholes and sinkhole lakes are especially common in north-central Florida, and in much of this area drainage is underground except for a few perennial streams that are entrenched into the limestone and that draw water from it.

The climate is temperate and humid. Annual rainfall ranges from about 38 inches 'at the southern tip of Florida to about 60 inches in western Florida, and the rainfall has a fairly even distribution during each year. The humid conditions, coupled with subdued topography and prevalent sandy soils, result in high rates of recharge to the limestone and in a relatively high water table where the limestone is at or near the ground surface. Consequently, almost all the

---
* Publication authorized by the Director, U.S. Geological Survey.

limestone, considered in aggregate volume, is below the water table. In spite of cavernous conditions, which represent much of the porosity and permeability, airfilled caves are confined locally to certain outcrop areas.

The authors have drawn upon their own experiences in the region (Stringfield, 1964), (Stringfield, in press), (Herrick and LeGrand, 1964), and upon the fine contributions of many other workers of the Coastal Plain. The extensive literature on hydrology of limestone terranes contains many good ideas that formed a basis for the present study.

Roger Waller, B.B. Hanshaw and William Back, USGS, reviewed this manuscript.

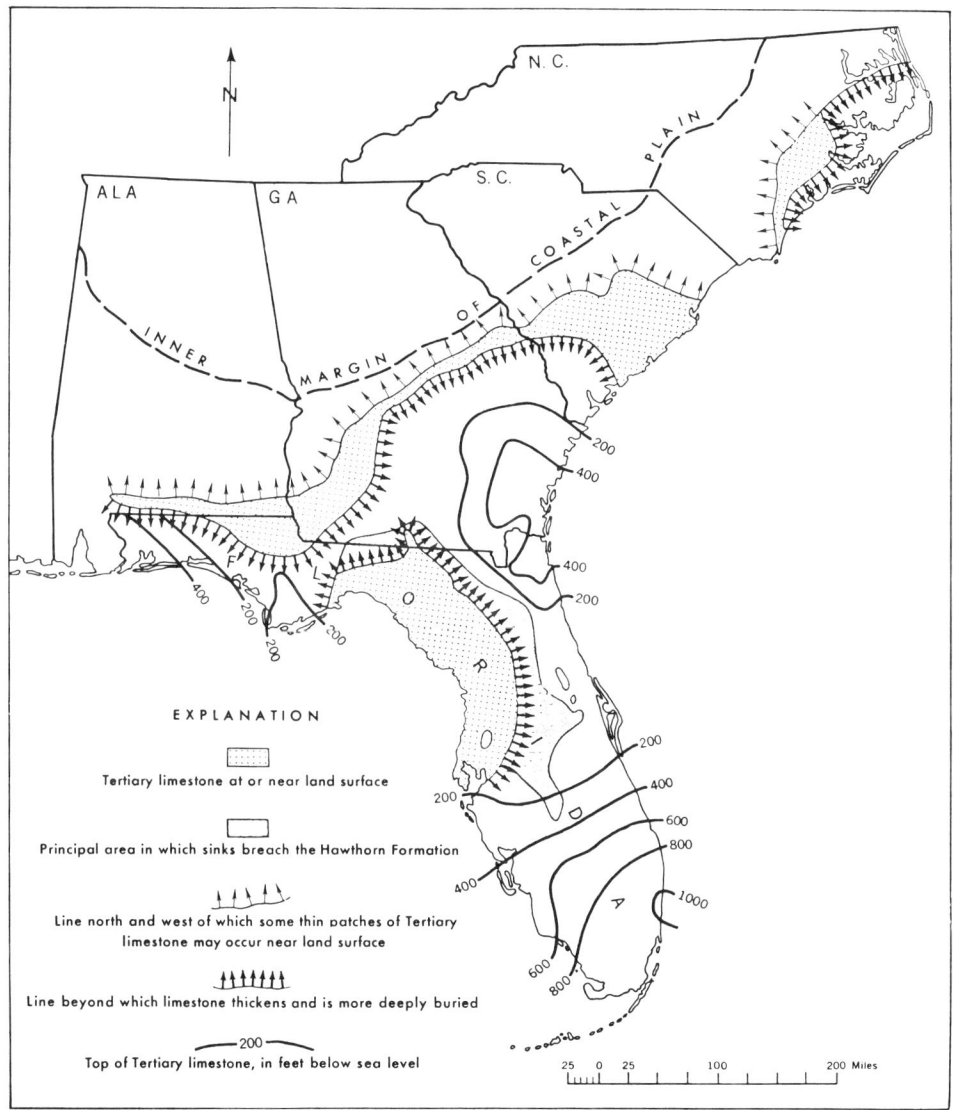

Fig. 1 — Map showing distribution of the Tertiary limestone at the land surface and beneath it.

Geologic and Hydrologic Setting

The limestone aquifer system, chiefly of Eocene age and to a lesser extent of Oligocene and Miocene age, has a widespread surface and nearsurface occurrence (fig. 1). The limestones tend to have a gentle homoclinal dip toward the coast, although broad gentle flexures (fig. 2) have

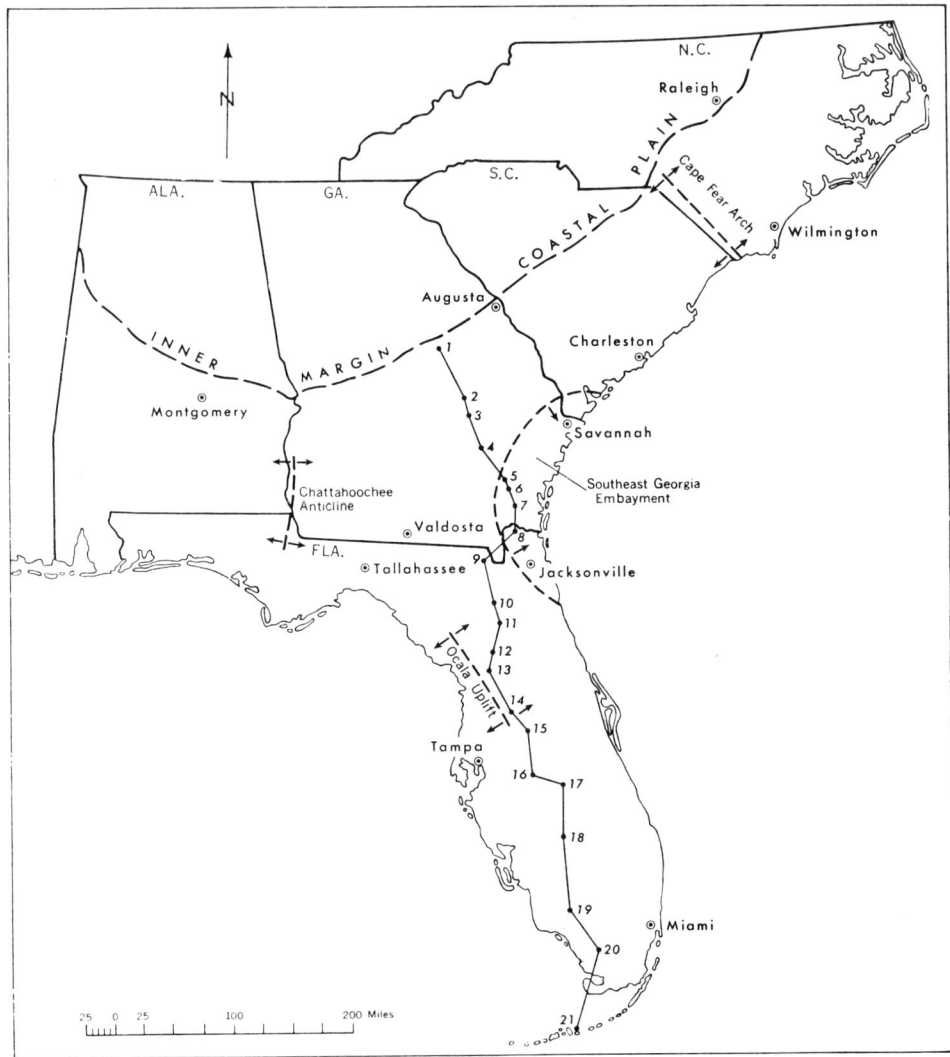

Fig. 2 — Major structural features of the Tertiary limestone. Line of wells 1-21 relates to cross section in figure 3.

modified the homoclinal dip (LeGrand, 1961, p. 1559). The formations commonly increase in thickness and are generally covered by overlying materials in coastal areas (fig. 3).

The limestones range greatly in texture and permeability. Some of the limestones in South Carolina have a chalky texture and a low permeability. Yet, most of the limestone in the region is at least partly indurated, and bedded crystalline limestone and dolomite are not uncommon. Marked differences in permeability, even within formations, occur locally. Much of the good permeability has been derived by circulation and solution within the limestone. Permeable zones are especially prominent in the upper part of the limestone unit but are not restricted to

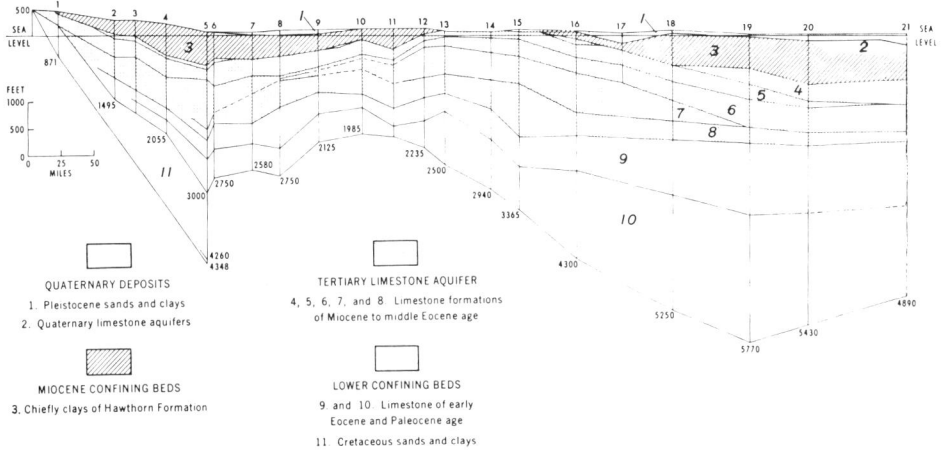

Fig. 3 — Cross section from the inner margin of Coastal Plain in Georgia to southern Florida (modified from Toulmin, 1952).

the upper part. Stewart (1959, p. 14) obtained information on 40 cavities penetrated by wells in Polk County, Florida; he reported cavities to be more numerous near sea level and at about 450 feet below sea level than at other depths. This suggests that solutional processes are not necessarily related to the top of the limestone but are more closely related to the top of the zone of saturation at certain stages in geologic history.

The Tertiary limestone aquifer system includes as many as seven geologic formations in Florida (Stringfield, 1964, p. 164) ranging in age from middle Eocene to middle Miocene. Farther northward and inland, as the aggregate system becomes thinner, only one or two formations are present. Unconformities separate several of the limestone formations as the sea advanced and withdrew during the aggregate depositional period. As a result of these changing conditions during geologic history, differences in lithology and texture are inevitable, and some clay and sand beds are locally present; yet, in aggregate, the entire unit may be considered as limestone, acting as a single hydrologic system.

Underlying the limestone aquifer system unconformably are formations ranging from lower Eocene limestones in some coastal areas to Cretaceous sands and clays along the inner margin of the Coastal Plain (Toulmin, 1952). Although some water passes between the underlying rocks and the limestone aquifer, the limestone aquifer system is in general much more permeable and is hydrologically distinctive.

Bold outcrops are rare, and even in interstream areas where the limestone is considered to crop out, a thin veneer of post-Miocene sands and clays or residual weathered clays commonly covers the limestone (LeGrand, 1961, p. 1564). Coastward, or downdip, the limestone is covered by Miocene formations, such as the Yorktown Formation in eastern North Carolina and the widespread Hawthorn Formation farther southward. The Miocene formations contain much clay, resulting in a confining unit for the artesian water in the homoclinal limestone aquifer.

Hydrologic principles of limestone terranes

Some principles necessary for an understanding of the evolutionary changes in limestone are set forth in this section. Special emphasis is placed here on limestone as a conduit or medium for the transmission of water, because only where acting as a conduit is it in contact with moving water, and therefore, vulnerable to solution. In acting as a conduit, a limestone terrane may be under either water-table or artesian conditions; in both cases, three things are essential. It must have (1) an area of intake, it must (2) transmit, and (3) discharge water. If any one of these three requirements is not met, the limestone body is hydrologically inert and cannot act as a conduit for water.

An understanding of the geologic history and the post-depositional processes is helpful in determining the permeability of limestone terranes. If the geologic history of a limestone terrane is long and varied, the permeability of the limestone has very likely undergone a considerable change. The original, or primary, permeability is provided by the original porosity, the permeability depending on the size and degree of intercommunication of the pore spaces existing in the unconsolidated or poorly consolidated deposits. As consolidation occurs the original porosity tends to be destroyed, and future circulation of water must be through interconnecting openings which have been enlarged by the solvent action of the water. A network of secondary permeability is thus established. This later permeability is the predominant hydrologic property of consolidated limestones; where water moves readily into and out of the limestone the openings tend to enlarge, and the permeability increases; thus, where ground water is in continual movement through pore spaces, it dissolves and carries away soluble material. This removal increases circulation and produces further permeability in the rock. The rate at which solution occurs thus depends largely upon the solubility of the material forming the deposits, the chemical character of the ground water, and the rate of circulation. Solution is especially great where water containing much carbon dioxide circulates rapidly through relatively pure limestone.

Where limestone crops out in large areas as shown by karst regions-water-table conditions, rather than artesian conditions, control the development of landforms. In these areas, intake, or recharge, facilities from precipitation are readily available. As concerns discharge, topographic relief is generally sufficient to facilitate removal of water by springs in the major valleys along the zone approximating local base level. Ground water, always controlled by gravity, takes the most direct natural path to discharge areas along perennial streams in inland areas and along sea coasts. In limestone terranes attention must be focused on the perennial stream because it is the linesink, or trough-like depression, in the hydraulic pressure surface and because it represents a base level below which the land does not appreciably subside as a result of solution. Along coastal areas the base level is sea level.

Theoretically, ground water moves in arcuate paths following lines of flow that have their origin at the top of the zone of saturation in recharge areas (Swinnerton, 1949, p. 665). The flow lines curve downward for some distance and then rise to an outlet or point of discharge. Diagrammatically, these lines can be represented by a family of curves the spacing of which is closest near the area of the outlet and becomes wider along the top of the zone of saturation as the distance from the outlet increases. In uniformly permeable material where geologic structure has only subordinate influence on the direction of movement, the water may be expected to have the arcuate pattern of flow. However, even in the initial stages of ground-water circulation in limestone and other carbonate rocks, lines of flow are modified. The circulation in the discharge area, together with the greater velocities in that area, may result in an enlargement of the outlet and a consequent shallowing of the more arcuate paths. Solution openings along the more direct paths will become larger than those along the less direct paths and will permit progressively larger flows at the expense of other passageways.

As stated by Rhoades and Sinacori (1941, p. 794), the initial flow occurring both at great and shallow depths causes solution which is quantitatively more pronounced in the upper zone because of greater circulation in that zone. Progressive concentration of flow in the upper part of the zone of saturation produces master conduits and causes eventual diminution of flow and solution at deeper levels.

Following enlargement of these openings, the limestone transmits water more readily than before, and also, the larger openings tend to further increase circulation. These interdependent tendencies are fundamental to the development of karst topography.

The above considerations deal chiefly with limestone bodies in which water-table conditions prevail. The movement of water in some limestone bodies is artesian. For example, east of the outcrop area in Georgia and the Carolinas, the limestone is a homoclinal artesian aquifer with a coastward dip slightly greater than that of the land surface; in the area downdip from the outcrop area no surface discharge points are available, and the relative impermeability of the confining clays tends to retard subsurface leakage. As a result, the poor discharge facilities of this type of limestone aquifer tend to slow the circulation of water within it.

The nature of underground circulation depends largely upon the geologic structure and the position of pertinent rocks with reference to ground-water discharge areas. Insofar as circulation is concerned, a limestone formation or body at any one place possesses at least one of the following four types of hydrologic zones:

ZONE 1. Limestone at or near the surface: The water table lies in the limestone. Water from precipitation moves vertically downward to the water table and then tends to move laterally toward a surface stream. (Example—karst regions, such as central Kentucky, and areas where the Tertiary limestone of the southeastern States is near the surface.)

ZONE 2. Limestone buried beneath an impermeable bed, forming a homoclinal artesian system: Water moves under artesian pressure from an elevated intake area, through the limestone, toward a lower discharge area. (Example—Tertiary limestone in coastal areas of Georgia).

ZONE 3. Limestone with no significant discharge facilities: It may be (*a*) a homoclinal artesian system so deeply buried beneath impermeable beds that almost no water can escape. (Example—the limestone unit of southern Florida, and Castle Hayne limestone in the vicinity of Cape Hatteras North Carolina.) It may also be (*b*) limestone lying below the streams controlling the base level of erosion and so faulted or folded as to nearly preclude the discharge of water. (Example—deep-lying Paleozoic limestones of the North-Central States).

ZONE 4. Limestone denuded and elevated above subjacent valley—sufficiently impervious locally to preclude a zone of saturation. No subsurface discharge of water. This zone is absent in many areas.

The classification above is not concerned primarily with thickness and areal distribution of formations. Rather it is based on unit volumes of limestone in which the circulation of water is controlled by specific geologic frameworks in relation to the ease or difficulty to which water can circulate in the limestone and discharge from it. Generalizations about the hydrology of limestone formations should be tempered by the realization that most limestone formations cross at least two of the four structural zones on which the classification is based. Only Zone 1 is involved in the development of karst topography. Zone 3 (*b*) is subjacent to Zone 1 in areas where the limestone extends to appreciable depths below the bottom of the largest stream. The top of Zone 3 (*b*) represents the highest level of the zone in which circulation is negligible and may vary from about 50 feet to a few hundred feet below the largest stream, depending upon local geologic controls. Where slightly inclined homoclinal artesian systems occur, as is the case in coastal areas of the Tertiary limestone, the zones lie in tandem, extending from Zone 1 at the surface, which passes into Zone 2, where potable artesian water occurs and into Zone 3 (*a*) which contains impotable artesian water that cannot escape (fig. 4). The presence of potable water is not an accurate criterion for separating Zones 2 and 3 (*a*). Actually, discharge capabilities, which control circulation of water, in Zone 2 are only slightly better than in Zone 3 (*a*). Consequently, the line between these zones is arbitrary.

Hydrology has had an influence on almost all economic considerations of limestone. It is believed, therefore, that zoning in the above manner will facilitate the evaluation of economic problems relating to the uses of limestones. Perhaps the most important features of the above classification is the distinction made between zones in which outlets for discharge are available and those in which there are no significant outlets for discharge. The major effect of limestones with poor discharge facilities is the presence within them of water more highly mineralized than is commonly used for drinking. This occurs in parts or all of many limestone formations which become consolidated before the buried sea water is removed and before potable water can enter

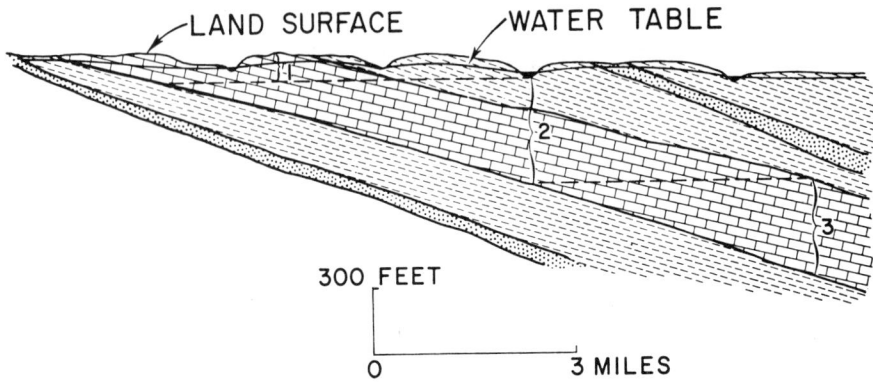

Fig. 4 — Generalized cross section showing three zones in which water moves at different rates within the limestone.

openings in them. The only parts of the limestone that will be purged of bad water are those having conditions favorable for discharging water.

Under water-table conditions the zone of greatest permeability tends to develop in the zone of greatest circulation and solution, which, as has already been noted, is commonly just below the water table. Figure 5 shows schematically the degree to which circulation and solution occur at various positions below land surface; the diagram is generalized and is not based on statistical evaluation of specific data. Topography and position of limestone below ground are important; limestones that allow at least moderate circulation and are entrenched by perennial streams tend to develop solution openings and to increase circulation. Deep-lying limestones that have impeded circulation are almost hydrologically inert; yet, the circulation and the discharge from some limestones as deep as several hundreds of feet may be adequate to cause a gradual increase in permeability in an artesian system if the water is not completely saturated with respect to calcium carbonate. Structural geology is very important, not as a separate entity but, in relation to topography and recharge-discharge facilities.

Thus far, attention has been focused on conditions at the present or at a particular time in the past; although the zone of greatest circulation and solution commonly represents the zone of greatest permeability, this common zone does not stay fixed. Of the following 2 types of changes in positions of limestone with respect to zones of circulation and solution within it, the first is of a smaller scale in space and time than that of the second:

1. As perennial streams progressively lower their beds into the limestone and as permeability increases in the zone, the water table is lowered, resulting in this zone ultimately lying above the water table. If solutional and mechanical weathering progress, this former zone of great circulation, solution, and permeability after seemingly rising above the water table may eventually be eroded away. Yet, as circulation and solution continue below the water table, permeability will develop below the water table in a new zone. Thus, although the zone of solution will likely be lowered, its relation with respect to land surface and with respect

to position of the water table may stay about stationary if stream incision and surface lowering of interstream areas are in equilibrium.

2. In addition to the gradual lowering of the water table that is related to concomitant lowering of stream level and lowering of land surface in interstream areas, appreciable changes in position of the water table have occurred as a result of changes in sea level. Elevation of limestone above sea level may expose it to meteoric weathering and to subaerial and solution erosion; all of the limestone may be removed in some places, the upper part in some places,

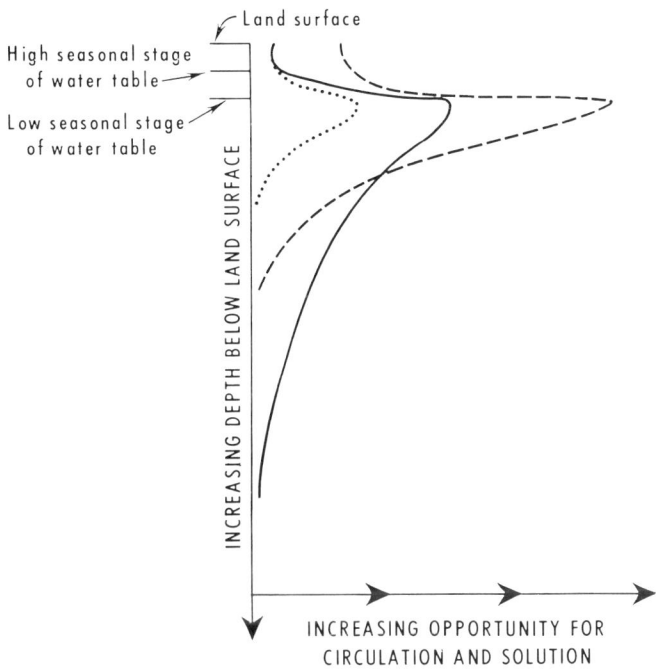

Fig. 5 — Schematic diagram showing the degree to which circulation of water and solution occur at various positions below land surface. Solid line represents a common condition and dashed and dotted lines represent less common but not unusual conditions.

and the remaining part in some places may have permeability increased greatly by solution. If uninterrupted deposition of long duration produced thick conformable deposits over limestone, ground-water circulation in the limestone may have been sluggish and insignificant, resulting in little or no solution. If deposits over the limestone were thin and if the limestone were elevated so that ground water could circulate in it, solutional openings and increased permeability could develop in some places. Development of permeability through solution action in many cases must wait for fractures to develop following compaction, consolidation, and perhaps recrystallization; in such cases, the absence of fractures may either prevent or retard the development of permeability even if the limestone is entrenched in a perennial stream. Folding and faulting may have occurred, hindering ground-water circulation in many cases but aiding it in others. If after a period of emergence and subjection to meteoric circulation of water and solution, and if the limestone unit is then submerged and covered with later marine deposits, later emergence may place that zone of great permeability at a greater depth than that of the present greatest circulation in water-table aquifers.

In the Tertiary limestone unit of the Southeastern States, erosional unconformities between several of the limestone formations are common, indicating that at certain times solutional development under water-table conditions was followed by marine submergence and a cover of more limestone. To what extent the former zones of greatest permeability have been removed by subaerial and marine erosion and to what extent they have been preserved is not clear. Our present state of knowledge does not indicate that buried limestone formations everywhere contain permeable zones in their upper parts, but there appears to be generally more permeable zones in the upper parts. Fluctuations of sea level during Pleistocene time have influenced positions of the water table and corresponding positions of greatest circulation and solution, especially in coastal areas underlain by limestone. Stringfield and LeGrand (1965, in press) point out that in Florida the water table stood at higher and lower levels during Pleistocene time in response to changes in sea level, and that the zone of greatest circulation of water also fluctuated correspondingly. High stages of sea level result in higher positions at which ground-water discharges and higher positions of greatest circulation and solution than those during low stages of sea level. Fluctuations of sea level may not cause changes in ground-water levels in inland areas where perennial streams are not also affected.

CIRCULATION OF WATER IN THE TERTIARY LIMESTONE

A map of the piezometric surface of water in the limestone aquifer in Florida and Georgia (fig. 6) indicates the general direction of movement of water, as well as areas of recharge and discharge. The value of this regional piezometric map was demonstrated by Stringfield (1936), who mapped the Florida part, and by Warren (1944), who mapped the Georgia part. The current map is based on compilations for Florida (Healy, 1962) and of Georgia (Stewart and Counts, 1958), (Stewart and Croft, 1960). The major features are essentially the same as those of the early maps except in areas affected by heavy withdrawal of water.

In general, water enters the aquifer where the piezometric surface is relatively high and is discharged where the surface is low. The highest part is as much as 240 feet above sea level in the area of outcrop of the aquifer extending northeast across Georgia. The water moves from that general recharge area toward the south and the Atlantic Coast. Callahan (1964, p. 24) describes the recharge areas in Georgia and indicates that the boundaries are not clearly defined. Much natural discharge occurs offshore. In Florida west of the Suwannee River the general movement of the water is south, toward the Gulf of Mexico. Recharge occurs in areas where the aquifer is near the surface (except in stream valleys) and through sinkholes which penetrate the overlying Hawthorn Formation. There is discharge to the Gulf through springs and submarine outcrops in the coastal belt. Discharge from the aquifer into the Suwannee and Santa Fe Rivers causes a large valley in the piezometric surface where these surface streams flow in channels cut into the limestone. In central and north-central Florida, high areas of the piezometric surface show recharge through filled sinkholes now occupied by lakes. Where the aquifer is at or near the surface in north-central Florida there is large recharge, but the large ground-water flow to the Gulf of Mexico and to the large springs, such as Silver Springs and Rainbow Springs, causes the piezometric surface to stand relatively low.

In the lake region of south-central Florida where the piezometric surface is as much as 100 feet above sea level, the water moves laterally in all directions from this recharge area. Part of the water on the north, northeast, and west is discharged through large springs. To the south, southeast, and southwest the water moves many miles to the coast and beyond. Where the edge of the continental shelf is near the southeast coast, some of the water may discharge where the aquifer is exposed to the ocean if the artesian pressure is large enough to overcome the back pressure of the ocean water. On the western coast of south Florida, where the edge of the continental shelf is many miles offshore, there probably is no groundwater discharge at the edge of the shelf, but there may be discharge by upward leakage through relatively permeable beds.

A regional piezometric map of the water in the limestone formations in South Carolina and North Carolina has not been made except near the Georgia and South Carolina State Line;

yet, the general features of groundwater movement are known. Moderately low topographic relief in the outcrop areas of the limestone results in low water-table gradients from interstream areas to streams that cut into the limestone. Much of the limestone is soft and relatively impermeable, seemingly still retaining much of its original permeability, whereas some limestone, especially near stream valleys, had developed degrees of later permeability. Toward the coast where the limestone is confined beneath younger clays and sands, the pattern of flow is not clear;

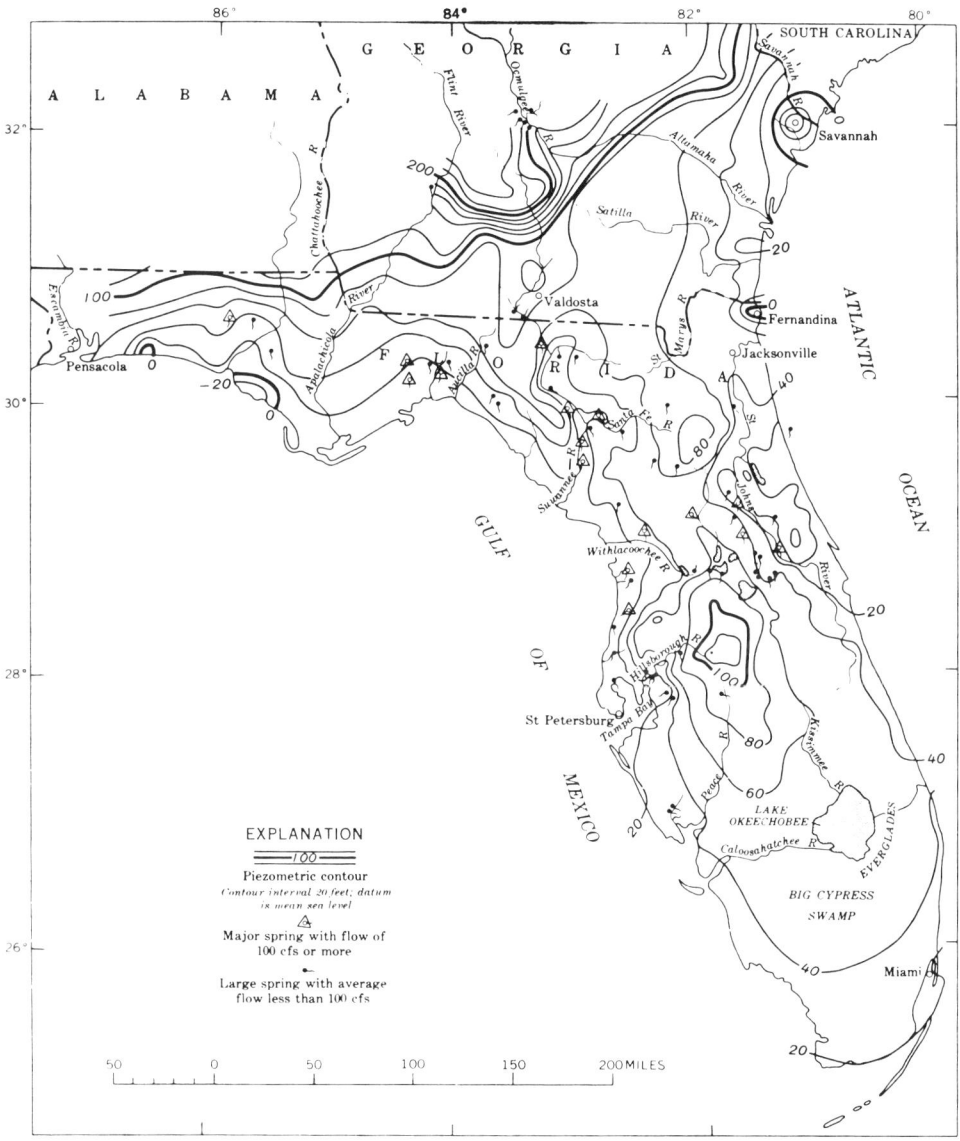

Fig. 6 — Piezometric map of water in the principal limestone artesian aquifer (modified from Stringfield, 1964, see text for source of data).

some of the water takes arcuate paths to streams and estuaries, although there is an overall low artesian gradient toward the east. Some of the confined limestone is very permeable, but no great areas of concentrated discharge have been noted.

A glance at figure 6 may suggest that movement of water in the limestone is too complex for analysis, but such is not the case. Admittedly, a simple homoclinal hydraulic gradient does not occur except locally, but changes in direction and rate of movement of water are explainable. In the absence of a prepared piezometric map, reasonable deductions can generally be made about the behavior of water, but a suitable piezometric map may reveal important specific aspects of water movement. Some general and gross features of the hydrogeologic framework must be known, including gross permeability within the limestone, positions of intake or recharge to the limestone, and the extent of cover. Areas may be mapped (1) in which the water is under water-table conditions, (2) in which water moves under confined conditions in the artesian system, and (3) in which water discharges from the artesian system.

PALEOHYDROLOGY OF THE TERTIARY LIMESTONE

Limestone terranes are noted for their great differences in permeability. Since aspects of permeability are mostly acquired after deposition—poor primary permeability as a result of consolidation and good secondary permeability as a result of solution action — it is proper to ask what features of limestone hydrology are related to present hydrogeologic conditions and what features have been retained from earlier times. If overlying deposits are thick and if the limestone is never elevated into a ground-water circulation system, no solution will occur and no secondary permeability will develop; on the other hand, lack of overlying deposits and early elevation into a ground-water circulation system will lead to development of secondary permeability and to removal of some of the limestone formation. In addition to the present hydrogeologic setting and the present ground-water circulation system in the Southeastern States, consideration should be given to the immediate post-depositional period of the limestone and to the subsequent pre-Recent Period.

Many of the limestone formations have been exposed to meteoric weathering and to subaerial and solutional erosion at some time in their history because oscillations of sea level have been common throughout the Coastal-Plain history. Much of the limestone is of middle and late Eocene age; this limestone was submerged beneath marine waters and beneath later marine deposits during part of late Eocene, Oligocene, and early Miocene time and emerged to be subjected to solution and erosion during part of this interval of time. Not until middle Miocene time when the Hawthorn Formation was deposited in Florida, Georgia, and a part of South Carolina, and not until late Miocene when the Yorktown Formation was deposited in North Carolina, did a blanket of relatively impermeable clay and marls cover the Tertiary limestone formations. This blanket converted much of the water-table circulation system that prevailed during emerged intervals to the present great artesian system. At the close of Miocene time the Hawthorn Formation, which still covers the limestone in a large part of the area, was more extensive than at present; it covered the area which is now occupied by the Flint River valley in southwest Georgia and most, if not all, of Florida east of the Apalachicola River.

The distribution of the cavities and their general decrease in size and numbers with distance from recharge and discharge areas suggest that the present pattern of circulation developed chiefly during the low stands of the sea in Pleistocene time. With little or no circulation in the deeper parts of the aquifer at the present time, it is believed that the most active solution was during the lowest stand of the Pleistocene sea.

The range in sea level was as much as 770 feet during Pleistocene time, if the estimates of 270 feet above the present level (Cooke, 1945) and 500 feet below the present level (Donn and others, 1962; Richards and Craig, 1963) are correct. With a low stand of the sea (as much as 500 feet below the present level), it is conceivable that the water level in part of the limestone in north-central Florida was as much as a few hundred feet lower than at the present time after circulation and solution had adjusted to the low sea level. Such changes in sea level during

Pleistocene time resulted in corresponding changes in positions of discharge zones for ground water and in positions of greatest circulation and solution. The rise of sea level led to filling of some sinks and solutional valleys with sand and to a general smoothing of topography, at least below present elevations of about 200 feet. As a result of the complex Pleistocene history, it is not easy to distinguish between permeability and storage features in the limestone that developed from those being developed now.

CONCLUSIONS

Limestone formations of Tertiary age, having a widespread distribution in southeastern United States, represent collectively a valuable aquifer system. A water-table circulation system occurs in surface and near-surface parts of the limestone formations; a homoclinal artesian system occurs coastward where the limestone is confined beneath relatively impervious beds of Miocene and younger age.

The limestone ranges greatly in permeability, but the more permeable zones indicate that circulation of water through them has been a major factor in removing some of the limestone through solution. This circulation is greatest in the upper part of the zone of saturation.

A piezometric map of water in the limestone aquifer system in Florida and Georgia indicates (1) the direction of flow, (2) the chief areas of recharge where the limestone crops out in interstream areas and where sinkholes breach overlying beds, and (3) chief discharge areas in valleys of some streams such as the Suwannee, Flint, St. Johns, as well as offshore. In North Carolina and in much of South Carolina, where a regional piezometric map is not available, the specific pattern of flow is not clear; in general, however, the moderately low topographic relief in the outcrop areas results in low water-table gradients from interstream areas to streams that cut into the limestone, and coastward where relatively impermeable materials confine the water, there is a gross overall artesian gradient toward the coast.

Development of permeability in the Tertiary limestone system is both current and historical. Where water-table conditions occur, or have occurred in the past, circulation of water and solution activity tend to be greatest in the upper part of the zone of saturation. As to both water-table and artesian condition, the ease or difficulty at which water can discharge from the limestone controls circulation and solution. The current discharge areas in general are easy to determine, but former discharge areas and areas of greatest circulation and solution must be inferred from a study of the geologic and hydrologic history.

Only a small part of the total Tertiary limestone unit has retained any semblance of the original or immediate post-depositional porosity and permeability; this part includes the chalky-textured marls of North Carolina and South Carolina and a few other unconsolidated calcareous deposits, which have never been deeply covered and compacted and which are positioned so that circulation of water in them and discharge of water from them have been limited. Erosional unconformities at the top of some of the limestone formations indicate periods of emergence at which times circulation of meteoric water and solution occurred before marine submergence and overlying deposition of additional calcareous materials. Oscillations of sea level have been common, and these changes in sea level have led to corresponding changes in positions at which water could discharge and in positions at which the greatest circulation and solution have occurred. Changes in sea level during Pleistocene time, ranging from about 270 feet above present sea level to about 500 feet below, were especially effective in changing positions of appreciable circulation and, therefore, in developing permeability at different positions in the limestone. The overall increase in permeability has increased the opportunity for the flushing of salty water that was either trapped in the limestone during Miocene or earlier time or entered during Pleistocene submergence. In much of the region salty water has been completely flushed from the aquifer, and in other places, especially along certain coastal areas, partial or nearly complete flushing has occurred.

A study of the Tertiary limestone aquifer indicates that the development of porosity and permeability can be related to (1) the present hydrogeologic conditions, (2) past hydrogeologic conditions, or (3) to both the present and past. A reconstruction of paleohydrology of a limestone region, together with analysis of the present hydrogeologic framework, leads to a better understanding of the distribution of permeable zones than is otherwise possible.

REFERENCES CITED

CALLAHAN, J.T., 1964, The yield of sedimentary aquifers of the Coastal Plain, Southeast River Basins: *U.S. Geol. Survey Water-Supply Paper* 1669-W, 56 pages.
COOKE, C.W., 1945, Geology of Florida: *Florida Geol. Survey Bull.*, 29, 339 pages.
DONN, W.L., FERRAND, W.R., and EWING, Maurice, 1962, Pleistocene ice volumes and sea level lowering: *Jour. Geology*, v. 70, no. 2, pp. 206-214.
HEALY, H.G., 1962, Piezometric surface and areas of artesian flow of the Floridan aquifer in Florida, July 6-17, 1961: *Florida Geol. Survey Map Ser.* 4.
HERRICK, S.M., and LeGRAND, H.E., 1964, Solution subsidence of a limestone terrane in southwest Georgia: *Internat. Assoc. Sci. Hydrology Bull.*, v. 9, no. 2, pp. 25-36.
LeGRAND, H.E., 1961, Summary of geology of Atlantic Coastal Plain: *Am. Assoc. Petroleum Geologists Bull.*, v. 45, no. 9, pp. 1557-1571.
—, 1962, Geology and ground-water hydrology of the Atlantic and Gulf Coastal Plain as related to disposal of radioactive wastes: *U.S. Geol. Survey Report TEI-805* (Open-File), 169 pages.
RHOADES, Roger, and SINACORI, M.N., 1941, Pattern of ground-water flow and solution: *Jour. Geology*, v. 49, no. 8, pp. 785-794.
RICHARDS, H.G., and CRAIG, J.R., 1963, Pleistocene mollusks from the continental shelf off Argentina: *Philadelphia Acad. Natl. Sci. Proc.*, v. 115, no. 6, pt. 2, pp. 127-152.
STEWART, H.G., Jr., 1959. Interim report on the geology and ground-water resources of northwestern Polk Country, Florida: *Florida Geol. Survey Inf. Circ.* 23, 83 pages.
STEWART, J.W., and CROFT, M.G., 1960, Ground-water withdrawals and decline of artesian pressures in the coastal counties of Georgia: *Georgia Geol. Survey Mineral Newsletter*, v. 13, no. 2, pp. 84-93.
STEWART, J.W., and COUNTS, H.B., 1958, Decline of artesian pressure in the Coastal Plain of Georgia, northeastern Florida, and southeastern South Carolina: *Georgia Geol. Survey Mineral Newsletter*, v. 11, no. 1, pp. 25-31.
STRINGFIELD, V.T., 1936, Artesian water in the Florida peninsula: *US Geol. Survey Water-Supply Paper* 773-C, pp. 115-195.
—, 1964, Relation of surface-water hydrology to the principal artesian aquifer in Florida and southeastern Georgia: *U.S. Geol. Survey Prof. Paper* 501-C, pp. 164-169.
—, 1965, (in press) Artesian water in Tertiary limestone in the Southeastern States: *U.S. Geol. Survey Prof. Paper* 517.
SWINNERTON, A.C., 1949, Hydrology of limestone terranes, *in* Meinzer, O.E. ed., Hydrology IX, Physics of the Earth: New York, Dover Publications, Inc., Chap. 14, pp. 656-677.
TOULMIN, L.D., 1952, Volume of Cenozoic sediments in Florida and Georgia, pt. 2 of Murray, G.E., ed.: Sedimentary volumes in Gulf Coastal Plain of United States and Mexico, *Geol. Soc. America Bull.*, v. 63, no. 12, pt. 1, pp. 1165-1176.
WARREN, M.A., 1944, Artesian water in southeastern Georgia: *Georgia Geol. Survey Bull.*, 49, 140 pages

32

Copyright ©1969 by Academic Press, Inc.
Reprinted from pages 53-56, 57, 58, 65, 70, 80-81, and 86-89 of *Flow Through Porous Media*, R. J. M. De Wiest, ed., Academic Press, New York, 1969, 530p.

Porosity and Permeability of Natural Materials

Stanley N. Davis
Department of Geology
University of Missouri
Columbia, Missouri

1. Introduction 54
   1.1. Porosity 54
   1.2. Permeability 55
2. Dense Rocks 56
   2.1. Introduction 56
   2.2. Weathering and Fracturing 56
   2.3. Relation of Permeability and Porosity to Depth 59
   2.4. Importance of Rock Type 61
   2.5. Effects of Solution on the Permeability and Porosity
       of Dense Rocks 61
   2.6. Anisotropic Characteristics 63
   2.7. Log-Normal Distribution of Well Yields 64
3. Volcanic Rocks 64
   3.1. Introduction 64
   3.2. Origin of Porosity and Permeability 65
   3.3. Aquifer Tests of Permeable Basalt 67
   3.4. Anisotropy of Basalt 68
4. Indurated Sedimentary Rocks 69
   4.1. Introduction 69
   4.2. Shale 69
   4.3. Sandstones 72
   4.4. Carbonate Rocks 74
   4.5. Statistical Distribution of Permeabilities and Porosities . . 75
5. Nonindurated Sediments 77
   5.1. Introduction 77
   5.2. Alluvium 78
   5.3. Eolian Deposits 83
   5.4. Glacial Deposits 84
   5.5. Fine-Grained Materials of Diverse Origins 85
References 86

1. Introduction

Few properties of natural materials have received more attention from hydrogeologists than porosity and permeability. Despite the vast amount of work expended in measuring these two properties in countless samples, reliable values can rarely be estimated on the basis of general geologic knowledge alone. The purpose of this chapter is to review the extent of our general knowledge concerning porosity and permeability and to relate this knowledge to the more analytical discussions which follow in subsequent chapters.

1.1. Porosity

Porosity, as used in this chapter, is defined as the ratio of pore volume to the total volume of a given sample of material. This definition is adequate for many purposes. Microscopically and submicroscopically, nevertheless, a complete gradation exists from large pores easily accessible to fluids to very small openings in minerals that are caused by minor lattice imperfections. The smallest openings contain fluids that are rendered essentially static by the close proximity of atomic force fields at the surfaces of the solids (Kemper et al., 1964). Whether or not these submicroscopic openings are measured in the laboratory as part of the void space depends on the methods of the measurement employed. Techniques using heat, centrifuging, liquid saturation under a vacuum, and other conditions abnormal to near-surface rocks will certainly displace static fluid that might be best considered part of the solid rock.

A large number of terms have been introduced into the literature to describe various physical measures related to porosity. Moisture equivalent, effective porosity, specific retention, drainage coefficient, coefficient of storage, and specific yield are some of the terms used to qualify variables such as degree of saturation, forces applied to the sample, length of test, degree of interconnection of pores, and fluid chemistry. Besides simple porosity, the only other closely related term used in this chapter will be specific yield, which will be defined as the volume of water drained by gravity from an initially saturated sample divided by the total volume of that sample. Ideally, specific yield should be measured using samples having column lengths of more than 3 m to avoid excessive capillary effects. Drainage time should be measured in many years for clay-size material. As a result of these inconvenient requirements, few reliable measurements of specific yield of fine-grained materials have been made. Modern laboratories tend to rely on centrifuging to speed drainage and to overcome capillary effects. Strain induced by centrifuging, however, makes the resultant specific yield values of fine grained materials open to question. This is particularly true of nonindurated silts and clays.

1.2. Permeability

Permeability is a measure of the ease with which fluids pass through a porous material. Permeability, or more properly, "intrinsic permeability," is a property of the solid material and is independent of the density and viscosity of the fluid. A common expression for permeability k is

$$k = -\frac{Q}{A} \frac{\mu}{\rho g} \left(\frac{\partial h}{\partial s}\right)^{-1}$$

in which Q is the volume of fluid discharged per unit time through a cross section having an area of A, μ is the viscosity of the fluid, ρ is the density of the fluid, g is the acceleration of gravity, and $\partial h/\partial s$ is the hydraulic gradient in the direction of flow s.

Permeability k has the dimensions of L^2. The darcy, which is almost universally used in the petroleum industry, will be used as the unit of permeability in this chapter. One darcy has the value of 0.987×10^{-8} cm². Units of hydraulic conductivity, which superimpose fluid properties on permeability and have the dimensions of LT^{-1} are most common in ground-water work. The "meinzer" is one of the popular units of hydraulic conductivity in the United States. It is related to a fixed viscosity and density of water (pure water at 60° F or 15.6° C). Media having a permeability of 1 darcy have a hydraulic conductivity of 18.2 meinzers.

Analytical work with single-phase flow of fluids almost universally assumes that permeability is constant with respect to time. Several important exceptions should always be kept in mind. Largest changes take place with changes of state or dissolution of minerals, as for example the slow formation of limestone caverns, the more rapid dissolution of gypsum, or the extremely rapid melting of granular ice as warm rain water percolates through a snow bank. Significant changes also take place in response to stresses associated with declining artesian pressures or with external compaction of sediments. If fluid chemistry changes, sedimentary materials will respond dynamically. Effects of partial or complete hydration which will cause clays, colloids, and some organic material to expand and close pore spaces are the most important of the responses to changing fluid chemistry. Permeabilities are largest with air and other nonpolar fluids. Partial hydration by brines will cause values to lower. Full hydration using distilled water as the measuring fluid will cause a drastic reduction of permeability (Johnston and Beeson, 1945). These effects are greatest in fine-grained sediments and could account for permeability differences of several orders of magnitude in the sample.

Effects of measuring techniques on the permeability of fine-grained sediments and rocks is particularly striking. Young et al. (1964) have shown that increasing confining pressures of samples of argillaceous rocks in order to duplicate subsurface conditions produces a decrease in permeability of more than tenfold over results of unconfined tests. Permeability contrasts in some nonindurated materials may be even larger.

Natural material is commonly assumed to be homogeneous and isotropic with respect to permeability. In general, computational problems associated with heterogeneous aquifers are easier to deal with than those of nonisotropic aquifers. The measurement and/or estimation of the actual extent of heterogeneity is also much easier geologically than is the measurement of anisotropy. Further attention will be given to these matters in later sections of this chapter as well as in other parts of this volume.

[*Editors' Note:* Material has been omitted at this point.]

TABLE 1

Rocks of Low Permeabilities and Porosities

Rock name	Porosity (%)	Permeability (darcys)	Reference
Chert	—	1.9×10^{-7}	Stuart et al. (1954)
Diabase	0.1	—	Brace (1965)
Dolomite	6.3	1.0×10^{-3}	Murray (1960)
	0.4	—	Brace (1965)
Granite	0.3	—	Brace (1965)
Graywacke	—	3.0×10^{-6} [a]	Stuart et al. (1954)
Limestone	8.4	1.0×10^{-3}	Murray (1960)
Marble	0.3	—	Brace (1965)
	0.6 [b]	—	Manger (1963)
Mica schist	—	2.1×10^{-6}	Stuart et al. (1954)
Quartzite	—	1.9×10^{-6}	Stuart et al. (1954)
	0.6	—	Brace (1965)
Rock salt	0.6	7.3×10^{-6} [c]	Gloyna and Reynolds (1961)
Slate	—	1.3×10^{-6} [d]	Stuart et al. (1954)
	3.4 [e]	—	Manger (1963)

[a] Median of 5 samples.
[b] Mean of 100 samples.
[c] Representative value taken from several different determinations. Permeability varied widely with confining pressure on specimen and with the type of fluid used.
[d] Median of 9 samples.
[e] Mean of 19 samples.

TABLE 2
Porosity and Permeability of Originally Dense Rocks Modified by Fracturing and Weathering

Rock name	Type of test	Porosity (%)	Specific yield (%)	Permeability (darcys)	Reference
Quartz-mica schist, weathered	Laboratory	48[a]	20.6[a]	3.3×10^{-2} [a]	Stewart (1964)
	Field	—	0.61	0.97[b]	Stewart (1964)
Graywacke, fractured	Field	—	—	4.5×10^{-2} [c]	Lewis et al. (1966)
Metasediments fractured	Field	2.4 (estimated)	—	3.1×10^{-2} [d]	Lewis et al. (1966)
Iron formation	Field	—	—	5.5×10^{-4} (minimum) 5.8×10^{-2} (maximum)	Stuart et al. (1954)
Metabasalt	Field	—	2.0	1.9[e]	Meyer and Beall (1958)
Schist	Field	—	3.0	1.4[e]	Meyer and Beall (1958)
Marble	Field	—	0.4	18[f]	Meyer and Beall (1958)

[a] Median values for 21 samples.
[b] Assumed aquifer thickness of 70 ft.
[c] Median of 11 tests.
[d] Median of 6 tests.
[e] Assumed aquifer thickness of 200 ft.
[f] Assumed aquifer thickness of 160 ft.

TABLE 3
Porosity and Permeability of Volcanic Rocks

Rock name	Porosity (%)	Permeability (darcys)	Reference
Basalt			
very dense	0.80[a]	—	Schoeller (1962)
porous	11.4	—	Schoeller (1962)
moderately dense	7.7[b]	1.4×10^{-5}	
Obsidian	0.52	—	Schoeller (1962)
Phonolite	1.98[c]	—	Schoeller (1962)
Pumice	87.3	—	Schoeller (1962)
Tuff	31.0[a]	—	Schoeller (1962)
zeolitized	39[d]	4×10^{-5}	Keller (1960)
pumiceous	40[d]	1.15×10^{-2}	Keller (1960)
friable	36[d]	1.4×10^{-3}	Keller (1960)
welded	14[d]	3.3×10^{-4}	Keller (1960)

[a] Average of 8 samples.
[b] Unpublished test results, K. Shizaki (1966), Univ. of Hawaii, Honolulu.
[c] Average of 5 samples.
[d] Porosity and permeability are averages of several samples.

TABLE 4
Permeability and Porosity of Sedimentary Rocks

Rock name	Porosity (%)	Permeability (darcys) Vertical	Permeability (darcys) Orientation not given	Permeability (darcys) Horizontal	Reference
Arkose					
coarse-grained	10.9	3.8×10^{-4}		5.5×10^{-4}	Rima et al. (1962)
fine-grained	14.4	1.6×10^{-4}		1.6×10^{-3}	Rima et al. (1962)
medium-grained	25.6	5.5×10^{-4}		1.1×10^{-3}	Rima et al. (1962)
Chalk, Cretaceous	29.2[a]	—		—	Schoeller (1962)
Chert, Mississippian	3.8	—		—	Manger (1963)
Conglomerate,					
coarse grained	17.3	3.8×10^{-4}		4.9×10^{-4}	Rima et al. (1962)
Dolomite					
Mississippian	27.8		2.9×10^{-1}		Murray (1960)
Ordovician	0.4[b]	—		—	Manger (1963)
	11.9		1.6×10^{-2}		Murray (1960)
Limestone					
Cretaceous	4.6[c]	—		—	Schoeller (1962)
oolitic	21.6		3.4×10^{-1}		Archie (1952)
Pennsylvanian	6.3[d]	—		—	Manger (1963)
Permian	10.1		7.7×10^{-3}		Archie (1952)
Sandstone					
Cambrian	11.2[e]	—		—	Manger (1963)
Cretaceous	—	4.8×10^{-2}		1.5×10^{-4}	Young et al. (1964)
Cromwell	16.6	1.7×10^{-1}		4.1×10^{-1}	Muskat (1937)
Gilcrest	27.4	5.6×10^{-1}		8.0×10^{-1}	Muskat (1937)
Pennsylvanian	17.4[f]	—		—	Manger (1963)
Prue	11.4	4.7×10^{-4}		3.4×10^{-3}	Muskat (1937)
Wilcox	15.6	3.6×10^{-1}		7.6×10^{-2}	Muskat (1937)
	12.4	8.5×10^{-2}		8.8×10^{-2}	Muskat (1937)
Shale					
Cretaceous	—		4×10^{-6}		Gondouin and Scala (1958)
Pennsylvanian	—		9×10^{-8}		Gondouin and Scala (1958)
Oligocene and Miocene	21.1[g]	—		—	Manger (1963)
Silurian	5.2[h]	—		—	Manger (1963)
Siltstone	9.7	1.6×10^{-4}		1.2×10^{-4}	Rima et al. (1962)
	—	1.5×10^{-8}		—	Young et al (1964)

[a] Mean of 16 samples.
[b] Mean of 56 samples.
[c] Mean of 24 samples.
[d] Mean of 2109 samples.
[e] Mean of 24 samples.
[f] Mean of 587 samples.
[g] Mean of 9 samples.
[h] Mean of 5 samples.

TABLE 5
Porosity and Permeability of Nonindurated Material

Description of sample	Sample number	Porosity (%)	Permeability (darcys) Horizontal	Permeability (darcys) Vertical	Median diameter (mm)	Reference
Clay						
marine	1	48.5	—	1.6×10^{-5}	0.0005	MacGary and Lambert (1962)
silty, depth 1558 ft	2	41.1	—	1.1×10^{-4}	0.003	Johnson and Morris (1962)
946 ft, lacustrine?	3	40.5	2.8×10^{-4}	1.1×10^{-5}	0.0021	Johnson and Morris (1962)
838 ft, lacustrine?	4	44.1	1.1×10^{-5}	1.1×10^{-5}	0.0036	Johnson and Morris (1962)
Sand						
Ohio River alluvium	5	—	25.2	26.4	—	Gallaher (1964)
marine	6	41.0	55.0	38.5	0.5	MacGary and Lambert (1962)
coarse, depth 1174 ft, alluvium	7	41.0	—	2.2	0.5	Johnson and Morris (1962)
medium, depth 767 ft, alluvium	8	42.9	18.2	0.6	0.32	Johnson and Morris (1962)
fine, alluvium	9	51.3	25.3	26.4	0.2	MacGary and Lambert (1962)
	10	46.3	13.2	16.5	0.2	MacGary and Lambert (1962)
depth 1284 ft	11	40.1	—	1.48	0.17	Johnson and Morris (1962)
depth 911 ft	12	42.4	1.05	0.27	0.14	Johnson and Morris (1962)
Silt						
sandy, depth 472 ft	13	39.4	—	3.8×10^{-2}	0.059	Johnson and Morris (1962)
clayey, depth 1527 ft	14	34.1	—	5.5×10^{-3}	0.04	Johnson and Morris (1962)
loess	15	50.0	0.22	0.33	0.02	MacGary and Lambert (1962)

Orientation of remaining samples was not specified in the literature cited

TABLE 5 (*continued*)

Description of sample	Sample number	Porosity (%)	Permeability (darcys)	Median diameter (mm)	Reference
Clay					
kaolinite under confining load of $\frac{1}{2}$ kg cm	16	58.8	2.38×10^{-3}	—	Olsen (1966)
Oligocene subsurface, Mississippi	17	33.3	1.1×10^{-5}	—	Unpublished data, U.S. Geol. Survey, Denver, Colorado
Gravel					
sand, silt; end moraine (median 7 samples)	18	—	2.2	—	Poth (1963)
silt, clay; colluvium	19	63.4	5.5×10^{-5}	0.13	Cohen (1965)
Sand					
aquifer Arkansas River alluvium	20	33.8	19.8	—	Tanaka and Hollowell (1966)
point bar, Humboldt River, Nevada	21	37.4	27.5	—	Tanaka and Hollowell (1966)
	22	38.3	77	0.82	Cohen (1965)
Patuxent formation, Cretaceous (median of 23 field tests)	23	—	36.8	—	Bennett and Meyer (1952)
from sand dune	24	35.8	28.0	—	Brown and Newcomb (1963)
Silt					
Miocene subsurface, Mississippi	25	33.7	3.85×10^{-5}	0.045	Unpublished data, U.S. Geol. Survey, Denver, Colorado
sandy overbank deposit, Humboldt River, Nevada	26	49.2	2.2×10^{-4}	0.02	Cohen (1965)
overbank deposit, Arkansas River	27	47.5	5.5×10^{-2}	—	Tanaka and Hollowell (1966)
	28	49.8	1.6×10^{-2}	—	Tanaka and Hollowell (1966)

REFERENCES

Archie, G. E. (1952). Classification of carbonate reservoir rocks and petrophysical considerations, *Am. Assoc. Petroleum Geologists Bull.* **36**, 278–298.

Atwater, G. I. (1966). The effect of decrease in porosity with depth on oil and gas reserves in sandstone reservoirs, Unpublished paper presented to the School of Earth Sci., Stanford Univ., California, Jan. 10, 1966.

Bedinger, M. S. (1961). Relations between median grain size and permeability in the Arkansas River Valley, Arkansas, U.S. Geol. Surv. Prof. Paper 424-C, pp. 31–32.

Bennett, R. R., and Meyer, R. R. (1952). Geology and ground-water resources of the Baltimore area, *Maryland Dept. of Geol., Mines, and Water Resources Bull.* **4**.

Bennion, D. W., and Griffiths, J. C. (1966). A stochastic model for predicting variations in reservoir rock properties, *Trans. Am. Inst. Mining Met. Engrs.* **237**, 9–16.

Brace, W. F. (1965). Some new measurements of linear compressibility of rocks, *J. Geophys. Res.* **70**, 398–398.

Brace, W. F., Paulding, Jr., B. W., and Scholz, C. (1966). Dilatancy in the fracture of crystalline rocks, *J. Geophys. Res.* **71**, 3939–3953.

Brown, S. G., and Newcomb, R. C. (1963). Ground-water resources of the coastal sand-dune area north of Coos Bay, Oregon, U.S. Geol. Surv. Water-Supply Paper 1619-D.

Bulnes, A. C. (1946). An application of statistical methods to core analysis data of dolomitic limestone, *Trans. Am. Inst. Mining Met. Engrs.* **165**, 223–240.

Cardwell, W. T., Jr., and Parsons, R. L. (1945). Average permeabilities of heterogeneous oil sands, *Trans. Am. Inst. Mining Met. Engrs.* **169**, 34–42.

Cohen, P. (1965). Water resources of the Humboldt River Valley near Winnemucca, Nevada, U.S. Geol. Survey Water-Supply Paper 1795.

Collins, R. E., (1961). "Flow of Fluids Through Porous Materials." Reinhold, New York.

Csallany, S., and Walton, W. C. (1963). Yields of shallow dolomite wells in northern Illinois, Illinois State Water Surv. Rept. Invest. 46.

Cushman, R. V., Krieger, R. A., and McCabe, J. A. (1965). Present and future water supply for Mammoth Cave National Park, Kentucky, U.S. Geol. Surv. Water-Supply Paper 1475-Q, pp. 601–647.

Davis, S. N. (1964). Silica in streams and ground water, *Am. J. Sci.* **262**, 870–891.

Davis, S. N. (1966). Initiation of ground-water flow in jointed limestone, *Natl. Speleological Soc. Bull.* **28**, 111–118.

Davis, S. N., and De Wiest, R. J. M. (1966). "Hydrogeology." Wiley, New York.

Davis, S. N., and Hall, F. R. (1959). Water quality of eastern Stanislaus and northern Merced Counties, California. Stanford Univ. *Geol. Sci.* **6**, 112.

Davis, S. N., and Turk, L. J. (1964). Optimum depth of wells in crystalline rocks, *Ground Water* **2**, 6–11.

Dingman, R. J., Fergusun, H. F., and Martin, R. O. R. (1956). The water resources of Baltimore and Harford Counties, *Maryland Dept. of Geol., Mines, and Water Resources Bull.* **17**.

Frommurze, H. F. (1937). The water-bearing properties of the more important geological formations in the Union of South Africa, Union of South Africa Geol. Surv. Memoir No. 34.

Gallaher, J. T. (1964). Geology and hydrology of alluvial deposits along the Ohio River between the Uniontown Area and Wickliffe, Kentucky, U.S. Geol. Surv. Hydrologic Atlas 129.

Gibbs, H. J., and Holland, W. Y. (1960). Petrographic and engineering properties of loess, U.S. Bur. of Reclamation Eng. Monographs No. 28.

Gloyna, E. F., and Reynolds, T. D. (1961). Permeability measurements of rock salt, *J. Geophys. Res.* **66**, 3913–3921.

Gondouin, M., and Scala, C. (1958). Streaming potential and the SP log, *Trans. Am. Inst. of Mining and Met. Engrs. Paper* 8023.

Greenkorn, R. A., Johnson, C. R., and Shallenberger, L. K. (1964). Directional permeability of heterogeneous anisotropic porous media, *Trans. Soc. of Petrol. Engrs.* **231**, 124–132.

Jahns, R. H. (1943). Sheet structure in granite, its origin and use as a measure of glacial erosion in New England, *J. Geol.* **51**, 71–98.

Johnson, A. I., and Morris, D. A. (1962). Physical and hydrologic properties of water-bearing deposits from core holes in the Los Banos-Kettleman City area, California, U.S. Geol. Surv. Open-File Rept. Denver, Colorado.

Johnson, A. I., and Sniegocki, R. T. (1967). Comparison of laboratory and field analyses of aquifer characteristics at an artificial recharge well site, Paper presented at *Intern. Assoc. Scientific Hydrology Symp. on Artificial Recharge of Aquifers, Haifa, Israel.*

Johnson, C. R., and Greenkorn, R. A. (1963). Correlation of lithology and permeability for a Virgilian sandstone reservoir in central Oklahoma, *Geol. Soc. Am. Proc.* **73**, 180–181.

Johnston, N., and Beeson, C. M. (1945). Water permeability of reservoir sands, *Trans. Am. Inst. Min. Met. Engrs.* **160**, 43–55.

Johnston, P. M. (1962). Geology and ground-water resources of the Fairfax Quadrangle, Virginia, U.S. Geol. Surv. Water-Supply Paper 1539-L.

Keller, G. V. (1960). Physical properties of tuffs of the Oak Spring Formation, Nevada, U.S. Geol. Surv. Prof. Paper 400-B.

Kemper, W. D., Maasland, D. E. L., and Porter, L. K. (1964). Mobility of water adjacent to mineral surfaces, *Soil Sci. Soc. Am. Proc.* **28**, 164–167.

Laney, R. L. (1965). A comparison of the chemical composition of rainwater and ground water in western North Carolina, p. 187–189, U.S. Geol. Surv. Prof. Paper 525-C.

Law, J. (1944). A statistical approach to the interstitial heterogeneity of sand reservoirs, *Trans. Am. Inst. Mining Met. Engrs.* **155**, 202–222.

LeGrand, H. E. (1954). Geology and ground water in the Statesville area, North Carolina, North Carolina Dept. of Conservation and Development, Div. of Mineral Resources Bull. 68.

Levorsen, A. I. (1954). "Geology of Petroleum." Freeman, San Francisco, California.

Lewis, D. C., Kriz, G. J., and Burgy, R. H. (1966). Tracer dilution sampling technique to determine hydraulic conductivity of fractured rock, *Water Resources Res.* **2**, 533–542.

MacGary, L. M., and Lambert, T. W. (1962). Reconnaissance of ground-water resources of the Jackson Purchase region, Kentucky, U.S. Geol. Surv. Hydrologic Atlas 13.

Manger, G. E. (1963). Porosity and bulk density of sedimentary rocks, U.S. Geol. Surv. Bull. 1144-E.

Masch, F. D., and Denny, K. J. (1966). Grain size distribution and its effect on the permeability of unconsolidated sands, *Water Resources Res.* **2**, 665–677.

Maxwell, J. C. (1964). Influence of depth, temperature, and geologic age on porosity of quartzose sandstone, *Bull. Am. Assoc. Petrol. Geologists* **48**, 697–709.

Mégnien, Claude. (1964). Observations hydrogéologiques sur le sud-est du bassin de Paris, Mém. Bur. de Rech. Géol. et Minières No. 25.

Meyer, G., and Beall, R. M. (1958). Water resources of Carroll and Frederick Counties, Maryland Dept. Geol., Mines, and Water Resources Bull. 22.

Murray, R. C. (1960). Origin of porosity in carbonate rocks, *J. Sediment. Petrol.* **30**, 59–84.

Muskat, M. (1937). "The Flow of Homogeneous Fluids Through Porous Media." McGraw-Hill, New York.

Norris, S. E. (1963). Permeability of glacial till, U.S. Geol. Surv. Prof. Paper 450-E, pp. 150–151.

Norris, S. E., and Fidler, R. E. (1965). Relation of permeability to particle size in a glacial-outwash aquifer at Piketon, Ohio, U.S. Geol. Surv. Prof. Paper 525-D, pp. 203–206.

Olsen, H. W. (1966). Darcy's law in saturated kaolinite, *Water Resources Res.* **2**, 287–295.

Parker, G. G. (1951). Geologic and hydrologic factors in the perennial yield of the Biscayne aquifer, *Am. Water Works Assoc. J.* **43**, 817–835.

Piersol, R. J., Workman, L. E., and Watson, M. C. (1940). Porosity, total liquid saturation, and permeability of Illinois oil sands, Ill. Geol. Surv. Rept. Invest. No. 67.

Poth, C. W. (1963). Geology and hydrology of the Mercer Quadrangle, Mercer, Lawrence, and Butler Counties, Pennsylvania, Pennsylvania Topographic and Geol. Surv. Ground-Water Rept. W-16.

Rima, D. R., Meisler, H., Longwill, S. (1962). Geology and hydrology of the Stockton Formation in southeastern Pennsylvania, Pennsylvania Topographic and Geol. Surv. Ground-Water Rept. W-14.

Samuelson, W. J. (1965). Engineering geology of the Norad Combat Operations Center, Colorado Springs, Colorado, *Engineering Geol.* **2**, 20–30.

Schoeller, H. (1962). "Les eaux souterraines." Masson, Paris.

Seaber, P. R., and Hollyday, E. F. (1966). Statistical analysis of regional aquifers, A paper presented to the 1966 Ann. Meeting, Geol. Soc. of Am., San Francisco, California.

Solov'yev, P. Ye. (1961). Change in the physical properties of ordinary chernozems under the effect of forest vegetation, *Dokl. Soil Sci.* Section No. 2, pp. 29–33. (English transl. by Am. Inst. of Biol. Sciences).

Stewart, J. W. (1964). Infiltration and permeability of weathered crystalline rocks Georgia Nuclear Laboratory, Dawson County, Georgia, *U.S. Geol. Surv. Bull.* **1133-D**.

Stuart, W. T., Brown, E. A., and Rhodehamel, E. C. (1954). Groundwater investigations of the Marquette Iron-Mining District: Michigan Geol. Surv. Div. Tech. Rept. 3.

Tanaka, H. H., and Hollowell, J. R. (1966). Hydrology of the alluvium of the Arkansas River, Muskogee, Oklahoma to Fort Smith, Arkansas, U.S. Geol. Surv. Water-Supply Paper 1809-T.

Tolman, C. F. (1937). "Ground Water." McGraw-Hill, New York.

Walker, E. H. (1956). Ground-water resources of the Hopkinsville Quadrangle, Kentucky, U.S. Geol. Surv. Water-Supply Paper 1328.

Walters, K. L. (1963). Highly productive aquifers in the Tacoma area, Washington, U.S. Geol. Surv. Prof. Paper 450-E, pp. 157–158.

Walton, W. C. (1962). Selected analytical methods for well and aquifer evaluation, Illinois State Water Surv. Bull. 49.

Walton, W. C., and Stewart, J. W. (1961). Aquifer tests in the Snake River Basalt, *Trans. Am. Soc. Civil Engrs.* **126**, 612–632.

Waring, G. A. (1965). Thermal springs of the United States and other countries of the world; a summary, U. S. Geol. Surv. Prof. Paper 492.

Warren, J. E., and Price, H. S. (1961). Flow in heterogeneous porous media, Trans. Soc. Petrol. Engrs, *Am. Inst. Mining, Met. Petrol. Engrs.* **222**, 153–169.

Yokota, J. (1963). Experimental studies on the design of grouting curtain and drainage for the Kurobe No. 4 Dam, *Rock Mech. Engineering Geol.* **1**, 104–119.

Young, A., Low, P. F., and McLatchie, A. S. (1964). Permeability studies of argillaceous rocks, *J. Geophys. Res.* **69**, 4237–4245.

AUTHOR CITATION INDEX

Adams, F. D., 4
Allen, D. N., de G., 240, 241
American Society of Civil
  Engineers, 277
Appel, C. A., 140
Aravin, V. I., 180
Archie, G. E., 419
Aris, R., 132
Atwater, G. I., 419

Back, W., 4, 139, 344
Baker, H. D., 205, 241
Baker, M. N., 4
Barnes, B. S., 277
Beall, R. M., 241
Bedinger, M. S., 419
Beeson, C. M., 421
Bennett, R. R., 296, 420
Bennion, D. W., 420
Bentall, R., 139
Bergstrom, R. E., 369
Bermes, B. J., 249
Biot, M. A., 12
Biswas, A., 4
Bittinger, M. W., 240
Bogomolov, G. B., 345
Boulton, N. S., 139, 161, 240
Brace, W. F., 420
Brater, E. F., 277, 345
Bredehoeft, J. D., 4, 139, 240
Bridges, P. M., 241
Brown, D. W., 295
Brown, E. A., 422
Brown, R. H., 180, 240
Brown, S. G., 420
Bruce, G. H., 240
Bruce, W. A., 200, 240
Bulnes, A. C., 420
Burgy, R. H., 421

Callahan, J. T., 410
Cambefort, H., 161
Cardwell, W. T., Jr., 420
Carslaw, H. S., 240, 382
Casagrande, A., 277
Chamberlin, T. C., 369
Cherry, J. A., 4

Cheung, Y. K., 357
Chow, V. T., 4
Clark, G., 382
Clark, W. O., 396
Cohen, P., 420
Collins, R. E., 420
Cooke, C. W., 410
Cooper, H. H. Jr., 139, 240
Couette, M., 22
Counts, H. B., 382
Craig, J. R., 410
Crank, J., 240
Croft, M. G., 410
Csallany, S., 420
Cuevas, J. A., 382
Cushman, R. V., 420

Dachler, R., 13, 37, 83
Darcy, H., 4, 22, 36,126
Darms, D. A., 205
Darton, N. H., 295
Davis, G. H., 295, 382, 383
Davis, S. N., 4, 420
De Glee, G. J., 180, 369
Denny, K. J., 241
De Wiest, R. J. M., 4, 13, 132,
  180, 240, 420
Dingman, R. J., 420
Domenico, P. A., 4, 382
Donn, W. L., 410
Dorn, W. S., 240
Douglas, J. Jr., 205, 240
Duke, H. R., 240, 241
Dupuit, J., 13

Eakin, T. E., 295
Emrich, G. H., 369
Eshett, A., 240
Ewing, M., 410

Farvolden, R. N., 295
Fayers, F. J., 240
Ferguson, G. E., 296
Fergusun, H. F., 420
Ferrand, W. R., 410
Ferrandon, J., 83
Ferris, J. G., 4, 369

Fidler, R. E., 241
Fiedler, A. G., 102, 109
Fiering, M. B., 240
Forchheimer, P., 13
Foster, J. W., 369
Fowle, F. E., 107
Freeze, R. A., 4, 139, 240, 295,
  357
Fried, J., 13
Frommurze, H. F., 420
Fuller, M. L., 88, 295, 369

Gabrysch, R. K., 241
Gallaher, J. T., 420
Garvin, W. W., 241
Geraghty, J. J., 344
Gibbs, H. J., 420
Girinsky, N. K., 180
Glebov, P. D., 180
Gloyna, E. F., 420
Goldman, M. I., 89
Gondouin, M., 420
Gottlieb, I., 252
Grace, H., 383
Graham, 22
Gray, 161
Green, J. H., 295, 382, 397
Greenkorn, R. A., 420, 421
Gregory, H. E., 27
Gregory, J. W., 87
Griffiths, J. C., 420
Groustein, W. C., 84

Hall, C. W., 369
Hall, F. R., 420
Hall, H. N., 241
Hanshaw, B. B., 4
Hantush, M. S., 132, 139, 180,
  240, 241, 251, 369, 382
Hard, H. A., 91, 99, 126, 296
Hay, R., 87
Hazen, A., 22, 33
Healy, H. G., 410
Hedberg, H. D., 87
Herrick, S. M., 410
Hill, R. T., 295
Hoag, L. N., 213

Author Citation Index

Holland, W. Y., 420
Hollowell, J. R., 422
Hollyday, E. F., 422
Horton, R. E., 4
Hubbert, M. K., 344, 357
Hunt, G. W., 397
Hursh, C. R., 277
Hvorslev, M. J., 139

Ince, S., 5
Ineson, J., 161
Inglis, G., 95
Irmay, S., 83

Jackson, A. S., 205
Jacob, C. E., 132, 139, 161, 181, 241, 246, 251, 295, 369, 382, 383
Jaeger, J. C., 240, 382
Jahns, R. H., 421
Javandel, I., 181
Johnson, A. I., 181, 383, 421
Johnson, C. R., 420, 421
Johnson, D. W., 90
Johnston, N., 421
Johnston, P. M., 421
Joint Committee on Water Problems of the California Legislature, 278
Jones, G. H., 278
Jumikis, A. R., 383

Karplus, W. J., 241, 246
Keller, G. V., 421
Kemper, W. D., 421
King, F. H., 89, 98, 295, 344
Kirkham, D., 84
Klute, A., 241
Knowles, D. B., 180
Kohler, M. A., 278
Korn, G. A., 205, 252
Korn, T. M., 252
Krieger, R. A., 420
Kriz, G. J., 421
Kuhn, T. S., 5

Lambert, T. W., 241
Laney, R. L., 421
Larson, T. E., 369
Law, J., 421
LeGrand, H. E., 296, 410, 421
Levorsen, A. I., 421
Lewis, D. C., 421
Liakopoulos, A. C., 241, 357
Linsley, R. K., Jr., 278
Lohman, S. W., 139, 295, 383
Longenbaugh, R. A., 240
Longwill, S., 422
Lonnquist, C. G., 140
Lonsdale, J. T., 295
Love, S. K., 296
Low, P. F., 422
Lueger, O., 22
Luthin, J. N., 140

Maasland, D. E. L., 421
Maasland, M., 84, 357
McCabe, J. A., 420
McCann, G. D., 205
McCombs, J., 109
McCracken, D. D., 241
McLatchie, A. S., 422
MacNeal, R. H., 203
MacRobert, 161
Manger, G. E., 241
Martin, R. O. R., 420
Mathews, 161
Maxey, G. B., 295, 382
Maxwell, J. C., 241
Mayer, P., 357
Mégnien, C., 241
Meinzer, O. E., 13, 88, 89, 91, 99, 126, 296, 344, 369
Meisler, H., 422
Mendenhall, W. C., 296
Meneley, W. A., 344
Meyboom, P., 296
Meyer, G., 241
Meyer, O. E., 22
Meyer, R. R., 420
Mifflin, M. D., 296
Miller, R. E., 296
Millman, J., 247
Montague, K. E., 200
Morris, D. A., 181, 383, 421
Morse, R. A., 241
Morton, K. W., 241
Moston, R. P., 181
Murray, R. C., 241
Muskat, M., 13, 37, 84, 126, 161, 241, 278
Myatiev, A. N., 181

Narasimhan, T. N., 181
Neill, J. C., 256
Nelson, R. W., 140
Nettleton, E. S., 99, 309
Neuman, S. P., 140, 181
Newcomb, R. C., 420
Noren, D., 278
Norris, S. E., 241
Norvatov, A. M., 344

Olmstead, F. H., 295
Olsen, H. W., 422
Osgood, W. R., 84

Papadopulos, I. S., 139, 240, 241
Parker, G. G., 296, 422
Parsons, R. L., 420
Paschkis, V., 205, 241
Patten, E. D., Jr., 241
Patten, E. P., 140
Patterson, O. L., 200
Paulding, B. W., Jr., 420
Paulhus, J. L. H., 278
Peaceman, D. W., 205, 240, 241
Peck, R. B., 383

Piersol, R. J., 422
Pinder, G. F., 140
Plotnikov, N. A., 345
Plummer, A. K., 278
Poiseuille, 22
Poland, J. F., 296, 383, 397
Polubarinova-Kochina, P. Ya., 5, 181, 241
Popov, O. V., 344
Porter, L. K., 421
Poth, C. W., 422
Pratt, W. E., 90
Price, H. S., 422
Prickett, T. A., 140, 241

Rachford, H. H., Jr., 205, 240, 241
Ralston, A., 212, 241
Rappleye, H. S., 397
Remson, I., 140
Reynolds, O., 62
Reynolds, T. D., 420
Rhoades, R., 410
Rhodehamel, E. C., 422
Rice, J. P., 240
Richards, H. G., 410
Richtmyer, R. D., 211, 241
Riesbol, H. S., 278
Rima, D. R., 422
Roessel, B. W. P., 278
Rorabaugh, M. I., 139, 140, 369
Rouse, H., 5, 278
Rowe, P. R., 197
Rubin, J., 241
Russell, W. L., 296

Samsioe, A. F., 84
Samuelson, W. J., 422
Sauer, S. P., 140
Saul'yev, V. K., 241
Sayre, A. N., 4, 296
Scala, C., 420
Scheidegger, A. E., 84
Schoeller, H., 422
Scholz, C., 420
Scott, E. J., 241
Scott, V. H., 140
Seaber, P. R., 422
Seely, S., 247
Shallenberger, L. K., 420
Shannon, W. L., 277
Shaw, F. S., 140
Sheldon, J. W., 240, 241
Shepard, J. H., 107
Sinacori, M. N., 410
Skibitzke, H. E., 140, 241, 244, 383
Slater, R. J., 181
Slichter, C. S., 29, 37, 126
Small, J. B., 382
Smith, H. F., 369
Smithsonian Physical Tables, 126
Snider, L. C., 91
Sniegocki, R. T., 421
Snyder, F. F., 278

Author Citation Index

Solov'yev, P. Ye., 422
Sorby, H. C., 87
Southwell, R. V., 140, 197, 241, 246
Spanier, J., 241
Stallman, R. W., 180, 197, 241, 246
Stearns, N. D., 126
Steggenventz, J. H., 181
Stephenson, D. A., 382
Stewart, H. G., Jr., 410
Stewart, J. W., 422
Stoek, H. H., 87
Stringfield, V. T., 296, 410
Stuart, W. T., 422
Suter, M., 278, 369
Swenson, F. A., 296
Swinnerton, A. C., 410

Tanaka, H. H., 422
Taylor, D. W., 383
Terwilliger, P. L., 241
Terzaghi, C., 87, 89, 90
Terzaghi, K., 13, 383
Theis, C. V., 126, 132, 161, 241, 369, 383
Thiem, A., 13, 241
Thiem, G., 13
Thompson, D. G., 95, 102, 104
Thordarson, W., 296
Tibbetts, F. H., 397
Todd, D. K., 5, 278

Todd, J., 241
Tolman, C. F., 5, 296, 397, 422
Tóth, J., 296, 345, 357
Toulmin, L. D., 410
Tribus, M., 262
Turk, L. J., 420
Tyson, H. N., Jr., 205

Ubell, K., 345
U. S. Coast and Geodetic Survey, 397

Van Ness, B. A., 181
Varga, R. S., 241
Vaughan, T. W., 295
Veatch, A. C., 88, 94
Verruijt, A., 13
Versluys, J., 13, 84
Volynskii, B. A., 241
Vreedenburgh, C. G. F., 13, 84

Walker, E. H., 422
Walker, W. H., 262
Walters, K. L., 422
Walton, W. C., 5, 140, 181, 241, 251, 256, 262, 369, 420, 422
Waring, G. A., 422
Warren, J. E., 422
Warren, M. A., 410
Wass, C. A. A., 252
Watson, M. C., 422
Webster, R. A., 140

Weinberg, G., 213
Weir, W. W., 397
Weiss, B., Jr., 200
Wenzel, L. K., 126, 140, 161, 197
Werner, P. H., 278
West, W. J., 241
Whisler, F. D., 241
White, W. N., 296
Wieckowska, H., 345
Wilf, H. S., 212, 241
Willers, F. A., 241
Wilsey, L. E., 241
Wilson, G., 383
Wilts, C. H., 205
Winograd, I., 296
Winslow, A. G., 383
Wisler, Ch. o., 345
Witherspoon, P. A., 140, 181, 295, 357
Wood, L. A., 241, 383
Workman, L. E., 422
Wylie, C. R., 84

Yang, S. T., 84
Ye Bukhman, V., 241
Yokota, J., 422
Young, A., 422
Young, D., Jr., 213
Young, L. E., 87

Zienkiewicz, O., 357

SUBJECT INDEX

Alluvial deposits, 233, 261-264, 318, 389, 417
Alternating direction implicit procedure, 225
Analog simulation, 137, 203-210, 218, 243-267
Anisotropy, 12, 71-84, 158, 250, 354
Aquifer
   analog simulation, 203-210, 218, 243-267
   artesian, 85-113, 114-126, 297-307, 313-316, 317, 359
   barometric efficiency, 122-124
   bounded, 137, 182-192, 229
   carbonates, 28, 295, 298-410
   coastal, 201, 318, 399
   compressibility, 85-113, 373
   confined. See Confined aquifer
   elasticity, 85-113
   fluvial deposits, 233, 261-264, 318, 389, 417
   glacial deposits, 28, 233, 360
   hydrogeology, 291-293, 301-303, 309-311, 317-327, 364, 389, 400
   igneous and metamorphic rocks, 302, 319, 326
   leaky, 136, 163-168, 169-181, 251, 294, 358-369
   numerical simulation, 211-215, 217-242
   response to pumping. See Drawdown cone; Well hydraulics
   sandstone, 11, 90, 102, 292, 308
   seawater intrusion, 97
   unconfined. See Unconfined aquifer
   yield, 319, 324, 327
Aquitard, 110-111, 303-307
   compaction, 370-383, 384-397
   glacial till, 360
   hydrogeology, 303-307, 309-311, 364, 389
   land subsidence, 90, 370-383, 384-397
   response to pumping in adjacent aquifers, 147-162, 169-181
   storativity, 172
Artesian well, 91, 99-105, 297, 308, 317
Artificial recharge, 199
Atmospheric pressure, 44, 116

Bank storage, 139, 268-287
Barometric efficiency, 122-124
Baseflow, 268-287
Body force, 56
Boundary-value problems, 8, 137
   for one-dimensional consolidation, 372
   for radial flow to a well, 127, 142, 151, 163
   for steady-state regional groundwater flow, 294, 331
Bounded aquifer, 137, 182-192, 229

Caves, 408
Chemical hydrogeology, 3
Clay, 87
Coefficient of permeability. See Hydraulic conductivity
Coefficient of storage. See Storage coefficient
Columbia River basalt, 326
Compaction, 370-383, 384-397
Compaction recorder, 393-395
Compressibility, 9
   aquifer, 85-113, 114, 128
   aquitard, 371
   from compaction recorder installations, 393-395
   range of values, 373
   water, 107, 115, 128, 371
Computer models. See Numerical simulation
Cone of depression, 121, 142, 152, 167, 172, 176, 185, 188, 238, 259-264, 320, 385
Confined aquifer, 85-113
   analog simulation, 218, 243-267
   diffusion equation, 114-126
   numerical simulation, 217-242
   pumping test, 121, 135, 145, 159, 168, 178, 185, 319, 361
   radial flow to a well, 119-121, 141-146, 163-168, 169-181
   storativity, 116, 127, 150, 163, 202, 220, 248
   transmissivity, 121, 127, 142, 153, 163, 202, 220, 248
Confining bed. See Aquitard
Consolidation, 90, 104, 303, 370-383, 393-395
Constant-head boundary, 185, 223, 252

Dakota sandstone, 11, 90, 102-104, 292, 308-316
Darcy's law, 9, 15-20, 22, 38-70, 75, 118, 131, 141, 150, 220
Deforming coordinate system, 12, 127
Delayed yield, 136, 147-162
Diffusion equation, 114-126
Diffusivity, 142
Digital computers, 3, 211, 217, 346
Dijon, France, 10, 18-19
Discharge
   area, 341, 347, 355
   as baseflow, 272
   as evapotranspiration, 292
   in hydrologic budget, 91, 344
   specific, 41
Discharge area, 341, 347, 355
Drainage basin, 328

427

Subject Index

Drawdown cone, 121, 142, 152, 167, 172, 176, 185, 188, 238, 261-264, 320, 362-363, 385
Driving force, 56, 64
Dupuit-Forchheimer theory, 10-11

Effective stress, 9, 11, 86, 103, 115, 128
  compressibility measurement, 90, 104
  due to external loads, 98
  in land subsidence, 90, 370-383
Effluent stream, 292
Elasticity, 85-113, 114, 127, 371
Electrical analog models. See Analog simulation
Elemental control volume, 128
Elevation head, 43-54
Ellipsoid of hydraulic conductivity, 78-83
Equation of continuity, 23, 75, 117, 151, 202, 370
Equations of groundwater flow, 9, 11, 21-25, 75, 114-126, 127-132, 151, 220, 370
Equipotential line. See Flow net
Evapotranspiration, 292
Exitation-response apparatus, 253
Experimental apparatus, 20, 39, 272
Exploration for aquifers, 301-307

Falling-head permeameter, 17
Finite-difference methods, 138, 194, 211-215, 217-242, 346
Finite-element methods, 138
Floridan limestone, 295, 399
Flowing artesian well, 88, 292, 297, 308, 313-316, 317
Flow nets, 11
  anisotropic systems, 354
  heterogeneous systems, 350-353
  homogeneous, isotropic systems, 332-338, 348
  by numerical simulation, 347
  regional groundwater flow, 291, 332-338, 342, 348-354
  to a well, 187, 189
Flow system, local, intermediate, and regional, 340
Fluctuations in water level, 94-101, 195-197, 208, 389, 405
Fluid pressure, 43, 86, 115, 125, 376. See also Pressure head
Fluvial deposits, 233, 261-264, 318, 389, 417
Fractured rocks, 302, 319, 326, 415
Free surface, 156

Geometrically-similar media, 60
Glacial deposits, 28, 233, 360, 417
Gradation curve, 33
Grain-size curve, 33
Grand Island, Nebraska, 135, 143
Grand Junction, Colorado, 322
Great Basin, 293, 308
Groundwater discharge. See Discharge
Groundwater divide, 329, 347
Groundwater flow
  basin, 329, 347, 356, 382
  equations of, 9, 11, 21-25, 75, 114-126, 127-132, 220
  in fractured rocks, 302, 319, 326
  macroscopic, 55, 64
  microscopic, 54, 64, 74
  regional. See Regional groundwater flow
  steady-state. See Steady-state flow
  transient. See Transient flow
Groundwater recharge. See Recharge

Hantush-Jacob theory, 163-168, 169-181
Head
  elevation, 43-54
  hydraulic, 43-54
  pressure, 43, 86, 115, 125
Hele-Shaw model, 269
Heterogeneity, 12
  regional flow systems, 350-353
History of hydrology, 3-4
Hydraulic conductivity, 9, 54-70, 118, 163, 411-422. See also Permeability
  anisotropy, 71-84, 158, 250, 354
  of basalt, 415
  of crystalline rock, 414
  ellipsoid, 78-83
  equivalent, 80
  of fluvial deposits, 417-418
  of fractured rocks, 415
  of glacial deposits, 361, 417-418
  heterogeneity, 12
  of limestone, 405, 416
  from piezometer tests, 137
  from pumping tests, 121, 135, 145
  of sandstone, 416
  of shale, 414, 416
  tensor, 74
Hydraulic diffusivity, 142
Hydraulic gradient, 9, 41, 45, 93
Hydraulic head, 40, 43-54
  fluctuations, 94-101
Hydraulic potential, 9, 11, 43-54, 330, 347
Hydrogeology, 4, 291-293, 301-303, 309-311, 317-327, 364, 400
Hydrograph
  baseflow, 284
  groundwater, 286
  streamflow, 281-287
Hydrologic budget, 91, 344
Hysteresis, in compressibility, 90

Illinois State Water Survey aquifer model, 138
Image-well theory, 10, 182-192
Imaginary impermeable boundary, 329, 347
Imaginary wells, 185-192
Impermeable boundary, 186-192, 223, 252, 329, 347, 356
Induced infiltration, 274
Influent stream, 292
Intermediate flow system, 340
Intertial force, 56
Intrinsic permeability, 70, 118
Inverse problem, 135

Jacob-Hantush theory, 163-168, 169-181, 268-287

Subject Index

Karst, 302, 398-410
Kinematically-similar flow, 60

Land subsidence, 90, 294, 370-383, 384-397
Laplace's equation, 23, 75, 331
Leaky aquifer, 163-168, 169-181, 251, 294, 358-369
Leaky well function, 166, 169-181
Limestone, 28, 295, 399
Local flow system, 340
Los Angeles coastal plain, 198-216

Macroscopic concept of flow, 55, 64
Manometer, 40
Mechanical energy, 46-52
Method of images, 10, 182-192
Microscopic concept of flow, 54, 64, 74
Modulus of elasticity, 85-113, 127
Multiple-well systems, 137
Musquoduboit Harbour, Nova Scotia, aquifer model, 231-239

Nevada test site, 293
Nonsteady flow. See Transient flow
Numerical simulation
    aquifer, 211-215, 217-242
    finite-difference, 211-215, 217-242, 346
    finite-element, 138
    regional flow, 346
    transient, 217-242

Paleohydrology, 408
Partial penetration, 137
Permeability, 39, 54-70, 118, 163, 411-422. See also Hydraulic conductivity
Permeameter, 17
Piezometer tests for hydraulic conductivity, 137
Piezometric surface. See Potentiometric surface
Pore pressure, 374
Porosity, 9, 27-35, 39, 88, 303, 372, 409, 411-422
Potential, fluid, 9, 43-54
Potentiometric surface, 141, 147, 194, 365, 367, 407
Pressure
    atmospheric, 44, 116
    effective, 377
    fluid, 43, 86, 115, 125, 376. See also Pressure head
    pore, 374
Pressure head, 43, 86, 115, 125
Principal directions of anisotropy, 71-85
Pumping tests, 121, 135, 145, 159, 168, 178, 185-190, 319, 361

Radial flow to a well, 119-122, 141-146, 147-162, 163-168, 169-181
Recharge
    area, 341, 347
    artificial, 199
    boundary, 185
    in hydrologic budget, 91
    induced infiltration, 275
Recharge area, 341, 347, 355
Recovery of well after pumping, 144, 158

Refraction of flowlines, 350
Regional groundwater flow, 294, 299, 328-345, 346-357, 402
Relaxation, 137, 211
Resistance-capacitance network, 137, 203-210, 218, 224, 243-267
Reynolds number, 62
Roswell Basin, New Mexico, 317
Runoff. See Streamflow

Sacramento River seepage, 275
Sandstone, 28, 86, 102-104, 416
Santa Clara Valley, California, land subsidence, 386
Scale factors, electric analog, 249
Seawater intrusion, 97
Shale, 414, 416
Shear stress, 59
Specific discharge, 41
Specific permeability, 70, 118
Specific storage, 128, 371
Specific yield, 412
Springs, 292
Steady-state flow, 21-25
    flow nets, 11, 291, 332-338, 342
    Laplace's equation, 23, 75, 331
    numerical simulation, 346-357
    regional, 328-342, 346-357
Stepped pumping rate, 137
Storage coefficient. See Storativity
Storativity, 116, 127-128, 150, 163, 202, 220, 248, 381, 412
    from pumping test, 121, 319, 324
Stream-aquifer interactions, 185, 268-287
Streamflow
    baseflow, 268-287
    hydrograph, 281-287
    in hydrologic budget, 91, 344
    induced infiltration, 274
Streamline, 55
Stress, effective. See Effective stress
Stress-strain, 12
Structural geology, 291, 323, 326, 400
Subsidence. See Land subsidence

Texts on groundwater, 4
Theis curve, 173, 228
Theis solution, 141-146, 173, 228
Tidal effects on groundwater, 94-96, 122-124
Till, 28
Transformed section, 75-77
Transient flow
    consolidation, 90, 104, 303, 370-383
    equations, 114-126, 127-132, 220
    numerical simulation, 211-215, 217-242
    to a well, 117, 141-146, 147-162, 163-168, 169-181
Transmissibility. See Transmissivity
Transmissivity, 127, 142, 153, 163, 202, 220, 235, 248
    from pumping test, 121, 144, 319, 324, 360
Type curve, 135, 173, 228

Subject Index

Unconfined aquifer
   analog model, 198-216, 257-259
   radial flow to a well, 136, 147-162
   regional, 344
   specific yield, 412
   water-level fluctuations, 94-101, 195, 208, 286, 389, 405
Uniformity coefficient, 34
U.S. Geological Survey, 2, 11, 135, 292
U.S. Geological Survey aquifer model, 138
Unsteady flow. *See* Transient flow

Velocity
   macroscopic, 55, 64
   microscopic, 54, 64, 74
Viscosity, 58-60
Void ratio, 372
Volcanic rocks, 326, 415

Water compressibility, 107, 115
Water table
   boundary condition, 330
   delayed response, 136, 147-162
   fluctuations, 94-101, 195-197, 208, 286, 389, 405
   free-surface, 156
   regional, 193, 330, 348, 404
Well
   artesian, 91, 99-105
   drawdown cone, 121, 142, 152, 167, 172, 176, 185, 188, 238, 259-264, 320
   flow artesian, 88, 292, 297, 308, 313-316, 317
Well function
   confined, 121, 144
   leaky, 166, 169-181
   unconfined, 153
Well hydraulics, 117, 135, 141-146, 147-162, 163-168, 169-181, 228
Well recovery, 144, 158

Yield
   aquifer, 319, 324, 327
   delayed, 136
   specific, 412

About the Editors

R. ALLAN FREEZE is a professor in the Department of Geological Sciences and an associate dean in the Faculty of Graduate Studies at the University of British Columbia in Vancouver, Canada. He received the B.Sc. in geological engineering from Queens University in 1961 and the Ph.D. in civil engineering from the University of California, Berkeley, in 1966. Before joining UBC, he was a research scientist with the Hydrologic Sciences Division of the Canada Inland Waters Branch in Calgary, Alberta, and a research staff member at the IBM Thomas J. Watson Research Center in Yorktown Heights, New York. He is the author of over fifty technical publications in the fields of hydrology, hydrogeology, soil physics, and engineering seepage. He is coauthor with J. A. Cherry of the textbook *Groundwater* published in 1979. Dr. Freeze was awarded the Horton Award by the American Geophysical Union in 1970 and 1972 (with J. A. Banner) and the Meinzer Award by the Geological Society of America in 1974. He served as editor of the journal *Water Resources Research* during the period 1976–1980. He is a Fellow of the Royal Society of Canada.

WILLIAM BACK is a research hydrogeologist with the U.S. Geological Survey, Reston, Virginia, and adjunct professor at George Washington University, Washington, D.C. He received the B.S. in geology from the University of California, Berkeley, the M. Pub. Admin., Conservation of Natural Resources, from Harvard University, and the Ph.D. in hydrogeology from the University of Nevada. Dr. Back has served as an advisor or consultant to the United Nations Development Programme, U.S. Agency for International Development, and Food and Agriculture Organization for the governments of Israel, Poland, Pakistan, Costa Rica, and Bolivia. He has authored or coauthored more than fifty articles, including chapters in *Advances in Hydroscience, Advances in Chemistry,* and *Advances in Groundwater Hydrology.* Dr. Back currently serves on the Karst Commission, International Association of Hydrogeologists, and on the Editorial Advisory Board for *Journal of Hydrology.* He was co-recipient with Dr. Bruce Hanshaw of the Meinzer Award by the Geological Society of America and was later selected to be that society's Birdsall Distinguished Lecturer in Hydrogeology.